Mesenchymal-Epithelial Interactions in Neural Development

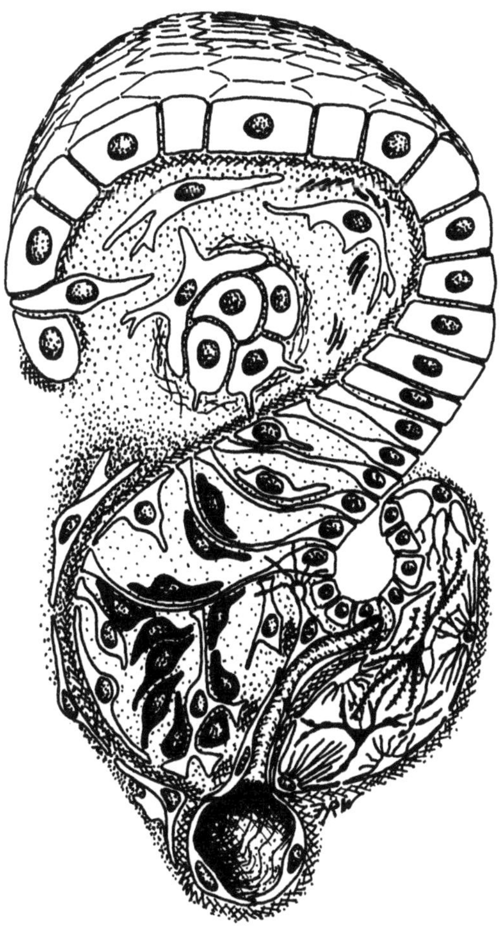

NATO ASI Series

Advanced Science Institutes Series

A series presenting the results of activities sponsored by the NATO Science Committee, which aims at the dissemination of advanced scientific and technological knowledge, with a view to strengthening links between scientific communities.

The Series is published by an international board of publishers in conjunction with the NATO Scientific Affairs Division

A Life Sciences **B Physics**	Plenum Publishing Corporation London and New York
C Mathematical and Physical Sciences	D. Reidel Publishing Company Dordrecht, Boston, Lancaster and Tokyo
D Behavioural and Social Sciences **E Applied Sciences**	Martinus Nijhoff Publishers Boston, The Hague, Dordrecht and Lancaster
F Computer and Systems Sciences **G Ecological Sciences** **H Cell Biology**	Springer-Verlag Berlin Heidelberg New York London Paris Tokyo

Series H: Cell Biology Vol. 5

Mesenchymal-Epithelial Interactions in Neural Development

Edited by

Joachim R. Wolff

Berlin Institute for Advanced Study
and
Universität Göttingen
Zentrum Anatomie
Kreuzbergring 36
3400 Göttingen, FRG

Jobst Sievers

Universität Kiel
Anatomisches Institut
Olshausenstr. 40–60, Haus N10
2300 Kiel, FRG

Martin Berry

Guy's Hospital Medical School
Department of Anatomy
London SE1 9RT, England

Springer-Verlag
Berlin Heidelberg New York London Paris Tokyo
Published in cooperation with NATO Scientific Affairs Divison

Proceedings of the NATO Advanced Research Workshop on Mesenchymal-Epithelial Interactions in Neural Development held at West-Berlin, FRG, 17–21 March, 1986

ISBN-13: 978-3-642-71839-7 e-ISBN-13: 978-3-642-71837-3
DOI: 10.1007/978-3-642-71837-3

Library of Congress Cataloging-in-Publication Data. NATO Advanced Research Workshop on Mesenchymal-Epithelial Interactions in Neural Development (1986 : Berlin, Germany) Mesenchymal-epithelial interactions in neural development. (NATO ASI series. Series H, Cell biology ; vol. 5) "Proceedings of the NATO Advanced Research Workshop on Mesenchymal-Epithelial Interactions in Neural Development held at West-Berlin, 17–21 March 1986"—T.p. verso. "Published in cooperation with NATO Scientific Affairs Division." Includes index. 1. Developmental neurology—Congresses. 2. Mesenchyme—Congresses. 3. Epithelium—Congresses. 4. Nerves—Growth—Congresses. 5. Brain—Growth—Congresses. 6. Extracellular matrix—Congresses. I. Wolff, Joachim R., 1935-. II. Sievers, J. (Jobst), 1948-. III. Berry, M. (Martin). IV. North Atlantic Treaty Organization. Scientific Affairs Division. V. Title. VI. Series. [DNLM: 1. Epithelium—physiology—congresses. 2. Mesoderm—physiology—congresses. 3. Nervous System—growth & development—congresses. WL 101 N2785m 1986] QP363.5.N37 1986 596'.03'33
86-31606

Softcover reprint of the hardcover 1st edition 1987

2131/3140-543210

Preface

This book is based on the NATO Advanced Research Workshop on "Mesenchymal-Epithelial Interactions in Neural Development" which was held in Berlin during March 1986. The idea that it may be the time for this workshop arose from a discussion among the organizers who met at a conference in Innsbruck.

During the twenties, Spemann demonstrated that, in vertebrates, the formation of neuroepithelium depends on the induction by specific mesodermal derivatives. This was about the first time that mesenchymal-epithelial interactions had been recognized as an important mechanism in the development of an epithelial organ. Since then two things had become clear:

1) The identity and role of molecules, elaborated at mesenchymal-neuroepithelial interfaces had been partially elucidated in spinal cord and peripheral nerve ontogeny but, until recently, had been ignored in studies on the development of the brain.
2) In many-non-neural epithelia, the surrounding mesenchyme regulates such fundamental developmental steps as, for example, cell proliferation, morphogenesis and cell differentiation, and there is no reason to exclude similar influences and interactions from brain development.

It seemed to us that the importance and implications of many of these findings had not been appreciated by many developmental neurobiologists and that an attempt should be made to initiate discussions and exchange of ideas among different groups of scientists involved in extracellular matrix (ECM) research. This turned out to be a formidable task because of the great diversity of disciplines contributing findings to this active field.

The plan of the workshop was to start with some basic aspects of the molecular and cell biology of the ECM, both in general and with specific reference to the nervous system. This section was kept relatively short, because several meetings on this topic had taken place in the last months before this workshop. The next section centered around mesenchymal-epithelial interrelationships in the morphogenesis of epithelial organs other than the nervous system and also covered examples of similar developmental interrelationships in the nervous system.

The next sections covered some developmental steps specific to the nervous system, i.e., neuronal and glial cell proliferation, growth and differentation, cell movement, growth and regeneration of neuronal processes, and the involvement of ECM and/or mesenchyme in these processes. In each case, examples from the development of non-neural organs preceded the discussion of these phenomena. The workshop ended with a session on intracerebral vascularization and the development of the blood-brain barrier. Until recently, this unique case of an "intraepithelial vascularization" has been largely disregarded, especially the cell biology and its molecular aspects.

The present book is a condensation of the contributions of the participants of this workshop. It is thus a heterogeneous compilation of reviews, short oral presentations, and posters from a diversity of scientific disciplines, methods, objectives and perspectives brought together under the heading of mesenchymal-epithelial interaction. Of necessity, this book is thus a proceedings report more than a state of the art book, a collection of several novel views, ideas and hypotheses more than a complete coverage of any of the individual sections. It contains accounts of all presentations except five. One of these missing articles is published at the same time in a scientific journal (K. Kratochwil et al., Normal epithelial branching morphogenesis in the absence of collagen I. Developmental Biology, in press). Other manuscripts couldn't be produced early enough to provide a quick publication.

Some aspects of the development of the nervous system in which mesenchyme and/or ECM are involved do not feature in this report, most importantly, neural crest development. This restriction was due to a combination of unfortunate circumstances (lack of time, coverage of the subject in other recent conferences, availability of speakers) and was not programmatic; yet findings and concepts from these areas were included in the discussion.

The NATO workshop and this book are a first attempt to focus attention on some yet undiscovered aspects of the development of the nervous system by drawing analogies from the development of other epithelial organs. It is meant to stimulate the search for basic principles in the development of epithelia including the brain which apparently need their surrounding mesenchyme to grow, differentiate and perform their functions. The nervous system is a very complex epithelium and its relationship to mesenchyme-derived structures are manyfold including an "intraepithelial" vascularization which is unique. Nevertheless, if this book succeeds in pointing out that there are many similarities between the development of non-neural epithelia and the nervous system, it has fulfilled its aspiration.

The Editors

Acknowledgements

This book presents the proceedings of a NATO Advanced Research Workshop which was held in Berlin from March 17–21, 1986. The Workshop was funded by the NATO Science Committee (Cell to Cell Signals in Plants and Animals). The editors wish to express their gratitude to the following people and institutions for their support: Dr. Silke Bernhard (Dahlem Conferences) who recommended suitable conference facilities; the Institute for Advanced Studies in Berlin and especially Barbara Golf, for considerable assistance in the organization; the Senate of the City of Berlin for providing additional financial support; and Ellen Raufeisen for skillful assistance in all aspects of administration and for preparation of the final manuscript.

Table of Contents

Contributors

Th. Bär, Zentrum Anatomie, Georg-August-Universität, Kreuzbergring 36, 3400 Göttingen, FRG

M. Berry, Department of Anatomy, Guy's Hospital, Medical School, London SE1 9RT, UK

P.S. Cooke, Department of Anatomy, University of California, San Francisco, CA 94143, USA

G.R. Cunha, Department of Anatomy, University of California, San Francisco, CA 94143, USA

A. & B. Delpech, Laboratory of Immunochemistry, Centre Henri-Becquerel, Rue D'Amiens, 76038 Rouen Cedex, France

S.S. Easter, Jr., Division of Biological Sciences, University of Michigan, Ann Arbor, MI 48109, USA

D. Edgar, Max-Planck-Institut für Psychiatrie, Abteilung Neurochemie, 8033 Martinsried, FRG

P. Ekblom, Max-Planck-Gesellschaft, Friedrich-Miescher-Lab., Spemannstr. 37–39, 4400 Tübingen, FRG

C. Gensburger, Centre de Neurochimie du CNRS, FRA 26 de L'INSERM, Faculté de Médecine, Strasbourg, France

D. Gospodarowicz, Department of Medicine and Ophthalmology, University of California, Medical Center, San Francisco, CA 94143, USA

U. Hahn, Medizinische Klinik, Abteilung Gastroenterologie, Klinikum Steglitz, Freie Universität Berlin, Hindenburgdamm 30, 1000 Berlin 45, FRG

B. Hausmann, Anatomisches Institut der Universität Kiel, Olshausenstr. 40–60, Haus N10, 2300 Kiel, FRG

R. Herken, Zentrum Anatomie, Universität Göttingen, Kreuzbergring 36, 3400 Göttingen, FRG

R.J. Keynes, Department of Anatomy, Downing St., Cambridge CB2 3DY, UK

P. Letourneau, Department of Anatomy, University of Minnesota, 4-135 Jackson Hall, Minneapolis, MN 55455, USA

A. Lumsden, Department of Anatomy, Guy's Hospital, Medical School, London SE1 9RT, UK

R.U. Margolis, Department of Pharmacology, New York University, Medical Center, 550 First Ave., New York, NY. 10016, USA

H.-J. Merker, Institut für Anatomie, Freie Universität Berlin, Königin-Luise-Str. 15, 1000 Berlin 33, FRG

A. Michler-Stuke, Zentrum Anatomie, Universität Göttingen, Kreuzbergring 36, 3400 Göttingen, FRG

F. W. Pehlemann, Anatomisches Institut der Universität Kiel, Olshausenstr. 40–46, Haus N10, 2300 Kiel, FRG

N. Ratner, Department of Anatomy and Neurobiology, Washington University, School of Medicine, Box 8108, St. Louis, MO 63110, USA

M. Rickmann, Zentrum Anatomie, Universität Göttingen, Kreuzbergring 36, 3400 Göttingen, FRG

W. Risau, Max-Planck-Institut für Entwicklungsbiologie, Spemannstr. 35, 7400 Tübingen, FRG

R. Schulz, Medizinische Klinik, Abteilung Gastroenterologie, Klinikum Steglitz, Freie Universität Berlin, Hindenburgdamm 30, 1000 Berlin 45, FRG

D. Schuppan, Medizinische Klinik, Abteilung Gastroenterologie, Klinikum Steglitz, Freie Universität Berlin, Hindenburgdamm 30, 1000 Berlin 45, FRG

M. Sensenbrenner, Centre de Neurochimie du CRNS, FRA 26 de L'INSERM, Faculté de Medecine, Strasbourg, France

J. Sievers, Anatomisches Institut der Universität Kiel, Olshausenstr. 40–60, Haus N10, 2300 Kiel, FRG

C. D. Stern, Department of Human Anatomy, South Parks Road, Oxford OX1 3QX, UK

P. Thorogood, Med. & Biol. Science Building, Department of Biology, Bassett Crescent East, Southampton, SO9 3TU, UK

B. P. Toole, Department of Anatomy and Cellular Biology, Tufts University, Schools of Medicine, Boston, MA 02111, USA

H. W. M. van Straaten, Anatomy and Embryology, Faculty of Medicine, Rijksuniversiteit Limburg, P. O. Box 616, 6200 MD Maastricht, NL

E. Wiertz-Hoessels, Anatomy and Embryology, Faculty of Medicine, Rijksuniversiteit Limburg, P. O. Box 616, 6200 MD Maastricht, NL

J. R. Wolff, Zentrum Anatomie, Universität Göttingen, Kreuzbergring 36, 3400 Göttingen, FRG

Y. Yamada, Laboratory of Developmental Biology, National Institute of Health, Bethesda, MD 70892, USA

Section 1

MOLECULAR AND CELL BIOLOGY OF THE EXTRACELLULAR MATRIX

COMPONENTS OF THE EXTRACELLULAR MATRIX (COLLAGENS, ELASTIN, GLYCO-PROTEINS AND PROTEOGLYCANS)

D. Schuppan and E.G. Hahn
Dept. of Gastroenterology
Klinikum Steglitz and Dept. of Anatomy
Freie Universität Berlin
West Berlin
Federal Republic of Germany

Introduction

Eukaryotic cells are surrounded by a complex assembly of macromolecules termed connective tissue or extracellular matrix (ECM)*. This specialized tissue comprises 1) the pericellular matrix of components interacting with structures of the cell membrane, 2) the classical interstitium, which occupies the intercellular spaces, when cells are separated from each other over longer distances and 3) basement membranes (BM), which support epithelial, endothelial and certain mesenchymal cells. The ECM is indispensable for the development of multicellular organisms. It is produced by the cells embedded in it, provides cohesiveness, and above all, directs these cells in that it tuns on, modifies or maintains their gene expression (for earlier studies on this subject refer to Slavkin and Greulich, 1975).

It was once believed that only a few molecules were involved in connective tissue organization and terms like "collagen" and "structural glycoprotein" still testify to that. With the availability of refined methods of purification, for structural analysis and for cell culture, and also with the aid of cDNA probes and novel antibody techniques, rapid progress has been made, especially during the last two decades, in our understanding of the ECM molecules. Much more is now known about their multiple interactions mediated by specialized structures, their supramolecular organization and their localization in tissues. Four major classes of components constitute the ECM, namely collagens,

---

* Abbreviations used in this paper:
ECM: extracellular matrix
BM: basement membrane(s)
PG: proteoglycan(s)
GAG: glycosaminoglycan(s)
HA: hyaluronic acid

NATO ASI Series, Vol. H5
Mesenchymal-Epithelial Interactions in Neural Development
Edited by J. R. Wolff et al.

elastin, glycoproteins and proteoglycans and each of these has been reviewed many times in recent years.

This chapter provides an orientation for those who are not familiar with the components of the ECM and sheds light on the growing importance of the ECM in cell biology. Considering the formidable complexity of the data, this review must neglect many basic aspects by referring to the literature, but on the other hand tries to highlight the major novel developments in connective tissue research which are of significance in cell biology and clinical-pathological research.

Collagen

Collagen represents roughly one third of the body protein and as late as 1969 there was little evidence that collagen isotypes existed different from what is now termed type I, the major fibril forming collagen. In subsequent years, however, other collagens designated types II, III, IV and V could be extracted from tissues and these collagens proved to be clearly distinct from type I, both biochemically and immunologically (Eyre, 1980; Bornstein and Sage, 1980; Miller and Gay, 1982; Miller and Rhodes, 1982). The less abundant collagens types VI-X, which have only recently been isolated, are still incompletely understood, but begin to reveal intriguing structural features. A common characteristic of all collagens is that they are proteins composed of three identical or similar polypeptide chains folded into a triple-helix, a structure, which endows the molecule with a particular stability. A prerequisite for triple-helix formation is the repetitive amino acid sequence $(\text{Gly-Xaa-Yaa})_n$ over most stretches of the polypeptide chains with a glycine at every third position, a high frequency of proline in positions Xaa or Yaa and of 4-hydroxyproline in Yaa (Fietzek and Kühn, 1976; Piez, 1984). The many collagen-specific post-translational modifications introduced into the nascent collagen molecule by specific enzymes during its intracellular biosynthesis such as hydroxylation of lysine - and of proline-residues and the glycosylation of hydroxylysine have been reviewed in detail (Kivirikko and Myllylä, 1982 and 1984). The abundant fibril-forming collagens type I and III of the interstitial ECM and tpye II of cartilage have been most intensively studied. They can be extracted and purified from tissues as native molecules after digestion with pepsin (Miller and Rhodes, 1982). The pepsin-resistant triple-helix can be dissociated at higher temperature (in the case of collagen I and II) or after additional reduction of intramolecular disulfide bonds (in collagen III) to yield the three constituent collagen α-chains of ca. 1050 amino acids and of a molecular weight of approx. 100,000.

As could be learned from cell culture studies, collagens I-III are synthesized as triple-helical precursor molecules, procollagens (Fig. 1), containing globular (non-collagenous) extension peptides at their N- and C-terminal ends (Fessler and Fessler, 1978; Prockop et al., 1979; Sage and Bornstein, 1982; Kivirikko and Myllylä, 1984). The

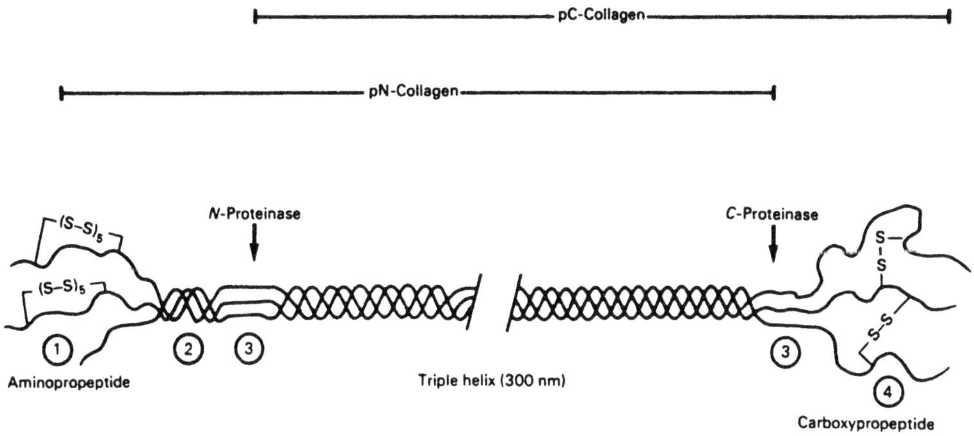

Figure 1: Structural model and processing of procollagen 1. Procollagen type I, as well as types II and III, consists of a 300 nm long triple helix and a compact aminoterminal and carboxyterminal propeptide. The aminopropeptide is composed of a globular domain with intrachain disulfide bonds (1), a short triple helical segment (2) and an aminoterminal region (3), whereas the carboxypropeptide contains a larger globular domain stabilized by interchain disulfide bonds (4) and a carboxyterminal telopeptide region (3). Specific N- and C-proteinases cleave the propeptides from the molecule at a single site in the telopeptide regions thus speeding up fibril formation. PN- and PC-collagen denote intermediate conversion products, found in tissues or cell culture media. Diagram taken from Timpl (1984b).

propeptides have a function in α-chain selection, which is of importance during the biosynthesis of correctly aligned triple-helical collagen in the rough endoplasmic reticulum of cells. During or shortly after export from the cell, the propeptides are then removed almost quantitatively by specific procollagen endopeptidases. This allows the formation of fibrils with an ordered lateral alignment of collagen triple-helices and a subsequent covalent crosslinking by the enzyme lysyl-oxidase (Fietzek and Kühn, 1976; Hofmann et al., 1980; Piez, 1984). The intermediate steps in fibril formation and the molecular packing of collagen molecules in fibrils are, however, still controversial (reviewed by Piez, 1984). Certain types of procollagen molecules do not, however, undergo extensive processing in the ECM, and this hold true especially for collagens type IV and VI. The propeptides of these collagen types appear to be necessary for their self-assembly into unique supramolecular structures (see below and Timpl, 1984b).

Although a great deal of homology exists between collagens I, II and III, amino acid sequence analysis has shown that collagen α-chains are genetically quite conservative and that there are only minor amino acid substitutions, when the sequences of e.g., $\alpha 1$ (I) from chicken or calf are compared, whereas much less homology exists between genetically distinct α-chains even from one species (Fietzek and Kühn, 1976). This allowed the production of collagen type specific antibodies (Timpl, 1976, 1984a), which have become valuable tools in cell biology.

Collagen genes are of unusual complexity (Boedtker and Aho, 1984) and a single α-chain may be coded for by as many as 35 exons. Recombinant DNA techniques have gained rapid entrance into connective tissue research and appear specially suited to elucidate the primary structure of large, cross-linked macromolecules.

The various types of collagen, their chain composition and major tissue localizations are listed in Table 1. An attempt to depict their molecular structure and extracellular processing is shown in Fig. 2.

Collagens I and III (Fig. 2a) are encountered in almost all interstitital connective tissues as the typical fibrillar collagen (Piez, 1984; Miller and Gay, 1982). Although it was long believed that they were constituents of two different fibril populations with type III as the delicate reticular collagen, there is now growing evidence of a population of mixed fibrils composed of both type I and type III collagens. Thus, Henkel and Glanville

Table I: Collagens and their localization

Type	Chain Composition	Localization
I	$\alpha1(I)_2\alpha2(I)$	major interstitial fibrils
	$\alpha1(I)_3$	skin, cornea, bone
II	$\alpha1(II)_3$	hyaline cartilage, vitreous
III	$\alpha1(III)_3$	interstitial fibrils
IV	$\alpha1(IV)_2\alpha2(IV)$	basement membranes
V	$\alpha1(V)_2\alpha2(V)$	minor interstitial fibrils,
	$\alpha1(V)\alpha2(V)\alpha3(V)$	around muscle cells,
	$\alpha1(V)_3$	intima of large vessels
VI	$\alpha1(VI)\alpha2(VI)\alpha3(VI)$	interstitial microfibrils
VII	$\alpha1(VII)_3$	anchoring fibrils
VIII	$\alpha1(VIII)_3$	descemets membrane
IX	$\alpha1(IX)\alpha2(IX)\alpha3(IX)$	cartilage, similar to type II
X	$\alpha1(X)_3$	hypertrophic cartilage

Ten different types of collagen, which are both chemically and immunologically distinct, have been described so far. Their ultrastructural correlation or typical localization in tissues is indicated. Each collagen molecule is composed of three identical or genetically different α-chains. The α1-trimer, $\alpha2(I)_3$, of type I collagen has been found as a trace component in several tissues or in cell culture media (Miller and Gay, 1982). For type V collagen at least three isomers of unknown biological function exist. It has not been unequivocally proven, whether type VI collagen is composed of three genetically distinct chains. For further details refer to Fig. 2 and the text with the citations given therein.

(1982) could isolate a substantial percentage of proteolytic peptides from fibrillar aortic collagen, which were derived from intermolecular covalent crosslinks between collagens I and III. The two collagens are, however, not entirely codistributed and type III is absent from, e.g., fully calcified bone and dentine (Veis, 1984), but it transiently reappears during bone remodeling (Becker et al., 1986b). Using indirect immunofluorescence, collagen III is more abundant in distensible structures such as arteries, upper dermis, intestinal and uterine walls (Fig. 5d). Its proportion is increased in the early phase of skin wound healing and of hepatic fibrosis (Gay and Miller, 1978). Contrary to the immunohistological findings, a biochemical analysis demonstrated an equal ratio of collagen I to III in all levels of the dermis (Epstein and Munderloh, 1978). Another remarkable feature is that a considerable proportion of collagen I and III retains its N-terminal propeptide after fibril assembly in the extracellular space. As shown by immuno-electron-microscopy (Fleischmajer et al., 1981), antibodies against the N-terminal propeptide of type I collagen decorate dermal fibrils with diameters up to 40 nm. On the other hand, antibodies against the N-terminal propeptide of type III collagen label fibrils with a diameter of up to 60 nm as well as do antibodies against triple-helical type III collagen, which suggests that the surface of these intermediate-sized fibrils may be coated by a layer of incompletely processed procollagen III molecules. It is generally believed that the N-terminal procollagen type III peptide has to be removed from the surface before collagen (III) fibrils can grow. The fact that a radioimmunoassay developed for this peptide (Rohde et al., 1979) detects highly elevated antigen levels in the serum of patients with active hepatic fibrosis strongly supports this concept (Hahn and Schuppan, 1985).

Collagen II (Fig. 2a) is the predominant fibrillar species in tissues exposed to compressive forces such as hyaline cartilage, nucleus pulposus of the vertebral disc and vitreous body of the eye (Miller and Gay, 1982).

Collagen IV (Fig. 2b), a major component of all basement membranes (BM) (Fig. 5a and c), possesses structural features, which are quite different from the fibrillar collagens. The procollagen IV molecule is considerably longer (Approx. 400 nm, Oberbäumer et al., 1982) than procollagens I-III (approx. 320 nm). Its N- or C-terminal domains are not removed by extracellular peptidases, but serve to assemble type IV collagen into a two- or three-dimensional network (Timpl et al., 1981). Four N-terminal ends are crosslinked in an antiparallel fashion forming a compact four-armed structure, named 7-S collagen, which can be extracted from tissues in homogeneous form after protease digestion (Risteli et al., 1980). The C-terminal ends are condensed to the globular hexameric NCl-domain (derived from two triple-helical procollagen IV molecules). It has been shown recently that the NCl-globule harbors the autoantigen of Goodpasture's syndrome, which involves pulmonary and renal BM (Butkowski et al., 1985) and cDNA clones for this part of the molecule are now available (Pihlajaniemi et al., 1985). The chain composition of collagen IV has long been disputed, but it is now generally agreed to be $\alpha1(IV)_2\alpha2(IV)$ (Trueb et al., 1984). The primary structure of type IV collagen is unusual in that it contains some short interruptions of the sequence $(Gly\text{-}Xaa\text{-}Yaa)_n$ (Schuppan et al., 1980), the continuity of which had long been considered a prerequisite for the folding of individual chains into a triple-helix. Such interruptions which could be visualized as hyperflexible triple-helical regions in the electron microscope after rotary shadowing the molecules (Glanville et al., 1982) may add to the flexibility of, e.g., vascular BM, which have to withstand changing blood pressure or vascular filling. Collagen IV has been the focus of much research, since BM are assumed to play a center part in morphogenesis, differentiation (Vracko, 1982), tumor metastasis (Barsky et al., 1983) or the microangiopathy of diabetes mellitus (Williamson and Kilo, 1977). Fragments of collagen IV and the BM glycoprotein laminin (see below) can be detected in serum and have been found elevated in rats with some tumors (Dickneite et al., 1985) or experimental diabetes mellitus (Risteli et al., 1982). In the human, however, significant rises in serum antigens have only been observed in fibrogenic liver diseases for 7-S collagen including the BM glycoprotein laminin (Niemelä et al., 1985) and the NCl-domain, the latter of which has been shown to reflect degradation of type IV collagen (Schuppan et al., 1986).

Collagen V (Fig. 2c) was first isolated by Burgeson et al. (1976) and Chung et al. (1976). Although a minor collagen, type V has attracted much attention, since its relative tissue content highly increased in atherosclerotic vessels (Ooshima, 1981; Morton and Barnes, 1982) and it appears in the early formation of granulation tissue (Kurita et al., 1985). Collagen V is produced by smooth muscle cells (Kumamoto and Fessler, 1980) and fibroblasts (Fessler et al., 1981b), where it might serve as a cellular exocytoskeleton (Gay et al., 1981). Its exact localization in tissues has long been disputed due to a lack of specific antibodies, but it now emerges that type V collagen is not a constitutent of BM. In immuno-electron microscopy, interstitial filaments were decorated by type V-antibody (Modesti et al., 1985) and light microscopy revealed fine insterstitial fibers, which are concentrated around smooth muscle, close to BM and in the arterial intima (Schuppan et al., 1986a and Fig. 5b). Accordingly, collagen V can form fine fibrils in vitro (Adachi and Hayashi, 1985) and recently, the partial amino acid sequence of the $\alpha2(V)$chain derived from cDNA studies demonstrated a remarkable homology of the C-terminal propeptide and part of the α-chain with the interstitial fibrillar collagens I-III (Myers et al., 1985). Nevertheless, the biochemistry of type V collagen is till incompletely understood and essentially nothing is known about its function. Its major chain composition is $\alpha1(V)_2$-

α2(V), but molecules with α1(V)$_3$ (Fessler et al., 1981b), α1(V)α2(V)α3(V) (Rhodes and Miller, 1982) and even an α4(V)chain (Fessler et al., 1985) have been described. The nature of chains 1α and 2α from cartilage, which show homology to α1(V) and α2(V) is still unclear (Burgeson and Hollister, 1979). Procollagen V as synthesized by cells is considerably larger than procollagens I-III and its extracellular processing by sequential removal of terminal peptides is quite complex (Kumamoto and Fessler, 1980; Fessler et al., 1981a, b, 1985). As shown in radiolabeled cell culture, noncollagenous peptides of molecular weight 33,000 are disulfide-linked to the terminal proα2(V) chain (Fessler et al., 1981a), which might be important in the supramolecular assembly of type V collagen. Type V collagen extracted from tissues without prior proteolysis contained an additional globule at one end of the molecule when visualized in the electron microscope (Broek et al., 1985) indicating that larger unprocessed forms are involved in its tissue organization.

Collagen VI (Fig. 2d) is widely distributed throughout the interstitial ECM (Von der Mark et al., 1984; Hessle and Engvall, 1984; Becker et al., 1986a, b). After its initial isolation from aortic walls (Chung et al., 1976) biochemical analysis (Odermatt et al., 1983) in combination with rotary shadowing electron microscopy (Furthmyr et al., 1983) lead to a structural model, in which two type VI protomers (length 105 nm with large globular domains at the ends) are aligned to an antiparallel dimer and the dimers then form a tetramer, which is further polymerized into a chain-like macromolecule (see Fig. 2d). Subsequent studies on the intra- and extracellular processing have basically supported this model (Engvall et al., 1986). Two or three genetically distinct polypeptide chains (molecular weight 140-260,000) are involved in the formation of the collagen VI triple-helix (Hessle and Engvall, 1984; Engvall et al., 1986). Type VI collagen has been shown to be the pepsin-generated fragment of GP (glycoprotein) 140, originally characterized from elastic tissue or fibroblast cell culture media and before thought to be a different molecular species (Heller-Harrison and Carter, 1984). In immuno-electron microscopy, antibodies to type VI collagen decorate microfibrils running mainly perpendicular to the major colagen fibrils of placenta and aortic media (Von der Mark et al., 1984). In spite of its low tissue content (below 1% relative to total collagen), type VI collagen turnover appears to be quite high as reflected by comparatively high levels of the serum antigen in humans (Schuppan et al., 1985).

Collagen VII (Fig. 2e) has been purified in minor quantities after limited pepsin digestion of amniotic membranes (Bentz et al., 1983). Collagen VII is a molecule composed of three identical α-chains. It contains the longest triple-helical segment (length 424 nm) observed for a collagen and an unusually large globular domain at its C-terminal end. Via smaller N-terminal domains two molecules are disulfide-linked forming a dimer of

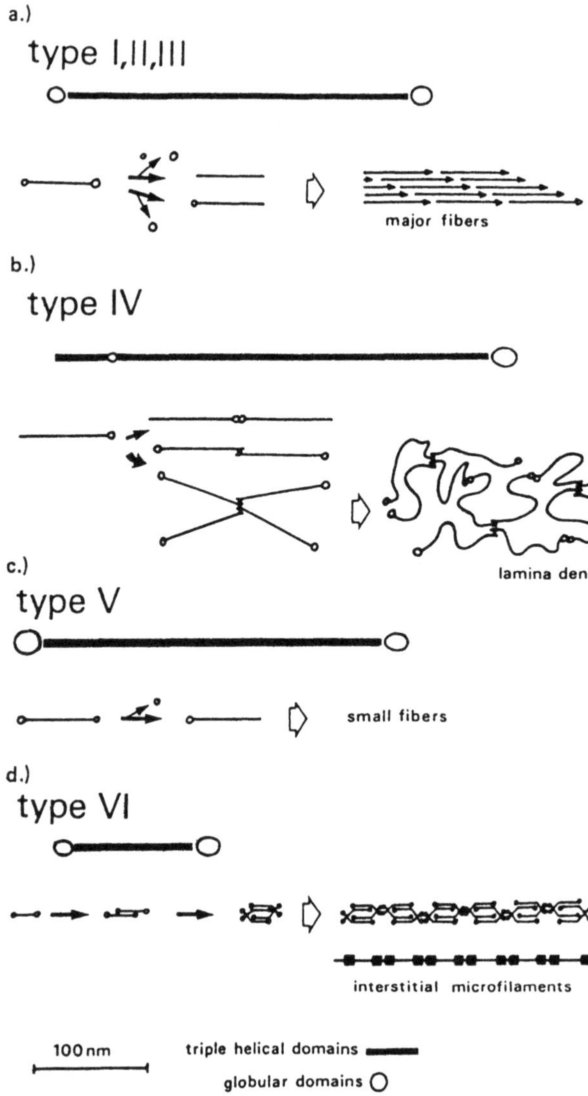

Figure 2: The collagen types and their conversion patterns. The triple helical stretches are depicted as black bars, the globular, mainly non-collagenous domains appear as open circles or ovoids according to their estimated sizes: a) the homologous collagens I, II and III form the major fibers after their propeptides are removed. The aminopropeptide is, however, partly retained in the ECM, possibly playing a role in the regulation of fibril diameter; b) collagen IV is condensated into a network via its terminal domains, which are not removed by proteolysis; c) type V collagen is similar in structure to collagens I, II and III, but contains an unusually large aminoterminal domain, which is not or only partly removed in the ECM; d) two antiparallel dimers of collagen VI form a tetramer and are finally organized into a chainlike macromolecule; e) collagen VII, which contains the longest triple helix hitherto observed and a large carboxyterminal globule, is disulfide-linked at its aminoterminus to a dimer of 780 nm length; f) little is known about collagen VIII, which might be a component of specialized tissues like Descemet's membrane; g) and h) collagens IX and X of cartilage undergo probably only little processing in the ECM. Although their primary structure is fairly well known, their macromolecular assembly is still unclear. For further details refer to the text and the literature cited therein.

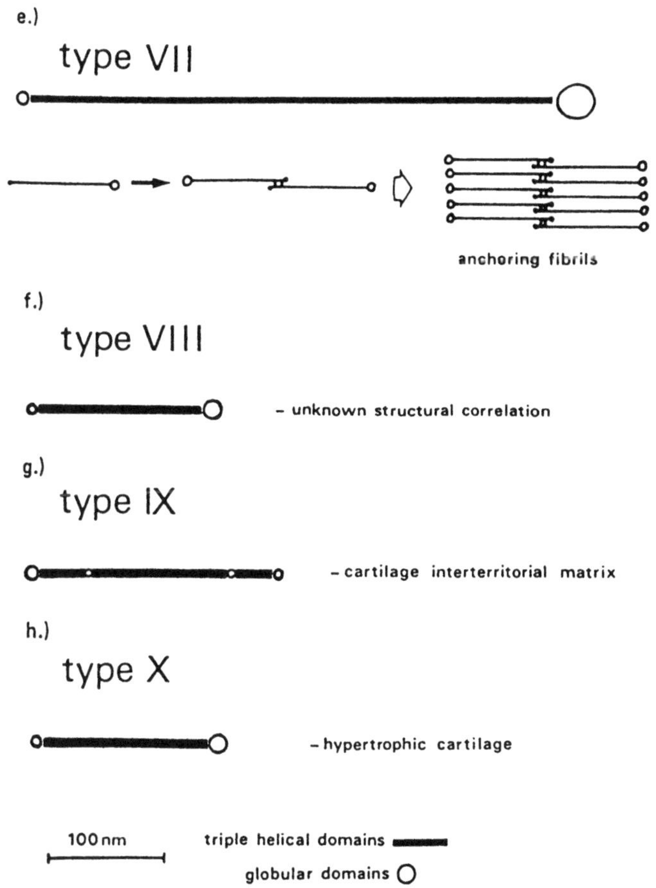

Figure 2 (cont.)

780 nm in length (Morris et al., 1986). Parallel aggregates of these dimers appear to be the structural basis of the anchoring fibrils beneath certain BM (Fig. 2e).

Collagen VIII (Fig. 2f) had been detected in cultures of endothelial cells (Benya, 1980; Sage et al., 1980) and was later found as a product of a variety of normal and malignant cells (Sage et al., 1984). A recent structural model based on chemical data shows that the molecule is a homotrimer composed of a short triple-helix with a smaller and a larger globular domain at the ends (Benya and Padilla, 1986). Although rabbit corneal endothelial cells may synthesize as much as 60% of their collagen as type VIII, it has not yet been isolated in chemical quantities to allow its tissue localization or characterization by microscopical methods. At present it is believed that type VIII collagen may have a function in some specialized BM such as descemets membrane of the eye (Benya and Padilla, 1986).

Collagen IX (Fig. 2g) has originally been extracted from hyaline cartilage as a complex mixture of pepsin-derived peptides (Reese and Mayne, 1981). The antigen is obviously codistributed with type II collagen in cartilagenous tissues. With the aid of a monoclonal antibody type IX collagen could be immuno-precipitated in an intact form and visualized in the electron microscope as a molecule with a short triple-helix and a prominent globular domain at one end (Irwin et al., 1985). The intact molecule, as observed by rotary shadowing, fulfilled the structural predictions from sequencing of cDNA clones (Ninomiya and Olsen, 1984). This and further studies (Van der Rest et al., 1985; Lozano et al., 1985) established the amino acid sequence of $\alpha 1(IX)$ as one of the three constituent polypeptide chains. No sequence homology was found with the known collagens I-IV and apart from a larger N-terminal and a small C-terminal noncollagenous domain, two interruptions of the triple-helical sequence, one of which harbors two disulfide links, was observed. The surprising finding that type IX collagen serves as a core protein in chicken sternal cartilage (Vaughan et al., 1985) illustrates well that only floating borders exist between different classes of ECM macromolecules.

Collagen X (Fig. 2h) was shown to be a distinct collagen type exclusively produced by hypertrophic cartilage, which is on its way to ossify (Schmid and Linsenmayer, 1982); Gibson and Flint, 1985). The molecule is a homotrimer with a length of 138 nm and a large terminal globular domain (Schmid et al., 1986). Due to its protease sensitivity its function might be to produce an ordered pattern of cartilage-removal prior to calcification (Schmid et al., 1986).

There are other less well defined proteins postulated to be new collagen types such as the low molecular weight fibroblast collagen predominantly seen in fetal cell culture (Sage, 1985), which await further characterization. Molecules like the subcomponent Clq of the first component of complement (Reid and Porter, 1981) or acetylcholinesterase (Mays and Rosenberry, 1981), which fulfill no structural role in the ECM, will be excluded from our considerations.

Elastin and Microfibrillar Protein

Elastic fibers only represent a relatively small proportion of ECM components, but are concentrated in specialized regions of the body, e.g., the bovine nuchal ligament. Elastic fibers render tissues such as skin, lung, and larger arteries pliable to mechanical stresses of distension and twisting (reviewed by Rosenbloom, 1982; Gosline and Rosenbloom, 1984). Elastin of several species has a characteristic amino acid composition with roughly one third of glycine, 10% of proline and 40% of hydrophobic amino acids (Sage and Gray, 1979). Structural analysis has been hampered for a long time by its insolubility. The isolation of a precursor molecule, tropoelastin, of Mr 72,000 from copper-deficient porcine arteries (Smith et al., 1968) has greatly speeded up research on the biosynthesis and extracellular organization of elastin. Polymerization of elastic requires the formation of two unique lysine-derived crosslinks, namely desmosine and isodesmosine, resulting in a random-coil structure with the properties of a rubber-like material (Gosline and Rosenbloom, 1984). The copper-enzyme involved in this step, lysyl-oxidase, is identical to that involved in collagen-crosslinking.

Less is known about the microfibrillar component, glycoprotein in nature, which surrounds growing and mature elastic fibers. Attempts to unambiguously identify the protein component(s) of elastic fiber microfibrils have so far been unsuccessful (Cleary and Gibson, 1983; Prosser et al., 1984).

Structural Glycoproteins

Glycoproteins have long been implicated as key elements in the structural organization of the ECM, but their characterization proved to be difficult due to their sensitivity towards tissue proteases, a tendency to aggregate and to an uncharacteristic amino acid composition (Anderson, 1976; Bach and Bentley, 1980). The identification of cold-insoluble globulin of human plasma (Mossesson and Umfleet, 1970), to be later renamed fibronectin, as a constituent of the cell surface and the ECM (Yamada and Weston, 1974), and the isolation of laminin from embryonal carcinoma cells (Chung et al., 1979) and a BM producing mouse tumor (Timpl et al., 1979) have greatly stimulated research activity and rapidly enhanced our understanding of these fascinating molecules (Hakamori et al., 1984).

Fibronection, the best known of the structural glycoproteins, is widely distributed in most interstitial connective tissues and mediates cellular adhesion to the ECM (Ruoslahti et al., 1981 and 1982; Furcht, 1983; Hakamori et al., 1984; Yamada et al., 1985). It is involved in the structural organization of granulation tissue and stimulates phagocytosis (Grinnel, 1984). Fibronection is a glycoprotein of molecular weight 440,000 composed of two homologous chains, which are disulfide-linked at the C-terminal end. The primary structures of plasma and cellular fibronection are slightly different from each other indicating that there might exist two or more genes or differential splicing of one parental gene transscript (Akiyama and Yamada, 1985). The generation of a number of smaller peptides with distinct biological activity after digestion of fibronectin by several proteases was a crucial step in the elucidation of this multidomain structure. The molecule carries regions, which interact noncovalently with heparin, cell surfaces, collagen, fibrin and staphylococcus aureus protein and which, in the case of fibrin and collagen, may be covalently crosslinked by a transglutaminase (Fig. 3). A major portion of the amino acid sequence of plasma fibronectin has been elucidated including stretches responsible for specific interactions (Petersen et al., 1983; Pierschbacher and Ruoslahti, 1984). A tetrapeptide with the sequence Arg-Gly-Asp-Ser, which was identified as a region promoting cell attachment (Pierschbacher and Ruoslahti, 1984), has recently been used to purify the cellular fibronection receptor, a membrane protein of molecular weight 140,000, by affinity chromatography (Pytela et al., 1985). The receptor is diffusely located on migrating neural crest cells, but assumes an ordered pericellular pattern contacting with cytoskeletal elements in stationary cells (Duband et al., 1986). There is, however, evidence that other cellular binding sites for fibronectin exist (McCarthy et al., 1986) and one of them is possibly represented by gangliosides (Thompson et al., 1986).

Vitronection, another distinct adhesion and spreading factor, with a molecular weight of 70,000, has been characterized by Hayman et al., (1983). It is present in serum and essentially codistributed with fibronectin on cell surfaces and in the ECM. The complete amino acid sequence of vitronectin as derived from protein and cDNA studies has revealed the triplet Arg-Gly-Asp similar to the cell-attachment region of fibronectin and a

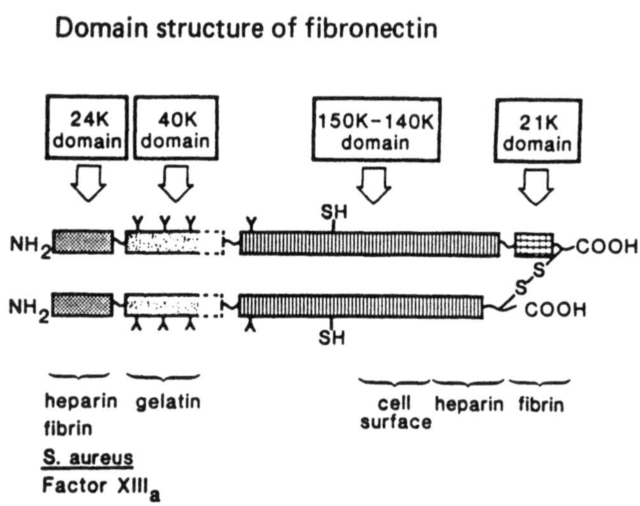

Figure 3: The two homologous subunits of hamster fibronectin are disulfide-linked at the carboxyterminus. Fragments produced after proteolysis, which retain specific biological activities are indicated on top. The domain f 21 kD is only present in the larger subunit. Each chain harbours binding sites for factor XIII, staphylococcus aureus, cell surfaces, gelatin (collagen) and two binding regions for heparin and fibrin. Y denotes carbohydrate side chains, SH a free sulfhydryl group. Taken from Hakamori et al. (1984).

basic glycosaminoglycan-binding site, but no other significant homology with fibronectin (Suzuki et al., 1985). Surprisingly, the first amino acids of vitronectin entail the entire sequence of somatomedin B.

Undulin has recently been isolated in substantial amounts from newborn monkey skins. It is a glycoprotein of moleuclar weight 700,000 composed of three constituent chains (Schuppan et al., 1986c) and immunlogically as well as biochemically distinct from the known ECM proteins. It has a remarkable affinity to collagen I and appears to be particularly sensitive to tumor proteases. Rotary shadowing electron microscopy visualizes undulin-monomer as a large globule linked to a smaller nodule via a 70 nm long arm or as intricately interconnected polymers connected specifically to type I collagen. Undulin is located around and between thick collagen fibers in immuno-electron microscopy and in indirect immunofluorescence exhibits an unprecedented pattern of uniform parallelly aligned fibers, which is limited to dense connective tissues (Fig. 5e and f). There are indications that undulin is produced by specialized mesenchymal cells and that it plays a major role in the supramolecular organization of collagen fibrils.

Several other glycoproteins located to the interstitial ECM have been described awaiting further characterization. These include hexabrachion, a large six-armed structure from cell surfaces (Erickson and Inglesias, 1984), glycoprotein 115 from chick blood vessels (Colombatti et al., 1985) and chick myotendinous antigen, which is found in myotendinous junctions (Chiquet and Farmbrough, 1984). The glycoproteins of bone have been reviewed by Butler (1984) and those of cartilage by Heinegard and Paulsson (1984). Furthermore, an attachment protein in cartilage, named chondronectin in analogy to fibronectin, has been described (Mollenhauer and Von der Mark, 1983).

Laminin, as another large multifunctional protein, has gained widespread attention due to its multiple biological activities (Timpl and Martin, 1982; Timpl et al., 1982; Hakamori et al., 1984; Kleinman et al., 1985). It is a cross-shaped molecule with a molecular weight of 900,000 and together with type IV collagen is a major component of all BM. Laminin possesses domains, which have a specific affinity for heparin, type IV collagen and the surface of mainly epithelial and endothelial cells (Fig. 4). The molecule has been assigned a central role in anchoring the cell to the various BM components including heparan sulfate proteoglycan and nidogen (see below) and thus is the focus of major interest in BM pathology (Martinez-Hernandez and Amenta, 1983). A putative laminin-receptor with a molecular weight of 68,000 has been isolated from muscle (Lesot et al., 1983) and fibrosarcoma cells (Malinoff and Wicha, 1983) and its expression is increased on the surface of (metastasizing) tumor cells (Hand et al., 1985). Laminin preferentially acts on epithelial cells regulating their growth, morphology and differentiation (Kleinman et al., 1985).

Whereas fetal sensory ganglia neurons attach equally well to both fibronectin and laminin (Baron Van-Evercooren et al., 1982), differentiated neurons prefer laminin as single substrate (Manthorpe et al., 1983). Accordingly, only laminin proved to be the active matrix factor promoting neurite-outgrowth (Lander et al., 1985). Analysis of the amino acid sequence of the three constituent laminin chains by the aid of cDNA probes has been initiated for the B1 and B2 chains (Barlow et al., 1984) and is close to completion for the rest of the molecule as will be discussed in a subsequent chapter.

Nidogen/Entactin. Nidogen was originally isolated as a fragment of molecular weight 80,000 from a BM producing mouse tumor (Timpl et al., 1983). Later, a supposedly intact form of molecular weight 150,000 could be characterized, which was visualized after rotary shadowing electron microscopy as a dumbell-shaped molecule composed of two differently sized globules interconnected by a short strand (Dziadek et al., 1985b). Nidogen tends to aggregate via the larger globular domain and is unusually sensitive towards proteases. As laminin, it is only found in BM and there is mounting evidence that nidogen is identical to entactin, a sulfated BM glycoprotein extracted previously from an endodermal cell culture (Carlin et al., 1981).

Due to their specific interaction repertoire (Timpl et al., 1984) mixtures of BM

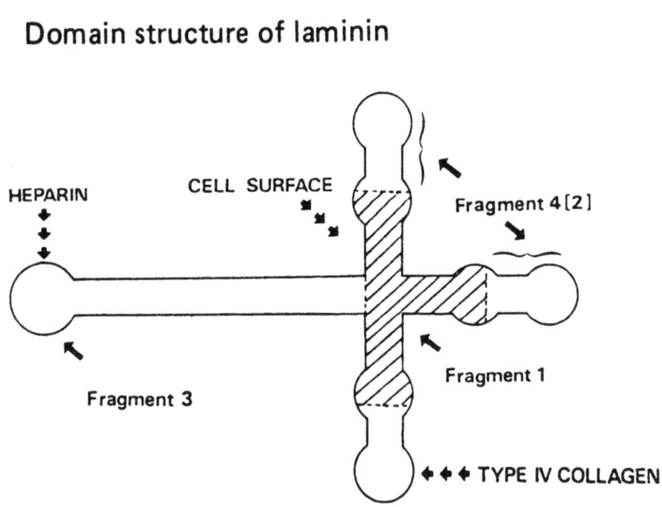

Figure 4: Laminin is a cross-shaped molecule composed of three constituent disulfide-linked chains (A, B1 and B2). The hatched area denotes the protease-resistant central domain (fragment 1) with the supposed cell binding region. Fragment 3 binds to heparin and fragment 4 (and subfragment 2) have an affinity for type IV collagen. Adapted from Timpl et al. (1982) and Hakamori et al. (1984).

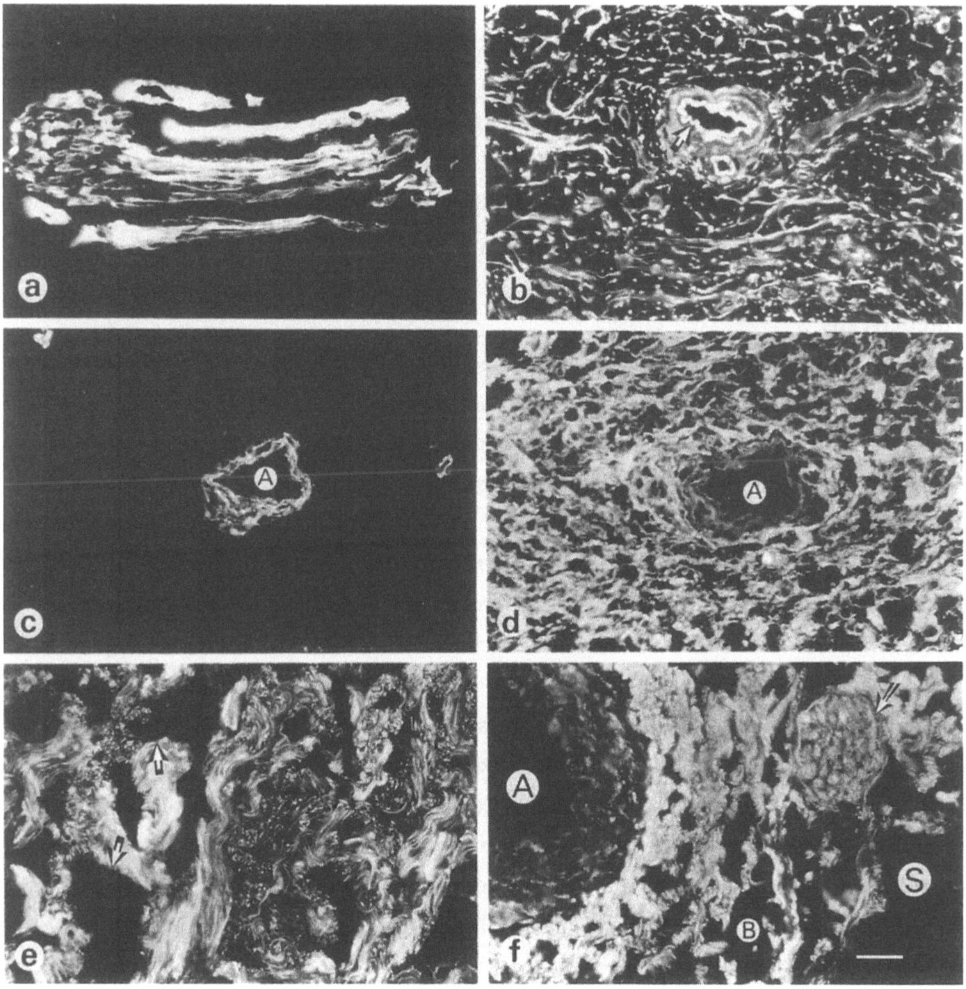

Figure 5: Immunolocalization of some ECM proteins in human tissue sections by indirect immunofluorescence. 4 µm cryostat sections were incubated with affinity-purified antibodies followed by rhodamine- or fluoresceine-conjugated second antibodies (in cooperation with Dr. J. Becker, Berlin); x 250; bar = 40 µm:

a) Type IV collagen (7-S domain) sharply delineates the axons of a myelinated nerve in desmodontium, whereas the surrounding ECM is completely negative. b) Fine interstitial fibers of type V collagen, which are partly sectioned transversely, in the oral mucosa. In an arteriole (arrow) the intima is predominantly labeled by the antibodies. c) and d) A section of dental pulp double-stained for type IV collagen (NCl domain) in c) and the aminopropeptide of type III collagen in d). Note that only the media of an arteriole (A) is stained for collagen IV, whereas procollagen III is amorphously distributed in the remaining stroma. e) in deeper layers of the dermis, undulin displays a unique pattern of uniform, undulating interstitial fibers. Fiber bundles are occasionally cut transversely thus appearing in a punctuate fashion. f) Undulin in a large portal tract of the liver. The media of an artery (A) is only faintly stained. The stroma of a myelinated nerve (arrow) is strongly positive. B = bile duct, S = sinusoidal space.

components reversibly assemble into a gel-like matrix (Kleinman et al., 1986). This co-operative assembly involves type IV collagen, heparan sulfate proteoglycan, laminin and nidogen/entactin and occurs spontaneously in a narrow concentration range for the individual components. These reconstituted matrices are ideal substrates to stimulate the growth and differentiation of certain cells (Kleinman et al., 1986). Based on electron microscopic studies Yurchenco et al. (1986) proposed several models for BM self-assembly.

Proteoglycans

Proteoglycans (PG) are compound macromolecules. They consist of a glycoprotein core to which chains of glycosaminoglycans (GAG) are covalently attached. Research on PG has been greatly expanded and a number of comprehensive reviews on their structure and function is now available (Silbert, 1982; Hascall and Kimura, 1982; Heinegard and Paulsson, 1984; Paulsson and Heinegard, 1984; Kuettner and Kimura, 1985; Poole, 1986) and most of the following data have been extracted from them. Five types of GAG are known, namely hyaluronic acid (HA), chondroitin sulfate, dermatan sulfate, keratan sulfate and heparan sulfate. They form unbranched chains of modified hexose sugars, which are highly negatively charged and only HA is not esterified with sulfate. The sulfated GAG may contain from 5-120 hexose residues per chain, whereas HA is remarkably larger (Table 2). HA proper is not covalently linked to core proteins, but is involved in the formation of aggregating PG (see below). Cartilage, which is exposed to extreme compressional forces, contains the highest proportion of GAG, mainly in the form of chondroitin sulfate, amounting to approx. 10% wet tissue weight, in contrast to the other tissues like skin or bone with less than 1% (Table 2). GAG display an unfolded structure with a high water-binding capacity, caused by the negative charges of the hexose building blocks and the resultant electrostatic repulsion. Consequently, GAG and PG fulfill a major function in structuring the solute space.

PG are polydispersed both with regard to the length of their GAG chains and to the size of the core proteins, which appear to be constantly processed by proteolytic events in the ECM (Heinegard and Paulsson, 1984). PG may roughly be divided into three structural groups, "small non-aggregating," "large non-aggregating," and "large aggregating PG." There is evidence that certain PG of similar size contain immunologically related core proteins (Heinegard and Paulsson , 1984).

<u>Small PG</u> (Fig. 6a), molecular weight 70-130,000, are composed of a core protein, which usually represents about 50% of the total weight, to which only 2-4 GAG side chains are linked. They are found in virtually all tissues of the body.

GAG Classes

	residues/chain	localization
HA (hyaluronic acid)	50-10.000	cartilage, fibroblast (early form)
CS (chondroitin sulfate)	20-60	cartilage (10% wet w.), cell surface
DS (dermatan sulfate)	30-80	skin, aorta (0.1-1% wet w.)
KS (keratan sulfate)	5-50	bone, cornea
HS (heparan sulfate)	10-120	skin, cell surface, BM

Table II: Five classes of glycosaminoglycans (GAG) exist. GAG are linear polymers consisting of two or three different alternating modified hexose sugars. All are highly negatively charged and carry sulfate groups with the exception of hyaluronic acid (HA). They occur in virtually all tissues intra-, peri- and extracellularly, but individually show preferred localizations. The sulfated GAG may be covalently linked to proteins thus forming proteoglycans (PG, see below). The values are taken from Heinegard and Paulsson (1984).

Large non-aggregating PG (Fig. 6b) are of high molecular weight and consist of a tiny protein core as in the case of mast cell heparan sulfate PG or of a large protein core with fewer GAG chains as in the case of the follicular fluid dermatan sulfate PG and the large BM heparan sulfate PG.

Aggregating PG (Fig. 6c) are complexes of HA, PG and link protein. Their core protein is loaded with 50-500 GAG chains, keratan sulfate/chondroitin sulfate for the cartilage PG and chondroitin sulfate/dermatan sulfate for the skin PG to yield molecules with a molecular weight of several million dalton. In addition, their core proteins carry a specific protein sequence for the attachment to HA, which is absent from the large non-aggregating PG. In the extracellular space, the complex is safely locked in position by core protein-specific link proteins (molecular weight ca. 50,000). Since several PG can complex with one chain of HA, giant aggregates can be formed.

Aside from their function as a molecular spacer PG/GAG fulfill specific biological and structural tasks. The heparan sulfate PG of BM are at present most intensively studied and their contribution to the assembly of a supramolecular complex has been mentioned above. Two classes of heparan sulfate PG of BM exist, as isolated from a mouse tumor, one of high density with a molecular weight around 130,000 and a low protein content and one of low density with a molecular weight above 400,000 and ca. 50%

Small PG (2-4 GAG chains)

	mol wt	tissue
1. KS-PG	72.000	cornea
2. CS-PG	76.000	cartilage, bone
3. CS/DS-PG	85.000	skin, sclera aorta etc.
4. HS-PG	130.000	liver, BM

a

Large PG/non-aggregating (4-200 GAG chains)

	mol wt	tissue
1. HS-PG	1.000.000	mast-cells

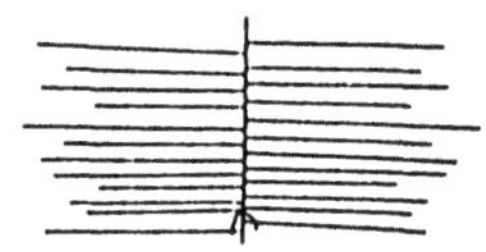

	mol wt	tissue
2. DS-PG	1.000.000	follic. fluid
3. HS-PG	450.000	BM

b

Large PG/aggregating (50-500 GAG chains)

	mol wt	tissue
1. KS/CS-PG	3.000.000	cartilage
2. CS/DS-PG	1.000.000	skin,glia, aorta

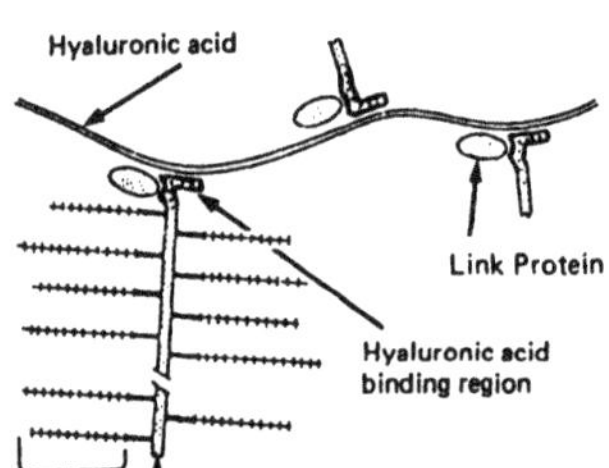

c

Figure 6: Classes of proteoglycans (PG). The molecular weights (mol wt) and tissue localization of some examples are given in the accompanying tables. Data have been condensed from Silbert (1982) and Heinegard and Paulsson (1984) and figures modified with permission. KS = keratan sulfate, CS = chondroitin sulfate, DS = dermatan sulfate, HS = heparan sulfate, GAG = glycosaminoglycan.
a) small PG with only 2-4 GAG chains covalently connected to the protein core. Carbohydrates are indicated by the short extensions. Two different GAG may be present in one PG molecule as is the case with the CS/DS-PG in skin. b) large, aggregating PG with more GAG chains and usually a larger protein core, e.g., the DS-PG of follicular fluid and the HS-PG of BM. An exception is the mast-cell HS-PG with only a tiny protein core. c) large, aggregating PG are supramolecular complexes of PG, which are non-covalently attached to hyaluronic acid via a specific binding region. This connection is further locked into position by link protein. The PG contain at least 50 GAG chains, either KS and CS or DS and CS.

protein, which are immunologically unrelated (Hassel et al., 1980; Fujiwara et al., 1984; Dziadek et al., 1985a). Similar molecular forms have been obtained from other tissues and their polydispersity with regard to molecular size may be explained by continuous extracellular degradation of the core protein to smaller fragments (Ledbetter et al., 1985). The high density form of heparan sulfate has been localized mainly to the laminae rarae of glomerular BM (Stow et al., 1985), but other cell membrane-associated forms exist. A heparan sulfate PG on the surface of dorsal root ganglia is required for the stimulation of Schwann cell proliferation (Ratner et al., 1985) and this stimulation is probably mediated via an interaction with the E3 domain of Schwann cell laminin (Edgar et al., 1984).

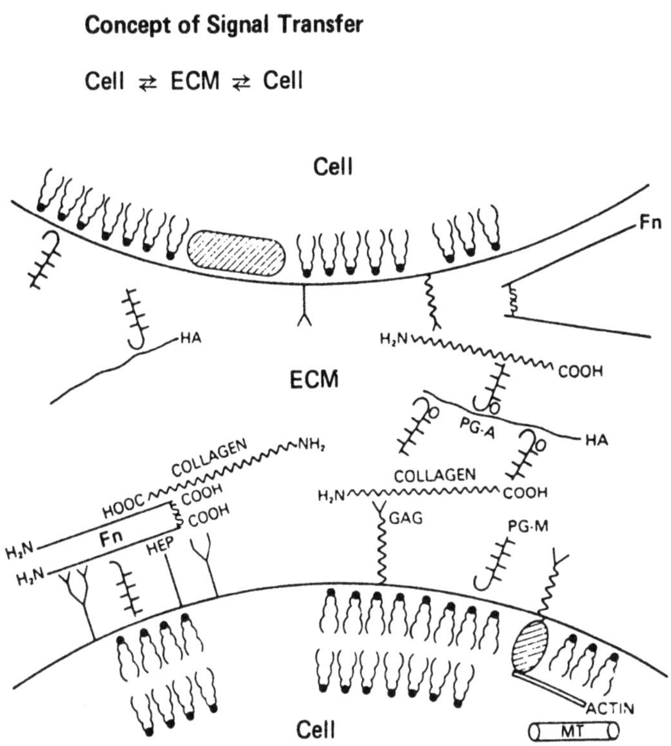

Figure 7: Schematic model for the signal-transfer between ECM and the cell (modified from Reddi, 1984). Cell surface receptors, which may be protein, heparin and heparan sulfate (hep) or proteoglycan-monomers (PG-M), interact with intracellular microfilaments, via actin bundles and microtubules (MT). Extracellularly, these receptors receive information from ECM macromolecules like fibronectin (Fn), collagen and proteoglycan-aggregate (PG-A). The matrix laid down by one cell thus influences and modulates the morphological and biochemical phenotype of a neighbouring cell as, e.g., in mesenchymal epithelial interactions.

In conclusion, it has become clear that the ECM is not simply a structural frame or an inert playground for cells, but that cellular behavior is greatly influenced by its composition in qualitative and quantitative terms. These influences are exerted upon the cells by specialized macromolecules with unique structural features, which form distinct supramolecular assemblies mediated by functional domains. It emerges that a continuous signal transfer directed from the nucleolus of the cell to the ECM and vice versa takes place, which involves the cytoskeleton and transmembrane receptors. Fig. 7 illustrates that the plasma membrane is not the ultimate barrier between the cell and the ECM and that adjacent cells may influence each other by the deposition of a specialized matrix. There are numerous such examples in development and organogenesis. We are now on the way to understand the mechanisms involved on a molecular level.

Acknowledgements. We are obliged to Ms. Herwig for typing the manuscript.

References

Adachi, E., and T. Hayashi (1985) In vitro formation of fine fibrils with a D-periodic banding pattern from type V collagen. Collagen Rel. Res. **5**: 225-232.

Akiyama, S.K., and K. Yamada (1985) Comparison of evolutionary distinct fibronections: Evidence for the origin of plasma and fibroblast cellular fibronectin from a single gene. J. Cell. Biochem. **27**: 97-107.

Anderson, J.C. (1976) Glycoproteins of the connective tissue matrix. Int. Rev. Connect. Tissue Res. **7**: 251-322.

Bach, P.R., and J.P. Bentley (1980) Structural glycoprotein, fact or artefact. Connect. Tissue Res. **7**: 185-196.

Barlow, D.B., N.M. Green, M. Kurkkienen, and B.L.M. Hogan (1984) Sequencing of laminin B chain cDNAs reveals C-terminal regions of coiled-coilalpha-helix. EMBO J. **3**: 2355-2362.

Baron-Van Evercooren, A., H.K. Kleinman, S. Ohno, P. Marangos, J.P. Schwartz, and M.E. Dubois-Dalcq (1982) Nerve growth factor, laminin and fibronectin promote nerve growth in human fetal sensory ganglia neurons. J. Neurosci. Res. **8**: 179-193.

Barsky, S.H., G.P. Siegal, F. Janotta, and L.A. Liotta (1983) Loss of basement membrane components by invasive tumors but not by their benign counterparts. Lab. Invest. **49**: 140-147.

Becker, J., D. Schuppan, H. Benzian, T. Bals, E.G. Hahn, C. Cantaluppi, and P. Reichart (1986a) Immunohistochemical distribution of collagens type IV, V, VI and procollagens type I and III in human alveolar bone and dentine. J. Historchem. Cytochem., in press.

Becker, J., D. Schuppan, E.G.Hahn, G. Albert, and P. Reichart (1986b) The immunohistochemical distribution of collagens tpye IV, V, VI and of laminin in the human oral mucosa. Archs. Oral Biol. **31**: 179-186.

Bentz, H., N.P. Morris, L.W. Murray, L.Y. Sakai, D.W. Hollister, and R.E. Burgeson (1983) Isolation and partial characterization of a new human collagen with an extended triple-helical structural domain. Proc. Natl. Acad. Sci. (USA) **80**: 3168-3172.

Benya, P.D. (1980) EC collagen: Biosynthesis by corneal endothelial cells and separation from type IV without pepsin treatment or denaturation. Renal Physiol. **3**: 30-35.

Benya, P.D., and S.R. Padilla (1986) Isolation and characterization of type VIII collagen synthesized by cultured rabbit corneal endothelial cells. J. Biol. Chem. **261**: 4160-4169.

Boedtker, H., and S. Aho (1984) Collagen gene structure: The paradox may be resolved.

Biochem. Soc. Symp. **49:** 67-84.
Bornstein, H., and H. Sage (1980) Structurally distinct collagen types. Ann. Rev. Biochem. **49:** 957-1003.
Broeck, D.L., J. Madri, E.G. Eikenberry, and B. Brodsky (1985) Characterization of the tissue form of type V collagen from chick bone. J. Biol. Chem. **260:** 555-562.
Burgeson, R.E., F. El Adli, I.I. Kaitila, and D.W. Hollister (1976) Fetal membrane collagens: Identification of two new collagen alpha chains. Proc. Natl. Acad. Sci. (USA) **73:** 2579-2583.
Burgeson, R.E., and W.E. Hollister (1979) Collagen heterogeneity in human cartilage: Identification of several new collagen chains. Biochem. Biophys. Res. Commun. **87:** 1124-1131.
Butkowski, R.J., J. Wieslander, B.J. Wisdom, J.F. Barr, M.E. Noelken, B.G. Hudson (1985) Properties of the globular domain of type IV collagen and its relationship to the Goodpasture antigen. J. Biol. Chem. **260:** 3939-3947.
Butler, W.T. (1984) Matrix macromolecules of bone and dentin. Collagen Rel. Res. **4:** 297-307.
Carlin, B., R. Jaffe, B. Bender, and A.E. Chung (1981) Entactin, a novel basal lamina-associated glycoprotein. J. Biol. Chem. **256:** 5209-5214.
Chiquet, M., and D.M. Farmbrough (1984) Chick myotendinous antigen II. A novel extracellular glycoprotein complex consisting of large disulfide-linked subunits. J. Cell Biol. **98:** 1937-1946.
Chung, A.E., R. Jaffe, I.L. Freeman, J.P. Vergnes, J.E. Bragniski, and B. Carlin (1979) Properties of a basement membrane related glycoprotein synthesized in culture by a mouse embryonal carcinoma-derived cell line. Cell **16:** 277-287.
Chung, E., R.K. Rhodes, and E.J. Miller (1976) Isolation of three collagenous components of probable basement membrane origin from several tissues. Biochem. Biophys. Res. Commun. **71:** 1167-1174.
Cleary, E.G., and M.A. Gibson (1983) Elastin-associated microfibrils and microfibrillar protein. Int. Rev. Connect. Tissue Res. **10:** 97-209.
Colombatti, A., G.M. Bressan, I. Castellani, and D. Volpin (1985) Glycoprotein 115, a glycoprotein isolated from chick blood vessels is widely distributed in connective tissue. J. Cell Biol. **100:** 18-26.
Dickneite, G., H.H. Sedlacek, R. Timpl, and F.R. Seiler (1985) Tumor growth and metastasis increases the serum concentration of basement membrane proteins. Behring Inst. Mitt. **78:** 159-166.
Duband, J.L., S. Rocher, W.T. Chen, K.M. Yamada, and J.P. Thiery (1986) Cell adhesion and migration in the early vertebrate embryo: Location and possible role of the putative fibronectin receptor complex. J. Cell Biol. **102:** 160-178.
Dziadek, M., S. Fujiwara, M. Paulsson, and R. Timpl (1985a) Immunological characterization of basement membrane types of heparan sulfate proteoglycan. EMBO J. **4:** 905-912.
Dziadek, M., M. Paulsson, R. Deutzmann, R. Timpl, S. Weber, and J. Engel (1985b) Structure and function of nidogen. In *Basement Membranes*, S. Shibata, ed., pp. 13-23. Elsevier, New York.
Edgar, E., R. Timpl, and H. Thoenen (1984) The heparin-binding domain of laminin is responsible for its effects on neurite outgrowth and neuronal survival. EMBO J. **3:** 1463-1467.
Engvall, E., H. Hessle, and G. Klier (1986) Molecular assembly, secretion, and matrix deposition of type VI collagen. J. Cell Biol. **102:** 703-710.
Erickson, H.P., and J.L. Inglesias (1984) A six-armed oligomer isolated from cell-surface fibronection preparations. Nature (Lond.) **311:** 267-269.
Epstein, E.H., and N.H. Munderloh (1978) Human skin collagen: Presence of type I and type III at all levels of the dermis. J. Biol. Chem. **253:** 1336-1337.
Eyre, D.R. (1980) Collagen: Molecular diversity in the body's protein scaffold. Science **207:** 1315-1322.
Fessler, J.H., K.J. Droege, K.G. Duncan, and L.I. Fessler (1985) Biosynthesis of collagen.

J. Cell. Biochem. **28:** 31-37.
Fessler, J.H., and L.I. Fessler (1978) Biosynthesis of procollagen. Ann. Rev. Biochem. **47:** 129-162.
Fessler, L.I., C.A. Kumamato, M.E. Meis, and J.H. Fessler (1981a) Assembly and processing of procollagen V (AB) in chick blood vessels and other tissues. J. Biol. Chem. **256:** 9640-9645.
Fessler, L.I., W.J. Robinson, and J.H. Fessler (1981b) Biosynthesis of procollagen [(proα1V)$_2$(pro 2V)] by chick tendon fibroblasts and procollagen (proα1V)$_3$ by hamster lung cell cultures. J. Biol. Chem. **256:** 9646-9651.
Fietzek, P.P., and K. Kühn (1976) The primary structure of collagen. Int. Rev. Connect. Tissue Res. **7:** 1-60.
Fleischmajer, R., R. Timpl, L. Tudermann, L. Raisher, M. Wiestner, J.S. Perlish, and P.N. Graves (1981) Ultrastructural identification of extension aminopropeptides of type I and III collagens in human skin. Proc. Natl. Acad. Sci. (USA) **78:** 7360-7364.
Fujiwara, S., H. Wiedemann, R. Timpl, A. Lustig, and J. Engel (1984) Structure and interactions of heparan sulfate proteoglycans from a mouse tumor basement membrane. Eur. J. Biochem. **143:** 145-157.
Furcht, L.T. (1983) Structure and function of the adhesive glycoprotein fibronectin, Mol. Cell. Biol. **1:** 53-117.
Furthmayr, H., H. Wiedemann, R. Timpl, E. Odermatt, and J. Engel (1983) Electron microscopical approach to a structural model of intima collagen. Biochem. J. **211:** 303-311.
Gay, S., and E.J. Miller (1978) *Collagen in the Physiology and Pathology of Connective Tissue*. Fischer, Stuttgart.
Gay, S., A. Martinez-Hernandez, R.K. Rhodes, and E.J. Miller (1981) The collagenous exocytoskeleton of smooth muscle cells. Collagen Rel. Res. **1:** 377-384.
Gibson, G.J., and M.H. Flint (1985) Type X collagen synthesis by chick sternal cartilage and its relationship to endochondral development. J. Cell Biol. **101:** 277-284.
Glanville, R.W., T. Voss, and K. Kuehn (1982) A comparison of the flexibility of molecules of basement membrane and interstitial collagens. In *New Trends in Basement Membrane Research*, K. Kuehn, H.H. Schoene, and R. Timpl, eds., pp. 69-77. Raven Press, New York.
Gosline, J.M., and J. Rosenbloom (1984) Elastin. In *Extracellular Matrix Biochemistry*, K.A. Piez and A.H. Reddi, eds., pp. 191-227. Elsevier, New York.
Grinnel, F. (1984) Fibronectin and wound healing. J. Cell Biocehm. **26:** 107-116.
Hahn, E.G., and D. Schuppan (1985) Ethanol and fibrogenesis in the liver. In *Alcohol Related Diseases in Gastroenterology*, H.K. Seitz and B. Kommerell, eds., pp. 124-153. Springer, Berlin.
Hakamori, S., M. Fukuda, K. Sekiguchi, and W.G. Carter (1984) Fibronectin, laminin, and other extracellular glycoproteins. In *Extracellular Matrix Biochemistry*, K.A. Piez and A.H. Reddi, eds., pp. 229-272. Elsevier, New York.
Hand, P.H., A. Thor, J. Schlom, C.N. Rao, and L. Liotta (1985) Expression of laminin receptor in normal and carcinomatous human tissues as defined by a monoclonal antibody. Cancer Res. **45:** 2713-2717.
Hascall, V.C., and J. Kimara (1982) Proteoglycans: Isolation and characterization. Methods Enzymol. **82:** 769-800.
Hassel, J.R., P. Gehron-Robey, H. Barrach, H.J. Wilczek, S. Rennard, and G.R. Martin (1980) Isolation of a heparan sulfate containing proteogylcan from basement membrane. Proc. Natl. Acad. Sci. (USA) **77:** 4494-4498.
Hayman, E.G., M.D. Pierschbacher, J. Öhgren, and E. Ruoslahti (1983) Serum spreading factor (vitronectin) is present at the cell surface and in tissues. Proc. Natl. Acad. Sci. (USA) **80:** 4003-4007.
Heinegard, D., and M. Paulsson (1984) Structure and metabolism of proteoglycans. In *Extracellular Matrix Biochemistry*, K.A. Piez and A.H. Reddi, eds., pp. 277-328. Elsevier, New York.
Heller-Harrison, R., and W.G. Carter (1984) Pepsin-generated type VI collagen is a

degradation product of GP 140. J. Biol. Chem. **259:** 6558-6564.
Henkel, W., and R.W. Glanville (1982) Covalent crosslinking between molecules of type I and type III collagen: The involvement of the N-terminal, nonhelical regions of the $\alpha 1(I)$ and $\alpha 1(III)$ chains in the formation of intermolecular crosslinks. Eur. J. Biochem. **122:** 205-213.
Hessle, H., and E. Engvall (1984) Type VI collagen. Studies on its localization, structure and biosynthetic form with monoclonal antibodies. J. Biol. Chem. **259:** 3955-3961.
Hofmann, H., P.P. Fietzek, and K. Kühn (1980) Comparative analysis of the sequences of the three collagen chains $\alpha 1(I)$, $\alpha 2$ and $\alpha 1(III)$. J. Mol. Biol. **141:** 293-314.
Irwin, M.H., H. Sandra, and R. Mayne (1985) Monoclonal antibody against chicken type IX collagen: Preparation, characterization, and recognition of the intact form of type IX collagen secreted by chondrocytes. J. Cell Biol. **101:** 814-823.
Kivirikko, K.I., and R. Myllylä (1982) Posttranslational enzymes in the biosynthesis of collagen: Intracellular enzymes. Methods Enzymol. **82:** 245-304.
Kivirikko, K.I., and R. Myllylä (1984) Biosynthesis of the collagens. In *Extracellular Matrix Biochemistry*, K.A. Piez and A.H. Reddi, eds., pp. 83-118. Elsevier, New York.
Kleinman, H., F.B. Cannon, G.W. Laurie, J.R. Hassel, M. Anmailley, V.P. Terranova, G.R. Martin, and M. DuBois-Dalcq (1985) Biological activities of laminin. J. Cell Biochem. **27:** 317-325.
Kleinman, H., M.L. McGarvey, J.R. Hassel, V.L. Star, F.B. Cannon, G.W. Laurie, and G. Martin (1986) Basement membrane complexes with biological activity. Biochemistry **25:** 312-318.
Kuettner, K.E., and J.H. Kimura (1985) Proteoglycans: An overview. J. Cell. Biochem. **27:** 327-336.
Kumamoto, C.A., and J.H. Fessler (1980) Biosynthesis of A, B procollagen. Proc. Natl. Acad. Sci. (USA) **77:** 6434-6438.
Kurita, K., Y. Hasimoto, T. Takai, T. Kaway, and T. Hayakawa (1985) Changes in collagen types during the healing of rabbit tooth extraction wounds. J. Dent. Res. **64:** 28-32.
Lander, A.D., D.K. Fuji, and L.F. Reichardt (1985) Purification of a factor that promotes neurite outgrowth: Isolation of laminin and associated molecules. J. Cell Biol. **101:** 898-913.
Ledbetter, S., B. Tyree, J.R. Hassel, and E.A. Horigan (1985) Identification of the precursor protein to basement membrane heparan sulfate proteoglycan. J. Biol. Chem. **260:** 8106-8113.
Lesot, H., U. Kühl, and K. Von der Mark (1983) Isolation of a laminin-binding molecule from muscle cell membranes. EMBO J. **2:** 861-865.
Lozano, G., Y. Ninomiya, H. Thompson, and B.R. Olsen (1985) A distinct class of vertebrate collagen genes encodes chicken type IX collagen polypeptides. Proc. Natl. Acad. Sci. (USA) **82:** 4050-4054.
Malinoff, H.L., and M.S. Wicha (1983) Isolation of a cell surface receptor protein for laminin from murine fibrosarcoma cells. J. Cell Biol. **96:** 1475-1479.
Manthorpe, M., E. Engvall, E. Ruoslahti, F.M. Longo, G.E. Davis, and S. Varon (1983) Laminin promotes neuritic regeneration from cultured peripheral and central neurons. J. Cell Biol. **97:** 1882-1890.
Martinez-Hernandez, A., and P.S. Amenta (1983) The basement membrane in pathology. Lab. Invest. **48:** 656-677.
Mays, C., and T.L. Rosenberry (1981) Characterization of pepsin-resistant collagen-like tail subunit fragments of 18S and 14S acetylcholinesterase from electrophorus electricus. Biochemistry **20:** 2810-2817.
McCarthy, J.B., S.T. Hagen, and L.T. Furcht (1986) Human fibronectin contains distinct adhesion- and motility promoting domains for metastatic melanoma cells. J. Cell Biol. **102:** 179-188.
Miller, E.J., and S. Gay (1982) Collagens - an overview. Methods Enzymol. **82:** 3-32.
Miller, E.J., and R.K. Rhodes (1982) Preparation and characterization of the different

types of collagen. Methods Enzymol. **82:** 33-64.

Modesti, A., T. Kalebic, S. Scarpa, S. Togo, G. Grotendorst, L. Liotta, and T. Triche (1985) Type V collagen in human amnion is a 12 nm fibrillar component of the pericellular interstitium. Eur. J. Cell Biol. **35:** 246-255.

Mollenhauser, J., and K. Von der Mark (1983) Isolation and characterization of a collagen-binding glycoprotein from chondrocyte membranes. EMBO J. **2:** 45-50.

Morris, N.P., D.R. Keene, R.W. Glanville, H. Bentz, and R.E. Burgeson (1986) The tissue form of type VII collagen is an antiparallel dimer. J. Biol. Chem. **261:** 5638-5644.

Morton, L.F., and M.J. Barnes (1982) Collagen polymorphism in the normal and diseased blood vessel wall: Investigation of collagen types I, III and V. Atherosclerosis **42:** 41-51.

Myers, J.C., H.R. Loidl, C.A. Stolle, and J.M. Seyer (1985) Partial covalent structure of the human α2(V) collagen chain. J. Biol. Chem. **260:** 5533-5541.

Niemelä, O., L. Risteli, E.A. Sotaniemi, and J. Risteli (1985) Type IV collagen and laminin-related antigens in human serum in alcoholic liver disease. Eur. J. Clin. Invest. **15:** 132-137.

Ninomiya, Y., and B.R. Olsen (1984) Synthesis and characterization of cDNA encoding a cartilage-specific short collagen. Proc. Natl. Acad. Sci. (USA) **81:** 3014-3018.

Oberbäumer, I., H. Wiedemann, R. Timpl, and K. Kühn (1982) Shape and assembly of type IV procollagen obtained from cell culture. EMBO J. **1:** 805-810.

Odermatt, E., J. Risteli, V. Van Delden, and R. Timpl (1983) Structural diversity and domain composition of a unique collagenous fragment (intima collagen) obtained from human placenta. Biochem. J. **211:** 295-302.

Ooshima, A. (1981) Collagen B chain: Increased proportion in human atherosclerosis. Science **213:** 666-668.

Paulsson, M., and D. Heinegard (1984) Noncollagenous cartilage proteins. Current status of an emerging research field. Collagen Rel. Res. **4:** 219-229.

Petersen, T.H., H.C. Thogersen, K. Skorstengaard, K. Vibe-Pedersen, P. Sahl, L. Sottrup-Jensen, and S. Magnasson (1983) Partial primary structure of bovine plasma fibronectin: Three types of internal homology. Proc. Natl. Acad. Sci. (USA) **80:** 137-141.

Pierschbacher, M.D., and E. Ruoslahti (1984) Cell attachment activity of fibronectin can be duplicated by small synthetic fragments of the molecule. Nature **309:** 30-33.

Pihlajaniemi, T.K., K. Tryggvason, J.C. Myers, M. Kurkinen, R. Lebo, M.C. Cheung, D.J. Prockop, and C.D. Boyd (1985) cDNA clones coding for the pro-α1(IV) chain of human type IV procollagen reveal an unusual homology of amino acid sequences in two halves of the carboxy-terminal domain. J. Biol. Chem. **260:** 7681-7687.

Piez, K.A. (1984) Molecular and aggregate structures of the collagens. In *Extracellular Matrix Biochemistry,* K.A. Piez and A.H. Reddi, eds., pp. 1-39. Elsevier, New York.

Poole, A.R. (1986) Proteoglycans in health and disease: Structures and functions. Biochem. J. **236:** 1-14.

Prockop, D.J., K.I. Kivirikko, L. Tuderman, and A. Guzman (1979) The biosynthesis of collagen and its disorders. N. Engl. J. Med. **301:** 77-85.

Prosser, I.W., M.A. Gibson, and E.G. Cleary (1984) Microfibrillar protein from elastic tissue: A critical evaluation. Aust. J. Exp. Biol. Med. Sci. **62:** 485-505.

Pytela, R., M.D. Pierschbacher, and E. Ruoslahti (1985) Identification and isolation of a 140 kD cell surface glycoprotein with properties of a fibronection receptor. Cell **40:** 191-198.

Ratner, N., R.P. Bunge, and L. Glaser (1985) A neuronal cell surface haparan sulfate proteogylcan is required for dorsal root ganglion neuron stimulation of Schwann cell proliferation. J. Cell. Biol. **101:** 744-754.

Reddi, A.H. (1984) Extracellular matrix and development. In *Extracellular Matrix Biochemistry,* K.A. Piez and A.H. Reddi, eds., pp. 375-412. Elsevier, New York.

Reese, C.A., and R. Mayne (1981) Minor collagens of chick hyaline cartilage. Biochemistry **20:** 5443-5448.

Reid, K.B.M., and R.R. Porter (1981) The proteolytic activation systems of complement. Ann. Rev. Biochem. **50:** 433-464.

Rhodes, R.K., and E.J. Miller (1982) Evidence for the existence of an 1(V)α2(V)α3(V) collagen molecule in human placental tissue. Coll. Rel. Res. **1:** 337-343.

Risteli, J., H.P. Bächinger, J. Engel, H. Furthmayr, and R. Timpl (1980) 7-S collagen: Characterization of an unusual basement membrane structure. Eur. J. Biochem. **108:** 239-250.

Risteli, J., K.E. Draeger, G. Regitz, and H.P. Neubauer (1982) Increase in circulating basement-membrane antigens in diabetic rats and effects of insulin treatment. Diabetologia **23:** 266-269.

Rohde, H., L. Vargas, E.G. Hahn, H. Kalbfleisch, M. Bruguera, and R. Timpl (1979) Radioimmunoassay for type III procollagen peptide and its application to human liver disease. Eur. J. Clin. Invest. **9:** 451-459.

Rosenbloom, J. (1982) Elastin: Biosynthesis, structure, degradation and role in disease processes. Connect. Tissue Res. **10:** 73-91.

Ruoslahti, E., E. Engvall, and E.G. Hayman (1981) Fibronectin: Current concepts on its structure and function. Collagen Rel. Res. **1:** 95-128.

Ruoslahti, E., E.G. Hayman, M. Pierschbacher, and E. Engvall (1982) Fibronectin: Purification, biochemical properties, and biological activities. Methods Enzymol. **82:** 803-831.

Sage, H. (1985) Low molecular weight fibroblast collagen: Structure, secretion, and differential expression as a function of fetal and cellular age. Biochemistry **24:** 7430-7440.

Sage, H., G. Balian, A.M. Vogel, and P. Bornstein (1984) Type VIII collagen: Synthesis by normal and malignant cells in culture. Lab. Invest. **50:** 219-231.

Sage, H., and P. Bornstein (1982) Preparation and characterization of procollagens and procollagen-collagen intermediates. Methods Enzymol. **82:** 96-127.

Sage, H., and W.R. Gray (1979) Studies on the evolution of elastin. I. Phylogenetic distribution. Comp. Biochem. Physiol. **64B:** 313-317.

Sage, H., P. Pritzl, and P. Bornstein (1980) A unique, pepsin-sensitive collagen synthesized by aortic endothelial cells in culture. Biochemistry **19:** 5747-5755.

Schmid, T.M., and T.F. Linsenmayer (1982) Metabolism of low molecular weight collagen by chondrocytes obtained from histologically distinct zones of the chick embryo tibiotarus. J. Biol. Chem. **257:** 12451-12457.

Schmid, T.M., R. Mayne, J.J. Jeffrey, and T.F. Linsenmayer (1986) Type X collagen contains two cleavage sites for a vertebrate collagenase. J. Biol. Chem. **261:** 4184-4189.

Schuppan, D., J. Becker, H. Boehm, and E.G. Hahn (1986a) Immunofluorescent localization of type V collagen as a fibrillar component of the interstitial connective tissue of human oral mucosa, artery and liver. Cell Tissue Res. **243:** 535-543.

Schuppan, D., M. Besser, R. Schwarting, and E.G. Hahn (1986b) Radioimmunoassay for the carboxyterminal cross-linking domain of type IV (basement membrane) procollagen in body fluids. J. Clin. Invest., in press.

Schuppan, D., C. Cantaluppi, E.G. Hahn, F. Schuppan, J. Becker, R. Fleischmajer, H. Wiedemann, and K. Kühn (1986c) Undulin, a type I collagen-associated glycoprotein from the connective tissue, submitted.

Schuppan, D., T. Rühlmann, and E.G. Hahn (1985) Radioimmunoassay for human type VI collagen and its application to tissue and body fluids. Analyt. Biochem. **149:** 238-247.

Schuppan, D., R. Timpl, and R.W. Glanville (1980) Discontinuities in the triple helical sequence Gly-X-Y of basement membrane (type IV) collagen. FEBS Let. **115:** 297-300.

Silbert, J.E. (1982) Structure and metabolism of proteoglycans and glycosaminoglycans. J. Invest. Dermatol. **79:** 31s-37s.

Slavkin, H.C., and R.C. Greulich, eds. (1975) In *Extracellular Matrix influences on gene expression*. Academic, New York.

Smith, D.W., N. Weissman, W.H. Carnes (1968) Cardiovascular studies on copper-deficient swine. XII. Partial purification of a soluble protein resembling elastin. Biochem. Biophys. Res. Commun. **31:** 309-315.

Stow, J.L., H. Sawada, and M.G. Farquhar (1985) Basement membrane heparan sulfate proteoglycans are concentrated in the laminae rarae and in podocytes of the ra renal glomerulus. Proc. Natl. Acad. Sci. (USA) **82:** 3296-3300.

Suzuki, S., A. Oldberg, E.G. Hayman, M.D. Pierschbacher, and E. Ruoslahti (1985) Complete amino acid sequence of human vitronectin deduced from cDNA. Similarity of cell attachment sites in vitronectin and fibronectin. EMBO J. **4:** 2519-2524.

Thompson, L.K., P.M. Horowitz, K.L. Bentley, D.D. Thomas, J.F. Alderete, and R.J. Klebe (1986) Localization of the ganglioside-binding site of fibronectin. J. Biol. Chem. **261:** 5209-5214.

Timpl, R. (1976) Immunological studies on collagen. In *Biochemistry of Collagen*, G.N. Ramachandran and A.H. Reddi, eds., pp. 319-375. Plenum, New York.

Timpl, R. (1984a) Immunology of the collagens. In *Extracellular Matrix Biochemistry*, K.A. Piez and A.H. Reddi, eds. pp. 159-190. Elsevier, New York.

Timpl, R. (1984b) Processed and unprocessed forms of procollagens. Biochem. Soc. Transactions **12:** 924-927.

Timpl, R., M. Dziadek, S. Fujiwara, H. Nowack, and G. Wick (1983) Nidogen: A new, self-aggregating basement membrane protein. Eur. J. Biochem. **137:** 455-465.

Timpl, R., S. Fujiwara, M. Dziadek, M. Aumailley, S. Weber, and J. Engel (1984) Laminin, proteoglycan, nidogen and collagen IV: Structural models and molecular interactions. In *Basement Membranes and Cell Movement*, CIBA Fdn. Symp. 108, pp. 25-43. Pitman, London.

Timpl, R., and G.R. Martin (1982) Components of basement membranes. In *Immunochemistry of the Extracellular Matrix*, vol. II, H. Furthmayr, ed., pp. 119-150. CRC Press, Boca Raton.

Timpl, R., H. Rohde, P. Gehron-Robey, S.I. Rennard, J.M. Foidart, and G.R. Martin (1979) Laminin - a glycoprotein from basement membranes. J. Biol. Chem. **254:** 9933-9937.

Timpl, R., H. Rohde, L. Risteli, U. Ott, P. Gehron Robey, and G.R. Martin (1982) Laminin. Methods Enzymol. **82:** 831-839.

Timpl, R., H. Wiedemann, V. van Delden, H. Furthmayer, and K. Kühn (1981) A network model for the organization of type IV collagen molecules in basement membranes. Eur. J. Biochem. **120:** 203-211.

Trueb, B., B. Gröbli, M. Spiess, B. Odermatt, and K.H. Winterhalter (1984) Basement membrane (type IV) collagen is a heteropolymer. J. Biol. Chem. **257:** 5239-5245.

Van der Rest, M., R. Mayne, Y. Ninomiya, N.G. Seidah, M. Chretien, and B.R. Olsen (1985) The structure of type IX collagen. J. Biol. Chem. **260:** 220-225.

Vaughan, L., K.H. Winterhalter, and P. Bruckner (1985) Proteoglycan Lt from chicken embryo sternum identified as type IX collagen. J. Biol. Chem. **260:** 4758-4763.

Veis, A. (1984) Bone and teeth. In *Extracellular Matrix Biochemistry*, K.A. Piez and A.H. Reddi, eds., pp. 329-374. Elsevier, New York.

Von der Mark, H., M. Aumailley, G. Wick, R. Fleischmajer, and R. Timpl (1984) Immunochemistry, genuine size and tissue localization of collagen VI. Eur. J. Biochem. **142:** 493-502.

Vracko, R. (1982) Role of basal lamina in the maintenance of orderly tissue structure. In *New Trends in Basement Membrane Research*, K. Kuehn, H.H. Schoene, and R. Timpl, eds., pp. 1-8. Raven, New York.

Williamson, J.R., and C. Kilo (1977) Current status of capillary basement membrane disease in diabetes mellitus. Diabetes **26:** 65-73.

Yamada, K.M., S.K. Akiyama, T. Hasegawa, E. Hasegawa, M.J. Humphries, D.W. Kennedy, K. Nagata, H. Urushihara, K. Olden, and W.T. Chen (1985) Recent advances in research on fibronectin and other cell attachment proteins. J. Cell. Biochem. **28:** 79-97.

Yamada, K.M., J.A. Weston (1974) Isolation of a major cell-surface glycoprotein from

fibroblasts. Proc. Natl. Acad. Sci. (USA) **71**: 3492-3496.
Yurchenco, P.D., E.C. Tsilibary, A.S. Charonis, and H. Furthmayr (1986) Models for the self-assembly of basement membrane. J. Histochem. Cytochem. **34**: 93-102.

STRUCTURE, EXPRESSION AND FUNCTION OF MOUSE LAMININ

Y. Yamada, A. Albini, I. Ebihara, J. Graf, S. Kato[1], P. Killen, H.K. Kleinman, K. Kohno[2], G.R. Martin, C. Rhodes, F.A. Robey[3], and M. Sasaki

Laboratory of Developmental Biology and Anomalies
National Institute of Dental Research
National Institutes of Health
Bethesda, MD
U.S.A.

Present addresses:
[1] Sagami Chemical Research Center, Sagami, Kanagawa, Japan
[2] First Department of Internal Medicine, Faculty of Medicine, Kyushu University, Fukuoka, Japan
[3] Center for Drugs and Biologics, Food and Drug Administration, Bethesda, MD, U.S.A.

Introduction

Basement membranes are thin extracellular matrices that separate epithelial and endothelial cells from underlaying stroma (Kefalides, 1973; Timpl and Martin, 1982). At the ultrastructural level, different zones are observed in the basement membrane including an electron dense zone, the lamina densa, and adjacent to it less electron dense zones, the lamina lucida. Basement membranes have diverse functions including forming barriers to the passage of proteins as well as of cells (Kefalides, 1979). They also play important roles in embryogenesis by delineating the developing tissues and by mediating the adhesion, growth, and differentiation of muscle, and peripheral nerve as well as of a variety of epithelial cells and organs.

All basement membranes contain type IV collagen, laminin, heparan sulfate proteoglycans, and nidogen. Type IV collagen is the major structural element, whereas the proteogylcans are thought to be involved in filtration (Hassell et al., 1985). Laminin (Mr = 800,000) is the major glycoprotein in basement membranes and is composed of one A chain (Mr = 400,000), one B1 (Mr = 200,000), and one B2 (Mr = 200,000) chain (Timpl et al., 1979; Chung et al., 1979). Electron microscopy after rotary shadowing reveals that the molecule is a cross-shaped structure with one long arm of 77 nm length and three short arms of 36 nm length (Engel et al, 1981). Two globular structures are observed at the end of each short arm and one large globule is observed at the end of the long arm. While the precise location of each chain in the molecule is still not fully established, the current model suggests that the cross is formed from a B1, a B2, and an A chain. It is speculated that the B1 and B2 chains form two of its short arms and bend toward the long

NATO ASI Series, Vol. H5
Mesenchymal-Epithelial Interactions in Neural Development
Edited by J.R. Wolff et al.

arm while the A chain forms a short arm and also extends down the entire long arm and forms the globule at the end of the long arm. Such a model is supported by the amino acid sequence analysis deduced from cDNA clones and by the amino acid sequence of a peptide (Barlow et al., 1984; Paulsson et al., 1985; Yamada et al., 1985). These studies show that both B chains contain fairly long α-helical segments at the carboxyl ends which are of the type that form coiled-coil structures in proteins such as myosin (McLachlan and Karn, 1982) and haemaglutinin (Bassel-Duby et al., 1985).

Laminin has been found to exert diverse biological activities including stimulating epithelial cell growth, migration, and proliferation (Kleinman et al., 1985). Laminin also promotes the adhesion of epithelial cells, myoblasts, Schwann cells, and certain tumor cells (Terranova et al., 1980; Kleinman et al., 1981). Laminin alters cell morphology, including inducing neurite outgrowth (Baron Van Evercooren et al., 1982; Rogers et al., 1985). Laminin also has structural functions. It binds to type IV collagen (Rao et al., 1982) and to the basement membrane specific heparan sulfate proteoglycan (Woodley et al., 1983; Del Rosso et al., 1981). In vitro studies, using fragments of laminin generated by proteolytic cleavage, suggest that the globular domains of the short arms contain binding sites for type IV collagen (Rao et al., 1982; Laurie et al., 1986), and that a fragment containing the three short arms, minus the globules, has the cell binding activity (Terranova et al., 1983). The region of laminin containing the globule binds to heparin, while the end of the long arm, plus adjacent segments, promotes neurite outgrowth (Edgar et al., 1984).

Laminin also mediates the attachment of carcinoma cells, increases the metastatic potential of tumor cells (Terranova et al., 1982) and causes the release of type IV collagenase (Turpeeniemi-Hujanen et al., 1986). Thus, laminin promotes the invasive phenotype of these cells. Laminin binds to cells through a specific membrane protein receptor, the laminin receptor (Mallinoff and Wicha, 1983; Rao et al., 1983; von der Mark and Kühl, 1985), as well as to gangliosides (Kennedy et al., 1983), heparan sulfate (Woodley et al., 1983; Del Rosso et al., 1981), and certain sulfated sugars (Roberts et al., 1985). There is a correlation between the metastatic potential of tumor cells and their numbers of laminin receptors (Terranova et al., 1983).

Some studies suggest that laminin is the earliest extracellular matrix protein to appear during embryonic development. The B chains appear at the 4-cell stage in the mouse embryo and the A chain at the 16-cell stage (Cooper and MacQueen, 1983). In mouse teratocarcinoma cells, the synthesis of all chains of laminin is increased more than 10-fold following treatment of the cells with retinoic acid and dibutyryl cyclic AMP (Strickland et al., 1980).

While major advances in our understanding of laminin have come from studies on the protein, its complex structure and size make information derived from its cDNA sequence helpful in elucidating the structure of the molecule. Here we report the isolation and characterization of cDNA clones for the B1, B2, and A chains of mouse laminin. Using these cDNA clones and a clone for the α1 chain of type IV collagen (Oberbäumer et al., 1985), we have measured the steady state levels of mRNA for basement membrane components. We have also determined the nucleotide sequence coding for the entire B1 chain and isolated the gene for the B1 chain. Antibodies against synthetic peptides from the sequence of various portions of the B1 chain have been used as probes to detect functional regions of the molecule.

Results and Discussion

Cloning cDNA

cDNA clones for the B1, B2, and A chains of the mouse laminin were initially isolated from two different cDNA libraries. One cDNA library was made from differentiated F9 cells poly(A) RNA using the modified Okayama-Berg method (Okayama and Berg, 1982; Yamada et al., 1985). The second library was constructed from poly(A) RNA isolated from the mouse EHS tumor and cloned into an expression vector, λgt11 (Young and Davis, 1983). The first library was screened with ^{32}P-labeled poly(A) RNA of larger than 28S from EHS tumor cells. Clones which showed strong hybridization to the differentiated F9 cell poly(A) RNA, but not to the undifferentiated RNA, were isolated and 20 several positive clones were characterized by DNA sequence analysis. Clone p24 was 5.3 Kb long and was identified as a B1 clone by comparison with the amino acid sequences from the B1 chain (Paulsson et al., 1985) and those deduced from other cDNA clones (Barlow et al., 1984). The second library was screened with polyclonal antibody against denatured laminin as described previously (Doege et al., 1986). Several hundred positives were obtained out of 2 X 10^5 recombinant phage and a dozen of these were purified and characterized by DNA sequence analysis. The deduced amino acid sequences were in agreement with the sequence of the B1 and B2 chains from cDNA cloning previously reported (Barlow et al., 1984) and the sequence of a peptide fragment of the A chain (R. Deutzmann, personal communication). To obtain cDNA clones further toward the 5' portion of mRNA, a synthetic oligonucleotide complementary to the nucleotide sequence about 100 nucleotides downstream of the 5' end of the cDNA clones was used as a primer and the extensions were cloned into λgt11. These cDNA libraries were screened with the most 5' portion of the cDNA clones and positive clones were identified as the laminin chains by DNA sequencing.

Some of the cDNA clones for the B1, B2, and A chains of laminin are shown in Fig. 1. Northern hybridization revealed that the sizes of mRNA for the B1, B2, and A chains are 6 Kb, 8 Kb, and 10 Kb, respectively (Yamada et al., 1985). The nucleotide sequences of the clones for the A and B2 chains have been determined in part. The sequence analysis of p7 coding for the B2 chain revealed that the B2 chain mRNA has about 2.5 Kb of 3' untranslated region. Since the B1 and B2 polypeptide chains are almost identical in size, the difference in their mRNAs appears to reflect a larger 3' untranslated region in the B2 chain mRNA. Based on mRNA size, we speculate that these B2 chain clones code for about 70% of the molecule, while the clones for the A chain code for only 20% of the protein.

The nucleotide sequence analyses of p24 and of λB1-11 which code for the entire B1 chain have been completed. The 5' portion of the mRNA has the presumptive initiation methionine codon which initiates an open reading frame of 5,383 bp encoding a polypeptide of 1,786 amino acids (Mr = 196,903). The initial methionine is followed by a stretch of 20 predominantly hydrophobic amino acids characteristic of a signal peptide. The nucleotide sequence of the 3' 1.6 Kb of p24 is the same as the cDNA sequence for the B1 chain reported previously (Barlow et al., 1984) except for one base which specifies a glutamine instead of a cysteine. The B1 chain has 13 Asn-X-Ser or Asn-X-Thr sequences, potential oligosaccharide attachment sites.

Laminin structure

A computer assisted analysis of the amino acid sequence of the B1 chain as deduced from its nucleotide sequence suggests that it is composed of a limited number of distinct

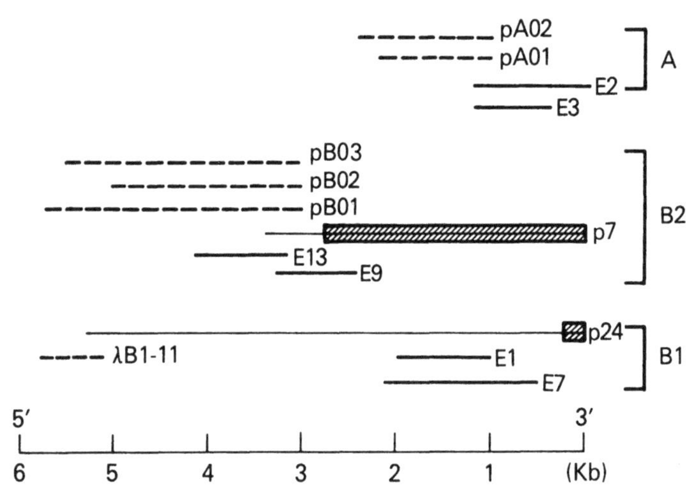

Figure 1: cDNA clones for the mouse laminin A, B1, and B2 chains. Clones E1, E3, E7, E9, and E13 were isolated from a cDNA library in the expression vector λgt11 (Yamada et al., 1985). Clones p7 and p24 were isolated from a cDNA library constructed by the modified Okayama-Berg method (Yamada et al., 1985). Clones λB1-11, pB01, pB02, pA01, and pA02 were isolated from a cDNA library using a specific primer (Sasaki, unpublished). Shaded boxes indicate the 3' untranslated sequence.

domains. Shown schematically in Fig. 2, these are numbered from 1 to 7 beginning at the carboxyl terminus of the B1 chain. Domain 1 (40 kD) has an extended α-helical sequence with heptad repeats of the type found in proteins such as myosin (Harington et al., 1984) and haemagglutinin (Bassy-Duby et al., 1985) whose chains are arranged in a coiled-coil structure. The adjacent sequence, Domain 2 (3 kD), has 33 amino acids with 6 cysteines plus several glycine and proline residues. This domain interrupts the α-helices, possibly serving as a site for interchain disulfide bond formation. Domain 3 (24 kD) is also α-helical but lacks the heptad repeat thought to be required for coiled-coil structures. Domains 4 (44 kD) and 6 (30 kD) are extraordinarily rich in cysteines. A striking feature of these domains is the occurrence of a homologous repeat of about 50 amino acids, each containing 8 regularly arranged cysteines. Domain 4 has 8 such repeats and domain 6 has 5 such repeats. Possibly these could form the rod-like segments adjacent to the globules in one of the short arms of laminin. Domains 5 (26 kD) and 7 (29 kD) contain a mixture of α-helical segments and random coil structure and could form the globules in one of the short arms of laminin visible in electron micrographs (Engel et al., 1981).

Fig. 2 shows a schematic diagram of the proposed alignment of the B1 chain in the laminin structure. According to this model, Domains 4 to 7 constitute a short arm. Domains 1 to 3 are located in the long arm of laminin. Such a model corresponds well to the electron microscopic observations on the laminin molecule. Assuming a distance of 0.15 nm per residence in the α-helix, the length of the structure formed by Domains 1 and 3 is calculated to be 82 nm as compared to an estimated length of the long arm of laminin

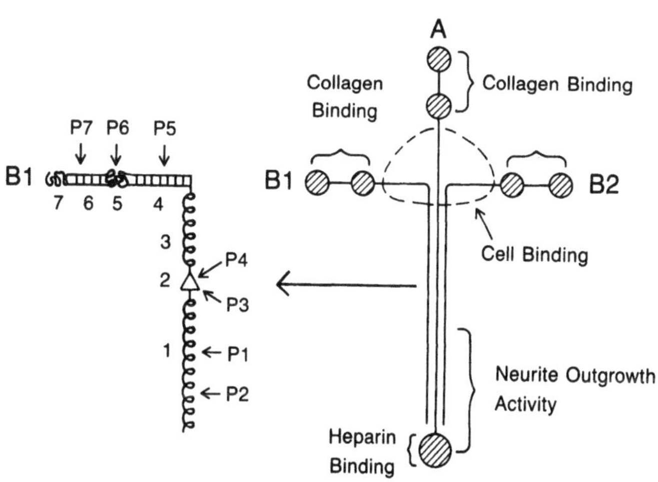

Figure 2: Schematic diagram of the laminin molecule. The numbers indicate the domains of the B1 chain. Arrows with P1 to P7 represent approximate locations of synthetic peptides.

of 77 nm from electron micrographs (Engel et al., 1981).

A computer search for homologous protein sequences through the National Biomedical Research Foundation data base revealed that a part of the repeat unit of the B1 chain in Domain 4 is homologous to several EGF-related proteins including EGF (Gray et al., 1983), TGFα (Derynck et al., 1984), vaccinia virus 19k (Stroobant et al., 1985) as well as coagulation factors X (Kurachi and Davie, 1982) and IX (McMullen et al., 1983). The sequence homology in the B1 chain and EGF is particularly interesting because both proteins share the ability to stimulate the growth of epithelial cells (Timpl et al., 1983; Carpenter and Cohen, 1979).

Structure of the B1 chain gene

cDNA clones for the B1 chain were used to screen a mouse genomic library. The B1 chain genes was isolated in a total of 8 overlapping clones spanning about 80 Kb of a contiguous DNA. Electron microscopic analysis of R-loop hybrids revealed that the gene is 63 Kb in size and contains at least 36 exons. The exon size is relatively homogenous averaging about 200 bp. Three different transcription initiation sites located close to one another were identified by S1-mapping and by primer extension analysis. The promoter region of the gene was characterized by DNA sequencing. There is a typical "TATA" box, but no "CAT" box. It is interesting to note that the promoter region contains tandem repeats of the sequence "5'-CCCGCC-3'" which is also found in the tk-genes (McKnight and Kingbury, 1982) as well as some other genes (Kohno et al., 1985; Reynolds et al., 1984). These sequences are known to be the site for binding to the transcription factor SP1 (Dynan and Tjian, 1983a, 1983b).

The presence of sequences homologous to EGF-like repeats suggests that these sequences arose in different proteins from a common ancestor. It has been speculated that proteins containing EGF-like repeats arose by exon shuffling (Doolittle, 1985; Patthy, 1985; Sudhof et al., 1985). This hypothesis is based on the findings that the repeat units are often coded for by a single exon. The size of several of the exons coding for the repeats of the B1 chain in Domains 4 and 6 have been determined by DNA sequence analysis. One of the exons codes for an entire repeat while in others the repeats are interrupted at various locations. Further analysis of exon size may help to elucidate how the repeat units of the B1 chain gene were assembled.

Expression of Basement Membrane Genes

Using cDNA clones for the B1, B2, and A chains of laminin and a cDNA clone for the α1 chain of type IV collagen, we measured steady state levels of mRNA for these basement membrane proteins in differentiating F9 cells and in 3T3 cells transformed with oncogenes as well as in various tissues.

It has been shown that mouse teratocarcinoma F9 cells produce a large amount of laminin and type IV collagen when they are induced to differentiate by exposure to retinoic acid and dibutyryl cyclic AMP (Strickland et al., 1980). RNA was isolated from F9 cells at various times after induction and analyzed by dot-blot hybridization. Very little mRNA for basement membrane components was detected at day 1. The levels of mRNA for A, B1, and B2 chains and the $\alpha 1$(IV) collagen were increased at day 2 and reached a maximum at day 6. The finding that the increases observed for each of these mRNA species was proportionate to the others suggests that their expression is coordinately regulated in differentiating F9 cells. Wang et al. (1985) have reported similar results using the $\alpha 1$(IV) and B1 chain probes. In addition, using a nuclear transcription assay, they found that there was an increased rate of transcription of these mRNAs.

It has been shown that the synthesis of type I collagen mRNA is down regulated in 3T3 cells transformed with oncogenes (Liau et al., 1985). We have studied the levels of mRNA for the basement membrane components in 3T3 cells and in sublines of 3T3 cells transformed with various oncogenes to see if the expression of these genes is altered (Fig. 3). The level of mRNA for the $\alpha 1$(IV) chain was not changed significantly in cells transformed by mos, ras, MMSV, and SSV. Similarly, the levels of laminin B1 and B2 chain mRNAs showed only a slight reduction in cells transformed by mos and SSV. However, the levels of mRNA for the B2 chain but not the B1 or $\alpha 1$(IV) chains, were increased about two-fold. The levels of mRNA for the $\alpha 2$(I) collagen were significantly reduced. These studies show distinct changes in transformed cells between the levels of mRNA expressed for type I collagen and those for basement membrane genes. Furthermore, the synthesis of the B1 and B2 chains is not coordinately regulated in certain transformed cells.

Little is known about the subunit composition of laminin in different tissues. Isoforms with different subunit composition may exist. For example, in mouse embryos, B chains are synthesized prior to the A chain (Cooper and MacQueen, 1983; Dziadek and Timpl, 1985). In culture, Schwann cells produce only B chains (Cornbrooks et al., 1983). Therefore, we examined the levels of mRNA for each chain of laminin in various mouse tissues (data not shown). As expected, spleen and liver, which contain very little basement membrane, had low levels of mRNA for basement membrane components. Heart, lung, and kidney in which basement membranes are more abundant had higher levels of mRNA for basement membrane proteins. Significant variations in the levels of mRNA for the laminin B1, B2, and A chains relative to the level of mRNA for the $\alpha 1$(IV) collagen were observed. The highest levels of A and B2 chain mRNA were found in heart, whereas B1 chain mRNA was highest in kidney. In the heart, the level of B2 chain mRNA is one third of that found in the kidney. The level of the A chain mRNA was surprisingly low in all tissues examined. Since RNA from induced F9 cells was used as an internal standard

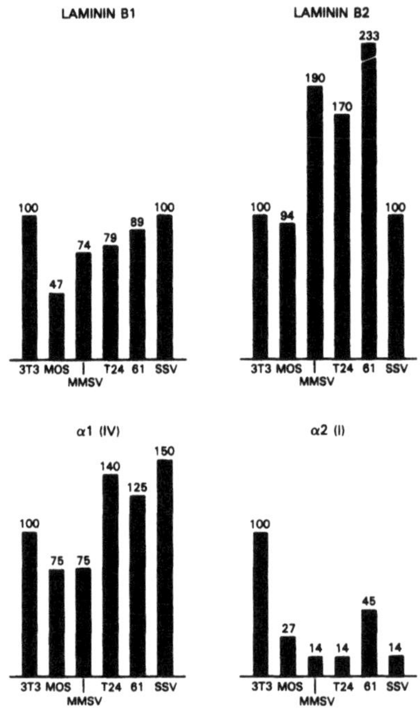

Figure 3: Steady state levels of mRNA for laminin-B1 and B2 chains, and α1 (IV) and α 2 (I) collagen chains. Results were obtained from densitometric tracing of RNA dot blot hybridizations. Numbers represent per cent of control cell lines. MOS, T24, and 61 represent 3T3 cells transformed by c-mos, c-Ha-ras-1 (Greig et al., 1985), and DNA from HS 242 tumor containing c-ras mutated in postion 61 (Aaronson S. unpublished), respectively. MMSV and SSV represent 3T3 cells transformed by Moloney murine sarcoma virus and Simian Sarcoma virus, respectively. cDNA probes used are p24 (Yamada et al., 1985), E9 (Kleinman et al., in preparation), PFAC (Oberbäumer et al., 1985) and AZ1002 (Schmidt et al., 1984) for B1 and B2 chains of laminin, α1(IV) and α2(I) collagen chains, respectively.

for all hybridization experiments, it seems likely that these low levels reflect actual steady state levels of mRNA for the A chain rather than differences in hybridization efficiency. These results suggest that genes for the laminin B1, B2, and A chains and for α 1(IV) collagen are not coordinately expressed in these tissues.

Activities of specific antibodies to the B1 chain

Several synthetic peptides of 20 amino acids in length were synthesized which correspond to various portions of the amino acid sequence of the B1 chain deduced from its nucleotide sequence. The location of these peptides in the B1 chain model is shown in Fig. 2. Antibodies against these laminin domain synthetic peptides were raised and their activities were examined (Table 1). The direct immunoprecipitations and indirect immunofluorescent assays showed that five (P1, P2, P4, P5, and P7) out of seven antibodies re-

Table 1: Activities of laminin domain specific antibodies

	P1	P2	P3	P4	P5	P6	P7
Reacts with Peptide -ELISA	+	+	+	+	+	+	+
Reacts with Native Laminin -Immunoptn -Indirect IF	+	+	0	+	+	0	+
Reacts with Denatured Laminin -Western Blot	+	+	0	+	+	0	+
Inhibits Cell Attachment	0	0	0	0	+	0	0

Active antibodies are designated by a + while inactive antibodies are designated by a 0. P1 to P7 in the Table indicate antibodies against synthetic peptides corresponding to various positions of the laminin B1 chain (see Fig. 2).
An enzyme linked immunosorbent (ELISA) assay was carried out by coating the peptides on microtiter wells and then reacting them with their respective antibodies at a 1:500 dilution, in accordance with Rennard et al. (1980).
Immunoprecipitation and indirect immunofluorescence were carried out using standard techniques (Bumol et al., 1982). For the immunoprecipitation, 20 µl of each antibody was coupled to protein A Sepharose and then incubated with ^{35}S-methionine labeled Engelbreth-Holm-Swarm (EHS) cell lysates (Hassell et al., 1985). For indirect immunofluorescence, a dilution of 1:20 was used in frozen sections of skin, EHS tumor, and kidney. Western blots were carried out according to established procedures (Towbin et al., 1979) with 50 µg per lane of laminin, and an antibody dilution of 1:300. Cell attachment assays using HT1080 cells were carried out as previously described (Klebe, 1974).
Various concentrations of antisera and of affinity purified antibodies were incubated first with laminin substrates and then, following a rinse to remove unbound antibodies, the cells were added in fresh media (Terranova et al., 1980). The number of attached cells was determined after trypsinization.

acted with native laminin. Immunostaining using western blotting revealed that these antibodies were specific for the B1 chain. The other two antibodies, P3 and P6, did not react with either native or denatured laminin though they reacted strongly with their own peptides. The epitopes for these antibodies appear to be hidden in the molecule.

Laminin is known to bind to cells through a specific receptor (Malinoff and Wicha, 1983; Rao et al., 1983; von der Mark and Kühl, 1985). In order to locate the domain for cell binding on laminin, the antibodies were tested to see if they could inhibit cell attachment to laminin. Antobody P5 showed strong activity in inhibiting cell attachment to laminin. Antibody P7 has some inhibitory activity in the presence of laminin on a type of IV collagen substrate, whereas the rest of the antibodies showed no inhibition. The location of peptide P5 whose antibody has the inhibitory activity for cell attachment corresponds well with the previously identified proteolytic fragment of laminin which has activity for cell binding. At the moment, it is not clear whether these sites are directly involved in cell binding or whether the antibody is causing steric blocking of the active site on laminin. More precise localization of active sites is in progress.

Conclusion

Interactions of cells with extracellular matrix components play an important role in the adhesion, migration, growth, and differentiation of a variety of cells and thus likely influence tissue dvelopment and function (Hay, 1982; Yamada and Akiyama, 1984). Laminin as the major component of basement membranes influences all these biological activities in a variety of cultured epithelial cells (Timpl and Martin, 1982; Kleinman et al., 1985). In this paper, we report on the structure, expression, and function of mouse laminin using molecular biological approaches.

We have isolated cDNA clones for the B1, B2, and A chains of mouse laminin. The nucleotide sequence of the cDNA clones coding for the entire B1 chain was determined. The analysis of the deduced protein sequence revealed that the B1 chain has at least seven distinct domains including those with α-helix, cysteine-rich repeats, and globules. The entire gene for the B1 chain was isolated. It is about 63 Kb in length and contains at least 37 exons. The promoter region of the gene and some of the exons encoding the repeats in domains 4 and 6 were characterized by DNA sequencing. Steady state levels of mRNA for the laminin B1, B2, and A chains, and the α1(IV) collagen were examined in cell lines and in tissues. The results suggest that the synthesis of these basement membrane components is not coordinately regulated in certain cells and in tissues. Antibodies were raised against synthetic peptides corresponding to various portions of the B1 chain in order to examine the structure and function of laminin. One of the antibodies specifically inhibited cell binding to laminin suggesting that the epitope recognized by this antibody is in a region of the molecule involved in cell attachment.

References

Barlow, D.P., N.M. Green, M. Kurkinen, and B.L.M. Hogan (1984) Sequencing of laminin B chain cDNAs reveals C-terminal regions of coiled-coil alpha-helix. EMBO J. **3**: 2355-2362.

Baron van Evercooren, A., H.K. Kleinman, S. Ohno, P. Marangos, J.P. Schwartz, and M.E. Dubois-Dalcq (1982) Nerve growth factor, laminin, and fibronectin promote neurite growth in human fetal sensory ganglia cultures. J. Neurosci. Res. **8**: 179-193.

Bassel-Duby, R., A. Jayasuriya, D. Chatterjee, N. Sonenberg, J.V. Maizel, and B.N. Fields (1985) Sequence of reovirus haemagglutinin predicts a coiled-coil structure. Nature **315**: 421-423.

Bumol, T.F., and R.A. Reisfeld (1982) Unique glycoprotein-proteoglycan complex defined by monoclonal antibody on human melanoma cells. Proc. Natl. Acad. Sci. (USA) **79**: 1245-1249.

Carpenter, G., and S. Cohen (1979) Epidermal growth factor. Annu. Rev. Biochem. **48**: 193-216.
Chung, A.E., R. Jaffe, I.L. Freeman, J.P. Vergnes, J. Braginski, and B. Carlin (1979) Properties of a basement membrane-related glycoprotein synthesized by a mouse embryonal carcinoma derived cell line. Cell **16**: 277-287.
Cooper, A.R., and H.A. MacQueen (1983) Subunits of laminin are differentially synthesized in mouse eggs and early embryos. Dev. Biol. **96**: 467-471.
Cornbrooks, C.J., D.J. Carey, J.A. McDonald, R. Timpl, and R.P. Bunge (1983) In vivo and in vitro observations on laminin production by Schwann cells. Proc. Natl. Acad. Sci. (USA) **80**: 3850-3854.
Del Rosso, M., R. Cappelletti, M. Viti, S. Vannucchi, and V. Chiarugi (1981) Binding of the basement-membrane glycoprotein laminin to glycosaminoglycans. An affinity-chromatography study. Biochem. J. **199**: 699-704.
Derynck, R., J.A. Jarrett, E.Y. Chen, D.H. Eaton, J.R. Bell, R.K. Assoian, A.B. Roberts, M.B. Sporn, and D.V. Goeddel (1984) Human transforming growth factor-beta complementary DNA sequence and expression in normal and transformed cells. Nature **316**: 701-705.
Doege, K., J.R. Hassell, B. Caterson, and Y. Yamada (1986) Link protein cDNA sequence reveals a tandemly repeated protein structure. Proc. Natl. Acad. Sci. (USA), in press.
Doolittle, R.F. (1985) The genealogy of some recently evolved vertebrate proteins. Trends Biochem. Sci. **10**: 233-237.
Dynan, W.S., and R. Tjian (1983a) Isolation of transcription factors that discriminate between different promoters recognized by RNA polymerase II. Cell **32**: 669-680.
Dynan, W.S. and R. Tjian (1983b) The promoter-specific transcription factor Sp1 binds to upstream sequences in the SV40 early promoter. Cell **35**: 79-87.
Dziadek, M., and R. Timpl (1985) Expression of nidogen and laminin in basement membranes during mouse embryogenesis and in teratocarcinoma cells. Dev. Biol. **111**: 372-382.
Edgar, D., R. Timpl, and H. Thoenen (1984) The heparin-binding domain of laminin is responsible for its effects on neurite outgrowth and neuronal survival. EMBO J. **3**: 1463-1468.
Engel, J., E. Odermatt, A. Engel, J.A. Madri, H. Furthmayr, H. Rohde, and R. Timpl (1981) Shapes, domain organizations and flexibility of laminin and fibronectin, two multifunctional proteins of the extracellular matrix. J. Mol. Biol. **150**: 97-120.
Gray, A., T.J. Dull, and A. Ullrich (1983) Nucleotide sequence of epidermal growth factor cDNA predicts a 128,000-molecular weight protein precursor. Nature **303**: 722-725.
Greig, R.G., T.P. Koestler, D.L. Trainer, S.P. Corwin, L. Miles, T. Kline, R. Sweet, S. Yokoyama, and G. Poste (1985) Tumorigenic and metastatic properties of "normal" and ras-transfected NIH/3T3 cells. Proc. Natl. Acad. Sci. (USA) **82**: 3698-3701.
Hassell, J.R., W.C. Leyson, S.R. Ledbetter, B. Tyree, S. Suzuki, M. Kato, K. Kimata, and H.K. Kleinman (1985) Isolation of two forms of basement membrane proteoglycans. J. Biol. Chem. **260**: 8098-8105.
Hay, E.D. (1982) Interaction of embryonic surface and cytoskeleton with extracellular matrix. Am. J. Anat. **165**: 1-12.
Kefalides, N.A. (1973) Structure and biosynthesis of basement membrane. Int. Rev. Connect. Tissue Res. **6**: 63-104.
Kefalides, N.A. (1979) Biochemistry and metabolism of basement membranes. Int. Rev. Cytol. **61**: 167-228.
Kennedy, D.W., D.H. Rohrbach, G.R. Martin, T. Momoi, and K.M. Yamada (1983) The adhesive glycoprotein laminin is an agglutinin. J. Cell Physiol. **3**: 257-262.
Klebe, R.J. (1974) Isolation of a collagen-dependent cell attachment factor. Nature **250**: 248-251.
Kleinman, H.K., R.J. Klebe, and G.R. Martin (1984) Role of collagenous matrices in the adhesion and growth of cells. J. Cell Biol. **88**: 473-485.
Kleinman, H.K., F.B. Cannon, G.W. Laurie, J.R. Hassell, M. Aumailley, V.P. Terranova,

G.R. Martin, and M. DuBois-Dalcq (1985) Biological activities of laminin. In The Role of Extracellular Matrix in Development, R.L. Trelstad, ed., pp. 123-143. Alan R. Liss, New York.

Kohno, K., M. Sullivan, and Y. Yamada (1985) Structure of the promoter of the rat type II procollagen gene. J. Biol. Chem. **260:** 4441-4447.

Kurachi, K. and E.W. Davie (1982) Isolation and characterization of a cDNA coding for human factor IX. Proc. Natl. Acad. Sci. (USA) **79:** 6461-6464.

Laurie, G.W., J.T. Bing, H.K. Kleinman, J.R. Hassell, M. Aumailley, G.R. Martin, and R.J. Feldmann (1986) Localization of binding sites for laminin, heparan sulfate proteoglycan, and fibronectin on basement membrane (Type IV) collagen. J. Mol. Biol., in press.

Liau, G., Y. Yamada, and B. de Crombrugghe (1985) Coordinate regulation of the levels of type III and type I collagen mRNA in most but not all mouse fibroblasts. J. Biol. Chem. **260:** 531-536.

Malinoff, A.L., and M.S. Wicha (1983) Isolation of a cell surface receptor protein for laminin from murine fibrosarcoma cells. J. Cell Biol. **96:** 1475-1479.

McKnight, S.L., and R. Kingsbury (1982) Transcriptional control signals of a eukaryotic protein-coding gene. Science **217:** 316-324.

McLachlan, A.D., and J. Karn (1982) Periodic charge distributions in the myosin rod amino acid sequence match cross-bridge spacings in muscle. Nature **299:** 226-231.

McMullen, B.A., K. Fujikawa, W. Kisiel, T. Sasagawa, W.N. Howald, E.Y. Kwa, and B. Weinstein (1983) Complete amino acid sequence of the light chain of human blood coagulation factor X: Evidence for identification of residue 63 as beta-hydroxy-aspartic acid. Biochemistry **22:** 2875-2884.

Oberbäumer, I., M. Laurent, U. Schwarz, Y. Sakurai, Y. Yamada, G. Vogeli, T. Voss, B. Siebold, R.W. Glanville, and K. Kuhn (1985) Amino acid sequence of the non-collagenous globular domain (NC1) of the α1(IV) chain of basement membrane collagen as derived from complementary DNA. Eur. J. Biochem. **147:** 217-224.

Okayama, H., and P. Berg (1982) High-efficiency cloning of full-length cDNA. Mol. Cell. Biol. **2:** 161-170.

Patthy, L. (1985) Evolution of the proteases of blood coagulation and fibrinolysis by assembly from modules. Cell **41:** 657-663.

Paulsson, M., R. Deutzmann, R. Timpl, D. Dalzoppo, E. Odermatt, and J. Engel (1985) Evidence for coiled-coil alpha-helical regions in the long arm of laminin. EMBO. J. **4:** 309-316.

Rao, C.N., I.M. Margulies, T.S. Tralka, V.P. Terranova, J.A. Madri, and L.A. Liotta (1982) Isolation of a subunit of laminin and its role in molecular structure and tumor cell attachment. J. Biol. Chem. **257:** 9740-9744.

Rao, C.N., S.H. Barsky, V.P. Terranova, and L.A. Liotta (1983) Isolation of a tumor cell laminin receptor. Biochem. Biophys. Res. Commun. **111:** 804-808.

Rennard, S.I., R. Berg, G.R. Martin, J.M. Foidart, and P. Gehron-Robey (1980) Enzyme linked immunoassay (ELISA) for connective tissue components. Analyt. Biochem. **104:** 205-214.

Reynolds, G., S.K. Basu, T.F. Osborne, D.J. Chin, G. Gil, M.S. Brown, J.L. Goldstein, and K.L. Luskey (1984) HMG CoA reductase: A negatively regulated gene with unusual promoter and 5' untranslated regions. Cell **38:** 275-285.

Roberts, D.D., C.N. Rao, J.L. Magnani, S.L. Spitalnick, L.A. Liotta, and V. Ginsburg (1985) Laminin binds specifically to sulfate glycolipids. Proc. Natl. Acad Sci. (USA) **82:** 1306-1310.

Rogers, S.L., J.B. McCarthy, S.L. Palm, L.T. Furcht, and P.C. Letourneau (1985) Neuron-specific interactions with two neurite-promoting fragments of fibronectin. J. Neurosci. **2:** 369-378.

Schmidt, A., Y. Yamada, and B. de Crombrugghe (1984) DNA sequence comparison of the regulatory signals at the 5' end of the mouse and chick alpha 2 type I collagen genes. J. Biol. Chem. **259:** 7411-7415.

Strickland, S., K.K. Smith, and K.R. Marotti (1980) Hormonal induction of differentiation in teratocarcinoma stem cells: Generation of parietal endoderm by retinoic acid and dibutyryl cAMP. Cell **21**: 347-355.

Stroobant, P., A.P. Rice, W.J. Gullick, K.J. Cheng, I.M. Kerr, and M.D. Waterfield (1985) Purification and characterization of vaccinia virus growth factor. Cell **42**: 383-393.

Sudhof, T.C., J.L. Goldstein, M.S. Brown, and D.W. Russell (1985) The LDL receptor gene: A mosaic of exons shared with different proteins. Science **228**: 815-822.

Terranova, V.P., D. Rohrbach, and G.R. Martin (1982) Role of laminin in the attachment of PAM 212 (epithelial) cells to basement membrane collagen. Cell **22**: 719-726.

Terranova, V.P., L.A. Liotta, R.G. Russo, and G.R. Martin (1982) Role of laminin in the attachment and metastasis of murine tumor cells. Cancer Res. **42**: 2265-2269.

Terranova, V.P., C.N. Rao, T. Kalebic, I.M. Margulies, and L.A. Liotta (1983) Laminin receptor on human breast carcinoma cells. Proc. Natl. Acad. Sci. (USA) **80**: 444-448.

Timpl, R., H. Rohde, P.G. Robey, S.I. Rennard, J.M. Foidart, and G.R. Martin (1979) Laminin - a glycoprotein from basement membranes. J. Biol. Chem. **254**: 9932-9933.

Timpl, R., and G.R. Martin (1982) Components of basement membrane. In *Immunochemistry of Collagen*, H. Furthmayr, ed., pp. 119-150, CRC Press, Cleveland, Ohio.

Timpl, R., J. Engel, and G.R. Martin (1983) Laminin - a multifunctional protein of basement membranes. TIBS **8**: 207-209.

Towbin, H., T. Stachlin, and J. Gordon (1979) Electrophoretic transfer of proteins from polycrylamide gels to nitrocellulose sheets: Procedure and some applications. Proc. Natl. Acad. Sci. (USA) **76**: 4350-4354.

Turpeenniemi-Hujanen, T., U.P. Thorgeirsson, C.N. Rao, and L.A. Liotta (1986) Laminin increases the release of type IV collagenase from malignant cells. J. Biol. Chem. **261**: 1883-1889.

von der Mark, K., and U. Kühl (1985) Laminin and its receptor. Biochem. Biophys. Acta 2: 147-160.

Wang, S., G.J. LaRosa, and J. Gudus (1985) Molecular cloning of gene sequences transcriptionally regulated by retinoic acid and dibutyryl cyclic AMP in cultured mouse teratocarcinoma cells. Dev. Biol. **107**: 75-86.

Woodley, D.T., C.N. Rao, J.R. Hassell, L.A. Liotta, G.R. Martin, and H.K. Kleinman (1983) Interactions of basement membrane components. Biochem. Biophys. Acta. **761**: 278-283.

Yamada, K.M. and S.K. Akiyama (1984) The interactions of cells with extracellular matrix components. In *Cell Membranes*, E. Elsan, W. Frazier, and L. Glaser, eds., pp. 77-148. Plenum, New York.

Yamada, Y., M. Sasaki, K. Kohno, H.K. Kleinman, S. Kato, and G. Martin (1985) A novel structure for the protein and gene for the mouse laminin B1 chain u. In *Basement Membrane*, S. Shibata, ed., pp. 139-146. Elsevier Science, Amsterdam.

Young, R.A., and R.W. Davis (1983) Efficient isolation of genes by using antibody probes. Proc. Natl. Acad. Sci. (USA) **80**: 1194-1198.

DISTRIBUTION OF ECM COMPONENTS IN EARLY EMBRYOS

R. Herken, Ch. Brandes, M. Fussek
Department of Anatomy
University of Göttingen
Göttingen
Federal Republic of Germany

Abstract: Extracellular matrix (ECM) components have been localized in preimplantation and early postimplantation mouse embryos at the ultrastructural level using a fixative supplemented with tannic acid. In the early blastocyst, tannic acid postive structures occurred in the blastocyst cavity. Moreover, positive reaction products were seen between polar trophoblast cells and cells of the inner cell mass. After implantation of the embryos (late day 4, early day 5) stained ECM components were located between the ectodermal and the endodermal cell layers. These structures occurred in the part of the embryo where polar trophoblast cells were in contact with endodermal cells. On late day 5 or early day 6 of development the tannic acid positive Reichert's membrane had formed. In the extraembryonic part of the developing egg cylinder, a fully developed basement membrane of the ectoderm could be seen, whereas in the embryo there were only patches of stained material which occurred between the ectodermal and endodermal cell layer.

In recent years, numerous extracellular matrix components have been identified in very early mammalian embryos. At the 2 cell stage, the basement membrane glycoprotein laminin could be detected in the embryo (Dziadek and Timp, 1985). Other authors, however, had identified laminin no earlier than in the 8 or 16 cell stage (Wu et al., 1983; Leivo et al., 1980). Heparan sulfate proteoglycan is another ECM constituent which seems to occur already in the 2 cell stage (Dziadek et al., 1985). The glycoprotein nidogen was found in 8 to 16 cell embryos (Dziadek and Timpl, 1985), type IV collagen in the early blastocyst (Leivo et al., 1980, Wu et al., 1983), and the glycoproteins fibronectin (Wartiovaara et al., 1979) and entactin (Wu et al., 1983) in the late blastocyst. The results of the localization of these ECM components in the embryo are summarized in Table 1. Apart from fibronectin all the other components found in preimplantation embryos are specific constituents of basement membranes.

One disadvantage in studying the localization of these ECM components at the ultrastructural level is that all these components cannot be made visible by conventional electron microscopical techniques. Localization is only possible at the ultrastructural level using immunohistochemical methods which are difficult to carry out due to prob-

NATO ASI Series, Vol. H5
Mesenchymal-Epithelial Interactions in Neural Development
Edited by J. R. Wolff et al.

Table 1: Localization of ECM components in early embryos.

Component	Stage of Development	Localization	Authors
Laminin	2-8 cells	patchy distribution on blastomers	Dziadek and Timpl, 1985
	8 cells	focal deposits on blastomers	Wu et al., 1983
	16 cells	outer trophoblast surface	Dziadek and Timpl, 1985
	16 cells	granular pattern lining intercellular contours	Leivo et al., 1980
			Wu et al., 1983
heparan sulfate proteoglycan	2-4 cells	punctate staining pattern on cell surface	Dziadek et al., 1985
Nidogen	8-16 cells	outer trophoblast surface	Dziadek and Timpl, 1985
Type IV collagen	early blastocyst	inner cell mass (ICM)	Leivo et al., 1980
	late blastocyst	inner aspect of trophectoderm	Leivo et al., 1980
Fibronectin	lage blastocyst	intercellular strands in ICM patches on mural trophectoderm	Wartiovaara et al., 1979
Entactin	late blastocyst	inner cell membrane of polar trophoblast	Wu et al., 1983

lems of both fixation and nonspecific binding of immunoglobulin to early embryos. This has led us to consider the possibility of localizing ECM components while, at the same time, avoiding the methodological problems of immunohistochemistry by the use of tannic acid fixation. Although, from the biochemical point of view, tannic acid fixation results in an unspecific staining of a variety of intracellular and extracellular structures (Review by Chaplin, 1985), one property of tannic acid fixation which makes it suitable for the ultrastructural localization of ECM components is the precipitation of glycoproteins and glycoaminoglycans, enabling the formation of complexes between carbohydrates and osmium, which results in an electron dense reaction product (Sannes et al., 1978).

Since all of the ECM components identified in early-stage embryos were glycoproteins or glycosaminoglycans which contain remarkable amounts of carbohydrates, we targeted these components for localization by tannic acid fixation. To begin with, we investigated the localization of ECM components in early mouse embryos using a fixative which was supplemented with tannic acid. Details of this fixation have been described previously (Herken and Hsu, 1983; Herken and Barrach, 1985).

Between the 2 and 8 cell stage of development, some tannic acid positive stained material occurred in the intercellular spaces between blastomers. Because most of this material was seen between the zona pellucida and the adjacent cells, we suspected that these stained components derived from the zona pellucida.

At the blastocyst stage of development (day 3), strong staining reactions were seen in the extracellular areas of the embryo. In the blastocyst cavity, tannic acid positive structures were found to be located predominantly near or at the inner surface of the mural trophoblast cells. The stained surface of the cells lining the blastocyst cavity continued into the area of the inner cell mass. However, this staining reaction was clearly stronger near to or at the surface of the mural trophoblast cells than that of the inner cell mass. Extracellular structures occurred between polar trophoblast cells and cells of the inner cell mass (Fig. 1), i.e., in those areas of the blastocyst which later form the egg cylinder.

One day later, i.e., on day 4 of development, shortly after implantation of the embryo, the inner cell mass cells had differentiated into endodermal and ectodermal cells. The polar trophoblast, however, had not yet formed the extraembryonic ectoderm and the ectoplacental cone. Stained ECM components were seen at this developmental stage between the endodermal and the ectodermal layer (Fig. 2). Staining appeared in those areas of the developing egg cylinder where the polar trophoblast cells were in contact with the endoderm and the ectoderm. Because, in later stages of development, this area corresponds to the transition zone between Reichert's membrane and the basement membrane of the extraembryonic ectoderm, it can be assumed that the stained

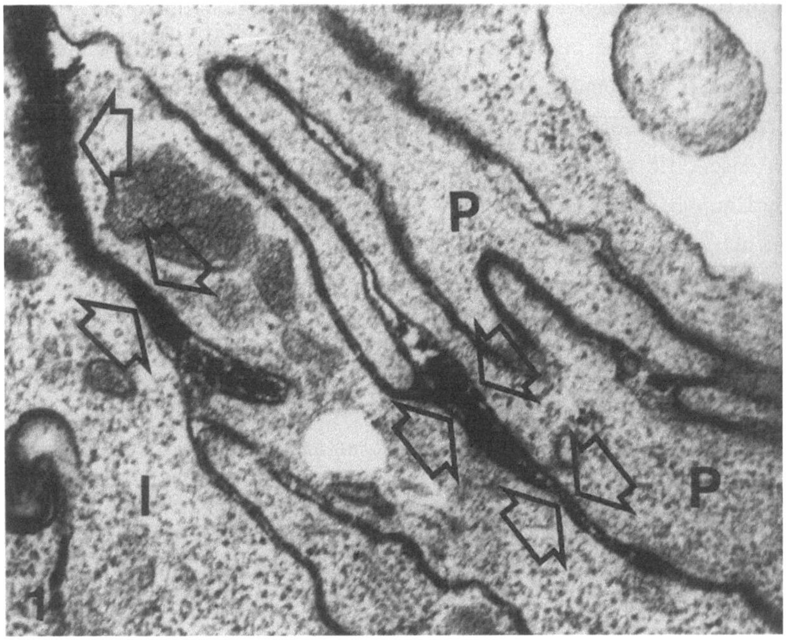

Figure 1: Tannic acid positive material (arrows) between polar trophoblast cells (P) and cells of the inner cell mass (I) of an early blastocyst. X 60,000.

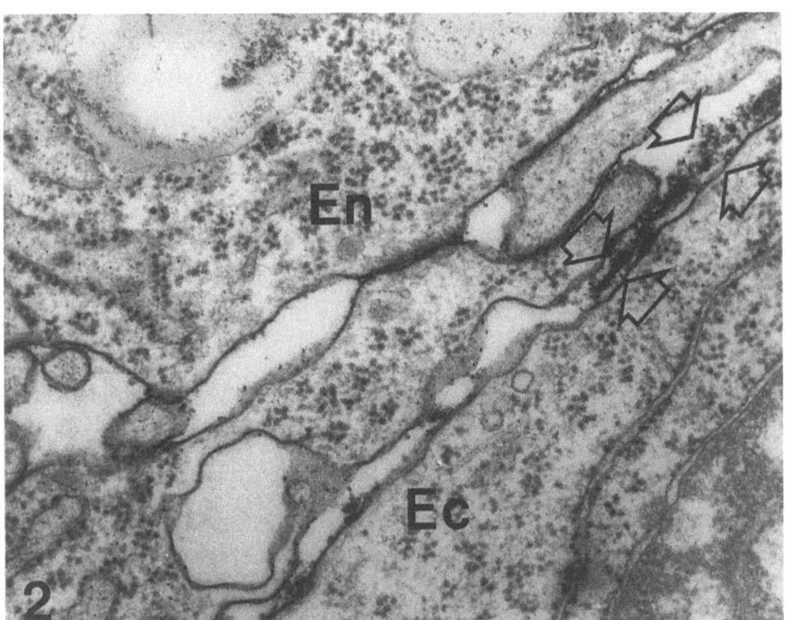

Figure 2: Tannic acid positive structures (arrows) between the endoderm (En) and the ectoderm (Ec) of an embryo shortly after implantation (late day 4, early day 5). X 60,000.

structures initiate the formation of Reichert's membrane as well as the formation of the basement membrane of the ectoderm.

On late day 5 or early day 6 of development, the typical tannic acid positive basement membrane of the ectoderm is formed in the extraembryonic part of the egg cylinder (Fig. 3), whereas in the embryonic part, only tannic acid positive patches are seen. Striking are the tannic acid positive structures at the surface of the extraembryonic endoderm orientated to the yolk sac cavity.

After having localized tannic acid positive ECM components in early embryos, we were interested to see exactly which of the known ECM components had been stained by this method. In a previous study (Herken and Barrach, 1985), we could demonstrate that tannic acid stains those parts of the basement membrane of the ectoderm which contain laminin and type IV collagen. Moreover, it seems likely that the appearance of a typical basement membrane in the embryo, which can be differentiated into lamina rara, lamina densa and lamina fibroreticularis, indicates that not only laminin and type IV collagen, but all ECM components have reached their final molecular organization. However, before the occurrence of a basement membrane in the embryo, the biochemical composition of tannic acid positive structures can only be identified using immunohistochemistry. Therefore, we started to investigate the appearance of laminin, type IV collagen and

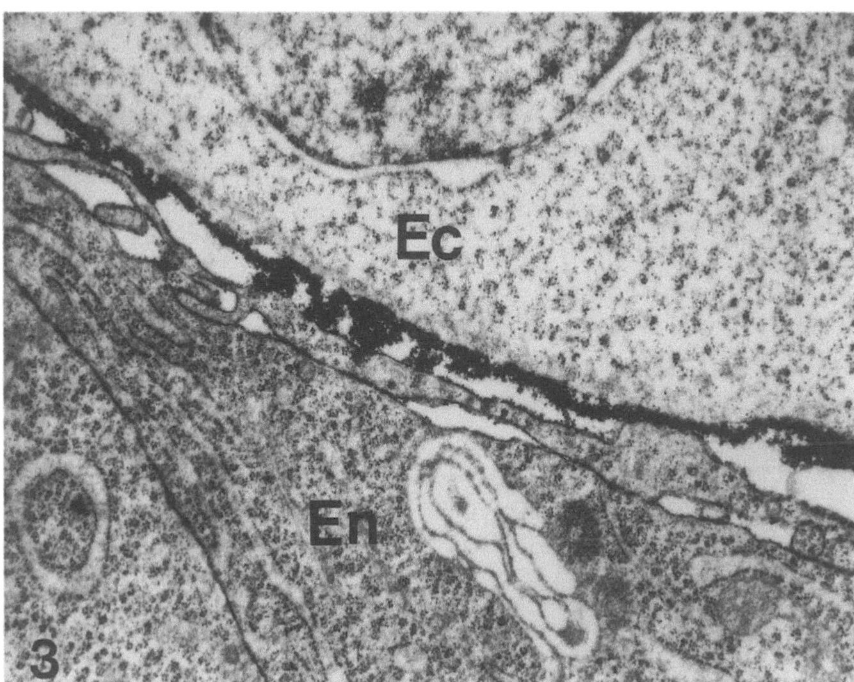

Figure 3: Tannic acid postive basement membrane between the extraembryonic ectoderm (Ec) and the endoderm (En). X 16,000.

heparan sulfate proteoglycan (BMI) in preimplantation embryos using immunoperoxidase staining on unfixed preimplantation embryos. Details of the staining technique used here have been described previously (Herken and Barrach, 1984, 1985).

We found no specific peroxidase positive staining in the embryo after staining for type IV collagen and heparan sulfate proteoglycan, although specific positive staining was seen between polar trophoblast cells and cells of the inner cell mass after staining for laminin (Fig. 4). The poor preservation of the cell structure was caused by the fact that the embryos were not fixed at all. The localization of the laminin staining can be correlated with the tannic acid positive structures seen between polar trophoblast and inner cell mass (Fig. 1). Moreover, Wu et al. (1983) found a similar staining pattern of immunostaining for laminin on fixed blastocysts.

Despite the specific staining for laminin, we found a strong nonspecific staining reaction on the surface of the embryos and between polar and mural trophoblast cells caused by nonspecific binding of the secondary antibody goat-anti-rabbit-IgG (Miles) which was used in our study (Fig. 5). Therefore, we declared all positive staining reactions which were located at the trophoblast surface of the embryos to be nonspecific

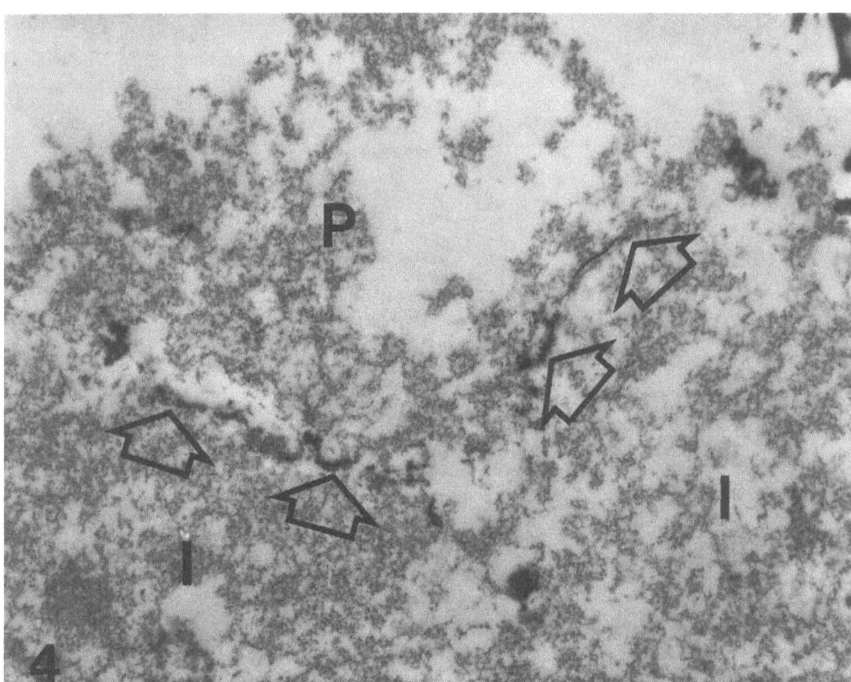

Figure 4: Immunoperoxidase staining for laminin on an unfixed blastocyst. Positive reactions (arrows) between polar trophoblast cells (P) and cells of the inner cell mass (I). X 160,000.

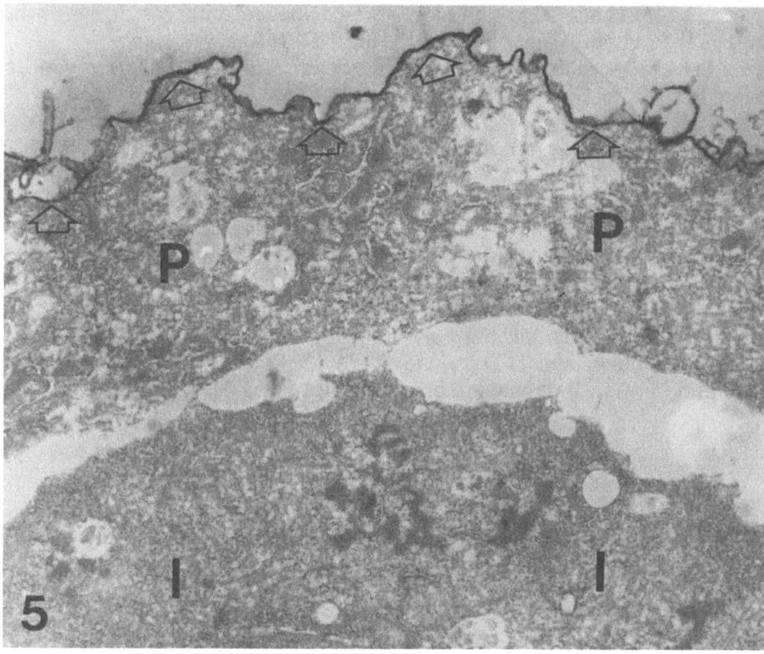

Figure 5: Immunoperoxidase stained control blastocyst treated with a goat-anti-rabbit immunoglobulin only. Striking is the positive reaction of the trophoblast surface (arrows). P = polar trophoblast, I = inner cell mass. X 9,500.

reactions for the reason that we had obtained the same staining pattern in the embryo even after the use of the secondary antibody only, i.e., without pretreatment with anti-laminin, anti-type IV collagen or anti-heparan sulfate proteoglycan antibodies. Further immunohistochemical studies with the use of secondary antibodies other than immunoglobulins have to be carried out to localize ECM components in early embryos.

Acknowledgements. The authors would like to thank Dr. Barrach for supplying us with anti-laminin, anti-type IV collagen and anti-heparan sulfate proteoglycan antibodies, W. Dresp, E. Heyder and B. Manshausen for their technical assistance, B. Rülke for her photographic work, and D. Landry-Lemme for editing the manuscript. This research was supported by a grant from the Deutsche Forschungsgemeinschaft (He 1327-2).

References

Chaplin, A.J. (1985) Tannic acid in histology: An historical perspective. Stain Technol. **60:** 219-231.

Dziadek, M., S. Fujiwara, M. Paulsson, and R. Timpl (1985) Immunological characterization of basement membrane types of heparan sulfate preteoglycan. EMBO J. **4:** 905-912.

Dziadek, M., and R. Timpl (1985) Expression of nidogen and laminin in basement membrane during mouse embryogenesis and in teratocarcinoma cells. Dev. Biol. **111:** 372-382.

Herken, R., and Y.-C. Hsu (1983) Development of mouse embryos grown in human cord serum (HCS) in vitro. Anat. Embryol. **168:** 137-150.
Herken, R., and H.-J. Barrach (1984) Immunhistochemische Lokalisation von Laminin und Kollagen Typ IV im Glomerulum der Niere. Verh. Ges. **78:** 167-169.
Herken, R., and H.-J. Barrach (1985) Ultrastructural localization of type IV collagen and laminin in the seven-day-old mouse embryo. Anat. Embryol. **171:** 365-371.
Leivo, I., A. Vaheri, R. Timpl, and J. Wartiovaara (1980) Appearance and distribution of collagens and laminin in the early mouse embryo. Dev. Biol. **76:** 100-114.
Sannes, P.L., T. Katsuyama, and S.S. Spicer (1978) Tannic acid-metal salt sequences for light and electron microscopic localisation of complex carbohydrates. J. Histochem. Cytochem. **26:** 55-61.
Wartiovaara, J., I. Leivo, and A. Vaheri (1979) Expression of the cell surface-associated glycoprotein, fibronectin, in the early mouse embryo. Dev. Biol. **69:** 247-257.
Wu, T., Y. Wan, A. Chung, and I. Damjanov (1983) Immunohistochemical localization of entactin and laminin in mouse embryos and fetuses. Dev. Biol. **100:** 496-505.

PRODUCTION AND FORMATION OF THE BASAL LAMINA

H.-J. Merker
Institut für Anatomie, Freie Universität Berlin and
Institut für Toxikologie und Embryonal-Pharmakologie
Freie Universität Berlin
Federal Republic of Germany

Introduction

The chemical composition of the basal lamina (BL), its morphology, localization and behaviour under experimental and pathological conditions have been investigated extensively over the last few years (Hawkes and Wang, 1982; Kühn et al., 1982; Lauri and Leblond, 1983; Martinez-Hernandez and Amanta, 1983; Porter and Whelan, 1984; Timpl and Martin, 1982; Piez and Reddi, 1984). However, the morphological basis for synthesis and secretion of BL-material and the actual formation of the BL in the extracellular matrix (ECM) have been somewhat neglected. In order to demonstrate and evaluate intracellular processes during BL production, data may be taken into consideration that were obtained for the production of the mesenchymal matrix. Since collagens, proteoglycans (PG) and glycoproteins (GP) are the main components of the BL as well as the matrix, these two parts of the intercellular space are, to a certain degree, comparable. Hence, the aim of the present study is to discuss 1) the morphology of BL production (synthesis, intracellular transport, secretion) in view of our knowledge of matrix production in general and 2) the processes of the actual BL formation in the extracellular space.

Except for hyaluronic acid, which is the only freely occuring GAG, all other components of the mesenchyme or the connective tissue contain varying amounts of protein. Consequently, the production begins with the formation of polypeptide chains at the ribosomes of the rough endoplasmic reticulum (ER). Therefore, all ECM-producing cells contain, depending on the synthesis rate, varying amounts of rough ER. As early as during the formation of the peptide chains, the first sugars are linked with certain amino acids. The binding of xylose to serine during chondroitin sulfate synthesis represents a well-investigated example in this connection. Xylose functions as starter for the further growth of the carbohydrate chains. Penetration of the synthesis products into the cavities of the ER is followed by vesicular transport into the Cis-components of the Golgi apparatus. In this organelle, important processing steps take place, for example, elonga-

NATO ASI Series, Vol. H5
Mesenchymal-Epithelial Interactions in Neural Development
Edited by J. R. Wolff et al.

tion of the carbohydrate chains and sulfation. At the trans-side of the Golgi apparatus, vesicles are cut off which contain the components of the ECM. The vesicles partly fuse to larger secretion granules. These migrate towards the cell membrane and their contents are secreted by exocytosis (Bornstein, 1974; Kivirikko and Risteli, 1976; Merker and Struwe, 1971; Mitchell and Hardingham, 1982; Palade, 1975; Piez and Reddi, 1984; Trelstad and Hayashi, 1979; Weinstock, 1972).

In the extracellular space of either the mesenchyme or the connective tissue, numerous other steps take place before the final state of maturation is reached. First, procollagens must be transformed into aggregatable collagens by means of collagenase which splits off telopeptides. This is followed by aggregation of tropocollagen molecules to primary filaments, filaments and cross-striated fibrils. The fibrils reach a thickness which is typical for each region, age, and species (Schwarz, 1957). These fibrils are bundled and form fibres that are arranged in a characteristic manner. The other structures, the glycosaminoglycans (GAG), PG and GP, are linked with one another, with collagen, basal lamina or the cell membranes. This linkage is the prerequisite for the functioning of the ECM (Kivirikko and Risteli, 1976; Manabe and Kikkawa, 1976; Morris et al., 1975; Schubert, 1966; Segrest and Cunningham, 1973; Silver et al., 1979; Smith, 1968).

During these extracellular processes, the individual components of the ECM influence each other. This phenomenon can best be demonstrated by the aggregation of collagen. Addition of highly sulfated GAG or PG, e.g., heparan sulfate, to cartilage cultures leads to the formation of thick segment-long-spacing or fibril-long-spacing collagen structures instead of the normal filaments (Merker et al., 1978). After the addition of highly sulfated GAGs to collagen solutions in the test tube, thicker fibrils are observed than after either the addition of hyaluronic acid or without any addition (Lilja and Barrach, 1981, 1983).

This rather brief discussion of synthesis, secretion and extracellular maturation of ECM-components does not consider a number of problems which have so far not been elucidated. They are, however, of significance for the interpretation of the mesenchymal functions.

These problems are:

1) Coupling of synthesis and secretion

Even in cells showing a high ECM production rate no or only very few secretion granules can be demonstrated. In the ECM-producing cells, secretion is coupled directly with synthesis, i.e., the synthesized components are immediately secreted. In contrast to other secreting cells, e.g., glandular cells, the synthesis product is not piled-up but continuously secreted without any special stimulus or signals. This is the so-called immediate type of secretion (Kelly, 1985). If secretion is inhibited, e.g., by colchicine, typical secretion granules appear (Feller and Merker, 1979).

Processing and intracellular transport are other steps of synthesis that can also be inhibited (Palade, 1975). According to the type of inhibition, some characteristic morphological findings are obtained: a dissociation of the polysomes, a ballooning of the endoplasmic cavities and of the Golgi apparatus and development of autophagic vacuoles (cytolysosomes).

2) Matrix-cell interaction

The ECM is undoubtedly able to influence its own production (Hawkes and Wang, 1982; Hay, 1981; Trelstad, 1984). We all know of examples that show an influence on quality, quantity and induction of ECM production. The inducing effect of the ECM can be demonstrated in a proliferating liver cell-line isolated from newborn mice. On glass or plastic, these cells do not produce an ECM. But on wet collagen, floating collagen or on gelatine ECM production can be demonstrated (Lilja and Merker, 1986).

3) The intracellular route

Unsolved is still the problem of the intracellullar secretion pathway of the ECM components, i.e., whether they are secreted jointly or individually. The available findings do not yet allow any final conclusions. The situation becomes even more confusing if we consider that the pathway for collagenase, which is necessary for the transformation of procollagens in the extracellular space, is not yet exactly understood either. The occurrence of cross-striated collagens in vacuoles or granules may indicate a common pathway, since a prerequisite for an aggregation of collagen is the enzymatic degradation of telopeptides (Deporter and Ten Cate, 1973; Weinstock, 1972). However, these findings may also be attributed to an inhibited secretion process. In this case, the lysosomal system is activated and tries to disaggregate the secretion product. Smith and Farquhar (1966) have demonstrated these processes in the hypophysis. Similar processes can be shown in cultured cartilage cells after colchicine application, which inhibits secretion (Feller and Merker, 1979).

4) Cell types producing ECM

According to the available findings, except for a few cell types (e.g., neurones), almost all cells are able to produce ECM components (Kraemer, 1982). Each cell type forms a quantitatively and qualitatively characteristic pattern of certain collagens and the corresponding GAG, PG and GP. Only the spatial allocation of the various minor collagens has not yet been fully elucidated. The interstitial ECM material is predominantly produced by corresponding local cells. But it is not exactly known yet whether other cells, epithelial, endothelial and muscle cells, are involved in this production.

5) The ECM pattern

Generally, the qualitative ECM pattern in the various regions is very typical and stable. However, under developmental, experimental and pathological conditions shifts from one genetic programme to another may occasionally occur. Such phenomena indicate the possibility of changes in gene activity in a specific cell type (Barrach et al., 1980, 1981).

In vivo examples are the menisci and other fibrous cartilages. They show a peculiar ECM pattern. The cells first produce much collagen type I, collagen type III and fibronectin and, in addition, much cartilage-specific PG but no collagen type II. After term, however, collagen type II is produced. This is reflected also by the electron microscopic pictures. Thick, typical type I fibrils containing many Ruthenium Red (RR)-positive granules are observed near the cells. Pericellularly, only thin filaments and densely packed RR-positive granules are located (Csato et al., 1986).

Blastemal cells of limb buds may serve as an in vitro example of a possible change in the qualitative ECM pattern. If isolated blastemal cells are cultured at high cell density, cartilaginous tissue containing collagen type II and cartilage-specific PG develop. If, however, these cells are cultured at low cell density in a monolayer culture system, they react like fibroblasts, i.e., they form collagen type I and III and unspecific PG. If these fibroblast-like cells are isolated again after 1 to 4 days and cultured at high cell density, they form a cartilage matrix. In contrast, isolated chondrocytes form a cartilage matrix if cultured in a monolayer culture system. After some days, they change again into fibroblast-like cells and synthesize collagen type I and III (Grundmann et al., 1980; Zimmermann et al., 1981). Hence, it can be postulated that a possibility exists to change the synthesis programme and the ECM pattern.

The findings described here for the production of the mesenchymal ECM and the ensuing analysis are, according to our experience, also true for the production of the BL. It is, however, not easy to demonstrate these processes. BL-producing cells have numerous other functions that can also be of secretory nature. Therefore, the structures that are responsible for BL production cannot be identified with certainty, i.e., they are masked. Moreover, the quantity of BL material that is produced by the individual cells is relatively small.

This is the reason why the morphological picture of the BM production is not very impressive. Therefore, a tumor that produces exclusively BL material in large amounts is especially suited for this type of investigation (Orkin et al., 1977). The picture of the tumor cells largely corresponds to that of ECM-producing cells of the mesenchyme and connective tissue. Thus, the morphological basis for the production of mesenchymal ECM and BL-ECM is identical (Fatemi and Leblond, 1985; Hogan et al., 1984; Merker and

Barrach, 1981).

As far as the site of formation is concerned, general agreement exists that the basal lamina components are formed by the cells resting on the BL (Kühn et al., 1982; Porter and Whelan, 1984). There is, however, one uncertainty: Is fibronectin a component of the basal lamina or is it just associated, and do epithelial cells also form fibronectin (Brownell et al., 1981)? Another problem occurs during embryonic development. In some organs, e.g., in the lung, BL components can also be demonstrated in the mesenchyme. According to own findings, at a certain stage, large amounts of laminin and collagen type IV are diffusely distributed in the mesenchyme of the embryonic mouse lung (Merker and Barrach, 1986).

Hence, based on these morphological findings the intracellular aspects of BL production are well understood. However, the problem of the actual BL production in the extracellular space is still unsolved and requires further investigation. These processes can be followed in embryonic tissues, e.g., in the dorsal part of the neural tube shortly after folding off (mouse embryo, day 9/10) and during early lung development (mouse embryo, day 12/13). Addition of tannic acid to the fixation solution considerably facilitates the demonstration of BL production and behaviour of BL material after secretion (Csat and Merker, 1983). Tannic acid obviously binds preferentially to glycoproteins (collagens, laminin, entactin, nidogen, fibronectin) and the glycoprotein-tannic acid bindings form electron-dense complexes together with osmium during postfixation (Simionescu and Simionescu, 1978).

BL production obviously starts with the differentiation of the responsible cells and synthesis and secretion (Fig. 1 and 2). The secreted material initially does not show any close or arranged spatial relationship to the cell membranes. It occurs in the form of vaguely demarcated plaques (0.5 - 1.5 µm) at a varying distance to the cell membrane. Higher magnification reveals a dense network of irregularly running filaments (..50 A) with incorporated granular structures (100 - 1000 A). Despite tannic acid or ruthenium red fixation the surface coat of the basal cell membrane is only weakly pronounced. In the following step of BL formation, the plaque-like BL material attaches to the cell membrane (Fig. 3). However, its arrangement does not yet resemble that of the typical lamina. Its demarcation from the mesenchyme and from the cell membrane is only weakly pronounced. Then, short segments develop that show the typical picture of a BL. A thick granular surface coat is now present in these areas. The Lamina rara is bridged by fine filaments. Almost at the same time the irregular plaques disappear. The typical BL segments increase in size. At the last stage of the process a BL is often still missing only in the regions of the intercellular gaps (Fig. 4). The gaps close from lateral direction. During the actual BL formation, a large number of free BL segments, of BL segments

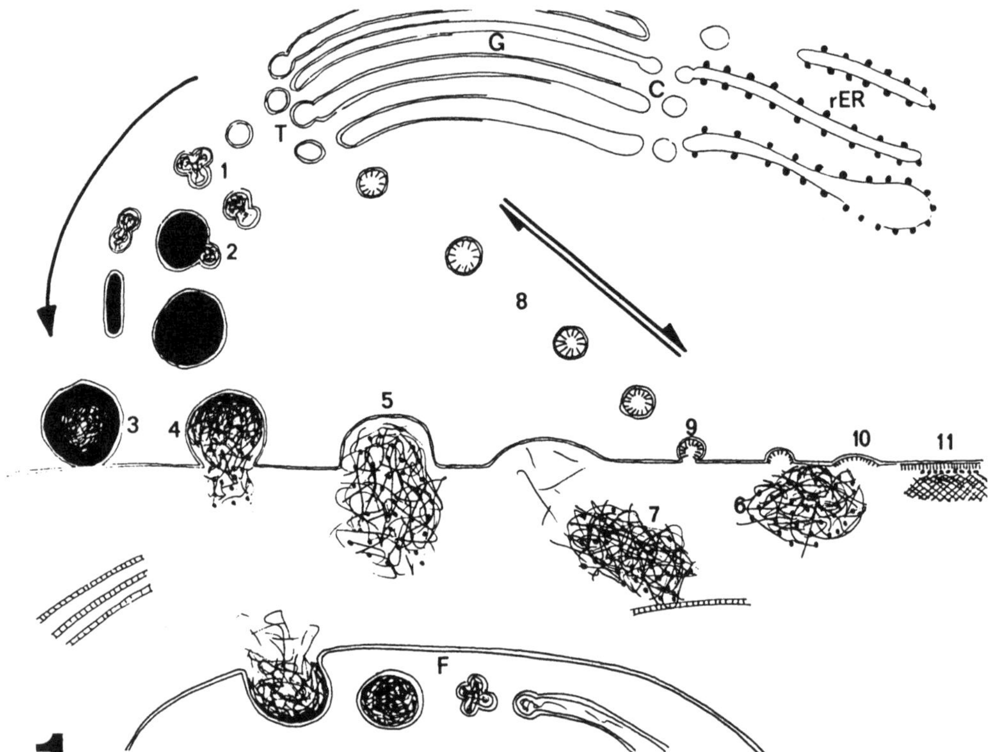
G
C
rER
T
1
2
8
3
4
5
9
10
11
7
6
F

1

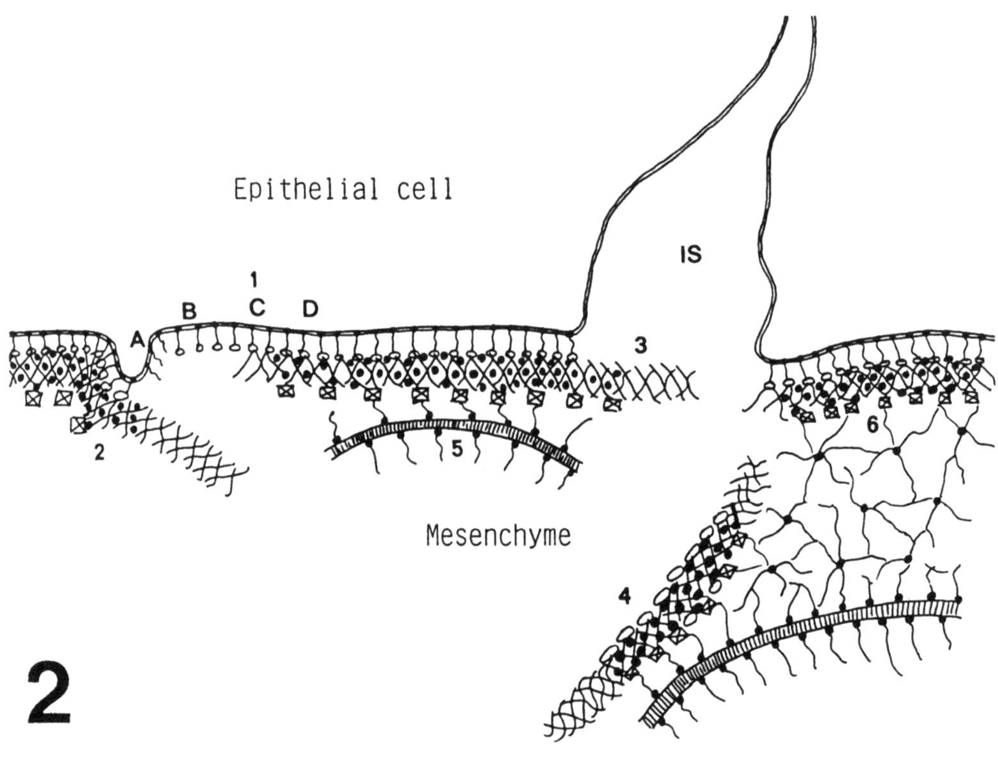
Epithelial cell
IS
1
B
C
D
A
3
2
5
6
Mesenchyme
4

2

Figure 1: Morphological basis for the synthesis, the intracellular transport and the secretion of BL components of an epithelial cell: synthesis in the rough endoplasmic reticulum (rER), transport by vesicles to the cis-side (C) of the Golgi apparatus (G), processing in the Golgi apparatus, development of vesicles containing secretion products at the trans-side (T), merging of the vesicles to prosecretion granules (1), densification of secretion granules (2), transport to the cell membrane (3), fusion and rupture of the membranes (3 and 4), release of the content of the secretion granules (4 and 5), localization of the secretion product in the vicinity of the cell membrane (6) or at a certain distance to it (7). After induction by secreted BL components vesicles are possibly transported from the Golgi apparatus to the cell membrane (8). The vesicle membranes contain receptors (laminin) that are incorporated into the cell membrane (10) by fusion (9). The receptors bind BL components thus giving rise to a BL (11). Fibrocytes (F) that also secrete collagen, proteoglycans and glycoproteins (fibronectin) are located in the mesenchymal space underneath the epithelium.

Figure 2: 1 = Formation of the BL at the cell membrane: A. The cell membrane receives receptors (e.g., for laminin). B. Laminin binds to the receptors; C. Laminin binds collagen type IV; D. The other BL components bind to the collagen skeleton. 2 = At one cell process the contact of the BL components gets lost and the BL grows away from the cell membrane through collagen self-assembly. 3 = At an intercellular space (IS) the binding and guidance by the cell membrane and the receptors cease and the BL continues to grow due to self-assembly. 4 = BL components may be bound by collagen, proteoglycans or fibronectin of the IS thus giving rise to free BL segments. 5 = The BL is additionally stabilized by accumulation of collagen and fibronectin at the mesenchymal side of the BL and the access of disaggregating enzymes is inhibited. 6 = Linkage of the BL with free proteoglycans of the IS, thus additional stabilization and inhibition of the turnover.

radiating from the actual BL and doublings of the BL are seen. After BL formation is completed, these aberrant forms no longer occur.

From these findings the following conclusions can be drawn and hypotheses postulated:

1) The production of BL components precedes actual BL formation.

2) The formation of first BL segments proceeds parallel to the occurrence and thickening of the surface coat. This behaviour speaks for a binding of BL components to the cell membrane as a triggering event, possibly via receptor-like molecules at the membrane surface, as other authors were able to demonstrate for some BL components (Bernfield et al., 1984; Liotta et al., 1984; Martin et al., 1984).

3) These binding molecules are obviously not present initially. In analogy to other receptor-like molecules, their formation is possibly only induced if ligands are present in the extracellular space.

4) The other BL components, especially collagen, bind to the first ligands. These components function as a kind of skeleton for the other BL elements.

5) After formation of the first BL segments collagen type IV is involved in the further

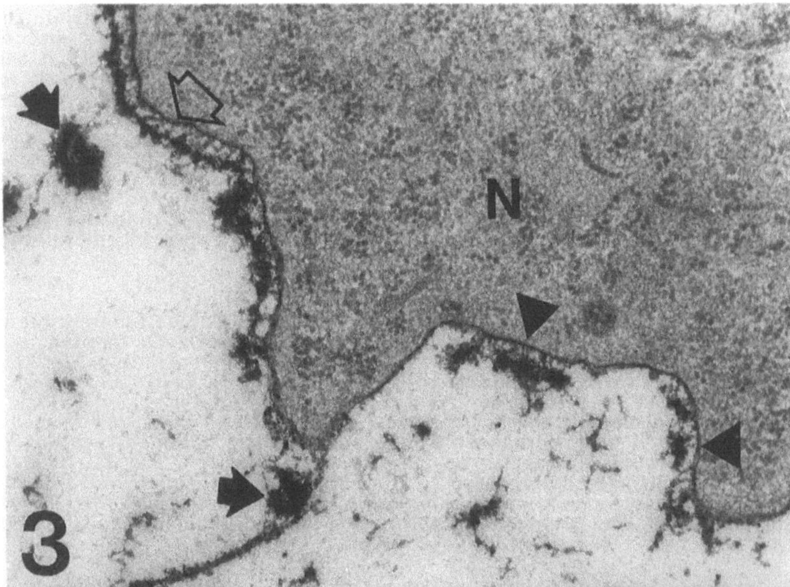

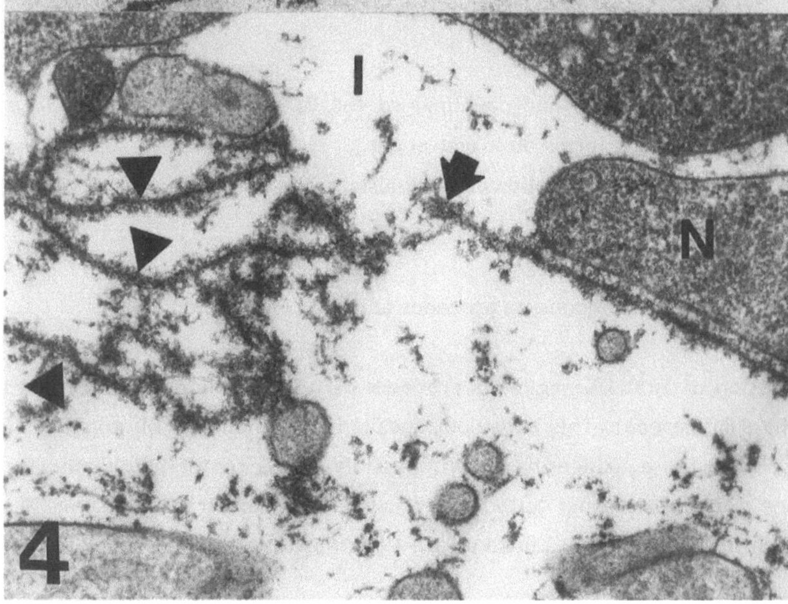

Figure 3: Dorsal region of the neural tube, mouse embryo day 9. Onset of BL formation. BL-like material freely occurring in the intercellular space (black arrows). First contacts of BL components with the cell membrane, occurrence of a surface coat (arrow heads). Completed BL segment (open arrows). N = neural epithelium.
X 55,000

Figure 4: Dorsal region of the neural tube, mouse embryo day 9. BL growth independent of the cell membrane (arrow) in the region of an intercellular space (I). Freely occurring BL segments (arrow heads). N = Neural epithelium.
X 42,000

growth of the BL in the form of a self-assembly process.

6) This capability of self-assembly plays a significant role in the bridging of intercellular spaces (Csato and Merker, 1983; Madri et al., 1984; Yurchenko et al., 1986).

7) Disturbance of the simultaneous binding to the cell membrane during self-assembly may lead to an aberrant course of the BL.

8) Although the cell membrane has a triggering function, BL formation may occasionally also start at the collagen, proteoglycan and glycoprotein structures, independent of the cell membrane, leading to free BL segments.

9) Superfluous BL material and aberrant BL segments are quickly dissolved, possibly by the mesenchyme.

Summary

The morphology of ECM production is well known. Biochemical findings as well as electron microscopic pictures help elucidate the synthesis of collagens, proteoglycans and glycoproteins at the ribosomes of the rough endoplasmic reticulum, the transport to the Golgi apparatus via a vesicle shuttle, the occurrence of the different processing events during this course, the preparation of the completed synthesis product for secretion in vesicles and granules and the exocytotic secretion mechanism. However, numerous processes must take place after secretion before the final maturation of the intercellular substance, for example: enzymatic transformation of collagen, aggregation of collagen modules to fibrils, their bundling to fibres that show a characteristic texture, interlinkage of ECM components and linkage of ECM components with BL and cell membranes. Obviously, a number of problems occurring in this connection have not yet been solved, as there are individual or joint secretion of the components, quantitative and qualitative regulation of the ECM production, the mechanism of matrix-cell interaction and the processes and site of hyaluronic acid production.

Since the basal lamina also consists of collagens, proteoglycans and glycoproteins, analogous events may be assumed for BL-producing cells. This hypothesis could be verified morphologically in BL-producing tumor cells, in parietal yolk sac cells of the rat and in embryonic epithelial tissue.

Few data are available on the actual formation of the BL in the extracellular space parallel to the cell membrane, It does not proceed simultaneously with the secretion of the BL material. The hypothesis is postulated that the secreted material initially induces the formation of receptor-like molecules. Subsequently, the BL components bind to the cell membrane via these receptors. The morphological equivalent of these processes is the occurrence of a surface coat that links with the developing BL. While the formation of receptors represents the first step and the localized binding of the BL components the second step, completion of the BL depends, in certain cases also independent of the cell membrane (intercellular gaps), upon the ability to self-assembly of collagen type IV. The occurrence of aberrant BL segments may also be attributed to this BL formation without binding to the cell membrane.

References

Barrach, H.-J., K. Grundmann, N. Hinz, and A. Felies (1980) Immunofluorescent microscopic investigations of intercellular substances during limb development. In *Teratology of the Limbs*, H.-J. Merker, H. Nau and D. Neubert, eds., pp. 273-293. Walter de Gruyter Verlag, Berlin.

Barrach, H.-J., K. Grundmann, T.E. Kwasigroch, N. Hinz, and A. Felies (1981) Comparison of the differentiation of muscle and connective tissue of mouse limb buds in culture and in vivo. In *Culture Techniques*, D. Neubert and H.-J. Merker, eds., pp. 135-159. Walter de Gruyter Verlag, Berlin.

Bernfield, M., S.D. Banerjee, J.E. Koda, and A.C. Rapraeger (1984) Remodelling of the basement membrane: Morphogenesis and maturation. In *Basement Membrane and Cell Movement*. Ciba Foundation Symposium 108, pp. 179-196, Pitman, London.

Bornstein, P. (1974) The biosynthesis of collagen. Ann. Rev. Biochem. **43**: 567-603.

Brownell, A.G., C.C. Bessem, and H.C. Slavkin (1981) Possible functions of mesenchyme cell derived fibronectin during formation of basal lamina. Proc. Natl. Acad. Sci. (USA) **78**: 3711-3715.

Csato, W., and H.-J. Merker (1983) Production and formation of the basement membrane in embryonic tissue of the mouse. Cell Tissue Res. **228**: 85-98.

Csato, W., H.-J. Merker, and H.-J. Barrach (1986) The intercellular matrix during pre- and postnatal development of the mouse menisci. In prep.

Deporter, D.A., and A.R. Ten Cate (1973) Fine structure localization of acid and alkaline phosphatase in collage-containing vesicles of fibroblasts. J. Anat. **114**: 457-461.

Fatemi, S.H., and C.P. Leblond (1985) Sulfatation and transport of basement membrane proteoglycans as visualized by ^{35}S-sulfate radioautography in the endodermal cells of the parietal yolk sac. Am. J. Anat. **173**: 127-145.

Feller, B., and H.-J. Merker (1979) Elektronenmikroskopische Untersuchungen zur Wirkung von Colchizin auf Knorpelgewebe aus Organkulturen von Extremitätenknospen (Mäuse-Embryonen). Acta Anat. **103**: 445-460.

Grundmann, K., B. Zimmermann, H.-J. Barrach, and H.-J. Merker (1980) Behaviour of epiphyseal mouse chondrocyte populations in monolayer culture. Virchows Arch. A, Pathol. Anat. and Histol. **389**: 167-187.

Hawkes, S., and J.L. Wang, eds. (1982) *Extracellular Matrix*. Academic Press, New York.

Hay, E., ed. (1981) *Cell Biology of the Extracellular Matrix*. Plenum Publ. Corp., New York.

Hogan, B.L.M., D. Barlowe, and M. Kurkinen (1984) Reichert's membrane as a model for studying the biosynthesis and assembly of basement membrane components. In *Basement Membrane and Cell Movement*, Ciba Foundation Symp. vol. 108, pp. 60-74. Pitman, London.

Kelly, R.B. (1985) Different types of secretion. Science **230**: 25-32.

Kivirikko, K.I., and L. Risteli (1976) Biosynthesis of collagen and its alterations in pathological states. Med. Biol. **54**: 159-186.

Kraemer, P.M. (1982) Perspectives in extracellular matrix research: The role of cell culture models. In *Extracellular Matrix*, S. Hawkes and J.L. Wang, eds., pp. 3-14. Academic Press, New York.

Kühn, K., H.H. Schöne, and R. Timpl, eds. (1982) *New Trends in Basement Membrane Research*. Raven Press, New York.
Lauri, G.W., and C.P. Leblond (1983) What is known of the production of basement membrane components? J. Histochem. Cytochem. **31**: 159-163.
Lilja, S. and H.-J. Barrach (1981) An electron microscopical study of the influence of different glycosaminoglycans on the fibrillogenesis of collagen type I and II in vitro. Virchow's Arch. A, Pathol. Anat. and Histol. **390**: 325-338.
Lilja, S., and H.-J. Barrach (1983) Normally sulphated and highly sulphated glycosaminoglycans (GAG) affecting fibrillogenesis of type I and type II collagen in vitro. Exp. Path. **23**: 173-181.
Lilja, S. and H.-J. Merker (1986) The behaviour of a proliferating hepatocyte line on different substrates. In prep.
Liotta, L.A., N.C. Rao, S. Barsky, and G. Bryant (1984) The laminin receptor and basement membrane dissolution: Role in tumor metastasis. In *Basement Membrane and Cell Movement*, Ciba Foundation Symp. 108, pp. 146-162. Pitman, London.
Madri, J.A., B.M. Pratt, P.D. Yurchenko, and H. Furthmayr (1984) The ultrastructural organization and architecture of basement membranes. In *Basement Membrane and Cell Movement*, Ciba Foundation Symp. 108, pp. 18-36. Pitman, London.
Manabe, T. and Y. Kikkawa (1976) Helical structure of human native collegan. Arch. Pathol. Lab. Med. **100**: 259-264.
Martin, G.R., H.K. Kleinman, V.P. Terranova, S. Ledbetter, and J.R. Hassel (1984) The regulation of basement membrane formation and cell matrix interaction by defined supramolecular components. In *Basement Membrane and Cell Movement*, Ciba Foundation Symp. 108, pp. 197-212. Pitman, London.
Martinez-Hernandes, A., and P.S. Amanta (1983) The basement membrane in pathology. Lab. Invest. **48**: 656-677.
Merker, H.-J., and K. Struwe (1971) Elektronenmikroskopische Untersuchungen zum Problem der Sekretion der bindegewebigen Interzellularsubstanz. Z. Zellforsch. **115**: 212-225.
Merker, H.-J., and H.-J. Barrach (1981) The morphology of basement membrane formation. Europ. J. Cell Biol. **26**: 111-120.
Merker, H.-J., and H.-J. Barrach (1986) The bahaviour of the components of the mesenchymal matrix and the basement membrane during development. In prep.
Merker, H.-J., S. Lilja, H.-J. Barrach, and Th. Günther (1978) Formation of an atypical collagen and cartilage pattern in limb bud cultures by highly sulfated GAG. Virchow's Arch. A, Path. Anat. and Histol. **380**: 11-30.
Mitchell, D., and T. Hardingham (1982) Monensin inhibits synthesis of proteoglycan, but not of hyaluronate in chrondrocytes. Biochem. J. **202**: 249-254.
Morris, N.P., L.I. Fessler, A. Weinstock, and J.H. Fessler (1975) Procollagen assembly and secretion in embryonic chick bone. J. Biol. Chem. **250**: 5719-5726.
Orkin, R.W., P. Gehron, E.B. McGoodwin, G.R. Martin, Th. Valentine, and R. Swarm (1977) A murine tumor producing a matrix of basement membrane. J. Exp. Med. **145**: 204-220.
Palade, G. (1975) Intracellular aspects of the process of protein secretion. Science **189**: 347-358.
Piez, K.A., and A.H. Reddi, eds. *Extracellular Matrix Biochemistry*. Elsevier, New York, 1984.
Porter, R., and I. Whelan, eds. *Basement Membranes and Cell Movement*, Ciba Foundation Symp. 108, Pitman, London, 1984.
Schubert, M. (1966) Structure of connective tissues, a chemical point of view. Fed. Proceed. **25**: 1047-1052.
Schwarz, W. (1957) Morphology and differentiation of the connective tissue fibres. In *Connective Tissue*, A Symposium by CION, pp. 144-156. Blackwell Sc. Publ., Oxford.
Segrest, J.P. and L.W. Cunningham (1973) Unit fibril models derived from the molecule topography of collagen. Biopolymers **12**: 825-834.

Silver, F.H., K.H. Langley, and R.L. Trelstad (1979) Type I collagen fibrillogenesis: Initiation via reversible linear and lateral growth steps. Biopolymers **18**: 2523-2535.

Simionescu, N., and M. Simionescu (1978) Galloylglucoses of low molecular weight as mordant in electron microscopy. I. Procedure and evidence for mordanting effect. J. Cell Biol. **70**: 608-621.

Smith, J.W. (1968) Molecular pattern in native collagen. Nature **219**: 157-158.

Smith, R.E., and M.G. Farquhar (1966) Lysosome function in the regulation of the secretory process in cell of the anterior pituitary gland. J. Cell Biol. **31**: 319-349.

book

Timpl, R., and G.R. Martin (1982) Components of the basement membranes. In Immunochemistry of the Extracellular Matrix, H. Furthmayr, ed., vol. 2, pp. 119-150. CRP-Press, Boca Raton.

Trelstad, R.L., ed. *The Role of Extracellular Matrix in Development*. Alan R. Liss, Inc., New York, 1984.

Trelstad, R.L., and K. Hayashi (1979) Tendon collagen fibrillogenesis: Intracellular subassemblies and cell surface changes associated with fibril growth. Dev. Biol. **71**: 228-242.

Weinstock, M. (1972) Collagen formation - observations on its intracellular packaging and transport. Z. Zellforsch. **129**: 455-470.

Yurchenko, P.D., E.D. Tsilibary, A.S. Charonis, and H. Furthmayr (1986) Models for the self-assembly of basement membrane. J. Histochem. Cytochem. **34**: 93-102.

Zimmermann, B., E. Kohnert-Stavenhagen, and K. Grundmann (1981) Das zeitabhängige Sekretionsverhalten von Chondrocyten in Monolayer-Kultur. Verh. Anat. Ges. **75**: 215-216.

CELL SURFACE AND EXTRACELLULAR MATRIX GLYCOPROTEINS AND PROTEOGLYCANS IN NERVOUS TISSUE

R.U. Margolis, J.A. Ripellino, and R.K. Margolis
Department of Pharmacology
New York University Medical Center
New York, NY
and
Department of Pharmacology
State University of New York
Downstate Medical Center
Brooklyn, NY
U.S.A.

There has recently been a greatly increased interest among neurobiologists in the roles of cell surface and extracellular matrix glycoproteins and proteoglycans in cell-cell interactions and other developmental processes. Our own efforts in this area have largely been directed towards studying their structure, localization, and metabolism in nervous tissue, since this information is essential for ultimately understanding their functional roles.

Proteoglycans are complex macromolecules in which one or more glycosaminoglycan chains are covalently linked to a protein moiety, and are the form in which all of the sulfated glycosaminoglycans are known to occur in tissues. Although recent studies on the biosynthesis of hyaluronic acid appear to have resolved a long-standing question by conclusively demonstrating that it does not occur as a covalent complex with protein (Prehm, 1983), since the structure (and many properties) of hyaluronic acid is similar to that of the other glycosaminoglycans, and because of its role in the formation of proteoglycan aggregates in many tissues, hyaluronic acid can best be considered together with the proteoglycans for the purposes of this discussion.

Chondroitin Sulfate Proteoglycans

The major glycosaminoglycans of nervous tissue are chondroitin sulfate, heparan sulfate, and hyaluronic acid. Chondroitin sulfate is composed of disaccharide repeating units consisting of glucuronic acid and N-acetylgalactosamine 4- or 6-O-sulfate. In brain, over

NATO ASI Series, Vol. H5
Mesenchymal-Epithelial Interactions in Neural Development
Edited by J. R. Wolff et al.

90% occurs in the form of chondroitin 4-sulfate. Approximately four chondroitin sulfate glycosaminoglycan chains with an average size of 19,000 daltons are linked to the protein core by a characteristic "linkage region" consisting of -glucuronosyl-galactosyl-galactosyl-xylosyl-O-serine (which is also present in sulfated glycosaminoglycans from other tissues).

The chondroitin sulfate proteoglycans of brain have a sedimentation coefficient of 6.5 S and a molecular size somewhat in excess of 300,000. They are composed of 56% protein, 24% glycosaminoglycans, and 20% of N- and O-glycosidically linked glycoprotein oligosaccharides. O-glycosidically linked oligosaccharides account for 28% of the glycoprotein carbohydrate in the chondroitin sulfate proteoglycan (as compared to only 9% in whole brain glycproteins), and the remainder is predominantly tri- and tetraantennary N-glycosidic oligosaccharides (over 65%), together with approximately 5% of biantennary oligosaccharides and less than 1% of high-mannose oligosaccharides (Kiang et al., 1981; Klinger et al., 1985). Most of the O-glycosidic oligosaccharides are unusual insofar as they are linked via mannose (rather than N-acetylgalactosamine) to serine and threonine residues in the protein core (Finne et al., 1979; Krusius et al., 1986). The proteoglycan also contains a small proportion of O-glycosidic mannose-linked keratan sulfate chains (with a molecular size of 3,000 to 10,000) composed of Gal(ß1-4)GlcNAc-6-O-SO_4(ß1-3) disaccharide repeating units (Krusius et al., 1986).

Monospecific antibodies to the chondroitin sulfate proteoglycan from rat brain were used in conjunction with the peroxidase-antiperoxidase technique for its localization by light and immunoelectron microscopy in developing and mature nervous tissue (Aquino et al., 1984a, b). To avoid potential difficulties in access of intact IgG molecules to antigenic sites in tissue sections and to minimize nonspecific binding through the Fc region, these studies employed $F(ab')_2$ fragments, which were further purified on an affinity column in which the proteoglycan was coupled to agarose.

The proteoglycan was found to be exclusively intracellular in adult cerebellum, cerebrum, brain stem and spinal cord. Some neurons and astrocytes (including Golgi epithelial cells and Bergmann fibers) show strong cytoplasmic staining. Although in the central nervous system there is heavy axoplasmic staining of many myelinated and unmyelinated fibers, not all axons stain. Staining was also seen in retinal neurons and glia (ganglion cells, horizontal cells, and Müller cells), but several central nervous tissue elements are consistently unstained, including Purkinje cells, oligodendrocytes, myelin, optic nerve axons, nerve endings, and synaptic vesicles. In sympathetic ganglion and peripheral nerve there is no staining of neuronal cell bodies, axons, myelin, or Schwann cells, but in sciatic nerve the Schwann cell basal lamina is stained, as is the extracellular matrix surrounding collagen fibrils. Staining was also observed in connective tissue surrounding the trachea and in the lacunae of tracheal hyaline cartilage. These findings are consistent with immunochemical studies demonstrating that antibodies to the chondroitin sulfate proteoglycan of brain also cross-react to varying degrees with certain connective tissue proteoglycans.

In contrast to the intracellular (cytoplasmic) localization of chondroitin sulfate proteoglycans in adult brain, immunoelectron microscopic studies in immature (7 day postnatal) rat cerebellum (Aquino et al., 1984b) demonstrate almost exclusively extracellular staining in the granule cell and molecular layers. Staining is also extracellular and/or associated with plasma membranes in the region of the presumptive white matter. Axons, which are unmyelinated at this age, generally do not stain, although faint intracellular staining is present in some astrocytes. At 10 and 14 days postnatal there is a significant decrease in extracellular space and staining, and by 21 days distinct cytoplasmic staining of neurons and astrocytes appears. This intracellular staining further increases by 33 days so as to closely resemble the pattern seen in adult brain. Analyses of the proteoglycans isolated from 7-day-old and adult brain demonstrated that they have essentially identical biochemical compositions, immunochemical reactivity, size, charge, and density. These findings indicate that the antibodies used in our studies recognize the same macromolecule in both early postnatal and adult brain, and that the localization of the chondroitin sulfate proteoglycan changes progressively from an extracellular to an intracellular location during brain development.

Hyaluronic Acid

As previously mentioned, it has recently become clear that hyaluronic acid, which contains a disaccharide repeating unit composed of glucuronic acid(ß1-3)N-acetylglucosamine(ß1-4).., does not occur as covalent complex with protein, and because of its highly conserved nature it has not yet been possible to raise high affinity antibodies to undegraded hyaluronic acid by conventional techniques. We have therefore made use of the properties of cartilage proteoglycan aggregates to develop a highly sensitive and specific biological probe for the localization of hyaluronic acid in tissue sections at both the light and electron microscopic levels (Ripellino et al., 1985).

In cartilage and certain other tissues, a specialized region of the core protein in the chondroitin sulfate proteoglycan monomers (called the hyaluronic acid binding region) is responsible for their alignment along a central filament of hyaluronic acid to form very large proteoglycan aggregates. The active binding site has a high affinity, with a K_d of approximately 10^{-6}M, for decasaccharide or larger oligomers of hyaluronic acid but a much lower affinity for shorter oligomers, and this interaction is highly specific, insofar as no other anionic biopolymer has been found which will compete with hyaluronic acid for binding. When a third constituent, called link protein, is present, the proteoglycan molcules are essentially locked into place on the hyaluronic acid, with a dissociation constant which is too low to measure conveniently (Hascall and Hascall, 1981).

In our studies, proteoglycan aggregates were isolated by CsCl density gradient centrifugation after extraction from a transplantable rat chondrosarcoma, and digested by clostripain to obtain the hyaluronic acid binding region as a complex with hyaluronic acid and link protein. For the histochemical visualization of hyaluronic acid, the hyaluronic acid binding region/link protein complex was biotinylated in the presence of hyaluronic acid to yield a product having a molar ratio of biotin to complex of approximately 1:1. The biotinylated probe was then separated from hyaluronic acid by gel filtration in the presence of 4 M guanidine HCl, and used in conjunction with avidin-peroxidase for the staining of hyaluronic acid in tissue sections. For control sections we have primarily relied upon the known ability of hyaluronic acid oligosaccharides of specified sizes to inhibit its binding to macromolecular hyaluronic acid. Hyaluronic acid oligosaccharides containing greater than 10 monosaccharide residues competely block staining of tissue sections by the biotinylated HA-binding complex, whereas oligosaccharides of smaller size do not inhibit staining. We have also demonstrated that treatment of tissue sections with Streptomyces hyaluronidase completely prevents staining by the HA-binding complex. This enzyme is specific for hyaluronic acid and shows no protease activity under the conditions employed.

Light microscopy of early postnatal rat cerebellum showed strong staining in the presumptive white matter, and to a lesser extent as a reticulum in the granular layer. During the first two weeks after birth, when the major developmental changes occur in rat cerebellum, there is also a band of intense staining which is initially present at the base of the molecular layer, near the apical tips of the maturing Purkinje cell dendrites (Fig. 1A). This band of staining progresses towards the pial surface as the Purkinje cell dendritic arbor shows normal branching and upward expansion, closely followed by the growing climbing fibers. In contrast to developing brain, in the mature cerebellum there appears to be intracellular staining of Purkinje cells as well as diffuse staining throughout the molecular layer, and the virtual disappearance of staining around myelinated axons (Fig. 1B).

Electron microscopy demonstrated that the staining in early postnatal brain is almost entirely extracellular (Fig. 2A), and in this respect closely resembles that which we previously found in immunoelectron microscopic studies on the localization of the chondroitin sulfate proteoglycan in developing brain (Aquino et al., 1984b). Also as in the case of the chondroitin sulfate proteoglycan, in more mature brain there is cytoplasmic staining of granule cells (Fig. 2B), and intra-axonal and membrane staining of parallel fibers. Although the possible occurrence of hyaluronic acid intracellularly will require further study, our results are consistent with the presence of some hyaluronic acid in the form of aggregates with chondroitin sulfate proteoglycan (Kiang et al.,1981), and with

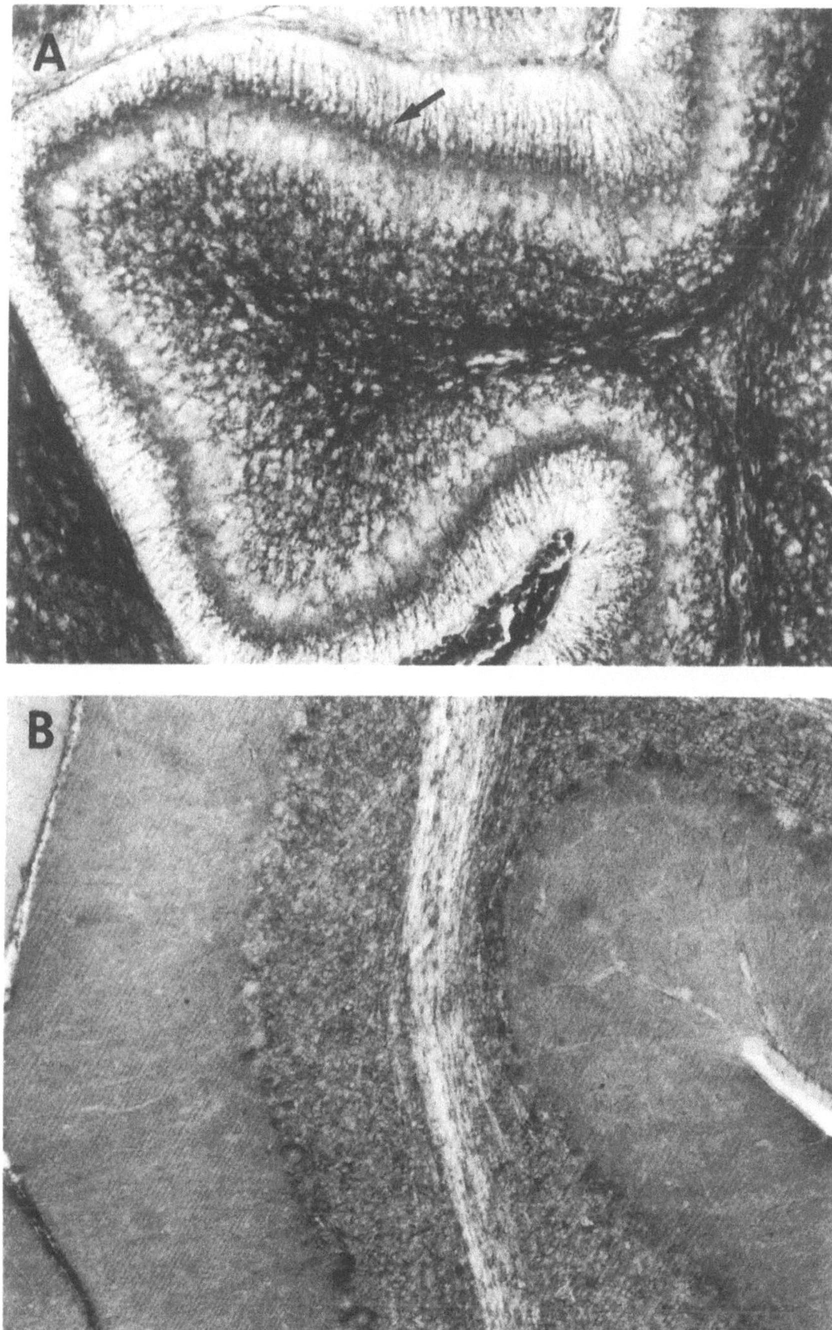

Figure 1: A: Staining of hyaluronic acid in sagittal section of 7-day rat cerebellum. Note prominent staining in the presumptive white matter, and dense band (arrow) above the Purkinje cells at the base of the moleculara layer. B: Localization of hyaluronic acid in adult cerebellum. X100.

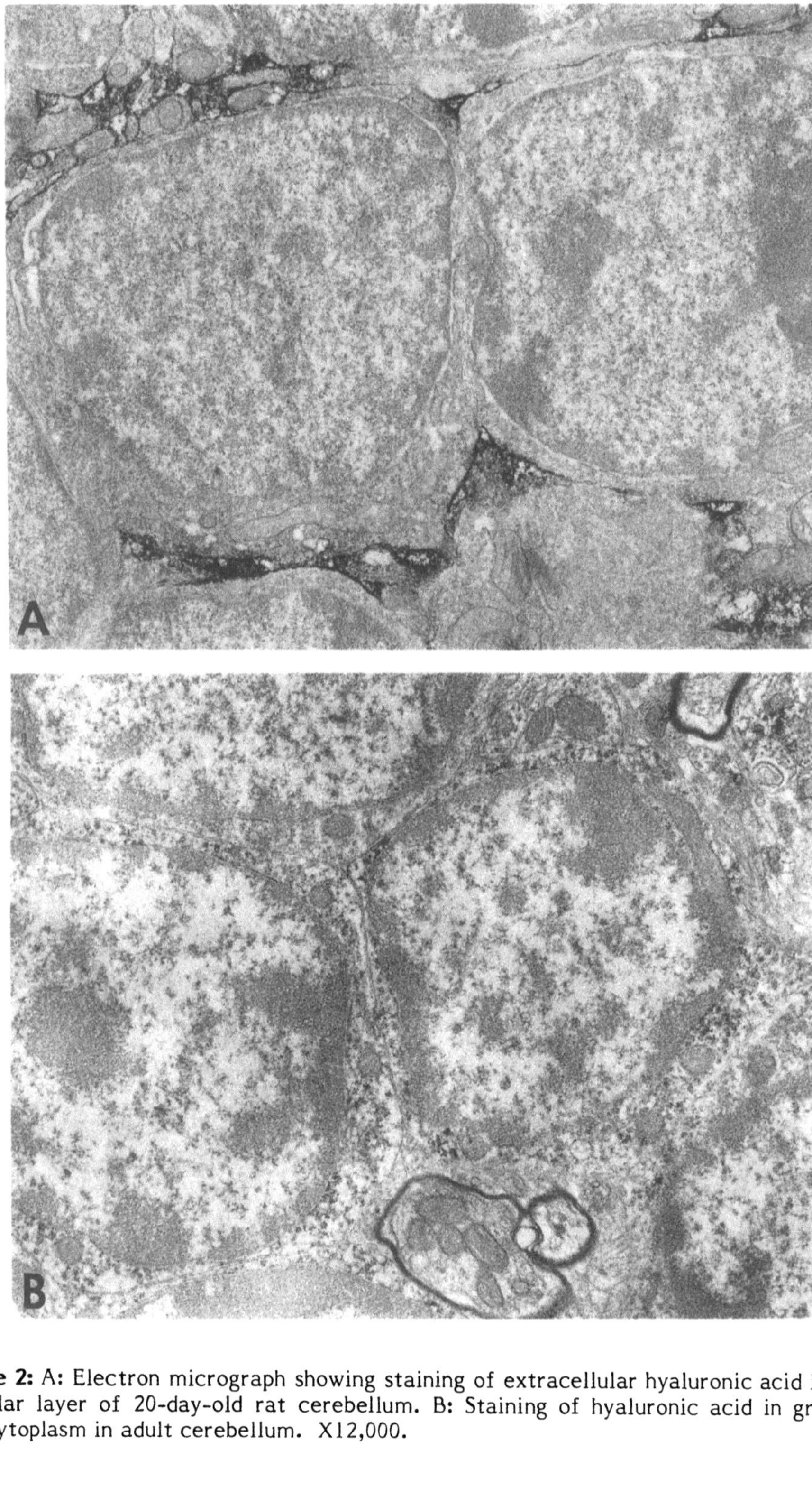

Figure 2: A: Electron micrograph showing staining of extracellular hyaluronic acid in the granular layer of 20-day-old rat cerebellum. B: Staining of hyaluronic acid in granule cell cytoplasm in adult cerebellum. X12,000.

preliminary evidence for the axonal transport of hyaluronic acid in rat brain (Gammon et al., 1985). Moreover biochemical analyses have previously detected the presence of significant amounts of hyaluronic acid in myelin-free axons isolated in bulk from rat and bovine brain (Margolis et al., 1975).

During early stages of brain development, it is likely that extracellular hyaluronic acid provides a highly hydrated and easily penetrable matrix through which neuronal migration may take place, before the intercellular space decreases to its much smaller dimensions in mature brain. It is also possible that extracellular hyaluronic acid and chondroitin sulfate proteoglycans must in some manner be removed or relocalized before various adhesive processes may begin. Aside from this type of passive or "facilitative" role, recent histochemical studies on the spatial and temporal distribution of extracellular matrix in developing cerebral cortex of normal and reeler mutant mice (Nakanishi, 1983) suggest that glycosaminoglycans and proteoglycans may also have a more direct involvement in relation to afferent axon targeting.

Heparan Sulfate Proteoglycans

The third major glycosaminoglycan present in nervous tissue is heparan sulfate, which is linked to a protein moiety by the same characteristic linkage sequence as described above for the chondroitin sulfate proteoglycans. Like the other glycosaminoglycans, heparan sulfate has a disaccharide repeating unit composed of a uronic acid (D-glucuronic acid and L-iduronic acid) and an amino sugar (N-acetyl- and N-sulfo-D-glucosamine). It bears a number of similarities with heparin, a product of mast cells which does not occur in nervous tissue parenchyma, and these two glycosaminoglycans can be considered as representing parts of a continous spectrum of related structures which vary mainly in their degree of sulfation and uronic acid composition.

Heparan sulfate proteoglycans were extracted from rat forebrain or microsomal membranes with deoxycholate, and purified by ion-exchange chromatography, affinity chromatography on lipoprotein lipase-agarose, and gel filtration (Klinger et al., 1985). The proteoglycan has a molecular size of approximately 220,000, and contains glycosaminoglycan chains of 14,000-15,000. It also contains essentially the same proportion (22%) of its carbohydrate in the form of glycoprotein oligosaccharides as does the chondroitin sulfate proteoglycan, and like the chondroitin sulfate proteoglycan these consist predominantly of N-glycosidically liked tri- and tetraantennary complex oligosaccharedes (60%, some of which are sulfated) and O-glycosidic oligosaccharides (33%).

In contrast to the rather restricted distribution of heparin in mast cells, heparan sulfate proteoglycans are ubiquitous components of the plasma membrane and basement membranes in many tissues, where they are thought to play important roles in the regulation of membrane permeability and in cell-cell interactions. It is also now becoming clear that a heparan sulfate proteoglycan accounts for a major portion of the internal contents of cholinergic synaptic vesicles from marine electric organ and mammalian brain (Stadler and Dowe, 1982; Carlson and Kelly, 1983; Walker et al., 1983). Although our studies indicate that most of the heparan sulfate proteoglycan of brain is an integral membrane component requiring detergents for extraction, a small but significant portion (approximately 20%) is probably loosely but specifically bound to the cell surface, since it can be released by incubating brain membranes with heparin. Chondroitin sulfate was ineffective in releasing heparan sulfate proteoglycans, and 2 M NaCl produced only a minimal effect (Klinger et al., 1985). These results are consistent with previous studies of other tissues, which have demonstrated the specific binding of heparan sulfate proteoglycans to the cell surface. We have prepared antibodies to the heparan sulfate proteoglycans of brain, and are currently using these for immunoelectron microscopic studies of their localization in developing and mature nervous tissue.

Cell Surface Glycoproteins

In addition to the proteoglycans discussed above, there is a very large number of cell surface glycoproteins in nervous tissue, many of which undoubtedly have important roles in cell-cell interactions during brain development. Most of these have not yet been identified even by relatively high resolution two-dimensional electrophoretic techniques, much less characterized in terms of their structures, localization to particular cell types or areas, or with regard to their developmental appearance and functional roles (for reviews of nervous tissue glycoproteins, see Margolis and Margolis, 1979, 1983). However, there is currently considerable interest in certain neuronal and neuronal-glial cell adhesion molecules which are also glycoproteins, as well as in laminin and related glycoproteins present in the extracellular matrix. It is therefore clear that this will be an area of increasingly intensive neurochemical and neurobiological investigation.

In our studies of nervous tissue glycoproteins, we have recently investigated the prevalence and localization of poly-N-acetyllactosaminyl oligosaccharides (Margolis et al., 1986), which are of particular interest insofar as they have been identified in glycoproteins and glycolipids which express a number of developmentally regulated and tumor-associated antigens (for a review, see Fukuda, 1985). They are characterized by the presence of up to six or more disaccharide repeating units having the structure (Galß1-4GlcNAcß1-3)$_n$, which in the case of glycoproteins are linked to a conventional $(Man)_3(GlcNAc)_2$ complex oligosaccharide core. These carbohydrate chains were originally detected in erythrocytes and their precursor cells, and have more recently been found in a number of other cell types including Chinese hamster ovary cells, thyroid, teratocarcinoma cells, and granulocytes. Susceptibility to several bacterial endo-ß-galactosidases is considered diagnostic of the presence of such oligosaccharide chains, although sulfation and branching at internal ß-galactosidic linkages is known to confer resistance to enzymatic hydrolysis.

Endo-ß-galactosidase treatment of glycopeptides derived from the trypsinate and membranes of PC12 pheochromocytoma cells and cultured sympathetic neurons demonstrated the presence of poly-N-acetyllactosaminyl units on tri- and tetraantennary oligosaccharides, some of which have a core fucose and a 2,6-substituted α-linked mannose residue. Nerve growth factor-induced differentiation of the PC12 cells led to a small but significant decrease in the proportion of these oligosaccharides. Glycoproteins containing poly-N-acetyllactosaminyl oligosaccharides were selectively labeled by treatment of PC12 cells with endo-ß-galactosidase to expose N-acetylglucosamine residues, followed by incubation with galactosyltransferalase and UDP-[^{14}C] galactose. SDS-polyacrylamide gel electrophoresis and fluorography revealed the presence of 15-20 distinct PC12 cell glycoproteins which contain these oligosaccharides, and have apparent molecular weights in the range of 25,000 to 250,000 Da. Treatment of PC12 cells with nerve growth factor (NGF) altered the relative labeling of several of the glycoprotein bands, with a time course similar to the effects of NGF on neurite outgrowth. However, most of the NGF effects on glycoprotein labeling are also seen in PC12 cells cultured in suspension, conditions under which neurite extension and certain other NGF effects do not occur, and they can be partially induced by long-term treatment with forskolin (which increases cellular levels of cyclic AMP), but not by epidermal growth factor.

One of the major components of PC12 cell membranes is a 230,000 Da cell surface glycoprotein whose concentration increases approximately three- to five-fold in the presence of NGF (McGuire et al., 1978; Salton et al., 1983). This surface glycoprotein has been named the NGF-inducible large external, or NILE, glycoprotein, and using indirect immunofluorescence it was found that polyclonal antisera against NILE glycoprotein purified from PC12 cells stained the PC12 cell surface in a uniformly distributed granular pattern (Salton et al., 1983). Immunofluorescent staining of primary cell cultures and tissue whole mounts demonstrated that immunochemically cross-reactive NILEF glycoprotein is expressed on the cell surfaces (somas and neurites) of all peripheral and central neurons examined, but was not found on a wide variety of other cell types (Salton et al., 1983; Stallcup et al., 1985). Recent findings indicate that the NILE glycoprotein is recognized by antisera against the L1 antigen (Bock et al., 1985; Sajovoc et al., 1986), and that it may be homologous or identical to a neuronal-glial cell adhesion molecule (Friedlander et al., 1986).

We have previously reported that the greater molecular size of the PC12 cell NILE glycoprotein as compared to the immunochemically cross-reactive species present in postnatal brain (205,000 Da) is due to the greater size of the PC12 cell tri- and tetraantennary complex oligosaccharides (Margolis et al., 1983). We now find that the largest of these oligosaccharides contain poly-N-acetyllactosaminyl units, and that after endo-ß-galactosidase treatment, most of the undegraded oligosaccharides which remain are of the same size as the single population of tri- and tetraantennary oligosaccharides present in the NILE glycoprotein of brain, which is not affected by endo-ß-galactosidase treatment. The presence of poly-N-acetyllactosaminyl oligosaccharides in the NILE glycoprotein of PC12 cells, but not that of postnatal brain, is probably in itself sufficient to provide a structural basis for the difference in size of these two related glycoproteins.

In preliminary studies of 1- and 30-day-postnatal rat brain we have not been able to detect the presence of poly-N-acetyllactosaminyl oligosaccharides, although such oligosaccharides were found in cultured neonatal sympathetic neurons, which can be considered to be a close non-tumor counterpart of NGF-treated PC12 cells. These results indicate that the occurrence of this oligosaccharide structure reflects the common embryonic origin of pheochromocytoma cells and normal sympathetic neurons, and that it is not necessarily related to the tumor origin of PC12 cells. Studies now in progress will attempt to determine whether poly-N-acetyllactosaminyl oligosaccharides are characteristic only of sympathetic or peripheral nervous tissue, or whether they might also be present in less-differentiated embryonic central nervous tissue.

The general neurobiological role of poly-N-acetyllactosaminyl oligosaccharides remains to be determined, although in other systems their presence and fine structure frequently appear to be related to specific stages of cell and tissue differentiation (Fukuda, 1985). Our preliminary finding that they do not occur in postnatal rat brain indicates that these oligosaccharides may be restricted to very early developmental periods, or to only certain types of cells, such as neurons of sympathetic origin but not central nervous tissue. It is of interest in this regard that the "extension enzyme", ß-galactoside:ß1-3 N-acetylglucosaminyl transferase, which adds N-acetylglucosamine to C-3 of a galactose residue in Galß1-4GlcNAc acceptors, is evidently present in brain, since as mentioned above, a small proportion of keratan sulfate chains (having the same repeating unit with sulfate at C-6 of the N-acetylglucosamine) occur in the chondroitin sulfate proteoglycan of brain (Krusius et al., 1986). It has also been reported that approximately 5-10% of the total carbohydrate in ß-N-acetylhexosaminidases A and B from bovine brain is in the form of biantennary poly-N-acetyllactosaminyl chains terminated by sialic acid (Overdijk et al., 1985). The apparent lack of poly-N-acetyllactosminyl oligosaccharides in most other brain glycoproteins could therefore be due to the relatively low proportion of acceptor proteins or the presence of structural features which make them resistant to endo-B-galactosidase (Margolis et al., 1986), rather than to the absence of the necessary glycosyltransferases.

Acknowledgements. This work was supported by Grants NS 09348 and NS 13876 from the

National Institutes of Health, and Grant BNS-8509528 from the National Science Foundation.

References

Aquino, D.A., R.U. Margolis, and R.K. Margolis (1984a) Immunocytochemical localization of a chondroitin sulfate proteoglycan in nervous tissue. I. Adult brain, retina and perpheral nerve. J. Cell Biol. **99:** 1117-1129.

Aquino, D.A., R.U. Margolis, and R.K. Margolis (1984b) Immunocytochemical localization of a chondroitin sulfate proteoglycan in nervous tissue. II. Studies in developing brain. J. Cell Biol. **99:** 1130-1139.

Bock, E., C. Richter-Landsberg, A. Faissner, and M. Schachner (1985) Demonstration of immunochemical identity between the nerve growth factor-inducible large external (NILE) glycoprotein and the cell adhesion molecule L1. EMBO J. **4:** 2765-2768.

Carlson, S.S., and R.B. Kelly (1983) A highly antigenic proteoglycan-like component of cholinergic synaptic vesicles. J. Biol. Chem. **258:** 11082-11091.

Finne, J., T. Krusius, R.K. Margolis, and R.U. Margolis (1979) Novel mannitol-containing oligosaccharides obtained by mild alkaline borohydride treatment of a chondroitin sulfate proteoglycan from brain. J. Biol. Chem. **254:** 10295-10300.

Friedlander, D.R., M. Grumet, and G.M. Edelman (1986) Nerve growth factor enhances expression of neuron-glia cell adhesion molecule in PC12 cells. J. Cell. Biol. **102:** 413-419.

Fukuda, M. (1985) Cell surface glycoconjugates as onco-differentiation markers in hematopoietic cells. Biochim. Biophys. Acta **780:** 119-150.

Gammon, C.M., J.F. Goodrum, A.D. Toews, A. Okabe, and P. Morell (1985) Axonal transport of glycoconjugates in the rat visual system. J. Neurochem. **44:** 376-387.

Hascall, V.C., and G.K. Hascall (1981) Proteoglycans. In *Cell Biology of the Extracellular Matrix*, E.D. Hay, ed., pp. 39-63. Plenum, New York.

Kiang, W.-L., R.U. Margolis, and R.K. Margolis (1981) Fractionation and properties of a chondroitin sulfate proteoglycan and the soluble glycoproteins of brain. J. Biol. Chem. **256:** 10529-10537.

Klinger, M.M., R.U. Margolis, and R.K. Margolis (1985) Isolation and characterization of the heparan sulfate proteoglycans of brain. Use of affinity chromatography on lipoprotein lipase-agarose. J. Biol. Chem. **260:** 4082-4090.

Krusius, T., J. Finne, R.K. Margolis, and R.U. Margolis (1986) Identification of an O-glycosidic mannose-linked sialylated tetrasaccharide and keratan sulfate oligosaccharides in the chondroitin sulfate proteoglycan of brain. J. Biol. Chem., in press.

Margolis, R.K., and R.U. Margolis (1983) Glycoproteins and proteoglycans. In *Handbook of Neurochemistry*, A. Lajtha, ed., vol. 5, pp. 177-204. Plenum, New York.

Margolis, R.K., L.A. Greene, and R.U. Margolis (1986) Poly-N-acetyllactosaminyl oligosaccharides in glycoproteins of PC12 pheochromocytoma cells and sympathetic neurons. Biochemistry, in press.

Margolis, R.K., R.U. Margolis, C. Preti, and D. Lai (1975) Distribution and metabolism of glycoproteins and glycosaminoglycans in subcellular fractions of brain. Biochemistry **14:** 4797-4804.

Margolis, R.K., S.R.J. Salton, and R.U. Margolis (1983) Structural features of the nerve growth factor inducible large external glycoprotein of PC12 pheochromocytoma cells and brain. J. Neurochem. **41:** 1635-1640.

Margolis, R.U., and R.K. Margolis, eds. (1979) *Complex Carbohydrates of Nervous Tissue.* Plenum, New York.

McGuire, J.C., L.A. Greene, and A.V. Furano (1978) NGF stimulates incorporation of fucose or glucosamine into an external glycoprotein in cultured rat PC12 pheochromocytoma cells. Cell **15:** 357-365.

Nakanishi, S. (1983) Extracellular matrix during laminar pattern formation of neocortex in normal and reeler mutant mice. Dev. Biol. **95:** 305-316.

Overdijk, B., E.P. Beem, G.J. Van Steijn, L.A.W. Trippelvitz, J.J.W. Lisman, J. Paz

Parente, P. Cardon, Y. Leroy, B. Fournet, H. Van Halbeek, J.H.G.M. Mutsaers, and J.F.G. Vliegenthart (1985) Structural analysis of the carbohydrate chains of ß-N-acetylhexosaminidases from bovine brain. Biochem. J. **232:** 637-641.

Prehm, P. (1983) Synthesis of hyaluronate in diffrentiated teratocarcinoma cells. Mechanism of chain growth. Biochem. J. **211:** 191-198.

Ripellino, J.A., M.M. Klinger, R.U. Margolis, and R.K. Margolis (1985) The hyaluronic acid binding region as a specific probe for the localization of hyaluronic acid in tissue sections. Application to chick embryo and rat brain. J. Histochem. Cytochem. **33:** 1060-1066.

Salton, S.R.J., C. Richter-Landsberg, L.A. Greene, and M.L. Shelanski (1983) Nerve growth factor-inducible large external (NILE) glycoprotein: Studies of a central and peripheral neuronal marker. J. Neurosci. **3:** 441-454.

Sajovoc, P., E. Kouvelas, and E. Trenkner (1986) Probable identity of NILE glycoprotein and the high-molecular-weight component of L1 antigen. J. Neurochem., in press.

Stadler, H., and G.H.C. Dowe (1982) Identification of a heparan sulphate-containing proteoglycan as a specific core component of cholinergic vesicles from Torpedo marmorata. EMBO J. **1:** 1381-1384.

Stallcup, W.B., L. Beasley, and J.M. Levine (1985) Antibody against nerve growth factor inducible large external (NILE) glycoprotein labels nerve fiber tracts in the developing rat nervous system. J. Neurosci. **5:** 1090-1101.

Walker, J.H., J. Obrocki, and C.W. Zimmermann (1983) Identification of a proteoglycan antigen characteristic of cholinergic synaptic vesicles. J. Neurochem. **41:** 209-216.

HYALURONECTIN AND HYALURONIC ACID DURING THE DEVELOPMENT OF RAT BRAIN CORTEX

Annie Delpech*, Bertrand Delpech, Nicole Girard, Philippe Bertrand and Claude Chauzy

* Laboratoire d'Histo-embryologie
Hôpital Charles Nicolle
et
Laboratoire d'Immunochimie
Centre Henri-Becquerel
Rouen
France

Introduction

Hyaluronic acid (HA) detection is particularly important during vertebrate embryonic development since it is believed to play a significant role in the differentiation and migration processes which occur in mesenchyme (Toole, 1973; Pratt et al., 1975; Toole, 1981) and in nervous tissue (Brunngraber, 1979). We report here results about HA detection during the development of rat cerebral cortex. HA location was compared to that of hyaluronectin (HN) which is characterized in vitro by its affinity for HA (Delpech and Halavent, 1981; Delpech, 1982). Recent results led to the conclusion that HN is the protein core of a small human brain proteoglycan (Delpech et al., 1986). Tissue HN detection was made by means of indirect immunohistochemical techniques using anti-HN antibodies. Tissue HA detection was made with the immune complex technique, using HN as a probe for HA. We describe in this paper the details of techniques, the preparation of reagents and the specificity criteria selected. In order to define precise topographic landmarks in the developing cerebral cortex, we refer to previous work on brain ontogenesis (Boulder Committee, 1970; Marin Padilla, 1978; Readler et al., 1980; Pinto-Lord and Caviness, 1979).

Materials and Methods

Tissue HN detection was made by means of indirect immunochemical techniques using either fluorescent or peroxidase labelled reagents with the avidin-biotin complex technique. Controls were made with (1) the preimmunization serum from the very rabbit which produced the anti-HN antibodies and (2) the monospecific anti-HN antiserum ab-

NATO ASI Series, Vol. H5
Mesenchymal-Epithelial Interactions in Neural Development
Edited by J. R. Wolff et al.

sorbed out with insolubilized HN. This absorption is the main control for tissue HN detection and must be carried out with insolubilized HN in order to prevent the binding of HN/anti-HN soluble immune complexes (IC) to tissue receptor for HN, that is, to tissue HA. This spontaneous binding precludes the use of soluble HN to absorb out the antibodies to HN (unless HA is added to HN) and was a key to set up a technique for the detection of tissue HA.

Tissue HA detection can be made with HN in two different ways:

(1) Incubation with HN followed by incubation with anti-HN antbodies: this allows the detection of tissue HN plus HN which has bound to tissue receptors for HN. A comparison should be made with results obtained in the absence of incubation with HN which hampers the interpretation.

(2) The IC technique allowed the characterization of only receptors for HN. HN was added to anti-HN antibodies in such a proportion that antibodies were saturated, which led to soluble immune complexes and blocked antibodies which could not bind to tissue HN. Then the antibodies of immune complexes were revealed with the classical techniques. The preparation is easy to perform (by just mixing solutions), but it must be controlled accurately in order to ascertain that (a) there are no more free antibodies to HN (which is particularly important since HN is also a nerve tissue component) and (b) there is not too large an excess of free HN which would compete with IC and would weaken the staining. The ELISA technique allowed the determination of the amounts of reagents to be used (Fig. 1). With HA coated plates it was easy to find which proportion of HN and antibodies gave the highest immune complex binding to the plate, and to verify that increasing the HN proportion led to a decrease of the immune complex binding. Then, using an HN coated plate it was verified that no antibody binding occurred with the selected IC solution. The optimum reaction was obtained with a solution containing 60 micrograms HN and 40 micrograms antibodies.

Specificity of IC binding to nerve tissue

Now the tissue receptor (which is presumably HA) has to be characterized. Like in the ELISA technique, the competition technique was used. In an investigation to see whether soluble molecules, mainly glycosaminoglycans (GAGs), could interfere with the IC binding to tissue, the addition of HA or HA derived oligosaccharides (Bertrand and Delpech, 1985) was found to be the only way to suppress IC binding (Girard et al., 1984, 1986). Moreover, this was obtained with a small amount of HA (10% of HN concentration in weight), whereas other GAGs were inefficient at concentrations 100 times higher. Finally, the destruction of the tissue target by Streptomyces hyaluronidase (Calbiochem, 50 TRU/ml, 90 minutes at 37°C, repeated once) suppressed IC binding to nerve tissue. Every specificity test was performed for the study of rat embryos.

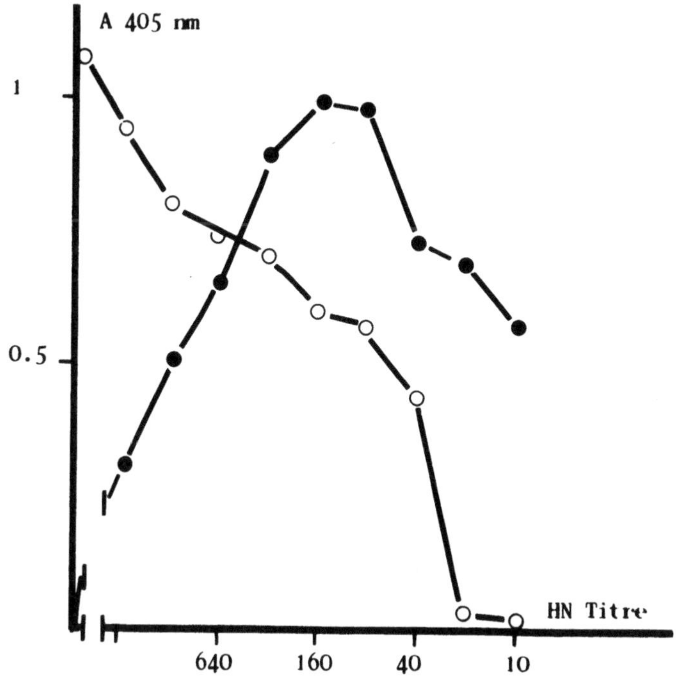

Figure 1: HN/anti-HN immune complexes titration. ELISA was carried out on (●) HA coated plates and on (○) HN coated plates. Anti-HN antibodies were 1 μg/ml.

Tissue preparation (Delpech and Delpech, 1984)

Rat embryos were studied from the morula stage (the day of positive vaginal smear was designated day 0 of pregnancy). All specimens were fixed in cold acetic alcohol, embedded in paraffin, and histochemical techniques were carried out on sections. Although the reagents were prepared from the human brain HN, mammals possess a similar antigen which cross-reacts with anti-human HN antibodies. The cross-reactivity was sufficient to obtain a good characterization of rat tissue HN.

Results

In the rat embryo, HA appeared on day 10 in the neural tube where weak positive staining was observed (Fig. 2a), whereas it was abundant in the surrounding mesenchyme. HN was not seen at this stage in the neural tube and was only present around neural tube and somites, whereas only traces were seen in the sclerotome (Fig. 2b).

The various zones encountered in the early developmental stages of the rat cerebral cortex are first described in order to correlate them with the histochemical results presented in this paper (Fig. 3). The ventricular zone (VL), which is the first zone to form, is characterized by a pseudostratified arrangement of neuroepithelial cells, precursors of all neurons and macroglia. On day 13, the marginal layer (ML) is the next re-

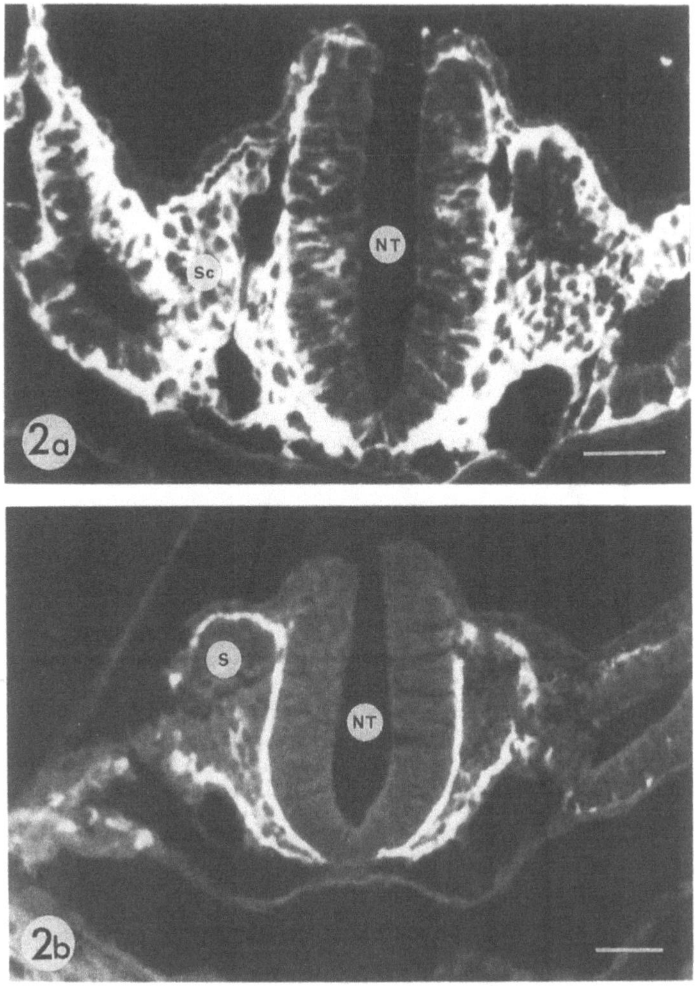

Figure 2: Transverse section of 10-day rat embryo. a: HA detection with fluorescence microscopy. Bright fluorescence is seen around neural tube (NT) and between sclerotomal cells (Sc). Fine strands are visible between neuroepithelial cells. Bar = 40 μm. b: HN detection with fluorescence microscopy. A thin intense fluorescent line surrounds the neural tube and somites (S). There is no staining within the neural tube. Bar = 40 μm.

gion to develop at the surface of the cortex. It is composed of the first developing neurons. The intermediate layer (IL), the presumptive cortical white matter, is the third zone to form, and it is composed mainly of migrating neurons. On day 16, the cortical plate (CP) and the subplate (SP) have developed between marginal and intermediate zones. The cortical plate is composed of undifferentiated bipolar cells. The subplate is composed of young neurons and contains large extracellular spaces.

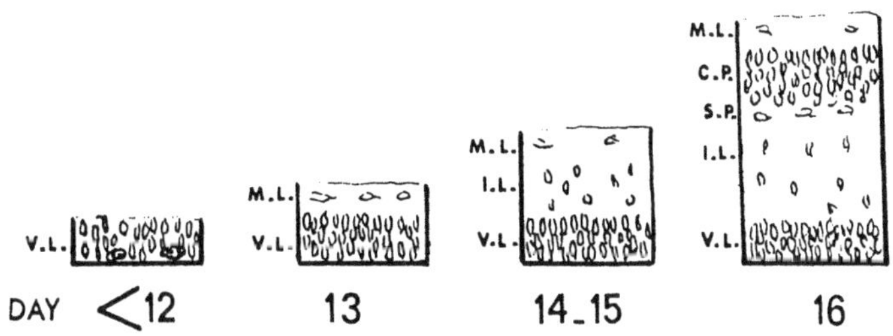

Figure 3: Schematic representation of the development of the rat cerebral wall, from the undifferentiated stage to the appearance of the cortical plate. ML: marginal layer; CP: cortical plate; SP: subplate; IL: intermediate layer; VL: ventricular layer.

At the end of day 11, HA is disseminated between neuroepithelial cells, throughout the neural tube (Fig. 4). Trace amounts of HN can also be detected. On day 15 (when the molecular layer is differentiated) HA and HN were seen only in the marginal layer (Fig. 5).

On day 16, when the cortical plate and the subplate appear, HA and HN were both observed in the subplate and the molecular layer (Fig. 6). This location remained identical during the following days. A greater magnification revealed that HN and HA were associated with extracellular filament-like structures (Fig. 7).

After birth, until 3 weeks, HN and HA locations were superimposable. They spread throughout the cortex, but the staining was weaker than it was in the embryo, suggesting a lower density of the molecules, i.e., HN, and HA sites accessible for HN. Around the 21st day after birth, HN density increases in the corpus callosum, whereas such an increase of HA was not seen with this technique, which suggests that HN progressively saturated HA. At the same time, some perineuronal staining was visible. The perineuronal site is one of the two characteristic HN locations staining in the adult brain cortex (Delpech et al., 1982). It is during the 5th week after birth that HN positive nodes of Ranvier were stained in the white matter: the nodes of Ranvier are the second characteristic HN location staining in the adult nervous system.

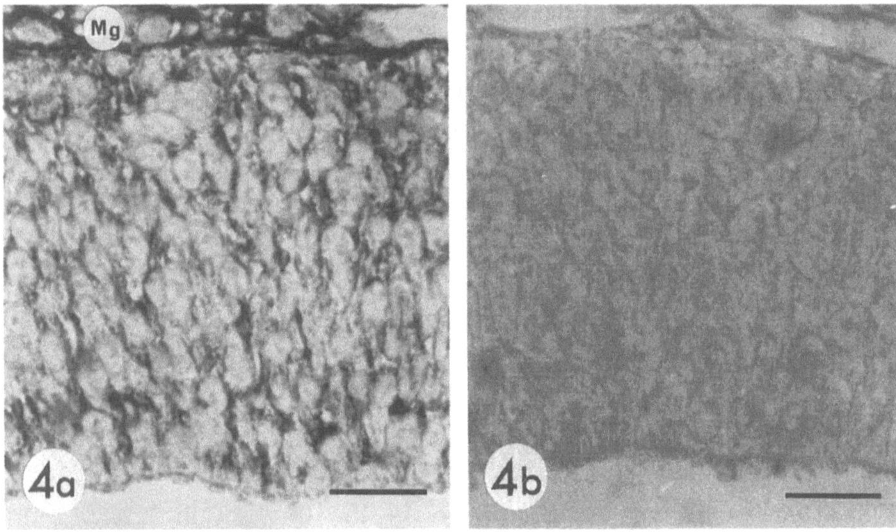

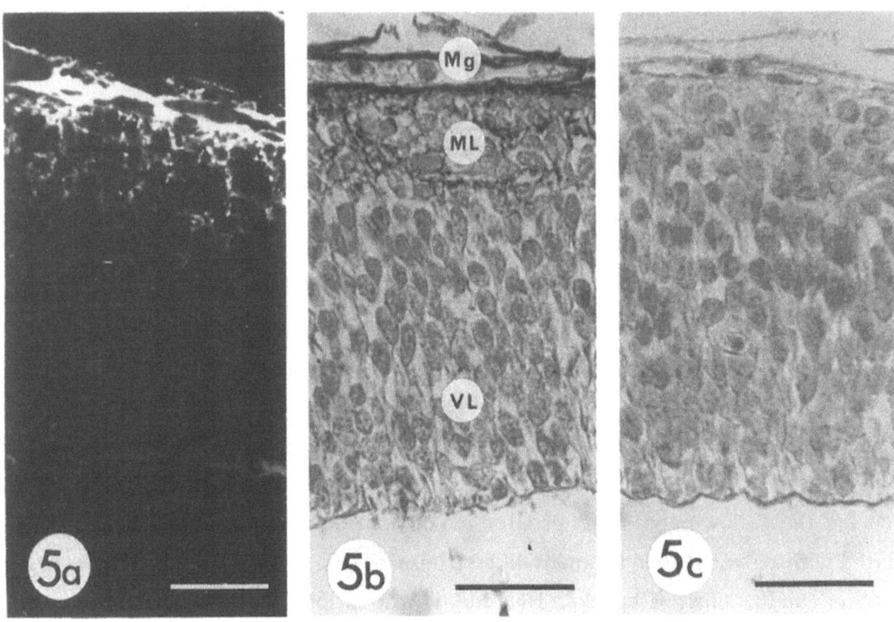

Figure 4: Cerebral cortex of the 11-day rat embryo. a: HA detection with immune complex peroxidase conjugate. Under meninges (Mg) HA is dispersed through the entire thickness of the cerebral wall between neuroepithelial cells; Bar = 20 µm. b: control incubation with immune complexes supplemented with HA. Bar = 20 µm.
Figure 5: Cerbral cortex of the 15-day rat embryo. a: HA detection with fluorescence microscopy. Under meninges (Mg), positive staining is located in the marginal layer (ML). The ventricular layer (VL) is devoid of staining. Bar = 40 µm. b: HA detection with immune complex peroxidase conjugate. Bar = 40 µm. The patterns are identical with both techniques. c: control incubation with immune complexes supplemented with HA. Bar = 40 µm.

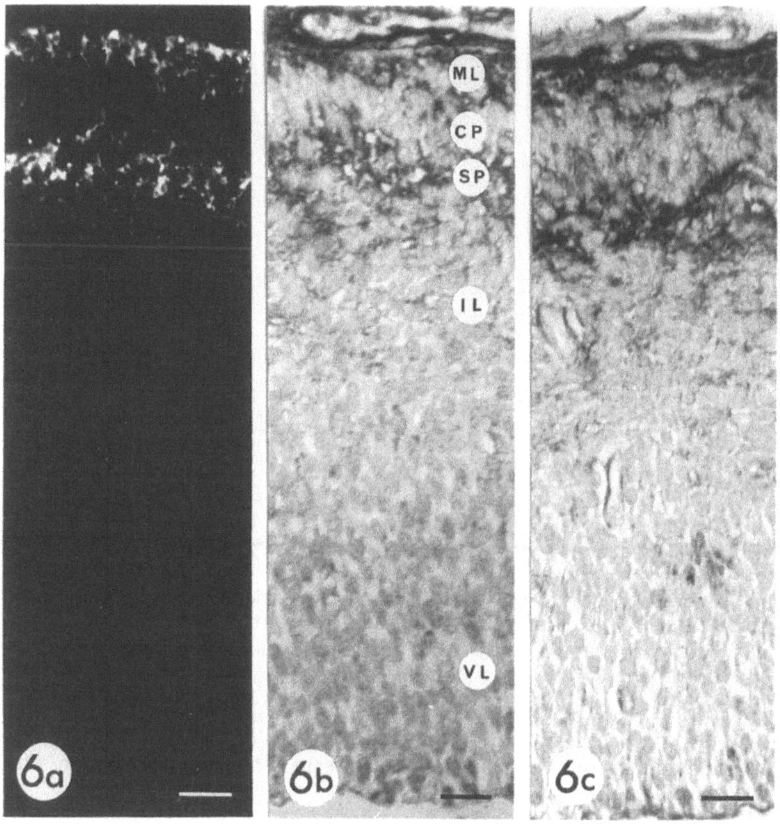

Figure 6: Cerebral cortex of 16-day rat embryo. a: HA detection with fluorescence microscopy. Positive staining is seen in the marginal layer (ML) and in the subplate (SP). The cortical plate (CP), the intermediary layer (IL) and the ventricular layer (VL) are negative. Bar = 20 µm. b: HA detection with immune complex peroxidase conjugate. The pattern is identical to that observed with the fluorescence technique. Bar = 20 µm. c: HN detection with anti-HN antibodies peroxidase conjugate. HN pattern is identical to HA pattern. Bar = 20 µm.

Very few results have been obtained in the human. It is interesting, however, to compare the above results to results obtained with a 4 month old human foetus. There was in this case an intense staining in the molecular layer, in the subplate and in the intermediary layer. This is very similar to observations made after birth in 3 day old rats.

Discussion

Two technical points about HA detection must be discussed. Although sensitive and specific, the technique can only reveal receptors for HN: if HA sites are saturated by molecules which have affinity for it, they are not detectable with this technique, unless the receptor can be cleared. The other point is that receptors, though free, could not be

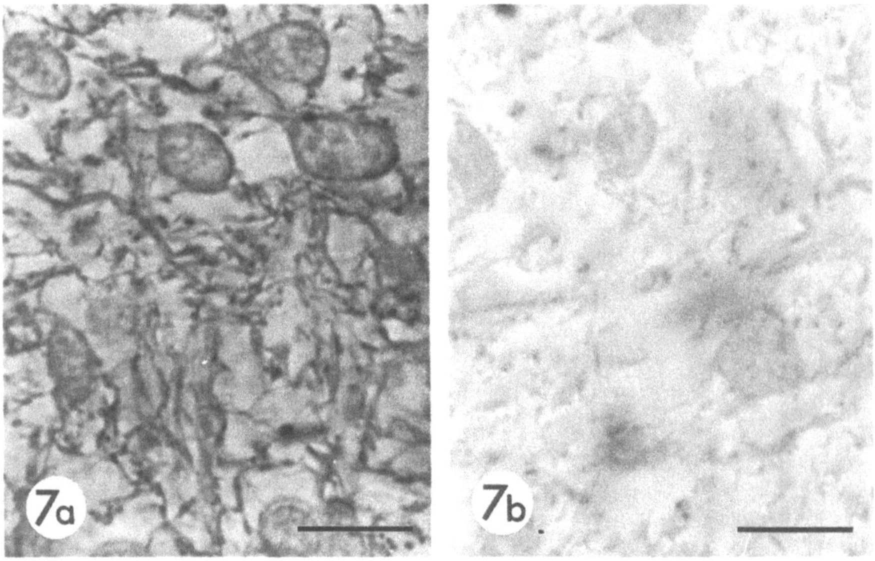

Figure 7: Subplate region of 18-day rat embryo. a: HA detection with immune complex peroxidase conjugate. The reticular pattern suggests that HA is associated with cellular processes. b: control incubation with immune complexes supplemented with HA. Bar = 10 µm.

Reached by the reagents due to proteoglycan steric hindrance. Although this was not seen for the nerve tissue, we could observe this for cartilage HA whose accessibility was made possible only after a short incubation with chondroitinase.

The immunochemical technique used for HA detection was based on the affinity of HN for HA, which allowed the use of anti-HN antibodies to detect HA. Its principle is similar to that of the techniques recently developed by Knudson and Toole (1985), who used the cartilage proteoglycan, and by Ripellino et al. (1985), who used a chondrosarcoma associated proteoglycan. The reaction was specific, since it was inhibited when the tissue section had been digested by Streptomyces hyaluronidase, or when immune complexes were supplemented with HA derived decasaccharide, before incubation on tissue section, whereas GAGs, other than HA, had no influence on the results.

Compared to autoradiographic methods, using enzyme digestion (Hay and Meier, 1974) or cetylpyridinium chloride fixation (Pratt et al., 1975), this technique was easier to perform and gave better results for morphological study. The distribution of HA was similar to that revealed by metachromatic techniques with alcian blue (Morris and Solursh, 1978). The reaction was easy to perform, using an indirect immunofluorescent or immunoperoxidase technique, and gave clear cut staining. After immunofluorescence detection, Evans blue counterstaining could be used to visualize tissue histology. The optimum reaction needed complexes made with defined amounts of both antigen and antibody, which were calculated from calibration curves obtained with the ELISA technique. The results of both ELISA and immunohistology demonstrated that it was necessary to add antigen in amounts, higher than that of antibodies, to get soluble complexes, which is in agreement with the principles of immunoprecipitation. The reagent could be kept at 4°C for two months without decay.

HA location in either mesenchyme or soft connective tissues has been well documented by metachromatic techniques. In the foetal nervous system, however, the immunohistochemical technique described in this work has permitted characterization of HA as a component of nervous extracellular matrix. It confirms that HN which was also detected in these areas by Delpech and Delpech (1984) is associated with HA in vivo. It is consistent with the description of Brückner and Biesold (1981) and Brückner et al. (1985) who characterized glycoproteins associated with glycosaminoglycans, in the same areas, by means of lectin binding combined with hyaluronidase digestion of tissues. It confirms and extends the observations of Derer and Nakanishi (1983) who described, using colloidal iron, polyanionic glycosaminoglycans in the mouse foetal cortex as early as 14 days post conception.

The immunohistochemical studies presented in this paper show the existence of regionally enriched HA and HN in the embryonic rat cerebral cortex. Before marginal layer formation, HA appeared as thin threads throughout the entire thickness of the neural tube. After the individualization of the marginal layer, HA and HN were concentrated in this zone. Another change in distribution occurred with the formation of the cortical plate. HA and HN were mainly located above and below the cortical plate, that is, in the marginal layer and in the subplate, with a fibrillary pattern.

What is the relationship between this latter pattern and cellular structures of the different layers? The ventricular layer and the cortical plate which are quite devoid of HA and HN are composed of closely packed undifferentiated cells with little developed extracellular spaces. The marginal layer, which was positive for HA and HN, is a sparsely populated layer, with extensive extracellular spaces, in which are the first post-mitotic preneurons which mature rapidly, the Cajal-Retzius cells (König et al., 1975; Readler et al., 1980; Derer, 1985). Moreover, in this region, axonal and dendritic outgrowth begins early and the earliest synaptic contacts were found (König et al., 1975). The subplate, in which the strongest stainings were seen, is a layer of relatively low cellularity with fully developed neurons, below the undifferentiated neurons of the cortical plate. Several points are common to both HA/HN rich layers: the presence of relatively large extracellular spaces; that of more differentiated cells and that of growing axons and dendrites.

We can conclude that the occurrence of HA and HN in the developing brain correlates spatially with areas containing large extracellular spaces and developing cell processes. This last point seems of the utmost importance since HA/HN is associated with cells, which are building up cell material (the processes), and with a growth movement, which is a preparatory step for synapse formation. It suggests that HA/HN plays a role between the process membrane and the extracellular matrix. According to this view, it could be hypothesized that HA/HN is also present at the synaptic membrane.

Acknowledgements. This work was supported by the University of Rouen. We are grateful to Dr. Derer, Dr. Brückner and Dr. Biesold for valuable discussion. We thank Mrs. Abdelouhab for skillful technical assistance.

References

Bertrand, P., and B. Delpech (1985) Interaction of hyaluronectin with hyaluronic acid oligosaccharides. J. Neurochem. **45:** 434-439.

Boulder Committee (1970) Embryonic vertebrate central nervous system, revised terminology. Anat. Record **166:** 257-261.

Brückner, G., and D. Biesold (1981) Wheat germ agglutinin indicates the commencement of neuronal differentiation in the embryonic brain. A comparative carbohydrate histochemical study. Acta histochem. **69:** 231-242.

Brückner, G., L. Muller, L. Wollweber, R. Samtleben, and D. Biesold (1985) Lectin binding sites and anionic components related to differentiation in the prenatal rat cerebral cortex. J. Hirnforsch. **26:** 615-634.

Brunngraber, E.G. (1979). In *Neurochemistry of Aminosugars*, E.G. Brunngraber, ed., pp. 128-177. C.C. Thomas Publ., Springfield, IL.

Delpech, B., and C. Halavent (1981) Characterization and purification from human brain of a hyaluronic acid-binding glycoprotein: Hyaluronectin. J. Neurochem. **38:** 978-984.

Delpech, A., N. Girard, and B. Delpech (1982) Localization of hyaluronectin in the nervous system. Brain Res. **245:** 251-257.

Delpech, A., and B. Delpech (1984) Expression of hyaluronic acid-binding glycoprotein, hyaluronectin, in the developing rat embryo. Develop. Biol. **101:** 391-400.

Delpech, B., P. Bertrand, B. Hermelin, A. Delpech, N. Girard, E. Halkin, and C. Chauzy (1986) Hyaluronectin. In *Frontiers of Matrix Biology*, L. Robert, ed. Karger Publish., Basel, Switzerland, in press.

Derer, P. (1985) Comparative localization of Cajal-Retzius cells in the neocortex of normal and reeler mutant mice fetuses. Neurosc. Letters **54:** 1-6.

Derer, P., and S. Nakanishi (1983) Extracellular matrix distribution during neocortical wall ontogenesis in "normal" and "reeler" mice. J. Hirnforsch. **24:** 209-224.

Girard, N., P. Bertrand, A. Delpech, and B. Delpech (1984) Caractérisation tissulaire de l'acide hyaluronique dans le cervelet par la hyaluronectine. Comptes Rendus Acad. Sc. (Paris) série III, **298:** 325-330.

Girard, N., A. Delpech, and B. Delpech (1986) Characterization of hyaluronic acid on tissue sections with hyaluronectin. J. Histochem. Cytochem. **34:** 539-541.

Hay, E.D., and S. Meier (1974) Glycosaminoglycan synthesis by embryonic inductors: Neural tube, notochord and lens. J. Cell Biol. **62:** 889-898.

Knudson, C.B., and B.P. Toole (1985) Fluorescent morphological probe for hyaluronate. J. Cell Biol. **100:** 1753-1758.

König, N., G. Roch, and R. Marty (1975) The onset of synaptogenesis in the rat temporal cortex. Anat Embryol. **148:** 73-87.

Marin-Padilla, M. (1978) Dual origin of the mammalian neocortex and evolution of the cortical plate. Anat. Embryol. **152:** 109-126.

Morris, G.M., and M. Solursh (1978) Regional differences in mesenchymal cell morphology and glycosaminoglycans in early neural-fold stage embryos. J. Embryol. exp. morphol. **46:** 37-52.

Pinto-Lord, M., and V.S. Caviness, Jr. (1979) Determinants of cell shape and orientation:

A comparative Golgi analysis of cell-axon interrelationships in the developing neocortex of normal and reeler mice. J. Comp. Neurol. **187:** 49-70.
Pratt, R.M., M.A. Larsen, and M.C. Johnston (1975) Migration of cranial neural crest cells in a cell-free hyaluronate-rich matrix. Develop. Biol. **44:** 298-305.
Readler, E., A. Readler, and S. Feldhaus (1980) Dynamic aspects of neocortical histogenesis in the rat. Anat. Embryol. **158:** 253-269.
Ripellino, J.A., M.M. Klinger, R.U. Margolis, and R.K. Margolis (1985) The hyaluronic acid binding region as a specific probe for the localization of hyaluronic acid in tissue sections. Application to chick embryo and rat brain. J. Histochem. Cytochem. **33:** 1060-1066.
Toole, B.P. (1973) Hyaluronate and hyaluronidase in morphogenesis and differentiation. Amer. Zool. **13:** 1061-1065.
Toole, B.P. (1981) Glycosaminoglycans in morphogenesis. In *Cell Biology of Extracellular Matrix*, E.D. Hay, ed., pp. 259-294. Plenum, New York.

Section 2

MESENCHYMAL-EPITHELIAL INTERACTION IN THE MORPHOGENESIS OF EPITHELIAL ORGANS

CONTROL OF EPITHELIAL POLARITY AND INDUCTION IN THE EARLY CHICK EMBRYO

Claudio D. Stern
Department of Human Anatomy
Oxford
U.K.

Abstract. The epiblast of the chick embryo at gastrulation is a simple, pseudo-stratified columnar epithelium. The tissue is polarised morphologically: the basal surfaces of the cells are covered by a basal lamina, and their apical surfaces display microvilli. The inter-cellular junctions appear near the apical ends of the cells, while the nuclei are located basally within the cytoplasm. The tissue is also polarised physiologically: sodium and water are transported from its apical to its basal side, the sodium pumps are located in the baso-lateral plasma membrane, and the tissue maintains a substantial trans-epithelial potential (about 30mV, basal side positive).

Experimental application of a potential across the epiblast, of similar magnitude but reverse orientation to the measured potential, leads to rapid and stable reversal of the morphological and physiological polarity of the epiblast.

A model to explain early morphogenetic events, based on these and other observations, is presented. Special reference is made to the role of the mesenchyme in induction of the embryonic axis and of the primitive neural axis.

Before gastrulation, the chick embryo consists of two layers of cells: the epiblast, dorsally, and the hypoblast, ventrally. Gastrulation sees the appearance of the primitive streak, which begins as a condensation of mesenchymal cells under the epiblast. The primitive streak imparts visible bilateral symmetry to the early embryo and constitutes the focal region for the formation of the third germ layer, or mesoderm. The nervous system subsequently forms coaxially with the primitive streak, as a result of "inductive" interactions between the mesoderm and the overlying epiblast. The cellular and molecular bases of these interactions are not understood at present, but considerable progress has been made recently, which may herald the solution of these age-old problems in the foreseeable future.

The epiblast is a "true" epithelium, in that it is polarised morphologically and physiologically. Morphologically, it is a pseudostratified, columnar epithelium. The apical (dorsal) aspect of the tissue displays some microvilli and other projections, intercellular junctions (tight, desmosomes and gap junctions) and a ring of actin microfilaments. The

NATO ASI Series, Vol. H5
Mesenchymal-Epithelial Interactions in Neural Development
Edited by J. R. Wolff et al.

basal (ventral) aspect of the cells is covered by a basal lamina composed primarily of hyaluronate, fibronectin, laminin, some collagen IV and V, and other glycoproteins and glycosaminoglycans. The nuclei appear near the basal ends of the cells, as does the Golgi apparatus.

The epiblast is also polarised physiologically. Studies using the vibrating probe on whole cultured chick embryos (Jaffe and Stern, 1979) have revealed a pattern of extracellular electric currents: cations enter the embryo through the whole of the epiblast and leave via the primitive streak. These findings are consistent with the notion that the epiblast is a polarised, ion-transporting epithelium, and that the primitive streak is the exit pathway for these ions from the interior of the embryo. The primitive streak could be acting either passively, as a zone of lower electrical resistance that allows the ions to flow between its cells, or actively, as a region where transport occurs in the opposite direction from the rest of the epiblast.

As a result of uni-directional transport of cations, the epiblast generates and maintains an electric potential of some 20mV across itself (Stern and MacKenzie, 1983). This potential was shown to be due primarily to uni-directional, apical-to-basal transport of sodium, with chloride probably being the major counterion (Stern and MacKenzie, 1983; Stern et al., 1985). ^{22}Na flux measurements revealed that sodium is transported across the epiblast at a rate of about 67 µmoles/cm^2/h (equivalent to about 38 pmoles/cell/h, or 2.3×10^{13} atoms/cell/h). Fluid is transported in the same direction as sodium, mainly driven by osmotic forces, at a rate of about 2 µl/cm^2/h (Stern et al., 1985). The uni-directional transport of sodium can be inhibited effectively using strophanthidin, a drug similar to ouabain, which interfaces with the sodium/potassium pump (Na/K-ATPase), which is located in the baso-lateral cell membrane of the epiblast cells as demonstrated by ^{3}H-ouabain binding studies (Stern, 1982; Stern and MacKenzie, 1983). The Na/K-ATPase drives sodium out of the cells and potassium into them actively, against an electrochemical gradient. Entry of sodium into the cells occurs passively, through different types of sodium channels. At least three types of sodium channels have been shown to be present in the epiblast cells in pharmacological and electrophysiological experiments (Stern et al., 1985): 1) a serum-stimulated, amiloride sensitive, Na^+/H^+ antiport, 2) a furosemide-sensitive, Na^+/Cl^- symport and 3) a third channel, not yet characterised, which may be a Na^+/K^+ antiport. There are no "fast" (tetrodotoxin (TXX)-sensitive) sodium channels, but it is likely that Na^+/Ca^{++} exchange channels are present. In order to maintain the trans-epithelial potential and uni-directional transport of ions and fluid, the epiblast is electrically tight (high resistance), with trans-epithelial resistances around 2KΩ/cm^2. This value is comparable to those of "classical" transporting epithelia, such as frog skin and amphibian bladder epithelia.

Epithelial Polarity Can Be Modulated Experimentally

If a sheet of epiblast, dissected free from hypoblast and any adhering mesoderm cells, is stretched over a small chamber so that both of its aspects are bathed by separate containers of fluid ("mini"-Ussing chamber), an Ag/AgCl electrode can be placed in each bath to measure or apply electric potentials across the tissue. An epiblast sheet set up in this way will generate a trans-epithelial potential, as it does in the embryo. If a potential similar in magnitude, but of reverse direction to that seen in the embryo, is applied across the two electrodes (about 25-30mV, apical side positive), the epiblast sheet reverses its morphological and physiological polarity. Morphologically, apical features appear towards the basal side of the cells and vice-versa. After stimulation, therefore, the original apical surface is covered by a basal lamina, which disappears from the basal side, nuclei and the Golgi apparatus become located towards the apical ends of the cells, and microfilament bundles and intercellular junctions move from an epical to a basal position. Physiologically, ^{22}Na transport occurs in a basal → apical direction, and this can be inhibited by strophanthidin, as in control tissues, ^{3}H-ouabain binding studies demonstrate that after stimulation the Na/K-ATPase is localised predominantly near the apical surfaces of the cells (Stern and MacKenzie, 1983).

The net rate of transport of sodium in the new direction, however, is less than the control rate in un-stimulated embryos. This appears to be due to the fact that some ^{3}H-ouabain binding sites (Na/K-ATPase sites) remain in the baso-lateral membrane after stimulation, from where they compete with those that have moved to the apical membrane.

The reversal of polarity is <u>rapid</u>, as it can be demonstrated after stimulation for 3 hr, <u>stable</u>, as the new polarity persists if no further stimulation is applied after this time, and <u>reversible</u>, as it can be returned to pre-experimental conditions if the direction of the applied voltage is reversed after 3 hr and stimulation applied for a further 3 hr in this new direction (Stern and MacKenzie, 1983).

Mechanism of Polarity Reversal

How does polarity reversal take place? One possibility is that the applied voltage moves the Na/K-ATPase sites along the plane of the membrane by electrophoresis or electro-osmosis. This mechanism was first suggested by Jaffe (1977) and has received experimental support in a number of systems. For example, Poo and Robinson (1977) demonstrated that surface Con.A receptors "cap" to one side of muscle cells if an electric field is applied to a culture of such cells, along the plane of the substrate. In support of this suggestion is the finding that stimulation of chick epiblast sheets for 3 hours at 4°C does not lead to overt reversal of polarity, but if the stimulation is then turned off and the tissue returned to 37°C its polarity is reversed (Stern and MacKenzie, 1983). It follows that stimulation at the lower temperature passively sets up some parameter(s) required for polarity reversal, but expression of the reversal requires some active collaboration by the cell.

These results suggest that the initial response to stimulation may be a passive translocation of the Na/K-ATPase sites from the baso-lateral membranes to the apical aspect of the cells. Thus, when stimulation is suspended, the tissue provides, as it were, its own voltage clamp in the reverse direction from the normal situation. However, it has been reported that intercellular junctions cannot open at low temperatures (Martinez-Palomo et al., 1980). In this case, the tight intercellular junctions should prevent the Na/K-ATPase sites from shifting from the baso-lateral to the apical domains at the lower temperature. Moreover, pulse-chase autoradiography experiments with labelled and unlabelled ouabain, respectively, showed that after stimulation at 4°C for 3 hr the labelled Na/K-ATPase pumps are located intracellularly rather than in the plane of the membrane (unpublished observations).

In conclusion, removal of Na/K-ATPase sites from the baso-lateral membrane takes place as a (passive?) initial response to stimulation. Re-insertion of the pumps into the apical membrane, on the other hand, is an active process and requires the cooperation of the cell. This conclusion implies that, if stimulation at the lower temperature is sufficient to set up the conditions for polarity reversal, the instructions for reversal are set up by re-distributing a cellular component other than the sodium pumps. At present it is not known what this component might be.

There remains one alternative attractive, though somewhat heretical, explanation of the results: it has been suggested (J.P. Kraehenbuhl, Lausanne, personal communication) that the 100Kd catalytic α-subunit of the sodium pump might also be a component of some sodium channels. The basis of this suggestion is that antibodies against this subunit of the sodium pump also decorate the apical surface in amphibian epithelial cells. This implies that, under certain conditions, channels and pumps could be interconvertible. If this is the case, it is not inconceivable that stimulation can convert at least some of the channels into pumps and vice-versa.

Control of Epithelial Polarity in the Embryo

As discussed above, the epiblast of the early chick embryo transports sodium and fluid in a uni-directional manner, from apical to basal aspects of the tissue, and generates a trans-epithelial potential of some 25mV. The magnitude of this potential is sufficient to control the morphological and physiological polarity of the epiblast. What could be the function of this reversal in the embryo?

^{3}H-ouabain autoradiography was also performed on older (primitive streak stage) embryos (Stern, 1982). The baso-lateral aspect of the cells in the epiblast in these embryos was labelled as in the younger embryos, but in the primitive streak itself the pattern of binding was reversed: the apical aspect of the primitive streak cells showed binding, while the baso-lateral membranes of the same cells did not. This finding correlates well with the pattern of extracellular electric current flow measured by Jaffe and Stern (1979): the primitive streak not only forms the pathway for current to leave passively, but it may actively pump sodium out of the embryo. Therefore, at some time during the formation of the primitive streak, some event takes place that acts as a stimulus that may reverse the polarity of the cells in this region.

I have put forward a simple model to interpret these findings (Stern, 1984a). In this model, cells in the epiblast prior to the appearance of the primitive streak pump sodium and fluid into the underlying intraembryonic cavity, where the sodium and fluid accumulate. As they continue to accumulate between the two germ layers, the electrochemical and hydrostatic gradient across the epiblast must reach a threshold, as the tissue can no longer continue to transport against an ever-steeper gradient. When this threshold is reached, the sodium pumps stop. This leads to a rise in intracellular sodium levels, which in turn triggers calcium release from intracellular pools, and this stimulates the apical cytoskeleton to contract. At the same time, intercellular junctions open, and some cells are forced by the contractions to enter the cavity, while allowing sodium and fluid to leak out between the cells, locally reversing the direction of the current. As a sub-threshold condition is restored, the tissue seals and resumes pumping in the original direction. The hypoblast begins to form nearest the future caudal end of the embryo, delimiting the intra-embryonic cavity and allowing sodium and fluid to accumulate within it. For this reason, the first mesodermal cells appear in the region covered by hypoblast. The cells that form in each contraction interact with the overlying epiblast in such a way as to make it more likely that subsequent contractions occur in the region where the cells are located. They could do this by secreting glycosaminoglycan- or protein-hydrolysing molecules that degrade the overlying basal lamina, and in this way autocatalyse their own formation.

The region where these cells appear becomes the primitive streak. By the time the primitive streak makes its appearance, currents have been flowing in the reverse direction from that in the rest of the epiblast for some time. The reversal in the direction of current flow could reverse the polarity of the cells in this region. Polarity reversal would restore equilibrium because the primitive streak can then provide a permanent outflow pathway for sodium to leave from the interior of the embryo. The main function of mechanisms of this sort would be, then, to prevent the appearance of supernumerary embryonic axes.

Polarity Reversal Is a Characteristic of Most Epithelia

Experimental reversal of epithelial polarity is not unique to the epiblast or even to embryonic tissues (Stern, 1986). The phenomenon was first described in adult porcine thyroid follicles in vitro by Mauchamp and his collaborators (Mauchamp and Fayet, 1974; Mauchamp et al., 1979) and since then has been studied by Nitsch and others (e.g., Nitsch and Wollman, 1980). Reversal of polarity in this epithelium can be brought about by controlling the serum concentration in the medium surrounding the follicles, and can be modulated by embedding the follicles in a collagen matrix. The role of the substratum in polarity reversal has been demonstrated elegantly by Crawford (1983) who showed that when cultured retinal pigment epithelium is presented with two substrates simultaneously, the cells either re-polarise or divide giving rise to two sheets of tissue enclosing a lumen between them, with the apical faces towards the lumen. Amemiya et al. (1979) have shown that the polarity of the sea urchin embryonic gut epithelium can be reversed by modulating the concentration of ions in the medium, and Schwartz et al. (1985) have found that rabbit kidney epithelial polarity can be reversed by altering the pH on one side of the epithelium.

The polarity of chick epiblast sheets can also be reversed by applying a pH gradient across the thickness of the tissue. In situ, the apical side of the epiblast faces the albumen, which is alkaline, at pH 9.5, while the basal side faces the interior of the embryo, which is about neutral. Exposing epiblast sheets to a pH gradient of about 3 units (pH6 apically, pH9 basally) also leads to reversal of polarity (unpublished observations).

The capacity for polarity reversal is therefore a general property of epithelia and can be brought about by a variety of different stimuli (pH, electrolyte concentration, serum concentration, electric potential, presence of a solid substrate, extracellular matrix fibres). What these stimuli all have in common is that they must be applied asymmetrically in the form of a gradient across the thickness of the tissue.

The Interconversion of Epithelia and Mesenchyme

Formation of the primitive streak involves the de-epithelialisation of some epiblast cells and their conversion into a mesenchymal morphology. Their behavior changes markedly as a result of this transition. They become fibroblastic in appearance, more highly motile, reduce their cell-cell and cell-substrate contacts, decrease their production of some substances, such as fibronectin, and increase their expression of others, like hyaluronidase (see Stern, 1984a, b for reviews). Moreover, they acquire the capacity to induce a primitive streak or neural axis (depending upon the age of the host and upon the position of the graft) if grafted ectopically, in contact with epiblast.

The transition from epithelium to mesenchyme is not irreversible and cannot be considered strictly as a differentiation event. Later in development, some of the mesoderm cells will become epithelial once again, and for some tissues the interconversion is repeated a few times in the course of development. For example, during somite formation, the mesoderm of the segmental plates condenses into a series of epithelial spheres. Each sphere consists of a simple epithelium with the apical poles of the cells facing inwards, towards the lumen or sclerocoele. A few hours later, the ventro-medial portion of the epithelium of the somite loses its epithelial nature once again, to form the sclerotome, a mass of mesenchymal cells that will later migrate to surround the notochord and become the osteogenic cartilage of the vertebrae and ribs. The dorso-lateral part of the somite epithelium (the dermomyotome) remains epithelial longer, until eventually it differentiates into dermatome (which will migrate to underlie the skin and give rise to the dermis) and myotome (which will migrate into the limb and spread around the trunk to give rise to skeletal muscle). Similarly, the mesoderm of the lateral plate will eventually split into two epithelioid sheets, the somatopleure and splanchnopleure, which constitute the lining of the coelom. The rod of mesoderm remaining between the somitic tissue and the lateral plate is the intermediate mesoderm, which will eventually become the epithelium of the mesonephric tubules that form the transitory embryonic kidney.

There are many examples of epithelial-mesenchymal transformation and vice-versa in embryonic development. The conversion of epthelium into mesenchyme has been better studied than the converse, although the latter should represent an ideal system for the study of the initial polarisation of cells to form a transporting epithelium.

Interactions Between Epithelia, Mesenchyme and Their Matrices

Little is known about the mechanisms that mediate epithelial-mesenchymal interconversion. Recently, Stoker and Perryman (1985) have succeeded in isolating a substance produced by a non-transformed cell line, MRC-5, of molecular weight around 50,000 daltons, which they have named "scatter factor". When this substance is added to cultures of epithelial cells they lose their epithelial characteristics and become fibroblastic, with a change in their cell cycle time. Scatter factor is produced by a number of different cell types, but neither transformed cells nor well-established cultures secrete it. Susceptible epithelial cells are not producers. It remains to be established whether newly-formed mesoderm cells become producers of scatter factor as a result of their transformation from an epithelium into a mesenchyme in a manner that would account for autocatalysis.

Another recent important finding comes from experiments on amphibian embryos (Smith et al., 1985). It was found that a substance isolated from chick embryos acts as a "vegetalising factor". When added to ectodermal cells, the factor stimulates them to become mesodermal. The substance appears to have a molecular weight similar to Stoker and Perryman's scatter factor (30,000 daltons), and it may well turn out that the two substances are one and the same. What is particularly interesting about the results of

Smith et al. is that the response of the ectoderm is concentration-dependent, as different structures are formed when different concentrations of the factor are applied.

In a previous study (Stern, 1984b) I have shown that mesoderm cells display elevated hyaluronidase activity with respect to the epiblast. Since the basal lamina intervening between epiblast and mesoderm contains a comparatively large amount of hyaluronate, this finding suggests that mesoderm cells degrade the overlying basal lamina. It is likely, therefore, that mesoderm cells produce other substances, directed against the remaining components of the basal lamina. Likely candidates are collagenase activities, and some sort of proteolytic activity that may degrade fibronectin and laminin. In preliminary enzyme assays, we have also found elevated ß-glucuronidase activity in mesoderm cells that may serve the function of attacking glycosaminoglycans and glycoproteins in the basal lamina (S. Bascomb, R. Bellairs and C.D. Stern, unpublished observations).

In summary, mesoderm cells express various activities that are consistent with a possible autocatalytic function, whereby they would stimulate their own formation. These substances include some that degrade basal lamina components, like hyaluronidase and ß-glucuronidase, and others, like scatter factor, that may induce epithelial-mesenchymal transformation more directly.

Less is known about the molecular bases of the transformation of mesenchyme into epithelium. It is possible that calcium and cell adhesion molecules such as the CAMs are involved in these transformations, since an increase in cell-cell adhesion must accompany the initial transition into an epithelium (e.g., during somite formation (Bellairs et al., 1981)). In support of this, Chernoff and Hilfer (1982) have found that calcium is likely to be implicated in somitogenesis, and more specifically in the process of contraction that accompanies somite formation. Sofar, neither N-CAM, nor L-CAM, nor NG-CAM have been found to differ in amount or distribution between the segmental plate mesoderm and the epithelial somite (M. Rickmann, J. Fawcett and R.J. Keynes, unpublished data).

Mesenchymal-Epithelial Interactions and Primary Induction in Neural Development

Mesenchymal-epithelial interactions are fundamental to the process of induction. In the early stages of development of amniote embryos, induction plays a role in the formation of the primitive streak and in the early development of the neural primordium. Both the hypoblast and the mesoderm have inducing ability for primitive streak mesoderm, while mesoderm from some regions (e.g., Hensen's node and the caudal portion of the primitive streak) can induce the formation of neural plates in the epiblast. It is not known whether the interactions that mediate induction of the primitive streak and of the neural anlage are the same in both cases. Smith et al. (1985) argue that in amphibian embryos there are several induction processes that differ from one another, but there is as yet no evidence that this is the case in amniotes.

It is important to realise that the process of induction itself is not yet understood. It is not known whether cells in the responding tissue seal their fates (i.e., mesodermal vs. epiblast, neural vs. epiblast) as a result of induction, as is often assumed. An alternative would be that the responding tissue is already a mosaic of cells committed to the two fates, and that induction consists of a process by which the two cell types sort out from each other under some influence from the inducing stimulus. This would account for results indicating that many different stimuli can trigger induction (acid pH, alkaline pH, coagulated organisers, methylene blue, alcohol, grains of sand, etc.); the inductive stimulus would be analogous to the grain of dust that initiates the crystallisation of a solute under appropriate conditions. By this alternative hypothesis, loss of "competence" by the responding tissue represents simply the depletion of one of the two cell types.

Sofar, no markers have been found that are common to all inducing tissues. The advent of monoclonal antibodies and of recombinant DNA technology may provide the means to look for such markers. It should be realised that isolation of the substances that mediate mesenchymal-epithelial interactions and represent the trigger for induction will not in itself represent the end of the quest to understand induction. Induction does not involve merely a series of genetic switches that results in all-or-none differentiation down a particular path, but also involves producing a new domain of spatial organisation. Neither genes nor gene products can hold information about spatial order. The solution to this problem will have to come from the study of cell interactions, developmental dynamics and a "field" approach to development. The study of mechano-chemical events that has begun to be undertaken recently (e.g., Odell et al., 1981; Oster et al., 1983, see review by Goodwin, 1985) may hold some of the answers.

References

Amemiya, S., K. Akasaka, and H. Terayama (1979) Reversal of polarity in ciliated cells of the isolated sea urchin Pluteus gut. J. exp. Zool. **210:** 177-182.

Bellairs, R., A.S.G. Curtis, and E.J. Sanders (1981) Cell adhesiveness and embryonic differentiation. J. Embryol. exp. Morph. **46:** 207-213.

Chernoff, E.A.G., and S.R. Hilfer (1982) Calcium dependence and contraction in somite formation. Tissue Cell **14:** 435-450.

Crawford, B.J. (1983) Some factors controlling cell polarity in chick retinal pigment epithelial cells in clonal culture. Tiss. Cell. **15:** 993-1005.

Goodwin, B.C. (1985) What are the causes of morphogenesis? BioEssays **3:** 32-36.

Jaffe, L.F. (1977) Electrophoresis along cell membranes. Nature **265:** 600-602.

Jaffe, L.F., and C.D. Stern (1979) Strong electrical currents leave the primitive streak region of chick embryos. Science **206:** 569-571.

Martinez-Palomo, A., I. Meza, G. Beaty, and M. Cereijido (1980) Experimental modulation of occluding junctions in a cultured transporting epithelium. J. Cell Biol. **87:** 736-745.

Mauchamp, J., and G. Fayet (1974) Three dimensional reorganization of thyroid cells in culture and iodine metabolism. Endocrinol. exp. **8:** 170.

Mauchamp, J., A. Margotat, M. Chambard, B. Charrier, L. Remy, und M. Michel-Bechet (1979) Polarity of three dimensional structures derived from isolated hog thyroid cells in primary culture. Cell Tiss. Res. **204:** 417-430.

Nitsch, L., and S.H. Wollman (1980) Ultrastructure of intermediate stages in polarity reversal of thyroid epithelium in follicles in suspension culture. J. Cell Biol. **86:** 875-880

Odell, G., G.F. Oster, B. Burnside, and P. Alberch (1981) The mechanical basis of morphogenesis. Develop. Biol. **85:** 446-462.

Oster, G.F., J.D. Murray, and A. Harris (1983) Mechanical aspects of mesenchymal morphogenesis. J. Embryol. exp. Morph. **78:** 83-125.

Poo, M.m., and K.R. Robinson (1977) Electrophoresis of Con.A receptors along embryonic muscle cell membrane. Nature **265a:** 602-605.

Schwartz, G.J., J. Barasch, und Q. Al-Awqati (1985) Plasticity of functional epithelial polarity. Nature **318:** 368-371.

Smith, J.C., L. Dale, und J.M.W. Slack (1985) Cell lineage labels and region-specific markers in the analysis of inductive interactions. J. Embryol. exp. Morph. **89 (Suppl.):** 317-331.

Stern, C.D. (1982) Localization of the sodium pump in the epiblast of the early chick embryo. J. Anat. **134:** 606-607,

Stern, C.D. (1984a) A simple model for early morphogenesis. J. theor. Biol. **107:** 229-242.

Stern, C.D. (1984b) Mini-review: Hyaluronidases in early embryonic development. Cell Biol. Int. Rep. **8:** 703-717.

Stern, C.D. (1986) The occurrence of reversible epithelia. Nature **318:** 728.

Stern, C.D., und D.O. MacKenzie (1983) Sodium transport and the control of epiblast polarity in the early chick embryo. J. Embryol. exp. Morph. **77:** 73-98.

Stern, C.D., S. Manning, and J.I. Gillespie (1985) Fluid transport across the epiblast of the early chick embryo. J. Embryol. exp. Morph. **88:** 365-384.

Stoker, M., and M. Perryman (1985) An epithelial scatter factor released by embryo fibroblasts. J. Cell Sci. **77:** 209-223.

CELL INTERACTIONS DURING KIDNEY DEVELOPMENT

P. Ekblom, E. Aufderheide, G. Klein, A. Kurz, A. Weller
Friedrich-Miescher-Laboratorium der Max-Planck-Gesellschaft
Tübingen
West Germany

Interactions between different tissue compartments are crucial for embryonic development. These interactions include long-range effects by serum factors (hormonese and growth factors), adhesive interactions between cells, and inductive interactions between two dissimilar cell populations located in close proximity to each other. In vitro cultures of embryonic tissues are well suited for analysis of these phenomena. In such cultures, the normal histoarchitecture is retained, but it is nevertheless possible to manipulate morphogenesis at least to some extent. Organ culture of embryonic tissues has been used to study the development of most parenchymal organs, such as the lung, heart, liver, many different glands, and the kidney. These studies have convincingly demonstrated the role of inductive interactions in development (see Wessells, 1977). We are using the developing kidney to study these phenomena.

The Developing Kidney As a Model System

The development of the mouse embryonic kidney in vitro has been extensively used to study cell-cell interactions and their developmental consequences, and several basic aspects of morphogenesis have been clarified using this model. As in many other organs, differentiation requires an interaction between a mesenchyme and an epithelium. The mesenchyme induces the epithelium to branch and to grow into the mesenchyme, but the epithelium also stimulates the mesenchyme. In the developing kidney, the mesenchyme responds to the signals from the epithelium by converting into an epithelium. Such a conversion is not seen during the development of other organs (lung, salivary, mammary, pancreas, tooth) (Fig. 1). The unusual property of the kidney mesenchyme to convert into an epithelium has given the researchers a unique opportunity to study the appearance of markers for the epithelial phenotype. It is very fortunate that the conversion can be induced also under in vitro conditions. Grobstein (1956) could show that several embryonic tissues in vitro can induce the differentiation of epithelial tubules in the nephrogenic mesenchyme. The availability of this model makes it possible to carefully study

NATO ASI Series, Vol. H5
Mesenchymal-Epithelial Interactions in Neural Development
Edited by J. R. Wolff et al.

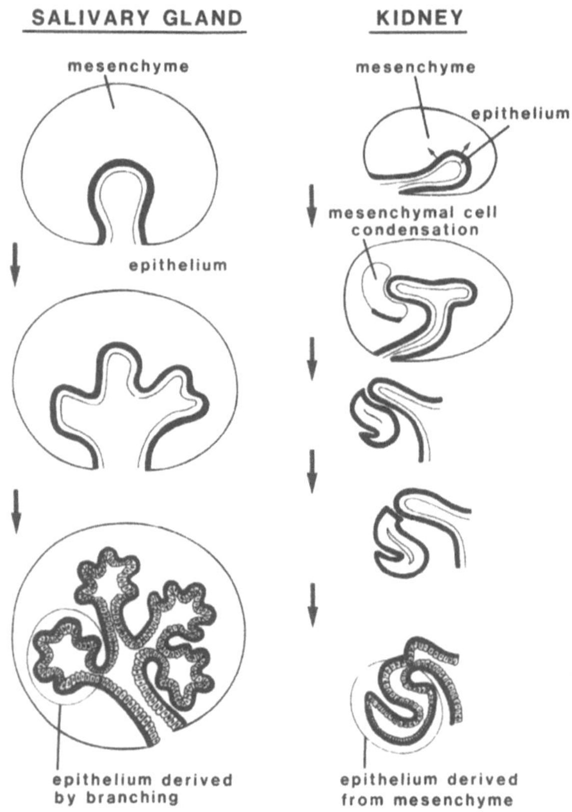

Figure 1: Branching epithelial morphogenesis in two embryonic organs. The mesenchyme stimulates the salivary epithelium to branch, but in the developing kidney, the mesenchyme also itself converts into an epithelium. In most other organs, no such conversion occurs.

various aspects of the induction process and the subsequent morphogenetic steps. It is peculiar that many embryonic tissues can serve as inductors of the nephrogenic mesenchyme. Embryonic neural tissues (brain, spinal cord) are particularly powerful inducers (Grobstein, 1955) and because the spinal cord is easy to dissect out from the embryo, it is the most frequently used inducer for the kidney mesenchyme during in vitro experiments (Saxén et al., 1968). The spinal cord can induce the differentiation of the mesenchyme across a membrane filter. In these transfilter cultures, the mesenchyme is cultured on top of the filter, and the inducer on the other side. After some 24 hours the inducer can be removed although no signs of morphogenesis at this stage are detectable (Fig. 2). However, if the pores of the filter are smaller than 0.1 µm, cell processes do not traverse the filter, and no induction of differentiation occurs. This suggests that induction of the kidney mesenchyme by the spinal cord requires close cell contacts (review: Lehtonen,

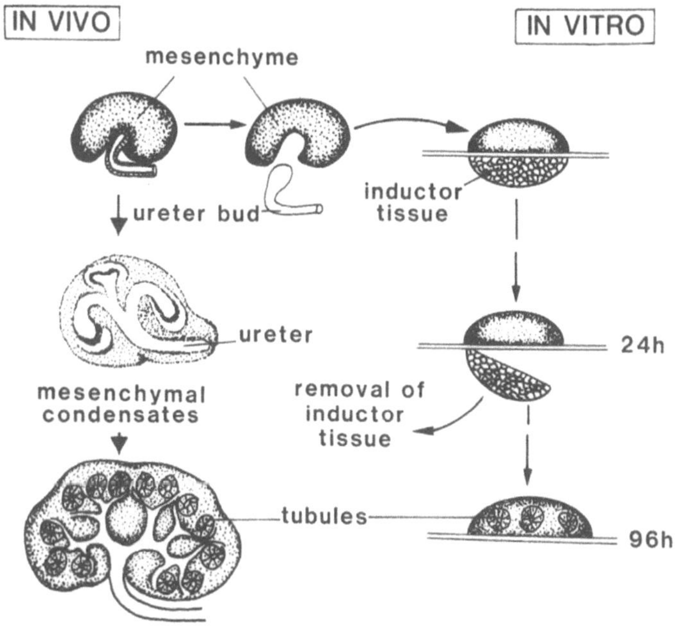

Figure 2: Comparison of kidney differentiation in vivo (left) and in vitro in a transfilter culture (right) according to Grobstein (1956).

1976). However, the molecular nature of the cell-cell contacts is not known, and it has not been directly shown that components of the inducer cell surface participate in induction of the mesenchyme.

The Response to Induction of the Nephrogenic Mesenchyme

Nothing is as yet known about the molecular nature of the inducer molecules in the kidney model. In contrast, we now have some clues about the molecular nature of the response to induction. In the in vitro model, induction is completed before overt morphogenesis is visible. Experiments with transfilter cultures of spinal cord and mesenchyme have shown that induction is completed at 24-30 hours, although there are no morphological changes at this stage. On day two, the induced cells begin to form aggregates, and on day three the aggregates will further develop into tubules composed of well polarized epithelial cells (Grobstein, 1956; Saxén et al., 1968). The polarized state of the cells is evident morphologically, but it can also be verified with the use of molecular markers for the apical side of the tubules. These markers become detectable on day three of development and they represent terminal differentiation of the cells (Ekblom and Thesleff, 1985). In the model system, several levels of differentiation can thus be distinguished. We can fairly well define the uninduced state, the induced state, and the subsequent differentiated state (Fig. 2).

Hence, it should be possible to use the model system to search for morphogenetically important gene products that are present before overt morphogenesis. In some other systems used to study induction, terminal differentiation occurs very soon after induction, and the gene products appearing in response to induction actually reflect terminal differentiation. This seems to be the case during induction of mesoderm which has been studied in Xenopus embryos. In that system, terminal differentiation is represented by the appearance of muscle specific characteristics (Nieuwkoop, 1977). Genes coding for muscle specific proteins are apparently directly activated during induction of the mesoderm (Gurdon et al., 1985; Dawid and Sargent, 1986). It is possible, however, that many controlling elements must be activated first. This is now being analysed by molecular cloning methods (Dawid and Sargent, 1986). The fact that terminal differentiation occurs so rapidly may be problematic in the mesoderm muscle induction-system. It may therefore turn out to be difficult to distinguish between early control events and the expression of the differentiated state.

In the developing kidney, terminal differentiation seems to occur several days after completion of induction. Kidney-specific antigens can be detected rather late, on day three and four of in vitro cultures. Well before that, a number of other molecular changes must occur, and some induction-dependent changes have been described at the protein level in the kidney system. In most cases, the evidence that the expression of a protein is dependent on embryonic induction is based on the detection of proteins by immunofluorescence, and several other approaches should now be used to verify the results. The induction-dependent changes seen during differentiation of the mesenchyme could be grouped into certain major categories:

1) an increased cell adhesion (extracellular matrix and cell surface),
2) an increased proliferation and an increased response to serum mitogens, and
3) changes leading to an attraction of blood vessels.

Cell Adhesion

The uninduced mesenchyme expresses interstitial collagens and fibronectin. Morphologically it resembles any other undifferentiated mesenchyme, and the matrix phenotype is also typical for embryonic mesenchyme. In vivo, the ureter bud will grow into the mesenchyme and it then induces the formation of mesenchymal condensates around the mesenchyme. As judged by immunofluoresence, these condensates do not express interstitial collagens (Fig. 3) or fibronectin, but laminin can now be detected in these areas. These studies suggest that induction changes the matrix phenotype of the cells. This is supported by data from the in vitro culture system (review: Ekblom and Thesleff, 1985). Since laminin was found in the early condensates before overt tubulogenesis, it was sug-

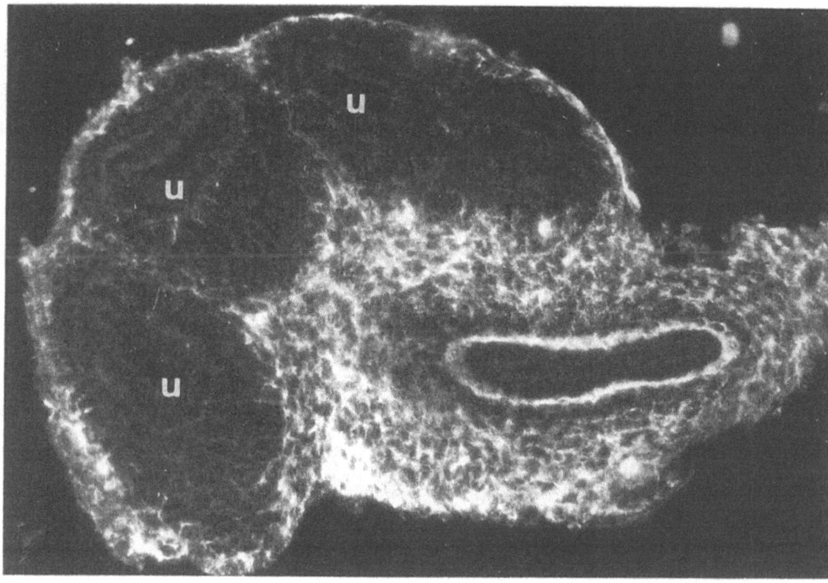

Figure 3: Type III procollagen expression during early in vivo development of the kidney. Note that type III colagen is expressed in the uninduced areas but not in the induced areas arund the tips of the ureter bds (u). (After Ekblom et al, 1982).

gested that it is a morphogenetically important protein (Ekblom et al., 1980). There is now an extensive amount of data which support this suggestion. It is known that laminin can promote differentiation, alter morphogenesis, and stimulate neurite outgrowth, to name a few particularly dramatic examples (reviews: Kleinman et al., 1985; Timpl and Dziadek, 1986).

In recent years it has become apparent that cell adhesion cannot be dependent merely on the presence of the correct matrix molecules, but it also requires cell-adhesion molecules located at the cell surface. In epithelial cells from a variety of different species, a cell adhesion molecule with an approximate size of 120 Kd has been detected (Edelman, 1985; Öbrink, 1986). All the 120 Kd CAMs seem to be identical or similar to uvomorulin, a cell surface glycoprotein which is important for cell compaction in the preimplantation mouse embryo (Kemler et al., 1977; Hyafil et al., 1981). It is now clear that this molecule is present on all epithelial cells, and its expression is apparently a very sensitive marker for the epithelial phenotype. In the developing kidney, the uninduced mesenchyme seems to express no uvomorulin, but when it is induced, uvomorulin gradually becomes detectable in the condensing areas. The appearance of immunoreactivity for uvomorulin during tubulogenesis has also been shown by protein blotting. Based on immunofluorescence data it would seem that laminin is detected slightly before uvomorulin. During tubulogenesis, laminin will rapidly become confined largely to the basal surface in the extracellular space, whereas no such polar expression is seen for uvomorulin. The expression studies on laminin and uvomorulin suggest that these proteins are important for the initial adhesion of the cells when they convert from mesenchyme to epithelium. No direct evidence for this proposal is, however, available. To study this, we are now trying to alter development by applying antibodies against laminin, uvomorulin and other adhesive proteins to the in vitro organ cultures. So far, data is available only for uvomorulin. It is noteworthy that anti-uvomorulin antibodies do not inhibit morphogenesis of the developing kidney, although the same antibodies can inhibit compaction during preimplantation development, histogenesis of the embryonic liver, and formation of cell polarity of kidney epithelial cells in vitro (Vestweber, 1985; Vestweber et al., 1985). Therefore, it is possible that other adhesive proteins such as N-CAM and laminin (or other basement membrane proteins) are more crucial for kidney tubulogenesis than uvomorulin.

Cell Proliferation and Serum Mitogens

Morphogenesis and growth of embryonic organs also require a constant increase in cell number. At the onset of kidney differentiation, the organ rudiment in the mouse consists only of some 20,000 cells, but in the adult, there are millions of nephric units, each composed of millions of cells. Growth of the embryonic kidney is very rapid. We have estimated that the doubling time during the early stages could be around 24 hours. Several factors, both local and systematic, undoubtedly contribute to growth, and much still remains to be learned about the factors controlling growth of the embryonic kidney.

It has been shown already 20 years ago that the inducer tissue exerts a direct positive effect on the thymidine incorporation of the nephrogenic mesenchyme (Vainio et al., 1965). These studies have been confirmed (Ekblom et al., 1983; Saxén et al., 1983). There seems to be an initial burst in proliferation at around 24 hr of in vitro culture. This proliferation may be caused by a local stimulus from the inducer tissue. However, for continued proliferation, serum is required. A variety of observations suggest that the effect of serum is due to the presence of transferrin, an iron carrier (Ekblom et al., 1983; Landschulz and Ekblom, 1985). Of the different serum mitogens tested (FGF, EGF, PGDF, insulin, transferrin), only transferrin can replace the effect of serum whereas the other mitogens are ineffective in the assay used. It is noteworthy that the uninduced mesenchyme does not respond to transferrin (Fig. 4). Therefore it has been postulated that induction of the mesenchyme leads to an increased expression of transferrin receptors (Ekblom et al., 1983). This possibility is now studied with the use of monoclonal antibodies to the receptor (Trowbridge et al., 1982) and with a mouse transferrin receptor cDNA (Stearne et al., 1985). For some other developmental organs, it has been convincingly shown that the epithelio-mesenchymal interaction leads to an expression of receptors for morphogenetically important serum factors (Kratochwil and Schwarz, 1976; Heuberger et al., 1982; Cunha et al., 1983).

The role of transferrin for organogenesis was first clearly noted for the developing kidney, but it is now becoming clear that transferrin is important for growth of a number of embryonic organs. It is required for tooth morphognesis (Partanen et al., 1984), it acts as a muscle trophic factor (Ii et al., 1982), and it could have important functions in the formation of neuromuscular junctions. Transferrin is mainly synthesized by the yolk sac and the liver in the embryo (Ekblom and Thesleff, 1985), but high levels of transferrin mRNA are also found in the developing muscle, and in adult brain (Levin et al., 1984). The importance of transferrin synthesis in the brain is unknown, but it is of interest that brain blood vessels express a high amount of the transferrin receptors (Jefferies et al., 1984), and that iron becomes incorporated into special sites of the brain (Hill et al., 1985). In the brain, transferrin is most likely not required for proliferation of cells, but iron-transferrin has some direct neurotrophic activity (Beach et al., 1983; Bloch et al., 1985). In embryonic chicken brain, postmitotic neurons are completely dependent on transferrin for survival (Aizenman et al., 1986).

In conclusion, it seems evdient that transferrin is important as a mitogenic factor for growth of many embryonic organs, and, in addition, it may have other vital non-

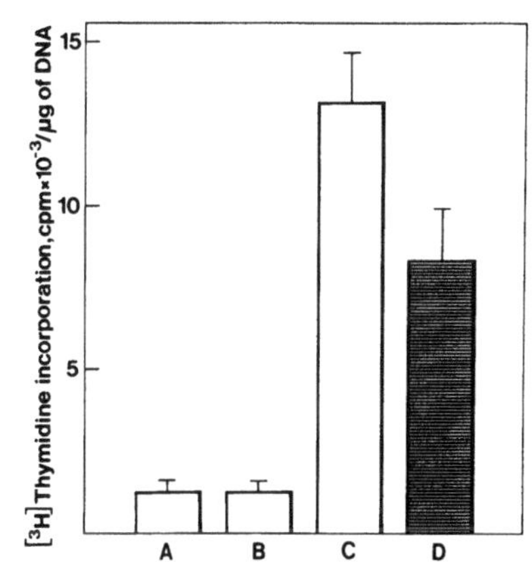

Figure 4: Transferrin-dependent proliferation on day two of in vitro cultures of the nephrogenic mesenchyme. (From Ekblom et al., 1983). (A) Non-induced cells do not respond to transferrin. (B) When induction is prevented by low-pore size filters, cells do not respond to transferrin. (C) Induced cells respond to transferrin. (D) Induced cells respond to 10% fetal calf serum. Note that transferrin is a stronger mitogen than serum. Bars: Standard error of triplicates.

mitogenic functions in embryonic and adult neurons.

Angiogenesis

Although the nephrogenic mesenchyme can convert into epithelium, it does not have the option to differentiate into endothelial cells (Ekblom, 1981; Bernstein, 1981). Blood vessels are nevertheless required early for the growing organ to meet the increased serum requirements. Furthermore, the main function of the kidney is to filter the blood in the glomeruli, and therefore a well developed vasculature is an integral part of the kidney structure. It has been found by transplantation procedures that the endothelial cells are not derived from the nephrogenic mesenchyme but from already differentiated endothelial cells which grow into the developing kidney (Sariola, 1985). The differentiating epithelial cells apparently secrete angiogenesis factors which induce the blood vessel ingrowth. Uninduced mesenchymes do not stimulate angiogenesis, suggesting that the secretion of angiogenesis factors is a differentiation-dependent event (Sariola et al., 1983). It is likely that such angiogenesis factors are secreted by many embryonic organs, since blood vessel development is so crucial for morphogenesis.

Future Prospects

The kidney mesenchyme responds to induction in a complex fashion, and many metabolic pathways are stimulated. By studying the changes that occur in response to induction, a number of morphogenetically important proteins have been identified. This approach led us to suggest a role for laminin in morphogenesis, and it was also possible to identify transferrin as an important stimulator of organogenesis. It is pleasing to note that these molecules are important for morphogenesis of many other organs as well.

Our list of changes that occur in response to induction of the nephrogenic mesenchyme is certainly incomplete, and many of the proposed changes must still be more carefully analyzed. In the studies so far performed, previously known proteins have been studied. In order to more completely characterize the induction-dependent changes, we have recently begun to use some other approaches as well. We are now trying to analyze the difference in the protein pattern between the induced and non-induced mesenchyme by two-dimensional gel electrophoresis. In addition, it should be possible to construct a cDNA library from the mRNA of the induced cells. This should make it possible to use differential hybridization procedures to detect specific cDNA molecules that would represent mRNA molecules found only in induced cells. The feasibility of this approach to study the repsonse to embryonic induction has recently been shown in another system (Dawid and Sargent, 1986). In the kidney model system, it may be a problem to obtain a sufficient amount of material for construction of a cDNA library, but the methods are now becoming increasingly sensitive, and therefore it may in the future be possible to apply such techniques in this model as well. It is likely that use of such techniques will unravel several important features of the early responses to induction.

References

Aizenman, Y., M.E. Weichsel, Jr., and J. De Vellis (1986) Changes in insulin and transferrin requirements of pure brain neuronal cultures during embryonic development. Proc. Natl. Acad. Sci. (USA) **83:** 2263-2266.

Beach, R.L., H. Popiela, and B.W. Festoff (1983) The identification of neurotrophic factor as a transferrin. FEBS Letters **156:** 151-156.

Bernstein, J., F. Cheng, and J. Roszka (1981) Glomerular differentiation in metanephric culture. Lab. Invest. **45:** 183-190.

Bloch, B., T. Popovici, M.J. Levin, D. Tuil, and A. Kahn (1985) Transferrin gene expression visualized in oligodendrocytes of the brain by using in situ hybridization and immunohistochemistry. Proc. Natl. Acad. Sci. (USA) **82:** 6760-6710.

Cunha, G.R., L.W.K. Chung, J. M. Shannon, O. Taguchi, and H. Fujii (1983) Hormone-induced morphogenesis and growth: Role of mesenchymal-epithelial interactions. Recent Progr. Horm. Res. **30:** 559-598.

Dawid, I.B., and T.D. Sargent (1986) Molecular embryology in amphibians: New approaches to old querstions. Trends in Genetics **2:** 47-50.

Edelman, G.M. (1985) Cell adhesion and the molecular processes of morphogenesis. Ann. Rev. Biochem. **54:** 135-169.

Ekblom, P. (1981) Formation of basement membranes in the embryonic kidney: An immunohistological study. J. Cell Biol. **91:** 1-10.

Ekblom, P., and I. Thesleff (1985) The role of transferrin and extracellular matrix components in kidney development. Modern Cell Biology **4:** 85-127.

Ekblom, P., K. Alitalo, A. Vaheri, R. Timpl, and L. Saxén (1980) Induction of a basement membrane glycoprotein: Possible role of laminin in morphogenesis. Proc. Natl. Acad. Sci. (USA) **77:** 485-489.

Ekblom, P., L. Saxén, and R. Timpl (1982) The extracellular matrix and kidney development. In *Membranes in Growth and Development*, G. Giebisch, J. Hoffman, and L. Bolis, eds., pp. 429-442, Liss, New York.

Ekblom, P., I. Thesleff, L. Saxén, A. Miettinen, and R. Timpl (1983) Transferrin as a fetal growth factor: Acquisition of responsiveness related to embryonic induction. Proc. Natl. Acad. Sci. (USA) **80:** 2651-2655.

Grobstein, C. (1955) Inductive interaction in the development of the mouse metanephros. J. Exp. Zool. **130:** 319-340.

Grobstein, C. (1956) Trans-filter induction of tubules in mouse metanephrogenic mesenchyme. Exp. Cell Res. **10:** 424-440.

Gurdon, J.B., S. Fairman, T.J. Mohun, and S. Brennan (1985) Activation of muscle-specific actin genes in Xenopus development by an induction between animal and vegetal cells of blastula. Cell **41:** 913-922.

Heuberger, B., I. Fitzka, G. Wasner, and K. Kratochwil (1982) Induction of androgen receptor formation by epithelium-mesenchyme interaction in embryonic mouse mammary gland. Proc. Natl. Acad. Sci. (USA) **79:** 2957-2961.

Hill, J.M., M.R. Ruff, R.J. Weber, and C.B. Pert (1985) Transferrin receptors in rat brain: Neuropeptide-like pattern and relationship to iron distribution. Proc. Natl. Acad. Sci (USA) **82:** 4553-4557.

Hyafil, F., C. Babinet, and F. Jacob (1981) Cell-cell interactions in early embryogenesis: A molecular approach to the role of calcium. Cell **26:** 447-454.

Ii, I., I. Kimura, and E. Ozawa (1982) A myotrophic protein from chick embryo extract, its purification, identity to transferrin, and indispensability for avian myogenesis. Dev. Biol. **94:** 366-377.

Jefferies, W.A., M.R. Brandom, S.V. Hunt, A.F. Williams, K.C. Gatter and D.Y. Mason

(1984) Transferrin receptor on endothelium of brain capillaries. Nature **312:** 162-163.

Kemler, R., C. Babinet, H. Eisen, and F. Jacob (1977) Surface antigen and early differentiation. Proc. Natl. Acad. Sci. (USA) **74:** 4449-4452.

Kleinman, H.K., J.R. Hassell, M. Aumailley, V.P. Terranova, G.R. Martin, and M. Dubois-Dalq (1985) Biological activities of laminin. J. Cell. Biochem. **27:** 317-325.

Kratochwil, K., and P. Schwarz (1976) Tissue interaction in androgen response of embryonic mammary rudiment of mouse: Identification of target tissue for testosterone. Proc. Natl. Acad. Sci. (USA) **73:** 4041-4045.

Landschulz, W., and P. Ekblom (1985) Iron delivery during proliferation and differentiation of kidney tubules. J. Biol. Chem. **260:** 15580-15584.

Lehtonen, E. (1976) Transmission and spread of embryonic induction. Med. Biol. **54:** 108-128.

Levin, M.J., D. Thil, G. Uzan, J.C. Dreufus, and A. Kahn (1984) High level of transferrin mRNA in fetal muscle and adult brain. Biochem. Biophys. Res. Commun. **122:** 212-217.

Nieuwkoop, P.D. (1977) Origin and establishment of embryonic polar axes in amphibian development. Curr. Top. Dev. Biol. **11:** 115-132.

Öbrink, B. (1986) Epithelial cell adhesion molecules. Exp. Cell Res. **163:** 1-21.

Partanen, A.M., I. Thesleff, and P. Ekblom (1984) Transferrin is required for early tooth morphogenesis. Differentiation **27:** 59-66.

Sariola, H. (1985) Interspecies chimeras: An experimental approach for studies on embryonic angiogenesis. Med. Biol. **63:** 43-65.

Sariola, H., P. Ekblom, E. Lehtonen, and L. Saxén (1983) Differentiation and vascularisation of the metanephric kidney on chorioallantoic membranes. Dev. Biol. **96:** 427-435.

Saxén, L., O. Koskimies, A. Lahti, H. Miettinen, J. Rapola, and J. Wartiovaara (1986) Differentiation of kidney mesenchyme in an experimental model system. Adv. Morphogen. **7:** 251-293.

Saxén, L., J. Salonen, P. Ekblom, and S. Nordling (1983) DNA synthesis and cell generation cycle during determination and differentiation of the metanephric mesenchyme. Dev. Biol. **98:** 130-138.

Stearne, P.A., G.A. Pietersz, and J.W. Goding (1985) cDNA cloning of the murine transferrin receptor: Sequence of trans-membrane and adjacent regions. J. Immunol. **134:** 3474-3479.

Timpl, R., and M. Dziadek (1986) Structure, development, and molecular pathology of basement membranes. Int. Rev. Exp. Pathol., in press.

Trowbridge, I.S., J. Lesley, and R. Schulte (1982) Murine cell surface transferrin receptor: Studies with an anti-receptor monoclonal antibody. J. Cell Physiol. **112:** 403-410.

Vainio, T., J. Jainchill, K. Clement, and L. Saxén (1965) Studies on kidney tubulogenesis: VI. Survival and nucleic acid metabolism of differentiating mouse metanephrogenic mesenchyme in vitro. J. Cell Comp. Physiol. **66:** 311-317.

Vestweber, D. (1985) Das Zelladhäsionsmolekül Uvomorulin, strukturelle und funktionelle Analyse. Dissertation, Tübingen.

Vestweber, D., R. Kemler, and P. Ekblom (1985) Cell-adhesion molecule uvomorulin during kidney development. Dev. Biol. **112:** 213-221.

Wessells, N.K. (1977) *Tissue Interactions and Development.* 276 p. W.A. Benjamin Inc., Menlo Park, CA.

CELL-MATRIX INTERACTION IN THE DIFFERENTIATION OF THE INTESTINAL EPITHELIUM

Ursula Hahn, D. Schuppan, E.G. Hahn, H.-J. Merker, and E.-O. Riecken
Klinikum Steglitz
Dept. of Gastroenterology and Dept. of Anatomy
Free University of West Berlin
West Berlin

Introduction

Epithelio-mesenchymal interactions are a prerequisite for the normal cytodifferentiation of all epithelial tissues during ontogenesis. Whereas the pioneering investigators reported on the necessity of the mere presence of mesenchymal tissue for epithelial cytodifferentiation (as reviewed by McLoughlin, 1963, and Grobstein, 1967), subsequent studies have focused on the biochemical nature of these events (Slavkin and Greulich, 1975; Yamada and Olden, 1978; Bernfield, 1981). It became obvious that the so-called mesenchyme is a very complex tissue and that further insight into its effects on epithelial tissues would primarily call for a scrutinizing look at its own proper structure and function. Especially the biochemical composition of the extracellular matrix evoked intensive interest. Consequently over the last years our knowledge of the distribution of the family of collagens, the non-collagenous glycoproteins and the proteoglycans within all vertebrate tissues has increased tremendously. Many reviews have dealt with the accumulating biochemical data on the structure and function of these macromolecules of the extracellular matrix (Porter and Whelan, 1984; Trelstad, 1984; Piez and Reddi, 1984; see also Schuppan, this volume).

In the intestine, an intimate morphological association between mesenchymal and epithelial cells in a period of intensive proliferation and differentiation of the epithelium has been clearly established. Close cell to cell contacts through gaps in the subepithelial basement membrane were identified ultrastructurally during a short period before and after birth in rats (Mathan et al., 1972). Later, basal processes of the epithelial cells in adult human and murine small intestinal mucosa were also observed (Partridge and Simpson, 1981). These findings point to a direct interaction of epithelial and connective tissue cells. More stringent evidence for this thesis came from work using techniques of recombination and grafting of tissues.

NATO ASI Series, Vol. H5
Mesenchymal-Epithelial Interactions in Neural Development
Edited by J. R. Wolff et al.

The Influence of Mesenchyne on the Fetal Endoderm

Isolated endoderm re-associated with fetal intestinal mesenchyme was grafted under the kidney capsule of adult rats and developed there into complete, organotypical intestinal mucosa. Moreover, intestinal fibroblasts exhibited powerful inductive properties compared to non-intestinal fibroblasts: while the former can induce heterotypic differentiation of gizzard endoderm into duodenal-like cells, including goblet and entero-endocrine cells, the latter are only able to induce homotypic differentiation. Intestinal mesenchyme of human origin apparently exerts an even stronger influence on endodermal cells than murine mesenchyme (Kedinger et al., 1981; Haffen et al., 1981; Haffen et al., 1983; Lacroix et al., 1984). Intestinal epithelial cell-lines, established from fetal or suckling rat intestine, resembling in some features intestinal crypt cells do not show self-differentiation in vitro (Quaroni et al., 1979; Negrel et al., 1983; Hahn et al., 1986a).

However, these primitive epithelial cells, when combined with fetal rat gut mesenchyme, will actually form villous structures lined by mature, absorptive enterocytes with a high and regular brush border as well as mucus, endocrine and Paneth cells (Kedinger et al., 1986).

Thus, the general importance of viable mesenchyme for intestinal cytodifferentiation has been well established.

Furthermore, the studies mentioned above indicate that a reciprocal effect of the epithelium on the fetal mesenchyme is also observed, since it will be transformed to the typical layers of smooth muscle cells and lamina propria in these grafting experiments, although this will have to be examined in more detail in the future. Another important issue remains: which of the aspects of differentiation of fetal endoderm into mature intestinal epithelium are really in every aspect comparable to the differentiation and maturation of the intestinal epithelial cells in the adult organ along the crypt-villus axis? In other words, is the adult connective tissue of the lamina propria also essential for intestinal cytodifferentiation?

The Extracellular Matrix of the Gut

When we investigate the mesenchymal influence on the epithelium, we would like to discriminate those events requiring living fibroblasts from those requiring distinct macromolecules of the extracellular matrix. As a first step, we decided to study the distribution of known matrix components within the entire human and murine gastrointestinal mucosa (Hahn et al., 1986b). In a comprehensive immunohistological study, we used antibodies to the interstitial collagens type I, III, V, VI; procollagen type III; basement membrane proteins collagen type IV; laminin; nidogen; fibronectin and undulin (for further information on these antigens see D. Schuppan, this volume). All along the entire alimentary canal, we found a very homogenous distribution of the basement membrane proteins.

Another basement membrane component, heparansulfate proteoglycan, has also been identified ultrastructurally together with laminin and collagen type IV (Lawrie et al., 1982). Interestingly, no apparent gradient of the amount of any one basement membrane component along the crypt-villus axis has been reported. Of fibronectin, an apparent gradient with decreasing amounts toward the villus tip was once oberved and it was claimed that this might be one of the causes of cellular exfoliation at the villus tip (Quaroni et al., 1978). This has not been verified, however, and no direct evidence has been presented that a biochemical change within the basement membrane along the crypt-villus axis actually evokes migration.

Except for undulin, all the other matrix proteins occur in the lamina propria and constitute the delicate network of interstitial fibrils and microfibrils formerly called reticulin. Anchoring fibrils project into a very dense layer of interstitial collagens immediately adjacent to the basement membrane on the lamina propria side. Undulin is restricted to the submucosa and thereby seems to define a very special fibroblast subpopulation. We concluded that even if the basement membrane proteins laminin, nidogen and collagen type IV do convey some kind of differentiation promoting signal to the epithelial cells, it could not be identified by this location. Basement membrane protein synthesis or basement membrane receptor synthesis and turnover may be another means by which the cells regulate their affinity towards the basement membrane. A receptor for laminin has recently been identified on human intestinal epithelial cells (Hand, 1985).

Intestinal Cell and Tissue Cultures

Numerous attempts to culture intestinal cells and tissues have been registered since the pioneering works of Browning and Trier (1969), Lichtenberger (1973), and Quaroni (1979), primarily for the purpose of studying epithelial cell functions in vitro. In reviewing the more recent experiments using intestinal mucosa of different species and varying strategies for cell culture conditions, however, some general features concerning epithelial-mesenchymal interactions have become evident.

1) In organ cultures of adult mucosa where the epithelium is explanted, resting on viable lamina propria, epithelial proliferation ceases after a few days. The epithelium resembles mature intestinal epithelium (for review see Trier, 1980).

2) In fetal organ culture, survival is much prolonged of moderately to highly differentiated intestinal epithelium together with the fetal mesenchyme (Howdle, 1983; Quaroni, 1985; Menard and Arsenault, 1985).

3) Rapidly proliferative epithelial cells of intestinal mucosa can only be established in vitro after having abolished all mesenchymal and fibroblastic contaminations. Exclusion of connective tissue from epithelial cells seems to permit the perpetual proliferation (Quaroni et al., 1979; Danes, 1980; Danes and Sutanto, 1982; Negrel et al., 1983; Terano et al., 1983; Siddiqui and Chopra, 1984).

4) A complete basement membrane, as far as it is documented, has only been found between a polarized, differentiated epithelium and a viable connective tissue in vitro.

In two studies on gastric mucous cells from adult guinea pig and fetal rabbit, a certain degree of polarization was accomplished although no well-defined basement membrane had been deposited (Logsdon et al., 1982; Rattner et al., 1985). In both situations, cells were grown on semipermeable collagen supports. It should be borne in mind that besides being influenced by the proteins within the support, the basal and the apical side of cells grown on "floating gels" have access to the liquid phase much more intensively than on solid supports which may exert a differentiation promoting effect.

Basement Membrane Protein Biosynthesis by Intestinal Epithelial Cells in Vitro

In a consecutive step, we have studied an in vitro model of fetal human and murine explant cultures in order to study the intra- and extracellular distribution of basement membrane proteins in intestinal epithelial cells. The methodology will be published in detail elsewhere (Hahn et al., 1986c). Briefly, intestinal fragments of timed pregnant mice and rats and of two human fetal intestines obtained from therapeutic hysterectomy were explanted in vitro. After 12-24h epithelial outgrowth was observed. Usually pure epithelial outgrowth with minimal mesenchymal proliferation at different sites of the explant that will not interfere with epithelial proliferation can be achieved by 50-80% of explants and only these were included in the study. The primary epithelial monolayer consists of a tightly packed monolayer of rather small, polygonal cells forming a thick rim at the leading edge, a phenomenon also described by others (Kondo et al., 1983; Quaroni, 1985). While this monolayer continues to expand for several days the pericentral zone begins to show an altered morphology: very little cytoplasm of the cells and the nucleus itself are hardly distinguishable because in this zone the epithelial cells have become very compacted. Following mitosis, many cells seem to be expelled into the medium instead of growing on the surface from this area. The borderline between the primary undifferentiated and the dense, pericentral zone is rather abrupt.

Histochemical staining for the brush border enzymes alkaline phosphatase, lactase and α-glucosidase, which are often used as marker enzymes of intestinal epithelial differentiation (Koldovsky, 1969; Grand et al., 1976), revealed that either all or the majority of the epithelial cells within the pericentral zone express brush border enzyme activity while the primary epithelial monolayer remains definitely negative.

Ultrastructural analysis of transverse sections confirmed the existence of a pericentral zone of differentiated epithelium: while the peripheral monolayer of epithelial cells consists of flat cells joined by gap junctions and carrying only very sparse microvilli, the pericentral epithelial cells are cuboidal, joined by tight junctions and extensive interdigitations of the lateral plasma membrane and are producing a rather high and regular brush border. Beneath these cells, a thin layer of mesenchymal cells has spread out from the original piece of mucosa and is now growing in between the primary epithelium and the culture dish, thereby lifting the epithelium off the plastic support.

The appearance of a continuous basement membrane coincides precisely with the transition of the primary undifferentiated monolayer to the differentiated, pericentral zone. Generally, it is observed in these explant cultures between polarized epithelium and viable mesenchyme and is lacking beneath the primary monolayer.

By immunofluorescence, this centrifugal gradient of epithelial differentiation is paralleled by changes of the apparent intracellular content of basement membrane proteins and the deposition of an extracellular matrix. Laminin and collagen type IV are detected intracellularly in the flat, peripheral cells. Within the differentiated zone, the intraepithelial content has decreased dramatically with the onset of the deposition of a veritable basement membrane. Fibronectin is found in surprisingly low amounts in the primary epithelial outgrowth compared to the production by intestinal epithelial cell lines (Quaroni et al., 1978) and is also not detected in the differentiated zone. Antibodies to procollagen type III, collagen type VI and undulin do not stain primary epithelial cells but decorate the intestinal mesenchymal cells proliferating from the mucosa fragment abundantly. Further characterization of the different types of mesenchymal cells originating from intestinal mucosa is under way (Hahn et al., 1986a). Principally, the scheme of intestinal differentiation in vitro was not altered when the explants were grown on glass coverslips coated with collagen type I, IV, laminin or fibronectin. Activation of brush border enzymes or polarization of epithelial cells was not induced by any single type of matrix molecule.

The most likely explanation for the apparent decrease of intracellular basement membrane proteins in the polarized cells is that, having deposited an authentic basement membrane in co-operation with the invading mesenchyme, the de novo biosynthesis is no longer required at that ratio and consequently drops below the level detectable by immunohistochemistry. This would also explain the lack of intracellular demonstration of basement membrane proteins in mature epithelia in vivo. Of course other reasons, like the lack of recognition of the respective antigen by the antibodies due to masking effects by other compounds, etc., have to be considered. For this reason we will attempt to quantify basement membrane biosynthesis by more sensitive methods.

Conclusions

The epithelial-mesenchymal interaction is a complex event and merits further refined research. It appears that induction of differentiation of intestinal epithelial cells requires viable mesenchyme and is not incited by any single matrix molecule. However, not all molecular candidates for such an event have been tested, and experiments with combinations of matrix components are under way. Preliminary results using co-cultures of intestinal fibroblasts and isolated epithelial cells confirm the results of the explant culture model described above. It will also facilitate using chemically defined medium, defined matrix proteins and defined cell types.

The process of simultaneous migration and differentiation of the adult intestinal epithelium requires unique regulatory mechanisms. We expect to unravel some of these complex interactions in such models and to be able to discriminate between cell-cell signals and cell-matrix-cell signals in the near future.

References

Bernfield, M.R. (1981) *Morphogenesis and Pattern Formation*. T.G. Connelly et al., ed. Raven Press, New York.

Browning, T., and J. Trier (1969) Organ culture of mucosal biopsies of human small intestine. J. Clin. Invest. **48:** 1423-1432.

Danes, S.B. (1980) Establishment of human colonic mucosal culture strains from colonoscopy specimens. Birth Defects **16:** 275-281.

Danes, S.B., and E. Sutanto (1982) Epithelial line from normal human colonic mucosa. J. Natl. Cancer Inst. **69:** 1271-1276.

Grand, R.J., J.B. Watkins, and F.M. Torti (1976) Development of the human gastrointestinal tract. Gastroenterology **70:** 790-810.

Grobstein, C. (1967) Mechanisms of organogenetic tissue interaction. Natl. Cancer Inst. Monogr. **26:** 279.

Haffen, K., M. Kedinger, P.M. Simon, and F. Raul (1981) Organogenetic potentialities of rat intestinal epitheliod cell cultures. Differentiation **18:** 97-103.

Haffen, K., B. Lacroix, M. Kedinger, and P.M. Simon-Assman (1983) Inductive properties of fibroblastic cell cultures derived from rat intestinal mucosa on epithelial differentiation. Differentiation **23:** 226-233.

Hahn, U., D. Schuppan, E.G. Hahn, and E.O. Riecken (1986a) Basement membrane metabolism of intestinal epithelial cells in vitro. In preparation.

Hahn, U., D. Schuppan, E.G. Hahn, and E.O. Riecken (1986b) The extracellular matrix of the gut. Submitted.

Hahn, U., D. Schuppan, E.G. Hahn, and E.O. Riecken (1986c) Intestinal epithelial cells produce basement membrane proteins in vitro. Gut, in press.

Hand, P.H., A. Thor, J. Schlom, C.N. Rao and, L. Liotta (1985) Expression of laminin receptor in normal and carcinomatous human tissues as defined by a monoclonal antibody. Cancer Res. **45:** 2713-2717.

Howdle, P.D. (1983) Organ culture in the study of the gastrointestinal tract in health and disease. Clin. Science **65:** 105-110.

Kedinger, M., P.M. Simon, J.F. Grenier, and K. Haffen (1981) Role of epithelial-mesenchymal interactions in the ontogenesis of intestinal brush-border enzymes. Dev.

Biol. **86:** 339-347.
Kedinger, M., P.M. Simon-Assman, B. Lacroix, A. Marxer, H.P. Hauri, and K. Haffen (1986) Fetal gut mesenchyme induces differentiation of cultured intestinal endoderm and crypt cells. Dev. Biol. **113:** 474-483.
Koldovsky, O., ed. (1969) *Development of the Functions of the Small Intestine in Mammals and Man.* Karger, Basel.
Kondo, Y., I. Rose, G.P. Young, and R.H. Whitehead (1983) Growth and differentiation of fetal rat small intestinal epithelium in tissue culture. Exp. Cell. Res. **153:** 121-134.
Lacroix, B., M. Kedinger, P. Simon-Assman and K. Haffen (1984) Early organogenesis of human small intestine: Scanning electron microscopy and brush border enzymology. Gut **25:** 925-930.
Lawrie, G.W., C.P. Leblond, and G.R. Martin (1982) Localization of type IV collagen, laminin, heparansulfate proteoglycan and fibronectin to the basal lamina of basement membranes. J. Cell. Biol. **95:** 340-344.
Lichtenberger, L., L.R. Miller, D.M. Erwin, and L.R. Johnson (1973) Effect of pentagastrin on adult rat duodenal cells in culture. Gastroenterology **65:** 242-251.
Logsdon, C.D., C.A. Bisbee, and M.J. Rutten (1982) Ultrastructural and transport properties of fetal rabbit epithelial cells cultured on floating collagen gels. In Vitro **18:** 233-242.
Mathan, M., J.A. Hermos, and J.S. Trier (1972) Structural features of the epithelio-mesenchymal interface of rat duodenal mucosa during development. J. Cell. Biol. **52:** 577-588.
McLoughlin, C.B. (1963) Mesenchymal influences on epithelial differentiation in cell differentiation. Symp. Soc. Exp. Biol. **17:** 359-377.
Menard, D., and P. Arsenault (1985) Explant culture of human fetal small intestine. Gastroenterology **88:** 691-700.
Negrel, R., P. Rampal, J.L. Nano, C. Cavenel, and G. Ailhaud (1983) Establishment and characterization of an epithelial intestinal cell line from rat fetus. Exp. Cell. Res. **143:** 427-437.
Partridge, B.T., and L.O. Simpson (1981) Basal processes on duodenal epithelial cells of man, mouse and rat. Pathology **13:** 463-472.
Piez, K.A. and A.H. Reddi (1984) *Extracellular Matrix Biochemistry.* Elsevier, New York.
Porter, R., and J. Whelan, eds. (1984) Ciba foundation symposium **108,** Pitman, London.
Quaroni, A., K.J. Isselbacher, and E. Ruoslahti (1978) Fibronectin synthesis by epithelial crypt cells of rat small intestine. Proc. Natl. Acad. Sci. (USA) **75:** 5548-5552.
Quaroni, A., J. Wands, R.L. Trelstad and K.J. Isselbacher (1979) Epitheloid cell cultures from rat small intestine: Characterization by morphologic and immunologic criteria. J. Cell. Biol. **80:** 248-265.
Quaroni, A. (1985) Development of fetal rat intestine in organ and monolayer culture. J. Cell. Biol. **100:** 1611-1622.
Rattner, D.W., S. Ito, M.J. Rutten, and W. Silen (1985) A rapid method for culturing guinea pig gastric mucous cell monolayers. In Vitro **21:** 453-461.
Siddiqui, K.M., and D.P. Chopra (1984) Primary and longterm epithelial cell cultures from human fetal normal colonic mucosa. In Vitro **20:** 859-868.
Slavkin, H.D., and R.C. Greulich, eds. (1975) *Extracellular Matrix Influence on Gene Expression.* Academic Press, New York.
Terano, A., T. Mach, J. Stachura, S. Sekhon, A. Tarnawski, and J. Ivey (1983) A monolayer culture of human gastric epithelial cells. Dig. Dis. Sci. **28:** 595-603.
Trelstad, R., ed. (1984) *The Role of the Extracellular Matrix in Development.* Alan R. Liss, New York.
Trier, J.S. (1980) Organ culture of the mucosa of human small intestine. Meth. Cell. Biol. **21b:** 365-384.
Yamada, K.M. and K. Olden, eds. (1978) *Cell Interactions and Development: Molecular Mechanisms.* Wiley Interscience, New York.

CONNECTIVE TISSUE AND DUCTULAR METAPLASIA OF HEPATOCYTES

R. Schulz, D. Schuppan, E.G. Hahn
Steglitz Medical Center, Department of Medicine
Free University of Berlin
Federal Republic of Germany

Introduction

The appearance of ductlike structures around portal tracts is a well known feature of many chronic liver diseases. These structures are called ductular proliferations, pseudoductules or pseudocholangioles and are thought to be one marker for the activity of the periportal fibrotic process (Popper et al., 1962; Scheuer, 1980). They are formed by small cells with a large nucleus and tiny basophilic cytoplasm, lining a round or flattened lumen at the margin of portal tracts and septa (Steiner and Carruthers, 1961). Their origin and function still remains unknown although they have been discussed for many years.

Some authors suggested that pseudoductules develop from pre-existing ductules (channels of Hering) (Schaffner and Popper, 1961; Grisham and Hartroft, 1961; Steiner and Carruthers, 1962; Carruthers et al., 1962; Popper et al., 1962). Others assumed that pseudoductules may originate from surrounding hepatocytes (Lucke, 1944; Leduc, 1955).

During the last few years, our knowledge and understanding of the structure and function of connective tissue has improved substantially (for reviews see: Bornstein and Sage, 1980; Hahn et al., 1980; Miller and Gay, 1982; Miller, 1984). There is more and more evidence for an important role of connective tissue in cell growth and differentiation (Kleinmann et al., 1981; Martin and Kleinmann, 1985).

Recently, we investigated the pericellular matrix of ductules and hepatocytes in various chronic liver diseases (Schulz et al., 1986) and here we describe and discuss the role of connective tissue changes in ductular metaplasia.

Material and Method

28 selected needle and wedge biopsies of human livers taken from patients of Steglitz Medical Center, West Berlin for diagnosis were stored at -70°C. Part of each biopsy served to establish the histological diagnosis. Only biopsies with an unequivocal clinical and histological diagnosis were investigated. Further information concerning disease, age and sex of the patients is given in Table 1. Three normal liver pieces taken from kidney

NATO ASI Series, Vol. H5
Mesenchymal-Epithelial Interactions in Neural Development
Edited by J. R. Wolff et al.

donors were used as controls.

Rabbit anti-human antibodies against collagen types I, III, VI, laminin fragment P1, plasma fibronectin and goat anti-human antibodies against collagen types IV (fragment 75 and NC1) and V were purified by cross-adsorption as described previously (Schuppan et al., 1985ab, 1986) (Table 2). All antibodies were tested for specifity by RIA-inhibition and showed no cross-reactivity.

Protein concentration of the antibody preparations as determined by the method of Lowry et al. (1951) was between 80 and 200 mg/l.

Unfixed cryostat sections of 4 µm thickness were prepared using a SLEE cryostat type HL and stained by the method of indirect immunofluorescence (Wick et al., 1976). Incubation with 5 µl of the first antibody for 30 mins and with 20 µl of the second antibody (rhodamine conjugated swine anti-rabbit IgG, DAKO R 156, Lot. No. 081 B, or rhodamin conjugated rabbit anti-goat IgG Cappel No. 180, Lot. No. 18915) was followed

Table 1: Summary of the examined biopsies (from (16)).

DIAGNOSIS	PATIENTS	AGE AND SEX
normal livers (kidney donors)	3	2 y,m / 18 y,f / 20 y,m
fatty liver	3	37 y,m / 43 y,m / 48 y,m
alcoholic hepatitis/ cirrhosis	5	38 y,m / 43 y,f / 47 y,m 48 y,f / 55 y,f
primary biliary cirrhosis	6	30 y,f / 40 y,f / 46 y,f 49 y,f / 73 y,f / 77 y,f
bacterial cholangitis	1	56 y,f
secondary biliary fibrosis	1	32 y,m
extrahepatic biliary atresia	2	case 1: f, 8 weeks / 10 weeks, 16 weeks case 2: m, 5 months
haemochromatosis	1	46 y,f
chronic active hepatitis	2	25 y,f / 41 y,m
idiopathic cirrhosis	1	39 y,f

Table 2: Human antigens used to produce specific antibodies (from (16)).

type	source	molecular weight (KD)	localisation
III	placenta	300	fibrillar, most tissues
IV(7S-L)	placenta	320	basement membranes
$V(AB_2)$	placenta, liver	320	interstitial, microfibrillar
VI	placenta	360	microfibrillar
laminin (P1)	placenta	300	basement membranes
fibronectin	plasma	540	interstitial, (basement membranes

by rinsing the sections in phosphate buffered saline for 30 mins each time.

The sections were counterstained with haematoxylin-eosin after photography of the immunofluorescence staining pattern.

Results

In normal livers, co-distribution of all proteins investigated was observed in the portal tracts as well as in the sinusoidal wall of the parenchyma, although the staining pattern differed distinctly.

Livers with periportal fibrosis are characterized by an increase of all examined proteins in this area. Hepatocellular islands, separated from the contiguous parenchyma by newly formed connective tissue, are delimited by a laminin and collagen type IV positive basement membrane on the septal matrix side (Fig. 1, Fig. 3). In comparison, periportal hepatocytes of normal liver never show a basement membrane-like structure. Also, a ring of collagen type V microfibrils is visible at the margin of the hepatocellular islands (Fig. 2), whereas in normal livers only bile ducts are surrounded by these fibrils. Hepatocytes apparently assume characteristics of bile ducts.

"Ductular proliferations" exhibit the same morphological appearance in immunofluorescence as hepatocellular islands (Fig. 3, 4). Because of their unusual shape they are cut longitudinally very often and show a direct transition to the neighbouring parenchyma (Fig. 1). The existence of hepatocytes and biliary epithelium together within the same cellular island can frequently be observed by the HE-counterstains (Fig. 4).

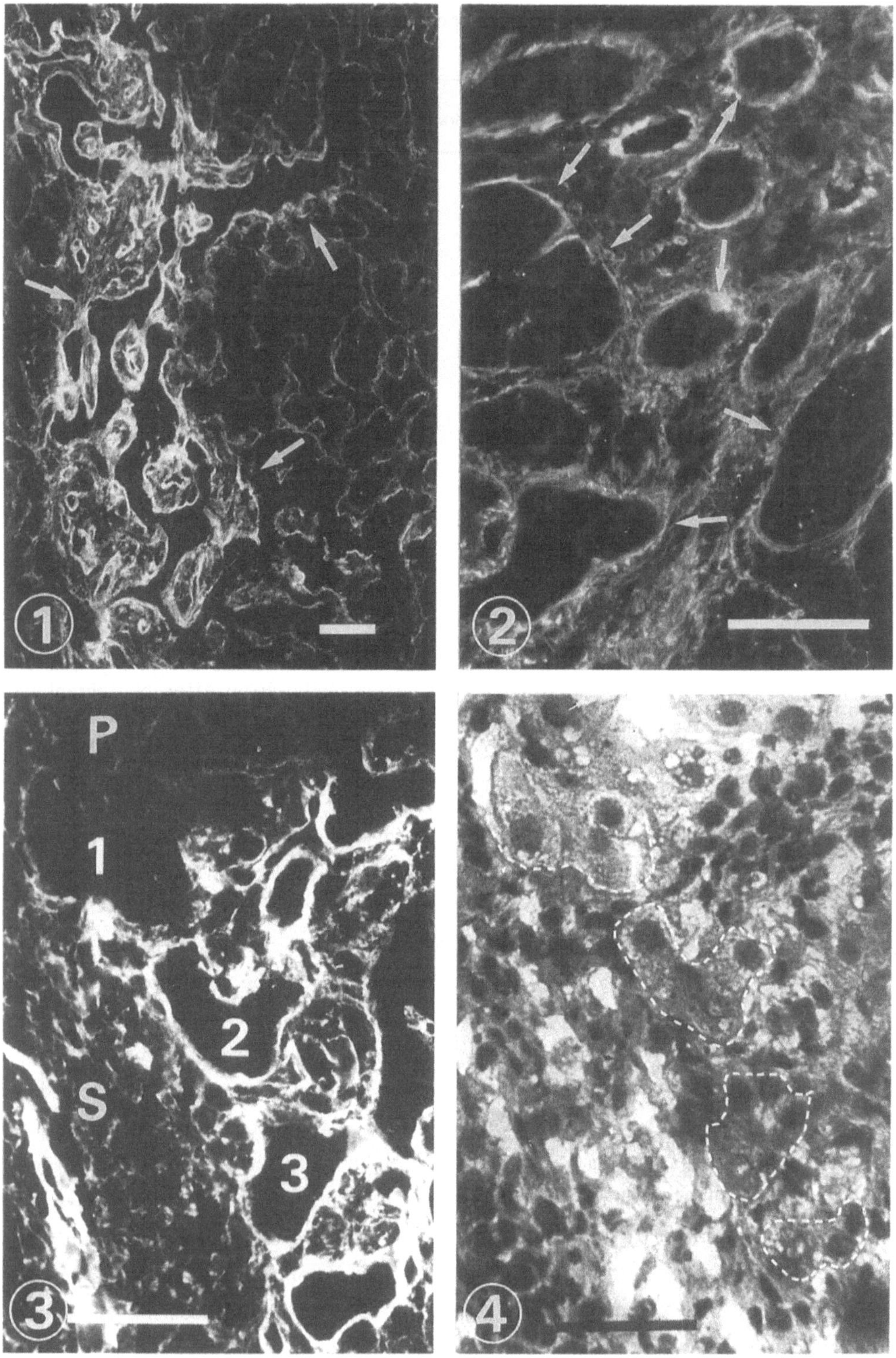
1
2
P
1
2
S
3
3
4

Legends to Figures:
S: septum P: parenchyma
The horizontal bar represents 50 µm.

Figure 1. Laminin: proliferating septum in the liver of a ten week old child with extrahepatic biliary obstruction. The arrows denote basement membranes around ductular proliferations and periportal hepatocytes. There is a direct transition of the pseudoducts into the parenchyma in the neighbourhood (original magnification: 200 X, sec. antibody rhodamine conjugated anti-rabbit IgG).
Figure 2. Collagen type V: border of septum in alcoholic fibrosis. The arrows denote Type V microfibrillar ring formation around hepatocellular islands and pseudoductules (original magnification: 500 X, sec. antibody rhodamin conjugated anti-goat IgG).
Figure 3. Collagen type IV: border of septum in primary biliary cirrhosis stage II-III. Partially (1) or completely separated cellular islands (2, 3) are delimited by a basement membrane (original magnification: 500 X, sec. antibody rhodamine conjugated anti-rabbit IgG).
Figure 4. HE-counterstain of the same section as shown in Fig. 3. The dotted line outlines an island containing one hepatocyte together with a bile duct-like formation (500 X).

Fibronectin forming intracellular dots is detectable in hepatocytes as well as in the epithelium of ductular proliferations. On the other hand, in normal livers fibronectin can only be recognized inside of the cytoplasm of hepatocytes.

These observations were most conspicuous in the rapidly progressive fibrotic liver diseases (alcoholic hepatitis, extrahepatic biliary atresia), but could also be recognized in other aetiologies.

Discussion

Our results are compatible with the assumption of a transformation of hepatocytes into biliary epithelium along with the formation of a characteristic fibrotic matrix in the periportal areas. As know from cell culture experiments and from the observation of developmental events the connective tissue can exert a stimulating or permissive effect on the differentiation of cells (Kleinman et al., 1981; Martin and Kleinmann, 1985). We postulate that the newly formed fibrotic matrix around hepatocellular islands acts like this and causes the transition of hepatocytes into biliary cells by an as yet unknown stimulus. Presumably, this reflects the reactivation of a normal developmental process, since during the ontogenesis of the liver the intrahepatic bile ducts are considered to derive from immature hepatocytes (Enzan et al., 1974) and their transformation takes place only in the presence of connective tissue (Shiojiri, 1984).

Our interpretations agree with recent observations of Vanstapel et al. (1984) and Butron Vila et al. (1984), who were able to demonstrate keratin filaments in periportal hepatocytes of fibrotic livers, whereas normal livers showed these filaments only inside of biliary epithelium. A relationship between bile duct cells and periportal hepatocytes under pathological conditions is also suggested by the expression of HLA-class I antigens including β_2-microglobulin on the cell membrane of either cell type (De Vos et al., 1985; Nagafuchi and Scheuer, 1985). On the contrary, in normal livers hepatocytes do not expose these antigens (De Vos et al., 1985; Nagafuchi and Scheuer, 1985).

In addition, Uchida and Peters (1983) reported the existence of transitional forms of hepatocytes and biliary cells within ductular proliferations using a light-, electron-microscopical and enzyme-histochemical approach. Further support comes from three-dimensional reconstructions (Jörgensen, 1971) and exact morphometrical studies (Jörgensen, 1973a), which showed that pseudoductules are similar in all hepatic diseases (Jörgensen, 1973b). While ducts and ductules of normal portal tracts have a cylindrical shape and do not communicate between each other, the pseudoductules under pathological conditions form plate-like tubes (Jörgensen, 1973b). These tubes build up a labyrinthine network by numerous connections between each other and therefore resemble the muralium architecture of liver parenchyma (Elias and Sherrick, 1969) much more than the ductular system of normal liver.

Our conclusion is that the transformation of hepatocytes into biliary structures in the periportal region could provide one mechanism for the lateral growth of fibrotic portal tracts and septa. Further studies are needed to investigate the role and influence of fibrotic connective tissue on hepatocytes. Cell culture experiments employing hepatocytes growing on fibrotic liver biomatrix could provide a tool to study the differentiation of these cells.

References

Bornstein, P., and H. Sage (1980) Structurally distinct collagen types. Annu. Rev. Biochem. **49**: 957-1003.

Butron Vila, M.M., J. Haot, and V.J. Desmet (1984) Cholestatic features in focal nodular hyperplasia of the liver. Liver **4**: 387-395.

Carruthers, J.S., R. Kalifat, and J.W. Steiner (1962) The ductular cell reaction of rat liver in extrahepatic cholestasis II. The proliferation of connective tissue. Exp. Mol. Pathol. **1**: 377-396.

De Vos,s R., C. de Wolf-Peeters, J.J. v.d. Oord, and V. Desmet (1985) Ultrastructural immunocytochemical demonstration of HLA class I antigens in human pathologic liver tissue. Hepatology **6**: 1071-1075.

Elias, H., and J.C. Sherrick (1969) Morphology of the liver. Chapter I, pp. 5-74. Academic Press, New York-London.

Enzan, H., T. Ohkita, H. Fujita, and S. Ijima (1974) Light and electron microscopic studies on the development of periportal bile ducts of the human embryo. Acta Path. Jap. **24**: 427-447.

Grisham, J.W., and W.S. Hartroft (1961) Morphologic identification by electron microscopy of "oval" cells in experimental hepatic degeneration. Lab. Invest. **10**: 317-332.

Hahn, E.G., U. Ott, and G.A. Martini (1980) Die Leberfibrose. Z. Gastroenterologie **18**: 453-469, 507-523.

Jörgensen, M. (1971) Three-dimensional reconstruction in histology. A short survey of available methods and description of a new technique. Acta Path. Microbiol. Scand. (A) **79**: 298-302.

Jörgensen, M. (1973a) A stereological study of intrahepatic bile ducts 1. Method and ap-

plication to normal livers. Acta Path. Microbiol. Scand. (A) **81:** 657-662.
Jörgensen, M. (1973b) A stereological study of intrahepatic bile ducts 2. Bile duct proliferation in some pathological conditions. Acta Path. Microbiol. Scand. (A) **81:** 663-669.
Kleinmann, H.K., J. Klebe, and G.R. Martin (1981) Role of collagenous matrices in the adhesion and growth of cells. J. Cell Biol. **88:** 473-485.
Leduc, E.H. (1955) Cell modulation in liver pathology. J. Histochem. Cytochem. **7:** 253-255.
Lowry, O.H., Rosebrough, A.L. Farr, and R.J. Randall (1951) Protein measurement with the folin phenol reagent. J. Biol. Chem. **193:** 265-275.
Lucke, B. (1944) I. The pathology of fatal epidemic hepatitis. II. The structure of the liver after recovery from epidemic hepatitis. Am. J. Pathol. **20:** 471-595.
Martin, G.R., and H.K. Kleinman (1985) The extracellular matrix in development and in disease. Sem. Liver Dis. **5:** 147-156.
Miller, E.J., and S. Gay (1982) Collagen: An overview. In *Methods in Enzymology*, L.W. Cunningham and D.W. Fredricksen, eds., vol. 82, pp. 3-32. 7 Academic Press, New York.
Miller, E.J. (1984) Chemistry of the collagens and their distribution. In *Extracellular Matrix Biochemistry*., K.A. Piez and A.H. Reddi, eds., pp. 41-81. Elsevier, New York.
Nagafuchi, Y., and P.J. Scheuer (1986) Hepatic $ß_2$-microglobulin distribution in primary biliary cirrhosis. J. Hepatol. **2:** 73-80.
Popper, H., F. Schaffner, and T. Barka (1962) Has proliferation of bile ductules clinical significance? Acta Hepatosplen. **9:** 129-139.
Schaffner, F., and H. Popper (1961) Electron microscopic studies of normal and proliferated bile ductules. Am. J. Pathol. **38:** 393-410.
Scheuer, P.J. (1980) *Liver Biopsy Interpretation*. Bailliere Tindal, London.
Schulz, R., D. Schuuppan, and E.G. Hahn (1986) The ductular metaplasia of hepatocytes: Characteristation of the pericellular matrix. Submitted.
Schuppan, D., M. Besser, and E.G. Hahn (1985a) Radioimmunoassay for the globular domain (NC1) of basement membrane collagen: Characterization and application to the metabolism of collagen type IV in fibrotic liver disease. J. Clin. Invest., in press.
Schuppan, D., T. Rühlmann, and E.G. Hahn (1985b) Radioimuunoassay for human type VI collagen and its application to tissue body fluids. Anal. Biochem. **149:** 238-247.
Schuppan, D., J. Becker, H. Boehm, and E.G. Hahn (1986) Immunofluorescent localisation of type-V collagen as a fibrillar component of the interstitial connective tissue of the human oral mucosa, artery and liver. Cell Tissue Res. **243:** 535-543.
Shiojiri, N. (1984) The origin of intrahepatic bile duct cells in the mouse. J. Embryol. Exp. Morph. **79:** 25-39.
Steiner, J.W., and J.S. Carruthers (1961) Studies on the fine structure of the terminal branches of the biliary tree. I. The morphology of normal bile canaliculi, bile preductules (ducts of Hering) and bile ductules. Am. J. Pathol. **38:** 639-661.
Steiner, J.W., and J.S. Carruthers (1962) Experimental extrahepatic biliary obstruction. Some aspects of the fine structural changes of bile ductules and pre-ductules (ducts of Hering). Am. J. Pathol. **40:** 253-270.
Uchida, T., and R.L. Peters (1983) The nature and origin of proliferated bile ductules in alcoholic liver disease. Am. J. Clin. Pathol. **79:** 326-333.
Vanstapel, M.-J., M. Butron, and V.J. Desmet (1984) Keratin immunohistochemistry in normal and pathological human liver. Hepatology **4:** 1037A.
Wick, G., S. Baudner, and F. Herzog (1976) *Immunfluoreszenz*. Medizinische Verlagsgesellschaft, Marburg.

EFFECTS OF AN INHIBITOR OF PROTEOGLYCAN BIOSYNTHESIS ON NEURON-INDUCED SCHWANN CELL PROLIFERATION AND BASAL LAMINA FORMATION BY SCHWANN CELLS

N. Ratner, C. Eldridge, R. Bunge, L. Glaser
Departments of Biological Chemistry and Anatomy and Neurobiology
Washington University School of Medicine
Division of Biology and Biomedical Sciences
St. Louis, MO
U.S.A.

It is now known that proteoglycans are components not only of extracellular matrices but are also found associated with the plasma membrane. Cell surface proteoglycans that have been described and characterized can be relatively small molecules (140kd, Carlstedt et al., 1983) or considerably larger (~400 kd, Woods et al., 1985) and contain mainly chondroitin, dermatan and/or heparan sulfate chains on a polypeptide chain which may be either peripherally or integrally associated with the cell membrane (Hook et al., 1984; c.f. Bernfield, this volume). These molecules contrast with extracellular matrix proteoglycans both in their location and size, since all matrix proteoglycans are extremely large (>500kd) (Caplan, 1984; Anderson and Fambrough, 1983; Hassell et al., 1985). We have studied the function of both cell surface and extracellular matrix proteoglycans in peripheral nerve tissues by the use of an inhibitor of proteoglycan biosynthesis, 4-methyl umbelliferyl-ß-D-xyloside.

These experiments have been performed using a tissue culture system in which primary cultures of purified embryonic rat neurons and Schwann cells can be separated and recombined. In the absence of serum in the culture medium Schwann cells adhere to neurites and proliferate in response to a signal on the neurite surface, eventually occupying all available neuritic outgrowth (Wood and Bunge, 1975; Moya et al., 1980); in the presence of a non-dialyzable serum factor and ascorbic acid these Schwann cells differentiate, form a basal lamina and elaborate myelin segments (Eldridge et al., 1985) (Fig. 1). The competitive inhibitor 4-methyl umbelliferyl-ß-D-xyloside has been added to neuron-Schwann cell preparations under both culture conditions. The xyloside acts as a competitive acceptor for the galactosyl transferase which normally adds two galactose

NATO ASI Series, Vol. H5
Mesenchymal-Epithelial Interactions in Neural Development
Edited by J. R. Wolff et al.

Interactions of Neurons and Schwann Cells *In Vitro*

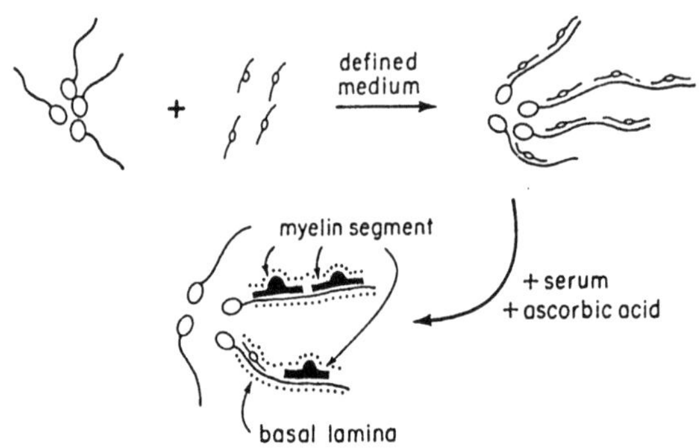

Figure 1: Dorsal root ganglion neurons from embryonic day 15 rat embryos are dissociated enzymatically and plated onto collagen substrata. In the presence of antimitotic agents, these cells survive, and grow neurites in the absence of non-neuronal cells. To these neuronal networks can be added, in the absence of antimitotic agents, embryonic or neonatal Schwann cells. In serum-free medium Schwann cells adhere to neurites and proliferate and within serveral weeks fill the neuritic outgrowth but do not assemble a basal lamina. If these cultures are shifted to medium containing serum and ascorbic acid, Schwann cells secrete and assemble a basal lamina and many Schwann cells myelinate neurites. Experimental procedures are described in Ratner et al. (1986a) and Eldridge et al. (1985).

residues onto a xylose of a nascent glycosaminoglycan chain prior to addition of repeating disaccharide units of the highly sulfated sugar chain (often 5-20kd), and therefore inhibits the formation of normal proteoglycans of the chondroitin sulfate (Robinson et al., 1975; Galligani et al., 1975) or haparan sulfate (Ledbetter and Hassel, 1986) type. In tissue culture, in this (Ratner et al., 1985) and other systems (Galligani et al., 1975; Schwartz et al., 1976; Schwartz, 1977; Lohmander et al., 1979; Thompson and Spooner, 1983; Kanwar et al., 1984), the addition of 4-methyl umbelliferyl-ß-D-xyloside results in the release of chondroitin sulfate chains which are not linked to protein. 4-methyl umbelliferyl-ß-D-xyloside at 1 mM inhibits neuron-induced Schwann cell proliferation by at least 80% (Ratner et al., 1985) as a result of its ability to interfere with the synthesis of one of more glycosaminoglycan containing proteins on the surface of neurons. Addition of the drug to fully nourished cultures of dorsal root ganglia neurons fully populated with Schwann cells results in the formation of a faulty basal lamina and few (and abnormal) myelin segments. It is important to note that protein synthesis is inhibited $<$ 10% in isolated Schwann cells in the presence of 1 mM 4-methyl umbelliferyl-ß-D-xyloside after 4 days, and that inactive isomers of 4-methyl umbelliferyl-ß-D-xyloside, as well as other ß-D sugars, have no effect on these neuron-Schwann cell interactions at the same concentration; neuron-Schwann cell interactions such as neuron-Schwann cell adhesion

(Ratner et al., 1985) and ensheathment of neurites by Schwann cells is normal (data not shown).

Tissue culture studies have established that neurons contain on their surface an intrinsic protein which is mitogenic for Schwann cells (Wood and Bunge, 1975; Salzer et al., 1980a, b; Ratner et al., 1985; Ratner et al., 1986b). This protein is probably also present in vivo since Aguayo and collaborators have shown that Schwann cell number in the rat can be regulated by axon number (Aguayo et al., 1976). We have shown that 4-methyl umbelliferyl-ß-D-xyloside treatment affects this neuronal protein, and not a Schwann cell surface protein, since prior drug treatment of neurons, but not Schwann cells, results in inhibition of Schwann cell proliferation (Ratner et al., 1985). We have also shown that digestion of the neuronal surface with the enzyme heparitinase, which degrades glycosaminoglycan chains of the heparan sulfate class, inhibits proliferation of Schwann cells, either in the whole cell system or when membranes are prepared from cells treated with heparitinase (Ratner et al., 1985).

To demonstrate that the neuronal mitogen is itself a heparan sulfate proteoglycan, it is necessary to isolate active mitogen from neurons and chemically characterize the isolated molecule. We have now achieved a part of this first goal, and can routinely purify mitogenic activity from cultures containing pure populations of dorsal root ganglion neurons. The purification of membrane fragments from primary cultures has been described previously (Ratner et al., 1985). All purification steps are designed to enrich for cell surface proteoglycans, which are presumed to 1) remain soluble after removal from membranes due to high negative charge on glycosaminoglycan chain(s), 2) pass through hydroxylapatite because of high negative charge density (Gorbonoff, 1984) and 3) adhere strongly to proteoglycan-binding proteins such as fibronectin and platelet factor IV (Lark and Culp, 1984). When proteins obtained from cultures of dorsal root ganglion neurons are purified by this protocol and iodinated using Bolton-Hunter reagent (which labels free amino groups) it can be seen that few polypeptides are present in the most purified fraction (Fig. 2). Approximately 2% of protein present in dorsal root ganglion membranes is recovered in the hydroxylapatite fraction following dialysis and centrifugation. Protein recovery has been determined either by Lowry protein assay (Markwell et al., 1978) after deoxycholate precipitation from tissue culture medium (Peterson, 1977) or by recovery of $[^{35}S]$-methionine labeled protein following 24h labeling of dorsal root ganglion neurons in 200 μCi/ml $[^{35}S]$-methionine as described (Ratner et al., 1985). It is more difficult to estimate protein in the most purified fractions because of the small amount of protein present. However, 1% of the $[^{35}S]$-methionine labeled protein in the hydroxylapatite fraction is present in the eluate from sepharose-fibronectin, indicating enrichment of 5,000-fold. Activity of fractions is measured by the ability of extract to stimulate incorporation of $[^{3}H]$-thymidine into DNA of normally quiescent, neonatal, rat Schwann

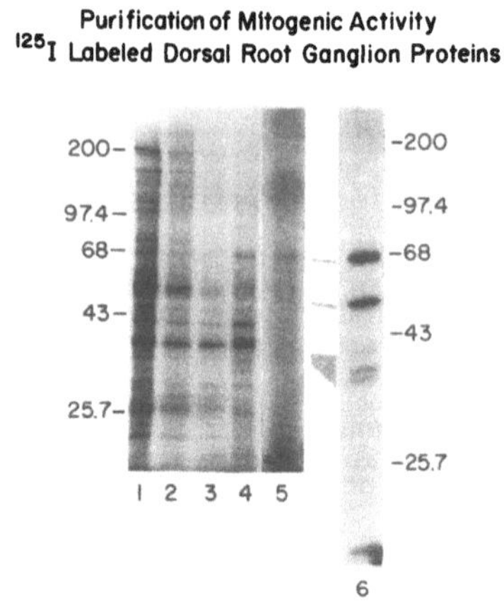

Figure 2: Purification of mitogen from dorsal root ganglion neurons. Membranes are prepared from dorsal root ganglion neurons grown in tissue culture (Ratner et al., 1985). 1-2 mg of membranes are extracted in 0.6 ml of detergent solution (30 mM octylglucoside, 0.5M NaCl, 10 mM EGTA, 10 mM Tris HCl pH 7.4, 0.16M sucrose, 1 mM PMSF, 17 mU/mL aparotinin). Following extraction for 30 min at 0°C samples are centrifuged 30 minutes at 100,000 x g. Solubilized protein is made 1mM in phosphate and passed over a hydroxylapatite column (0.2 ml hydroxylapatite/mg protein, based on OD_{280}). The flow through is collected, dialyzed against 1000 volumes of Dulbecco's minimal essential medium (DMEM) and insoluble material removed by centrifugation. The supernatant is diluted 3-fold with sterile water and applied to a 0.1 ml sepharose-fibronectine column (Lark and Culp, 1984), then eluted with DMEM. Alternatively, material is applied without dilution to Sepharose-platelet factor IV (Maresh et al., 1984) and eluted with 0.4M NaCl. (In this case, samples were dialyzed into DMEM prior to assay of mitogenicity.) Partially purified protein was dialyzed into phosphate buffered saline and iodinated using fresh Bolton Hunter reagent (New England Nuclear, Boston, MA) which had been dried, then resuspended in PBS and aliquoted into vials containing the various samples. Protein was iodinated for 1h at 0°C then the reaction stopped by adding 1/4 volume of 4x Laemmli sample buffer (Ratner et al., 1985). Samples were immediately electrophoresed on 10% acrylamide gels (Ratner et al., 1985) and bands detected by autoradiography on Kodak XAR film. Lane 1 = membranes, lane 2 = octylglucoside extract, lane 3 = flow through from hydroxylapatite, lane 4 = flow through after dialysis and centrifugation, lane 5 = eluate from platelet factor IV column, lane 6 = eluate from Sepharose-fibronectin column.

Hatched lines are drawn between bands of common mobility in the two most purified fractions, which are from different polyacrylamide gels.

cells (Brockes et al., 1979) as assayed by autoradiography (Ratner et al., 1985). Up to 30% recovery of activity present in solubilized extracts from neurons is recovered in the most highly purified fraction of mitogen eluted from sepharose-fibronectin. It has not yet been determined whether the proteins in this purified fraction are present in a complex, are related through proteolysis, or whether only one of these bands on a polyacrylamide gel is related to mitogenic activity. All these data are consistent with the notion that a proteoglycan is involved in neuron-induced Schwann cell proliferation, but the rigorous demonstration of a heparitinase-sensitive sulfated glycosaminoglycan in at least one of these molecules which corresponds with mitogenic activity, is still lacking.

Until now, we have been limited by the amount of membranes present in primary cultures of neurons, even after growth of several million neurons for over one month in the culture dish. Therefore, a new approach has been tried. Reasoning that neonatal rat brain contains few oligodendrocytes (the myelin-producing cell of the CNS) and relatively few astrocytes, and should therefore be a rich source of neuronal membrane, membrane fractions have been prepared from whole postnatal day 1-3 rat brains. It has previously been shown that neurite growth from embryonic CNS neurons from spinal cord and retina is able to stimulate Schwann cell proliferation in vitro (Ratner et al., 1986b) as are membrane fragments from adult white matter (DeVries et al., 1982, 1983; Cassel et al., 1982; Sobue et al., 1983; Meodor-Woodruff et al., 1984; Yoshimo et al., 1984; Sobue and Pleasure, 1985). We have found that mitogenic activity from neonatal rat brain copurifies with dorsal root ganglion derived activity using the purification procedure described above. In addition to co-purification, both mitogens are active in the absence of serum, which is a characteristic to date only of the authentic dorsal root ganglion neuronal membrane associated mitogen (Cassel et al., 1982; Sobue et al., 1983; Ratner et al., 1984). Chemical characterization of the mitogen should now be feasible.

If neuron-Schwann cell cultures are allowed to become fully populated with Schwann cells in defined medium and then are shifted to a medium containing serum and ascorbic acid (myelinating medium), Schwann cells assemble a basal lamina around each axon-Schwann cell unit. This basal lamina contains laminin, type IV collagen and heparan sulfate proteoglycan (Eldridge et al., 1985; Bunge et al., 1986). Within two weeks of adding myelinating medium to a populated neuron-Schwann cell culture, many hundreds of myelin segments form. When 1 mM 4-methyl umbelliferyl-ß-D-xyloside was added to such a fully populated neuron-Schwann cell culture at the time of the shift to myelinating medium, little myelin formed (Fig. 3). The morphology of the Schwann cells in these cultures was also altered. Instead of the usual elongated spindle shape, aligned with the axis of the axon, Schwann cells were flat and more stellate (Fig. 4). In addition, these cells appeared to be fewer in number than cells in control cultures, which could indicate the ongoing proliferation of Schwann cells in normal cultures or a requirement for Schwann cells to undergo one or several rounds of cell division prior to differentiating. This has not been investigated further. Fig. 4 also shows one of the rare typical myelin segment in a 4-methyl umbelliferyl-ß-D-xyloside treated culture. Myelin segments in drug treated cultures were measured using an eyepiece micrometer and were 70% the length of control segments (Control = 95 ± 23.2 μM, +4-methyl umbelliferyl-ß-D-xyloside = 66.7 ± 26.1 μM N = 100). Such abnormal segments often extended on only one side of the Schwann cell nucleus; normally the myelin-forming Schwann cell nucleus is located at the midpoint between the bordering nodes of Ranvier.

A limited examination by electron microscopy of xyloside treated neuron-Schwann cell cultures did not reveal manifest differences in basal lamina formation from control cultures, and extensive ensheathment of neurites by Schwann cells was observed. Basal lamina is apparently required for myelin formation by Schwann cells (Bunge et al., 1986).

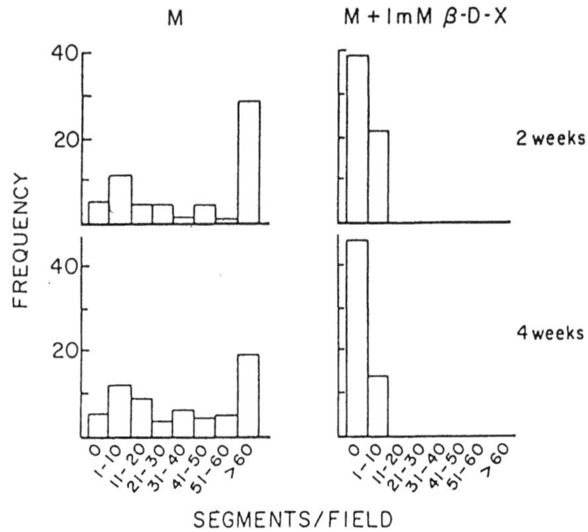

Figure 3: Myelin formation is inhibited by 4-methyl umbelliferyl-ß-D-xyloside. Cultures of neurons fully populated with Schwann cells were shifted from serum-free medium to myelinating medium for 4 weeks in the presence or absence of 1mM 4-methyl umbelliferyl-ß-D-xyloside (see Fig. 1). Fresh medium was added from frozen aliquots 3 times every week. At the end of the experiment cultures were fixed and stained with Sudan Black (Ratner et al., 1986a) and the number of myelin segments in 1.25 mm^2 fields counted as described (Eldridge et al., in preparation; Ratner et al., 1986a). Frequency is the number of fields in which a given number of myelin segments was observed.

A major component of Schwann cell basal lamina is a well characterized heparan sulfate proteoglycan (Hassel et al., 1985; Eldridge et al., 1986) whose synthesis should be inhibited by 4-methyl umbelliferyl-ß-D-xyloside. Therefore we further analyzed the basal lamina formed in the presence of the drug by labeling cultures with antibodies against the major basal lamina components: laminin, type IV collagen and heparan sulfate proteoglycan. Detection of antibody using fluorescent secondary antibodies showed that control cultures contained the characteristic pattern of basal lamina staining, in which tubes of basal lamina, homogenously stained with antibodies against all three proteins, outlined neuron-Schwann cell units. In the periphery of control cultures Schwann cells were present in a monolayer and resembled cells from drug-treated cultures. In these regions of the cultures laminin, type IV collagen and heparan sulfate proteoglycan were present diffusely. It is this irregular "immature" deposition of all three basal lamina components which was observed in 4-methyl umbelliferyl-ß-D-xyloside treated cultures, even after as long as 4 weeks in the presence of myelinating medium (Fig. 5).

When neurons and Schwann cells are cultured in certain serum-free media, they fail to assemble a basal lamina (Moya et al., 1980; Carey et al., 1983; Bunge et al., 1986). Punctate cell surface staining of Schwann cells with antibodies against laminin and heparan sulfate proteoglycan is observed; no staining with anti type IV collagen is seen

(Eldridge et al. in preparation). The basal lamina components interact in vitro in a specific fashion and it has been suggested that such interactions are involved in basal lamina deposition (Kleinman et al., 1983; Fujiwara et al., 1984; Yurchenko, 1986). Therefore, the assembly of any basal lamina components under conditions which should inhibit the production of a major basal lamina component, heparan sulfate proteoglycan, was surprising. Two possibilities were considered to explain this apparent discrepancy: 1) the anti-heparan sulfate proteoglycan antibody (Eldridge et al., 1986) was staining core protein in the matrix, or 2) sufficient leakage of intact proteoglycan was occuring to, in time, generate assembly of apparently normal basal lamina.

To investigate the molecular nature of the antigen stained with the anti-heparan sulfate proteoglycan antibody, cultures were labeled with either [^{3}H]-leucine or [^{35}S]-SO_4. Samples from the culture medium and the basal lamina (4M guanidine HCl extract) were analyzed by immunoprecipitation and SDS polyacrylamide gel electrophoresis followed by fluorography (Fig. 6). Anti-laminin antisera precipitated from 4-methyl umbelliferyl-ß-D-xyloside treated cultures a [^{3}H]-leucine labeled glycoprotein of mobility identical to intact laminin from control cultures (lanes 5 and 6). The amount of laminin in the 4-methyl umbelliferyl-ß-D-xyloside treated culture was lower than in control cultures, possibly due to the lower Schwann cell number in treated cultures. Total radioactivity in [^{3}H]-leucine labeled protein in 4-methyl umbelliferyl-ß-D-xyloside treated cultures was 50% of control, indicating that the total amount of basal lamina in treated cultures was half that of controls.

Secretion of anti-heparan sulfate proteoglycan immunoreactive material into the culture medium (lanes 10, 11) was severely decreased after xyloside treatment, yet a small amount of anti-heparan sulfate proteoglycan immunoreactive material was present in the basal lamina of xyloside treated cultures (lanes 7, 8, 9). This material differed qualitatively from intact heparan sulfate proteoglycan. While some apparently intact proteoglycan was present (lane 9), an equal amount comigrated with naked core protein (detected by heparinase digestion of the intact molecule, Eldridge et al., 1986). Apparently the small amount of relatively normal proteoglycan synthesized in the presence of 4-methyl umbelliferyl-ß-D-xyloside was assembled into the basal lamina as was the naked core protein synthesized in this condition. The presence of a small amount of heparan sulfate proteoglycan appears sufficient to organize the assembly of considerable amounts of basal lamina; that other basal lamina proteins are normal is implied by the immunoprecipitation of normal laminin from treated cultures and from the normal pattern of other basal lamina proteins separated on polyacrylamide gels (lanes 3 and 4). Kleinman et al. (1983) showed that precipitation of laminin is stimulated 9-fold by heparan sulfate proteoglycan when laminin is present in a 3-10 fold excess over proteoglycan. Since basal lamina formation is thought to be required for normal Schwann cell differentiation, we

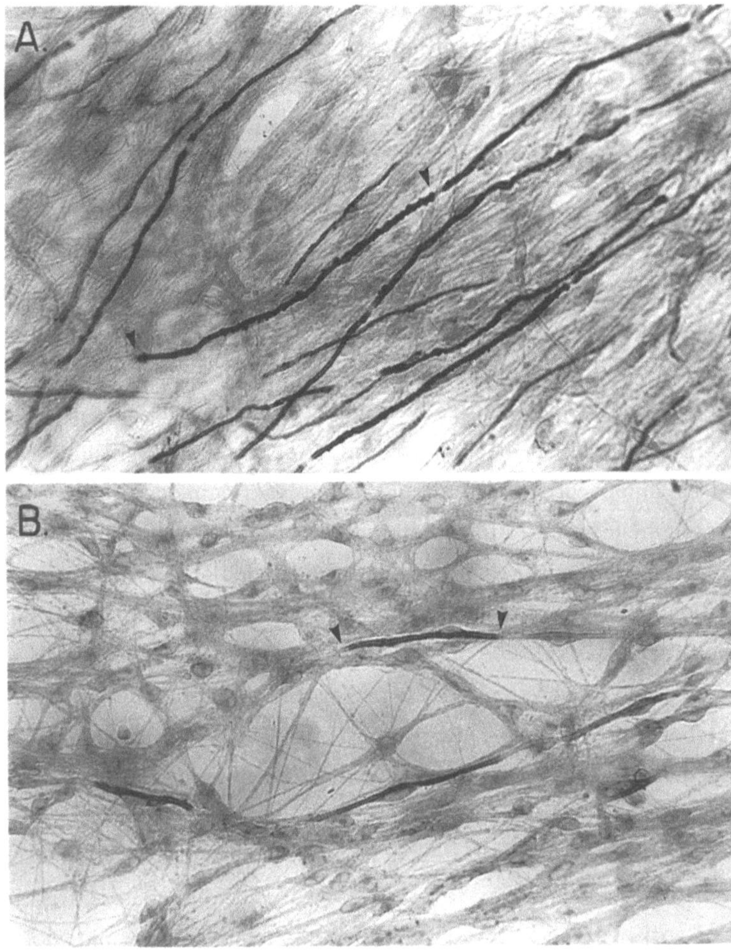

Figure 4: Morphology of Schwann cells in 4-methyl umbelliferyl-ß-D-xyloside-treated neuron-Schwann cell cultures. Cultures of neurons and Schwann cells grown, fixed, Sudan Black stained, and photographed using phase contrast optics on a Zeiss Universal Research microscope. Photographs were taken at an equal distance from the neuronal somata at the center of the culture. A = control, B = xyloside-treated. Arrows indicate nodes between myelin segments. X500.

Figure 5 (facing page): Assembly of basal lamina components in 4-methyl umbelliferyl-ß-D-xyloside treated neuron-Schwann cell co-cultures.

Fully populated neuron-Schwann cell co-cultures were switched to myelinating medium for 2 weeks in the presence or absence of 1 mM 4-methyl umbelliferyl-ß-D-xyloside. Living cultures were either double stained with rabbit antisera to mouse laminin (Timpl et al., 1979) and mouse monoclonal anti-heparan sulfate proteoglycan ascites fluid (Eldridge et al., 1986) or stained with antibody to the NCl domain of type IV collagen (Weber et al., 1984). These methods have been described (Ratner et al., 1986). a, d, g, j, k, l are phase contrast pictures; b, e, h, are stained with anti-laminin and photographed

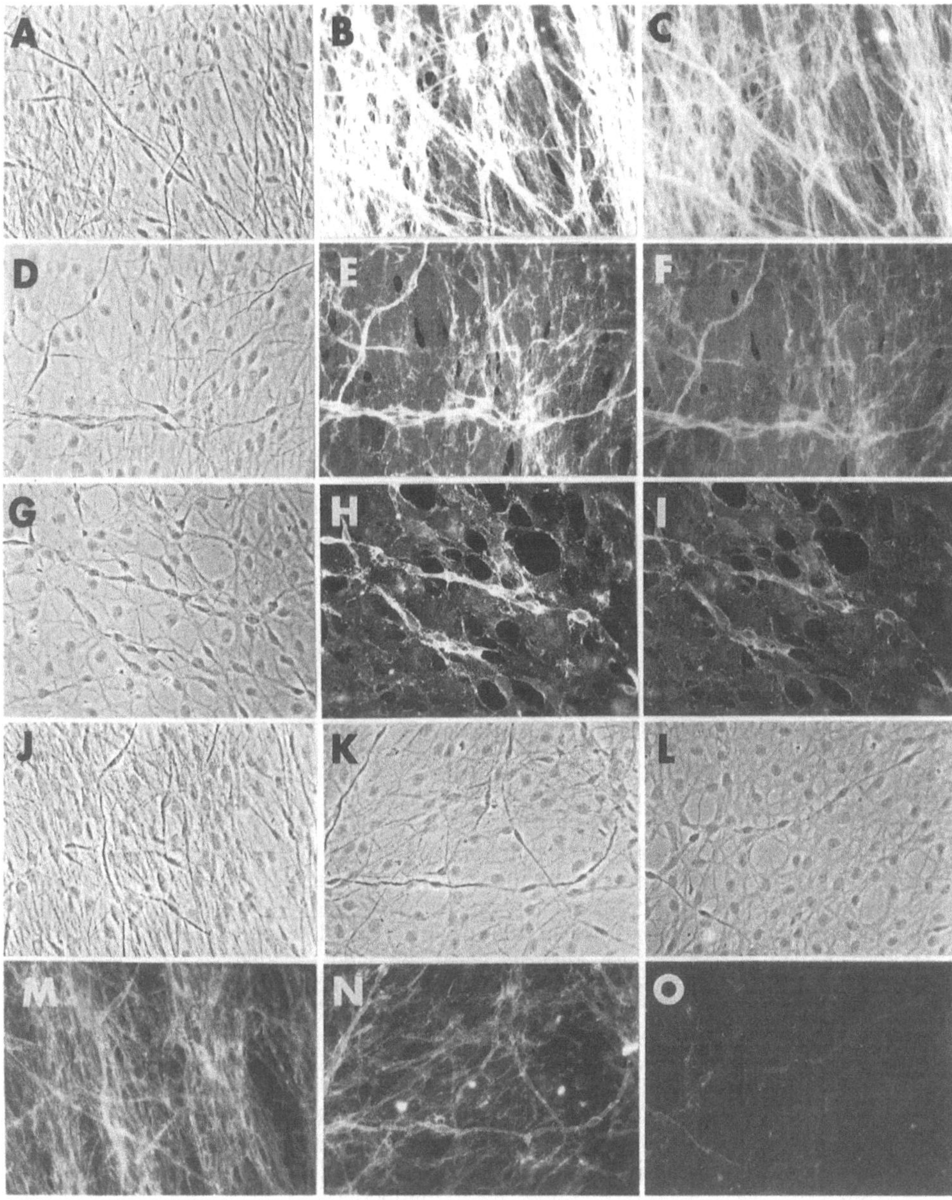

with fluorescein optics; c, f, i are stained with anti-heparan sulfate proteoglycan (Eldridge et al., 1986) and photographed with rhodamine optics; m, n, o are stained with anti-NCl domain of type IV collagen and photographed with fluorescein optics.

All fluorescence pictures for a given antibody were exposed for the same time and all prints were made using identical exposure and paper. a-c, j, m were taken in an area of a control culture close to the center of the culture showing well-assembled basal lamina. d-f, k, n were taken in the periphery of a control culture. g-i, l, o are representative of basal lamina throughout 4-methyl umbelliferyl-ß-d-xyloside treated cultures (X178).

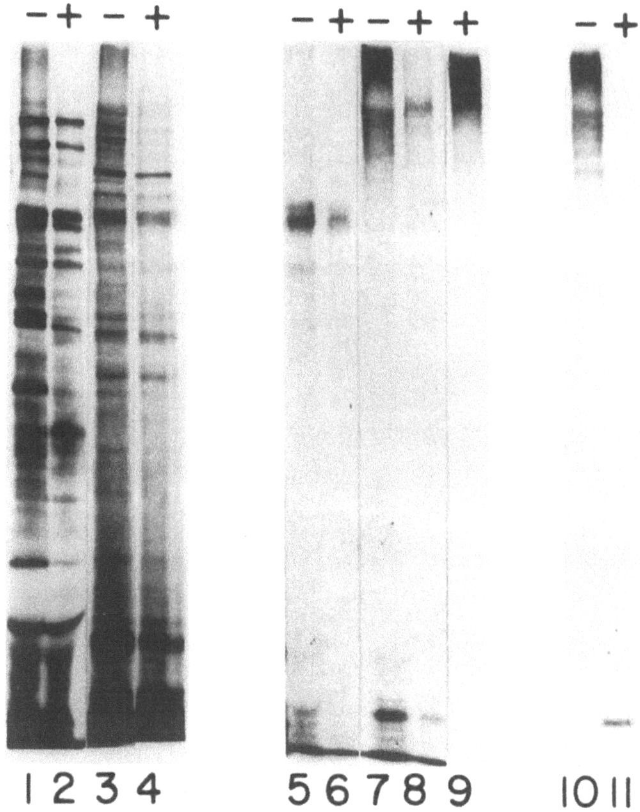

Figure 6: Immunoprecipitation of basal lamina components. Neuron-Schwann cell co-cultures were switched from serum-free medium to myelinating medium for a total of 1 week in the presence or absence of 1mM 4-methyl umbelliferyl-ß-D-xyloside. During the last 48h cultures were labeled either in 200 µCi/ml [^{3}H]-leucine or [^{35}S]-SO_4. At the end of the labeling period culture aliquots were removed for analysis of medium, then cultures were washed, cells were lysed, cell debris washed away, and basal lamina extracted in 4M guanidine-HCl 24h at 0°C. Extracts were dialyzed into immunoprecipitation buffer and precipitated. All steps were as described by Eldredge et al. (1986). Proteins were electrophoresed on 6% SDS polyacrylamide gels and fluorographed. Lane 1 = 12.5 µl of [^{3}H]-leucine labeled medium, lane 2 = 12.5 µl [^{3}H]-leucine labeled medium from a culture treated with 4-methyl umbelliferyl-ß-D-xyloside. Lane 3 = 50 µl [^{3}H]-leucine labeled basal lamina extract, underexposed (2X) relative to lane 4 to show co-migration of many basal lamina proteins. Lane 4 = 50 [^{3}H]-leucine labeled basal lamina extract from drug treated culture. Lane 5-8 = immunoprecipitation of 50 µl of [^{3}H]-leucine basal lamina extract. Lane 5 = control, anti-laminin immunoprecipitate. Lane 6 = 4-methyl umbelliferyl-ß-D-xyloside, anti-laminin immunoprecipitate. Lane 7 = control, anti-heparan sulfate proteoglycan immunoprecipitate. Lane 8 = 4-methyl umbelliferyl-ß-D-xyloside anti-heparan sulfate proteoglycan immunoprecipitate. Lane 9 = same as lane 8 but precipitated from [^{35}S]-SO_4 labeled basal lamina extract. Lane 10 = control medium, 50 µl, precipitated using anti-heparan sulfate proteoglycan. Lane 11 = 4-methyl umbelliferyl-ß-D-xyloside culture medium, precipitated as lane 9. Arrowhead points to the ß chain of laminin. Arrow points to the core protein of the heparan sulfate proteoglycan. + indicates treatment with 4-methyl umbelliferyl-ß-D-xyloside.

suggest it is the deficiency (in quantity and molecular composition of the proteoglycan) of the basal lamina which inhibits the formation of normal myelin segments. We cannot rule out the possibility that an as yet undefined proteoglycan is required for myelin formation.

Two proteoglycans, one a small protein on the neuronal surface and one a large component of the basal lamina produced by Schwann cells, are required for normal interaction of neurons and Schwann cells. In the peripheral nerve a small cell surface proteoglycan appears to play a functional role in the stimulation of Schwann cell proliferation, while the large matrix proteoglycan plays a role in establishing and maintaining a normal basal lamina required for Schwann cell differentiation.

Acknowledgments. We thank Laura Tynan for assistance with culture work, Peggy Bates for assistance with electron microscopy, and Susan Mantia for typing the manuscript. This work was supported by NIH grants NS19923 (to L. Glaser) and NS09923 (to R. Bunge). Nancy Ratner is a Life Sciences Research Foundation Fellow of the Burroughs-Wellcome Fund. Charles Eldridge was supported by NIH grant NS09923.

References

Aguayo, A.J., J.M. Peyronard, L.C. Terry, J.S. Romine, and G.M. Bray (1976) Neonatal neuronal loss in rat superior cervical ganglia: Retrograde effects on developing preganglionic axons and Schwann cells. J. Neurocytol. **5**: 137-155.

Anderson, M.J., and D.M. Fambrough (1983) Aggregates of acetylcholine receptors are associated with plaques of a basal lamina heparan sulfate proteoglycan on the surface of skeletal muscle fibers. J. Cell Biol. **97**: 1396-1411.

Brockes, M.P., K.L. Fields, and M.C. Raff (1979) Studies on cultured rat Schwann cell. I. Establishment of purified populations from cultures of peripheral nerve. Brain Res. **165**: 105-118.

Bunge, R.P., Bunge, M.B., and C.E. Eldridge (1986) Linkage between axonal ensheathment and basal lamina production by Schwann cells. Ann. Rev. Neurosci. **9**: 305-328.

Caplan, A.I. (1984) Cartilage. Sci. American (Oct.): 84-94.

Carey, D.J., C.F. Eldridge, C.J. Cornbrooks, R. Timpl, and R.P. Bunge (1983) Biosynthesis of type IV collagen by cultured rat Schwann cells. J. Cell Biol. **91**: 473-479.

Carlstedt, I., L. Coster, A. Malmstrom and L.-A. Fransson (1983) Proteoheparan sulfate from human skin fibroblasts. Isolation and structural characterization. J. Biol. Chem. **258**: 11629-11635.

Cassel, D., P.M. Wood, R.P. Bunge and L. Glaser (1982) Mitogenicity of brain axolemma membranes and soluble factors for dorsal root ganglia Schwann cells. J. Cell. Biochem. **18**: 433-445.

DeVries, G.H., J.L. Salzer, and R.P. Bunge (1982) Axolemma-enriched fractions isolated from PNS and CNS are mitogenic for cultured Schwann cells. Dev. Brain. Res. **3**: 295-299.

DeVries, G.H., L.N. Minier, and B.L. Lewis (1983) Further studies on the mitogenic response of cultured Schwann cells to rat CNS axolemma-enriched fractions. Dev. Brain Res. **9**: 87-93.

Eldridge, C.F., M.B. Bunge, and R.P. Bunge (1985) Serum ascorbic acid regulates myelin formation and basal lamina assembly by Schwann cells in vitro. Soc. Neurosci. Abstr. **11**: 986.

Eldridge, C.F., A.Y. Chiu, R.P. Bunge, J.R. Sanes, and C.J. Cornbrooks (1986) Basal lamina-associated heparan sulfate proteoglycan in the rat peripheral nervous system: Characterization and localization using monoclonal antibodies. J. Neuro-

cytol, in press.
Fujiwara, S., H. Wiedemann, R. Timpl, A. Lustig, and J. Engel (1984) Structure and interactions of heparan sulfate proteoglycans from a mouse tumor basement membrane. Eur. J. Biochem. **143:** 145-157.
Galligani, L., J. Hopwood, N.B. Schwartz, and A. Dorfman (1975) Stimulation of synthesis of free chondroitin sulfate chains by ß-D-xyloside in cultured cells. J. Biol. Chem. **250:** 5400-5406.
Gorbunoff, M.J. (1984) The interactions of proteins with hydroxylapatite. Analyt. Biochem. **136:** 425-432.
Hassell, J.R., W.C. Leyshon, S.R. Ledbetter, B. Tyree, S. Suzuki, M. Kato, K. Kimata, and H.K. Kleinman (1985) Isolation of two forms of basement membrane proteoglycans. J. Biol. Chem. **260:** 8098-8105.
Hook, M., L. Kjellen, S. Johanson, and J. Robinson (1984) Cell-surface glycosaminoglycans. Ann. Rev. Biochem. **53:** 847-869.
Kanwar, Y.S., V.C. Hascall, M.L. Jakubowski, and J.T. Gibbons (1984) Effect of ß-D-xyloside on the glomerular proteoglycans. I. Biochemical studies. J. Cell Biol. **99:** 715-722.
Kleinman, H.K., M.L. McGarvey, J.R. Hassell, and G.R. Martin (1983) Formation of a supramolecular complex is involved in the reconstruction of basement membrane components. Biochemistry **22:** 4969-4974.
Lark, M.W., and L.A. Culp (1984) Multiple classes of heparan sulfate proteoglycans from fibroblast substratum adhesion sites. J. Biol. Chem. **259:** 6773-6782.
Ledbetter, S.R., and J.R. Hassell (1986) ß-D-xyloside-mediated alteration in the synthesis of basement membrane proteoglycan. Arch. Biochem. Biophys. **246,** in press.
Lohmander, L.S., and V.C. Hascall (1979) Effects of 4-methyl umbelliferyl-ß-D-xyloside on chondrogenesis and proteoglycan synthesis in chick limb bud mesenchymal cell cultures. J. Biol. Chem. **254:** 10551-10561.
Maresh, G.A., E.A.G. Chernoff, and L.A. Culp (1984) Heparan sulfate proteoglycans of human neuroblastoma cells: Affinity fractionation on columns of platelet factor-4. Arch. Biochem. Biophys. **233:** 428-437.
Markwell, M.A., S.M. Aaas, L. Bieber and N.E. Tolbert (1978) A modification of the Lowry procedure to simplify protein determination in membrane and lipoprotein samples. Anal. Biochem. **82:** 206-210.
Meador-Woodruff, J.H., B.L. Lewis, and G.H. DeVries (1984) Cyclic AMP and calcium, as potential mediators of stimulation of cultured Schwann cell proliferation by axolemma-enriched and myelin-enriched membrane fractions. Biochem. Biophys. Res. Comm. **122:** 373-380.
Moya, F., M.B. Bunge, and R.P. Bunge (1980) Schwann cells proliferate but fail to differentiate in defined medium. Proc. Nat. Acad. Sci., (USA) **77:** 6902-6906.
Peterson, G.L. (1977) A simplification of the protein assay method of Lowry et al. which is more generally applicable. Analyt. Biochem. **83:** 346-356.
Ratner. N.L., L. Glaser, and R.P. Bunge (1984) PC12 cells as a source of neurite-derived cell surface mitogen, which stimulates Schwann cell division. J. Cell Biol. **98:** 1150-1155.
Ratner, N., R.P. Bunge, and L. Glaser (1985) A neuronal cell surface heparan sulfate proteoglycan is required for dorsal root ganglion neuron stimulation of Schwann cell proliferation. J. Cell Biol. **101:** 744-754.
Ratner, N., A. Elbein, S. Porter, M.B. Bunge, R.P. Bunge, and L. Glaser (1986a) Asparagine-linked sugars are not required for many neuron-neuron and neuron-Schwann cell interactions. J. Cell Biol., in press.
Ratner, N., P. Wood, L. Glaser, and R.P. Bunge (1986b) Further characterization of the neuronal cell surface protein mitogenic for Schwann cells. In *Neuron-Glial Communication in Development and Regeneration,* H. Althaus and W. Seifert, eds., Academic Press, London, in press.
Robinson, H.C., M.J. Brett, Tralaggen, D.A. Lowther, and M. Okayama (1975) The effect of d-xylose, ß-D-xylosides and ß-D-galactosides on chondroitin sulphate biosynthe-

sis in embryonic chicken cartilage. Biochem. J. **148:** 25-34.
Salzer, J.L., R.P. Bunge, and L. Glaser (1980a) Studies of Schwann cell proliferataion. III. Evidence for the surface localization of the neurite mitogen. J. Cell Biol. **84:** 767-778.
Salzer, J.L., A.K. Williams, L. Glaser, and R.P. Bunge (1980b) Studies of Schwann cell proliferation. II. Characterization of the stimulation and specificity of the response to a neurite membrane fraction. J. Cell Biol. **184:** 753-766.
Schwartz, N.B. (1977) Regulation of chondroitin synthesis. Effect of ß-xylosides on synthesis of chondroitin sulfate proteoglycan, chondroitin sulfate chains, and core protein. J. Biol. Chem. **252:** 6316-6321.
Schwartz, N.B., P.-L. Ho, and A. Dorfman (1976) Effects of ß-xylosides on synthesis of cartilage-specific proteoglycan in chondrocyte cultures. Biochem. Biophys. Res. Comm. **71:** 851-856.
Sobue, B., and D. Pleasure (1985) Adhesion of axolemmal fragments to Schwann cells: A signal- and target-specific process closely linked to axolemmal induction of Schwann cell mitosis. J. Neurosci. **5:** 379-387.
Sobue, G., B. Kreider, A. Asbury and D. Pleasure (1983) Specific and potent mitogenic effect of axolemmal fraction on Schwann cells from rat sciatic nerves in serum-containing and defined media. Brain Res. **280:** 263-275.
Timpl, R., R. Heilwig, P.G. Robey, S.I. Rennard, J.-M. Foidart, and G.M. Martin (1979) Laminin - glycoprotein from basement membranes. J. Biol. Chem. **254:** 9933-9937.
Thompson, H.A., and B.S. Spooner (1983) Proteoglycan and glycosaminoglycan synthesis in embryonic mouse salivary glands: Effects of ß-D-xyloside, and inhibitor of branching morphogenesis. J. Cell Biol. **96:** 1443-1450.
Weber, S.J., J. Engel, H. Wiedemann, R.W. Glanville, and R. Timpl (1984) Subunit structure and assembly of the globular domain of basement-membrane collagen type-IV. Eur. J. Biochem. **139:** 401-410.
Wood, P.M., and R.P. Bunge (1975) Evidence that sensory axons are mitogenic for Schwann cells. Nature, **256:** 662-664.
Woods, A., J.R. Couchman, and M. Hook (1985) Heparan sulfate proteoglycans of rat embryo fibroblasts. A hydrophobic form may link cytoskeleton and matrix components. J. Biol. Chem. **260:** 10872-10879.
Yanagashita, M., and V.C. Hascall (1984) Proeoglycans synthesized by rat ovarian granulosa cells in culture. Isolation, fractionation, and characterization of proteoglycans associated with the cell layer. J. Biol. Chem. **259:** 10260-10269.
Yoshino, J.E., M.P. Dinneen, B.L. Lewis, J.H. Meador-Woodruff, and G.H. DeVries (1984) Differential proliferative responses of cultured Schwann cells to axolemma- and myelin-enriched fractions. I. Biochemical studies. J. Cell Biol. **99:** 2309-2313.
Yurchenko, P.D. (1986) Basement membrane collagen, laminin and heparan sulfate proteoglycan self-assemble into a complex heteropolymer using domain-specific interactions. Fed. Proc. **45:** 1815a.

MECHANISMS OF MORPHOGENETIC SPECIFICATION IN SKULL DEVELOPMENT

Peter Thorogood
Department of Biology
Southampton University
Southampton
U.K.

The nature of the developmental transition from the linear complexity of the genotype, encoded in the sequence of nucleotides along the DNA molecule, to the three-dimensional complexity of the phenotype remains a fundamental challenge to contemporary developmental biologists. In this paper, I wish to consider one particular three-dimensional pattern or form - that of the vertebrate skull, and to address the question of how the embryo constructs such a form. In other words, what are the 'assembly rules' underlying the formation of the skull during development? Answering this question is clearly an awesome task if one tries to analyse development in terms of the finished structure. Instead, of course, we can rationalise the form into its component, blastemal parts as seen in the embryo. Thus, the developing skull can be rationalised into 'neurocranium' and 'viscerocranium' which fulfill, respectively, the traditionally observed functions of protection (of the brain and sense organs) and ingestion/respiration. Both components contain cartilaginous and 'membranous' (membrane bone) elements and it is the formation of the cartilaginous neurocranium which I wish to discuss. This structure has a basic pattern which is common to all vertebrates from living agnathans to Man. There are two paired elements - the trabeculae and the parachordals - which fuse together to form the embryonic cranial floor, and three pairs of capsular cartilages - olfactory, optic and otic - which surround and support the organs of smell, sight and hearing/balace respectively (de Beer, 1937).

This primordium has a fundamental role in that it functions as a structural template around which the bony parts of the skull subsequently form. Thus, change in the form of the cartilaginous neurocranium will presage changes in skull form, be they congenitally abnormal or of phylogenetic significance. The mesenchyme from which the cartilage differentiates arises from two sources - from primary mesoderm and from the neural crest (NC), mesenchyme from the latter sometimes being termed 'ectomesenchyme'. The relative contributions from these two sources to the tissues of the skull has

NATO ASI Series, Vol. H5
Mesenchymal-Epithelial Interactions in Neural Development
Edited by J. R. Wolff et al.

been defined and catalogued elsewhere (e.g., amphibians - Horstadius, 1950; e.g., birds - Noden, 1978; Le Lievre, 1978), and it has been found that although there are broad similarities amongst the vertebrates studied, proportionate contributions may vary from one vertebrate order to another and may even vary between related species. In fact this variation can render homologising the component parts of the skull problematic, as has recently been shown for the reptilian orbitosphenoid (Bellairs and Gans, 1983) and the avian columella (Noden, 1983). For instance, in the latter case various experimental manipulations can produce columellae which are morphologically normal, but whose composition, in terms of ectomesenchyme and mesodermally-derived mesenchyme, can vary dramatically. Clearly any explanation of skull morphogenesis has to take account of such variation in composition and derivation of what are, essentially, homologous skeletal elements.

There are at least two traditional ways of tackling the problems of morphogenesis and pattern formation. The first is to regard the system under study as a 'morphogenetic field' which can be experimentally perturbed by any one of a variety of manipulations (e.g., ablation, grafting, etc.); the subsequent effects on form and pattern are then interpreted in terms of the (often largely unknown) cellular response to the perturbation. The second is to identify those genetic/molecular factors which underlie behavioural and differentiative changes in cells and then define the distribution and activity of these factors in developmental time and space. The two approaches, respectively, perhaps typify the 'top-down' and 'bottom-up' approaches to biological problems, enunciated by Sydney Brenner (Brenner, 1975). The work described in this brief review falls clearly into the second of these two categories; parts A) and B) briefly review our current understanding of the developmental events which elicit a chondrogenic response from head mesenchyme and parts C) and D) describe a morphogenetic model which specifies where and when such events take place.

A) The Role of Tissue Interactions

It is now established that epithelio-mesenchymal interactions, occurring during very early craniofacial development, promote skeletongenic differentiation within head mesenchyme (e.g., Drews et al., 1972; Epperlein and Lehmann, 1975; Tyler and Hall, 1977; Bee and Thorogood, 1980). Thus, premigratory mesencephalic crest from avial embryos will form a mesenchyme readily in culture, but that mesenchyme will display no differentiation into cartilage or bone, or even collagen synthesis, in isolation. However, premigratory crest, cultured in combination with an appropriately-staged epithelium taken from a point along the migration pathway for crest cells at that particular axial level, will form cartilage and in some cases, bone (Bee and Thorogood, 1980). Such recombination experiments do not only reveal interactions to be an important component of skull formation but also demonstrate that the in vivo event of migration is not a prerequisite for normal differentiation. However, as we shall see later, it is a critical factor in morphogenesis and pattern formation.

I shall refer to such interactions as 'promoting' chondrogenesis or skeletogenic differentiation, because we know so little of the state of commitment/determination of the cells concerned. Use of the more traditional term 'inducing' cannot be justified as it implies an understanding of the commitment state, potency and lineage of these cells, which in reality simply does not exist yet. If considered in terms of 'instructive' or 'permissive' interactions (Wessells, 1977) then the available evidence suggests that such interactions are permissive in nature (discussed by Hall, 1982), although the question has not yet been analysed rigorously in an experimental sense and indeed, the frames of reference which we now use may turn out to be inadequate.

B) The Mechanism of Tissue Interaction

In order to categorize such chondrogenesis-promoting interactions as either contact-, diffusion- or matrix-mediated types (Saxen, 1980), one particular system - the formation of the optic capsule (the 'scleral' cartilage in birds) - was selected for further study. This structure is generated by an interaction between the pigmented retina of the newly-formed optic cup, and the periocular mesenchyme derived from the mesencephalic NC. Standard transfilter cultures of pigmented retina and premigratory crest (as the progenitor tissue of the mesenchyme) reveals that the crest component will differentiate chondrogenically but only under certain conditions. Such transfilter interactions will occur only if the Nuclepore filters used are of a port size large enough to allow crest cell processes to traverse the filter (see Figs. 1 and 2). Smaller port size filters, through which diffusion is not impeded but which cell processes cannot penetrate, prevent the interaction taking place although, in all respects other than differentiation, the cultured tissues grow perfectly normally (Smith and Thorogood, 1983). The prevention of the interaction by filters which allow free diffusion of large molecules indicates that pigmented retina and neural crest cells or their progeny interact either by direct plasmalemmal contact between the cells of the two populations or by contact between the cells of one population and an extracellular matrix (ECM) produced by the other, rather than by diffusible signals. Examination of the ultrastructure of the tissue interface during the interaction allows one to distinguish between these two possibilities, assuming that direct plasmalemmal contact would be manifest by crest/mesenchymal cell processes extending through discontinuities in the epithelial basal lamina to contact the cells of the pigmented retina. Evaluation of the in vitro tissue interface in transfilter cultures is complicated by the removal of the basal lamina during initial (enzymatic) tissue isolation, and the variation in the rate at which this matrix is regenerated by the epithelium.

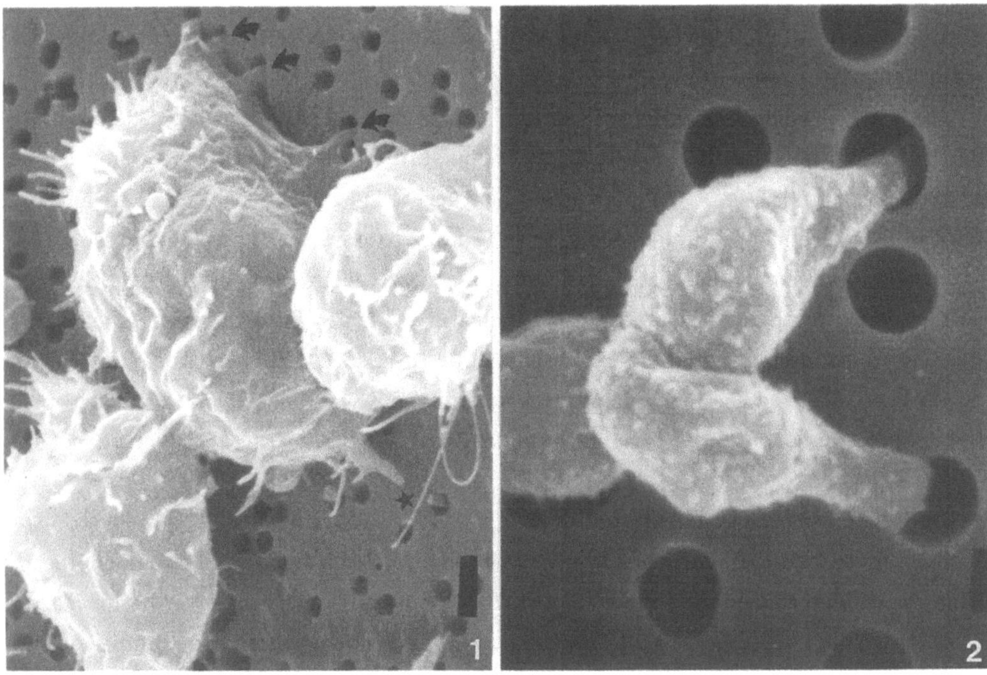

Figure 1: Scanning electron micrograph of chick mesencephalic NC cells on 0.8 µm Nuclepore filter, after 24h in culture. Note the numerous filopodial cell processes (arrows) extending into the pores of the filter (asterisk denotes a slender process artefactually broken during tissue processing). Bar equals 2 µm (micrograph kindly supplied by Dr. Linda Thomas).

Figure 2: Scanning electron micrograph of underside of 0.8 µm Nuclepore filter, similar to that shown in Fig. 1, and displaying the distal tips of neural crest cell processes extended through the pores. Bar equals 0.6 µm (micrograph kindly suppliedd by Dr. Linda Thomas).

However, a clear picture emerges from studying the in vivo interface. Examination of large areas of the pigmented retina/periocular mesenchyme interface reveals a complete absence of discontinuities in the basal lamina and of plasmalemmal contact betwen epithelial and mesenchymal cells, both during the interaction and after its completion (Thorogood and Smith, 1984; Thorogood et al., 1986a). This unequivocal result means that cell contact-mediated mechanisms can now be discounted along with the diffusion-mediated possibilities eliminated by the earlier transfilter culture results. Thus, circumstantial evidence indicates that a matrix-mediated event is the most likely causal mechanism.

Should this be the case, then a compositional change in the interface ECM should be detectable during and after the interaction. Immunocytochemical investigation, using a range of affinity-purified, polyclonal antibody probes to major ECM components, has revealed just such a change in matrix composition. Whereas collagen types I and IV, and matrix glycoproteins fibronectin and laminin, are present at the interface throughout the period studied, collagen type II is only found during the interaction and is no longer detectable following its completion (Thorogood and Smith, 1984; Thorogood et al., 1986b). In fact, collagen type II does reappear several days later when the mesenchyme begins to express overtly a chondrogenic phenotype. Type II is a collagen usually regarded as specific to cartilage (although it is also found in cornea and vitreous humor) and its transient occurrence at such an early stage, at the site of a chondrogenesis-promoting tissue interaction, is reminiscent of the avian notochord/sclerotome interaction. In that system, the somite-derived sclerotome cells undergo a permissive, matrix-mediated interaction with the embryonic notochord and differentiate chondrogenically to form the cartilaginous primordia of the vertebral bodies. The matrix in question is the perinotochordal sheath, produced by the notochordal cells themselves, and which contains, in addition to a number of typical basement membrane components (such as laminin and collagen type IV) and collagen type I, two matrix components characteristic of cartilage, namely cartilage-specific proteoglycan (Vasan, 1981) and collagen type II (von der Mark et al., 1976). Upon exposure to the perinotochordal sheath, or even to either one of these two components in vitro, sclerotome cells display a positive feedback response by synthesizing large amounts of collagen type II and cartilage-specific proteoglycan themselves and subsequently differentiating into chondrocytes (e.g., Kosher and Church, 1975; Lash and Vasan, 1978). Thus, matrix components, synthesized during a short window in development by the notochordal cells, have a causal role in the differentiation of the somite-derived sclerotome cells into cartilage.

That collagen type II might have a similar, if not identical, role in cranial chondrogenesis seems, at this stage, a distinct possibility. However, the appropriate in vitro experiments remain to be satisfactorily carried out before this can be established. Instead, our attentions have moved on to the possibility that collagen type II may have a morphogenetic role in addition to this putative cytodifferentiative one. That is to say that it, and possibly other matrix molecules, will be causally involved in the specification of chondrocranial form. Should this be the case then two, testable predictions can be made. Firstly, that type II will be expressed elsewhere in the developing head, wherever chondrogenesis-promoting tissue interactions are taking place. And secondly, that type II will be expressed at the time that such interactions occur. These predictions have been tested by immunocytochemical analysis of serial sections of embryonic heads of various appropriate, developmental stages, using affinity-purified, polyclonal antibodies raised against chicken collagen type II. This study revealed patterns of collagen type II expression not previously reported (see, for example, Figs. 3 and 4). In addition to its known locations in the cornea, vitreous humor and perinotochordal sheath, collagen type II was found to be present at the basal surfaces of some neuroepithelia - at sites where chondrogenesis-promoting tissue interactions are known to occur, and which result in the formation of the component parts of the cartilaginous neurocranium (see Figs. 5 and 6). Moreover, this pattern changes with developmental stage in that type II is only detectable at these sites during the period that the interaction is known to take place (Thorogood et al., 1986b). Thus, there are definable patterns of transient collagen type II expression in the early developing head, which map topographically, and correlate temporally, with the interactions generating the neurocranium. We have proposed elsewhere that the location and timing of this type II expression serves a morphogenetic function in that it specifies, to a responsive mesenchyme, precisely where and when a chondrogenesis-promoting tissue interaction will take place (Thorogood et al., 1986b). The derivation of that mesenchyme, be it from mesoderm or from NC, may well be of secondary importance, hence the occurrence of cartilage elements of mixed mesenchymal compo-

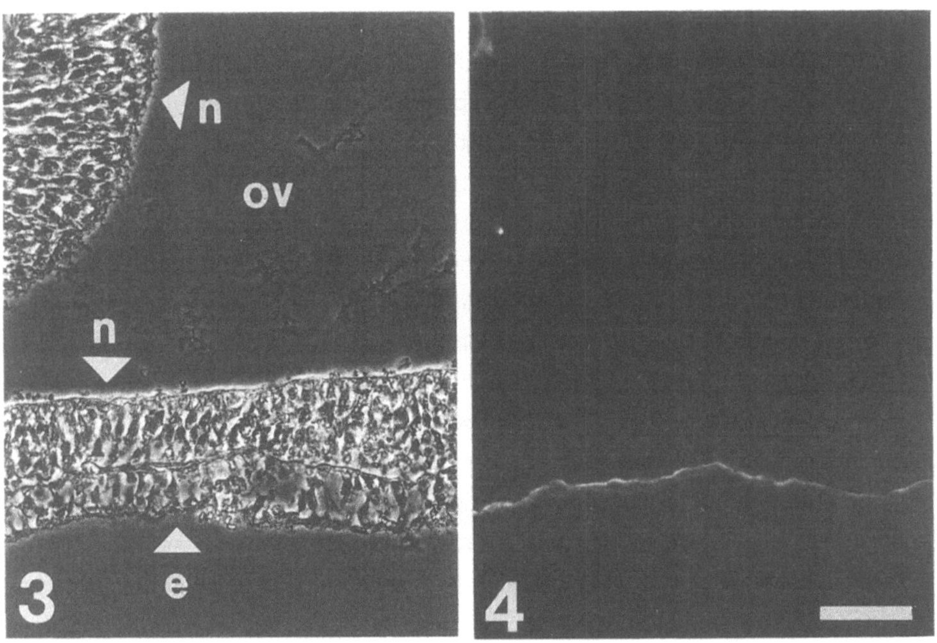

Figures 3 and 4: Phase contrast and immunofluorescent micrographs respectively, of transverse section through diencephalic region of stage 14 quail embryo, stained for collagen type II. The area shown is the basal region of the uninvaginated optic vesicle (ov); positive staining is seen around the basal aspect of the neuroepithelial wall (n) of the optic vesicle and of the ventral diencephalon where it abuts against the ventral ectoderm (e) of the early head. Bar equals 15 μm.

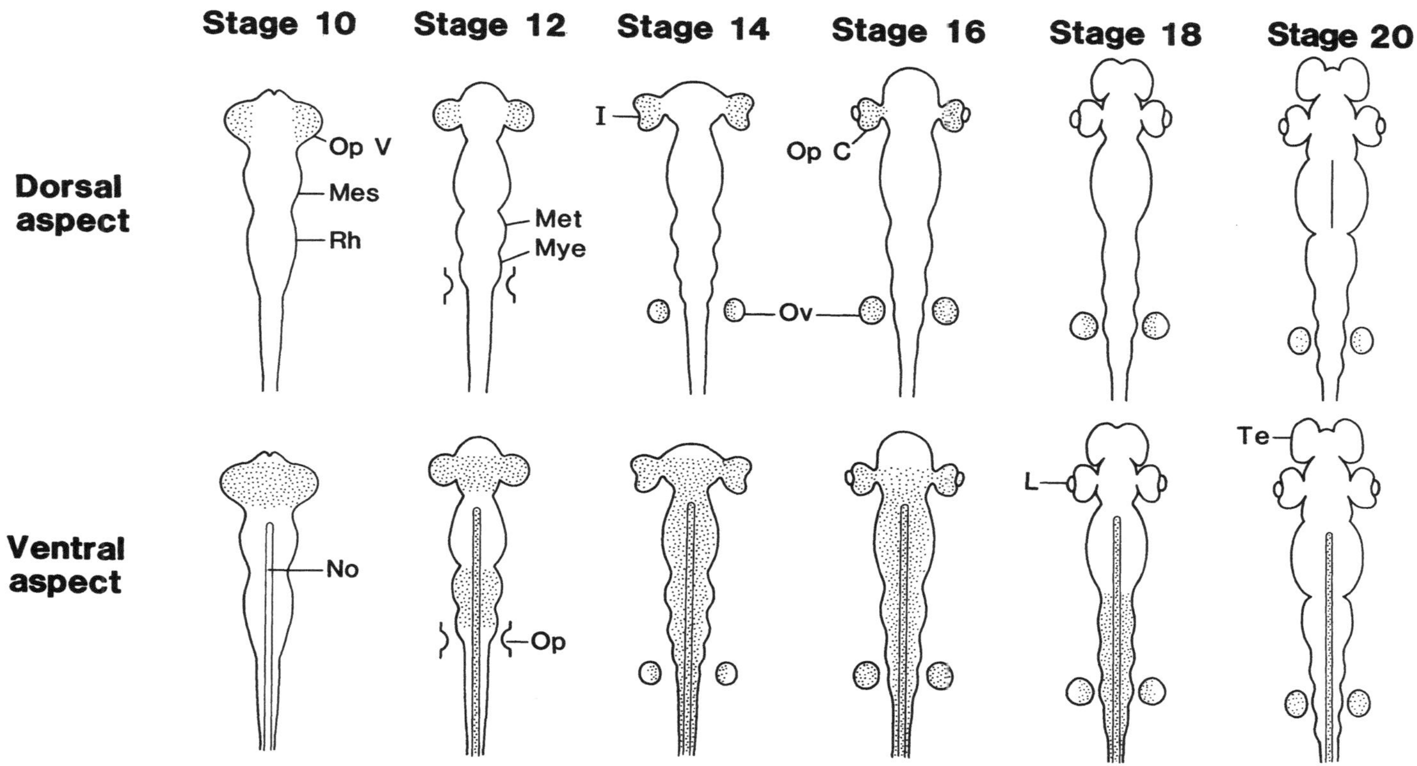

Figure 5: Diagram summarising the distribution of collagen type II around the brain and notochord of the early avian embryo. The sequence is staged developmentally from left to right, and for each stage both dorsal and ventral aspects are illustrated. Key: optic vesicle (OpV); mesencephalon (Mes); rhombencephalon (Rh); notochord (No); metencephalon (Met); Myelencephalon (Mye); invaginating otic pit (Op); invaginating optic vesicle (I); otic vesicle (Ov); optic cup (OpC); telencephalon (Te). Not drawn to scale (taken from Thorogood, 1986b).

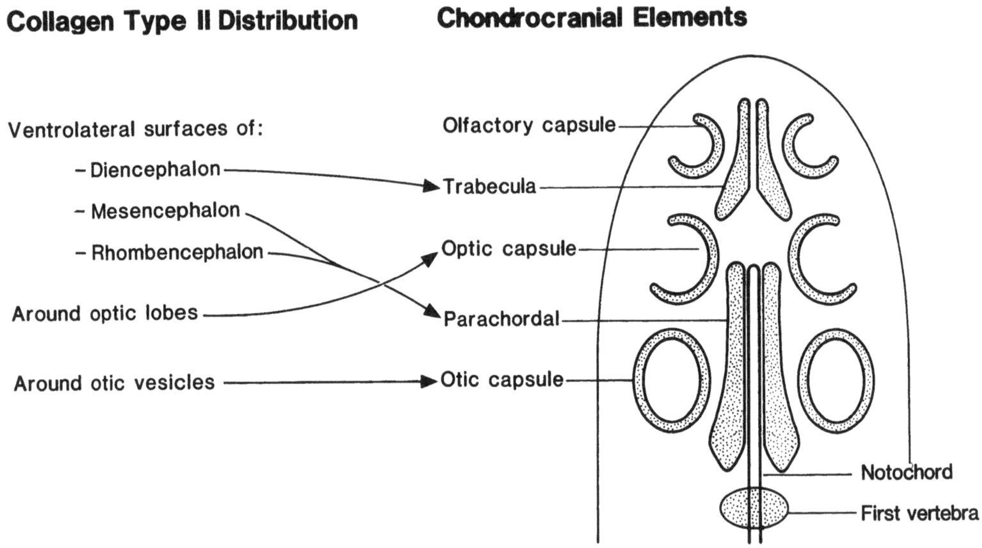

Figure 6: Diagram mapping the distribution of collagen type II with the elements of the cartilaginous neurocranium, which form subsequent to tissue interactions at those sites (the neurocranium has been idealised to a general vertebrate pattern and is not drawn to scale) (taken from Thorogood, 1986b).

sition (e.g., Noden, 1978, 1983). It must be stressed that these results are at best correlative in nature and the interpretation speculative. However, experimental testing of this interpretation is now possible following the formulation of the morphogenetic model described in Section C.

Currently we believe that type II is of (neuro)epithelial origin but without, as yet, any direct experimental evidence for such an assumption. Although typically synthesized by chondrocytes there are precedents for an epithelial origin - neural retina and corneal epithelium - in addition to embryonic notochord, both synthesize collagen type II (Smith et al., 1976; Linsenmayer et al., 1977). One further aspect needing clarification is the precise ultrastructural location of collagen type II at the tissue interfaces in question. TEM immunocytochemistry currently in progress will, it is hoped, reveal whether it is located in the lamina lucida or lamina densa of the basal lamina or more superficially in the reticular lamina.

C) A Morphogenetic Model

Although tissue interactions determine the cytodifferentiative fate of the cells, it is the parameter of cell migration (maybe supplemented by passive cell displacement) which is fundamentally important in skull morphogenesis. For instance, retinoid-induced perturbation of mammalian NC migration generates anomalies of skull form and pattern, leaving the range of cytodifferentiative potential relatively unscathed (Morriss and Thorogood, 1978). The reasons for cell movement being a major morphogenetic factor in this system are twofold. Firstly, it is the event of migration which creates the opportunities for epithelio-mesenchymal interactions to take place, and secondly, it is the time and position of arrest of migration which defines precisely where the ensuing skeletogenic differentiation will be expressed (Thorogood, 1981). Pursuing this line of enquiry, we have found that matrices derived from epithelium in the embryonic head can cause arrest of NC cell motility in culture. Thus, mesencephalic crest cell behaviour is dramatically changed on plastic surfaces 'conditioned' by pigmented retina ('conditioned' in this sense simply means the deposition of ECM by cultured epithelial cells); the cells become spindle shaped, many round up and even display some degree of aggregation (Thorogood and Smith, 1984; Yallup et al., 1986). It is particularly striking that this behavioural response closely parallels the events occurring in vivo when mesencephalic NC cells, having arrived in the periocular region, cease migrating and condense around the optic vesicle/optic cup as a prelude to differentiating into (scleral) cartilage several days later.

These observations on NC migration and cell motility, together with the results reviewed earlier on matrix-mediated tissue interactions, provide the basis for a morphogenetic model which we have dubbed the 'Flypaper Model'. This can best be summarised in a pointwise fashion:

i) The early developing head is viewed as a convoluted bag of folded epithelia enclosing an extensive extracellular compartment which is only sparsely populated by mesodermally-derived mesenchyme cells.

ii) Into these matrix-filled spaces NC cells migrate autonomously and opportunistically.

iii) Their migratory behaviour is, to some extent, controlled by ECM and in particular by epithelially-derived matrices (there is ample evidence from the extensive studies of trunk NC that components of the ECM can and do operate in this way, e.g., Weston et al., 1984). Certain matrices will serve a 'trapping' function causing localised arrest of cell migration - just as the pigmented retina-derived matrix does in vitro.

iv) At these sites, cells accumulate and undergo a matrix-mediated interaction with those epithelia and differentiate into cartilage. Thus, the pattern of cell trapping will epigenetically determine the form of the cartilaginous neurocranium. (Certain matrix molecules may serve a dual role functioning in both steps iii. and iv. This possibility emerges from preliminary observations that purified collagen type II, already shown to be present at such tissue interfaces, is very potent in eliciting, from cultured NC cells, the behavioural response previously seen on pigmented retina-conditioned substrata - B. Yallup, unpublished observations).

v) The spatial patterning of such 'trapping' events, and the ensuing tissue interactions, is the consequence of epithelial folding (of epiblast and, later, of ectoderm and neuroepithelium). In other words, topographical differences in invagination and evagination of head epithelia will create different patterns of trapping.

If, as suggested, the pattern of localised cell trapping determines the form of the cartilaginous neurocranium, then this form can be specified simply by using the parameters of epithelial folding and (rates and timing of) cell migration and movement. Quantitative changes therein will create different patterns of trapping and thus generate different neurocranial forms. Because the cartilaginous neurocranium serves as a structural template for the deposition of the bony skull, then one further epigenetic consequence will be change in overall skull form. Indeed, quantitative shifts in these two parameters may underlie the phylogenetic diversity of skull form and much of the congenital abnormality of skull morphogenesis.

D) Predictions Arising from the Model

The vigour of this model can be guaged by making a number of predictions which can be tested experimentally. Firstly, if the 'trapping' hypothesis is valid then:

i) Migrating neural crest cells should express a cell surface receptor for collagen type II. In fact, although collagen receptors are generally not well characterised at present, a receptor for type II has been isolated from chondrocyte membranes and partially characterised. Anchorin CII is a 31K integral membrane thought to mediate attachment of chondrocytes to their type II-rich matrix (Mollenhauer et al., 1984). Preliminary experiments carried out in collaboration with Klaus von der Mark and colleagues have revealed that in mesencephalic NC cultures, a sub-population of cells stain up with a polyclonal antibody raised against chondrocyte anchorin CII. Apart from the interesting possibility of cell heterogeneity within the crest, this result indicates that an appropriate surface receptor, necessary to mediate the trapping, is present on the surface of at lease some neural crest cells.

ii) The introduction of ectopic collagen type II should cause precocious arrest of migration and maybe even ectopic chondrogenesis. This set of experiments is still in progress and as yet there are no results to report.

A third prediction arises from the suggested parallel between the type of interaction considered here, promoting a chondrogenic response from head mesenchyme, and that matrix-mediated interaction between notochord and sclerotome cells which generates the vertebral primordia (discussed in Section B.).

iii) Other matrix components, thought or known to have a role in chondrogenesis-promoting tissue interactions elsewhere, will show patterns of distribution similar to that displayed by collagen type II during early craniofacial development. One such group of matrix components are the cartilage-specific proteoglycans which co-distribute with type II in the perinotochordal sheath. MZ15 is a monoclonal antibody prepared against a keratan sulfate-rich proteoglycan specific to cornea, cartilage and perinotochordal sheath in non-mammalian vertebrates (Smith and Watt, 1985). From the immunocytochemical work carried out to date it is clear that there is a large degree of co-distribution with collagen type II, for instance, both molecules are present at the interface between otic vesicle and periotic mesenchyme. Definition of the complete pattern of expression of this proteoglycan, to the level displayed in Fig. 5 for collagen type II, is currently in progress.

Thus out of the three predictions made here, the two which to date have been tested experimentally serve to corroborate the Flypaper Model. Clearly, further testing is needed to validate or repudiate the model as it stands. One particular task that we anticipate will be to identify unequivocally which cells have their collagen type II genes in a transcriptionally active state prior to the stage at which we can first detect the gene product immunocytochemically at the site of the interaction.

Acknowledgements. I am grateful to Dr. Bridget Yallup for her comments on the manuscript, and to Sandy Spanner and Lynette Potter for assistance with the illustrations. Much of the work described has been supported by the Medical Research Council of Great Britain.

References

Bee, J., and P. Thorogood (1980) The role of tissue interactions in the skeletogenic differentiation of avian neural crest cells. Devl. Biol. **78:** 47-66.

Bellairs, A. d'A., and C. Gans (1983) A reinterpretation of the amphisbaenian orbitosphenoid. Nature **302:** 243-244.

Brenner, S. (1975) Closing remarks: the genetic outlook. In *Cell Patterning*, CIBA Foundation Symposium 29 (New Series), pp. 343-345. Elsevier North Holland, Amsterdam.

De Beer, G. (1937) Development of the Vertebrate Skull. University Press, Oxford.

Drews, U., U. Kocher-Becker, and U. Drews (1972) Die Induktion von Klemenknorpel aus Kopfneuralleistenmaterial durch präsumptiven Kiemendarm in der Gewebkultur und das Bewegungsverhalten der Zellen während ihrer Entwicklung zu Knorpel. Wilhelm Roux' Archiv. **171:** 17-37.

Epperlein, H.H., and R. Lehmann (1975) Ectomesenchymal-endodermal interaction system (EEIS) of Triturus alpestris in tissue culture. 2. Observations on the differentiation of vesceral cartilage. Differentiation **4:** 159-174.

Hall, B.K. (1982) The role of tissue interactions in the growth of bone. In *Factors and Mechanisms Influencing Bone Growth*. A.S. Dixon and B.G. Sarnat, eds., pp. 205-215. A.R. Liss Inc., New York.

Horstadius, S. (1950) The Neural Crest: Its Properties and Derivatives in the Light of Experimental Research. University Press, Oxford.

Kosher, R.A., and R.L. Church (1975) Stimulation of in vitro chondrogenesis by procollagen and collagen. Nature **258:** 327-330.

Lash, J.W., and N.S. Vasan (1978) Somite chondrogenesis in vitro; stimulation by exogenous extracellular matrix components. Devl. Biol. **66:** 151-171.

Le Lievre, C. (1978) Participation of neural crest-derived cells in the genesis of the skull in birds. J. Embryol. exp. Morph. **47:** 17-37.

Linsenmayer, T.F., G.N. Smith, and E.D. Hay (1977) Synthesis of two collagen types by chick embryonic corneal epithelium in vitro. Proc. Natl. Acad. Sci. (USA) **74:** 39-43.

Mollenhauer, J., J.A. Bee, M.A. Lizarbe, and K. von der Mark (1984) Role of Anchorin CII, a 31,000-mol-wt membrane protein, in the interaction of chondrocytes with type II collagen. J. Cell. Biol. **98:** 1572-1578.

Morriss, G.M., and P. Thorogood (1978) An approach to cranial neural crest cell migration and differentiation in mammalian embryos. In *Development of Mammals*, M.H. Johnson, ed., vol. 3, pp. 363-412. Elsevier North Holland, Amsterdam.

Noden, D.M. (1978) The control of avian cephalic neural crest cytodifferentiation. I. Skeletal and connective tissues. Devl. Biol. **67:** 296-312.

Noden, D.M. (1983) The role of the neural crest in patterning of avian cranial skeletal, connective and muscle tissue. Devl. Biol. **96:** 144-165.

Saxen, L., P. Ekblom, and I. Thesleff (1980) Mechanisms of morphogenetic cell interactions. In *Development of Mammals*, M.H. Johnson, ed., vol. 4, pp. 161-202. Elsevier North Holland, Amsterdam.

Smith, G.N., T.F. Linsenmayer, and D.A. Newsome (1976) Synthesis of type II collagen in vitro by embryonic chick neural retina tissue. Proc. Natl. Acad. Sci. (USA) **73:** 4420-4423.

Smith, J.C., and F.M. Watt (1985) Biochemical specificity of Xenopus notochord, Differentiation **29:** 109-115.

Smith, L., and P. Thorogood (1983) Transfilter studies on the mechanism of epitheliomesenchymal interaction leading to chondrogenic differentiation of neural crest cells. J. Embryol. exp. Morph. **75:** 165-188.

Thorogood, P. (1981) Neural crest cells and skeletogenesis in vertebrate embryos. Histochem. J. **13:** 631-642.

Thorogood, P., and L. Smith (1984) Neural crest cells: The role of extracellular matrix in their differentiation and migration. In *Matrices and Cell Differentiation*, R.B. Kemp and J.R. Hinchliffe, eds., pp. 171-185. A.R. Liss Inc., New York.

Thorogood, P., B. Yallup, L. Smith, and T. Courtenay (1986a) The ultrastructure of an epitheliomesenchymal interface during a chondrogenesis-promoting tissue interaction. In prep.

Thorogood, P., J. Bee, and K. von der Mark (1986b) Transient expression of collagen type II at epitheliomesenchymal interfaces during morphogenesis of the cartilaginous neurocranium. Devl. Biol., in press.

Tyler, M.S., and B.K. Hall (1977) Epithelial influences on skeletogenesis in the mandible of the embryonic chick. Anat. Rec. **188:** 229-240.

Vasan, N.S. (1981) Analysis of perinotochordal materials. I. Studies on proteoglycan synthesis. J. Exp. Zool. **215:** 229-233.

Von der Mark, H., K. von der Mark, and S. Gay (1976) Study of differential collagen synthesis during development of the chick embryo by immunofluorescence. Devl. Biol. **48:** 237-249.

Wessells, N.K. (1977) Tissue Interactions and Development. W.A. Benjamin Inc., New York.

Weston, J.A., G. Ciment, and J. Girdlestone (1984) The role of extracellular matrixin neural crest development: A reevaluation. In *The Role of Extracellular Matrix in Development*, R.L. Trelstad, ed., pp. 433-460. A.R. Liss Inc., New York.

Yallup, B., L. Smith, and P. Thorogood (1986) Epithelially-derived extracellular matrix modulates neural crest cell motility. In prep.

INFLUENCE OF THE NOTOCHORD ON THE MORPHOGENESIS OF THE NEURAL TUBE

H.W.M. van Straaten and J. Drukker
Department of Anatomy and Embryology
National University of Limburg
Maastricht
The Netherlands

In embryogenesis a well known phenomenon of induction is the determination of the neural plate as a result of contact between notochordal tissue and ectoderm (Jacobson, 1978). The factors involved in subsequent determination and morphogenesis of the neural tube are less well understood; influences emanating from the mesoderm, ectoderm and notochord together with intrinsic factors are held responsible for this development. Based on the results of isolation and of experimental rotation of the rhombencephalic neural plate in amphibians, Källen (1965) summarized that the alar and basal parts of the neural tube are already determined at the stage of the neural plate, and that there is good evidence for the assumption that many later processes of neural differentiation are already determined at the neural plate stage as well. Similar experiments in chick embryos of 2 days (neural tube just closed) did not lead to a uniform conclusion, which is likely to be caused by differences in experimental procedures (Steding, 1962; Jacob et al.; 1976; Martin, 1977).

The notochord appears to be involved in proliferation of the neural tube. In rhombencephalic explants, proliferation hardly occurs without this structure, and a possible general mitotic stimulator originating from the notochord and acting on neural tissue was postulated (Källen, 1965). Such a role of the notochord was also illustrated in experiments with intracoelomic grafts of entoderm and notochord in chick embryos; only in the presence of the notochord did development occur (Wiertz-Hoessels et al., 1986). Takaya (1977) stated that in vitro it is not only the strong inductive activity of the notochord, but the scarcity of surrounding mesoderm that results in the production of a large mass of neural tissue. On the other hand, experiments of Bergquist (1959) and Burda (1968) pointed to an inhibitory action of the notochord on proliferation. From studies on amphibians (Weiss, 1955), and from cases of duplicitas anterior in birds (Watterson et al., 1955; v. Limborgh, 1956) Watterson (1965) concluded that "wherever a notochord con-

NATO ASI Series, Vol. H5
Mesenchymal-Epithelial Interactions in Neural Development
Edited by J.R. Wolff et al.

tacts the neural plate or early neural tube, even when an extra notochord is present, a floor plate reaction occurs, probably by inhibition of proliferation; wherever a somite contacts the neural tube the opposite occurs, namely a thickening of the neural tube wall, probably by enhancement of proliferation". In chick embryos, the latter influence could not be demonstrated experimentally by Martin (1971). It thus appears that at least the notochord is involved in the early development of the neural tube, but that its role with respect to proliferation is indistinct.

We are interested in the specific development of matrix cells into neuroblast in the neural tube and the role of the notochord in this development.

The matrix of the neural tube is constituted by a homogeneous cell population (Fujita, 1963); after these cells have become determined into neuroblasts during their last mitotic cycle they leave the matrix layer rapidly and their cytodifferentiation becomes evident (Fujita, 1963; Miki et al., 1981a). In the cervical region, the first post-mitotic cells are observed at 2 days (Lyser, 1964; Miki et al., 1981b). The first neuroblasts arise from the basal plate of the neural tube, though the dorsal part at least delivers some of these cells (Kanemitsu, 1977; Thors et al., 1982). At 4 days 95% of the prospective motoneurons, and part of the prospective interneurons, are present (Fujita, 1964; Kanemitsu, 1977; Hollyday and Hamburger, 1977). The cells are located then in the ventrolateral region, forming a longitudinal uniform column.

Abundant experimental and descriptive studies about the regulating influence of target organs on the numerical and cytological development of prospective motoneurons have been published (Houthoff and Drukker, 1977; Hamburger, 1980; Landmesser, 1980), but with respect to the earliest development of these cells no specific regulating factors are known. Martin (1977) and Landmesser (1980) stated that generation and differentiation of large numbers of motoneurons appear to be intrinsic properties of the neural tube, in that these processes are unaffected by the somite, the presence of the limb or by afferent or descending input.

Since the notochord is involved in the development of the early neural tube we expect that this influence will be expressed in the proliferation of matrix cells and consequently in the number of neuroblasts.

Supernumerary Notochord

In our first experiment (van Straaten et al., 1985a) the neural tube was flanked by an extra notochord in chick embryos of two days (stage 12-15, Hamburger and Hamilton, 1951, Fig. 1). From a donor embryo, a notochord fragment of about 500 µm of length was dissected. The host embryo was operated in ovo. In the region around somite 22, the notochord fragment was inserted in between the neural tube and somite. Thereafter, the eggs were incubated further up to 4 days. The insertion of an additional notochord always

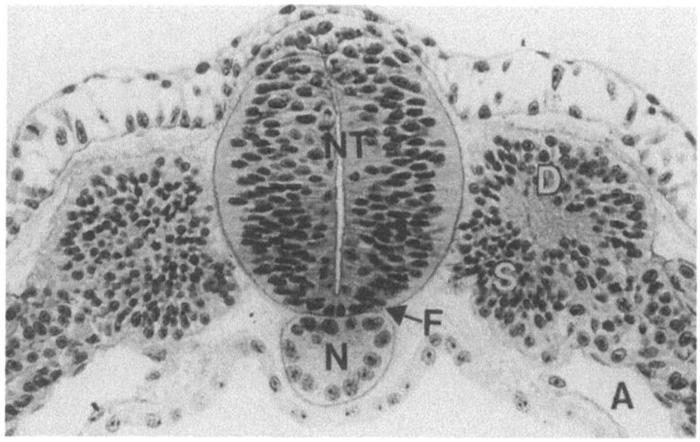

Figure 1: Transverse section (5 µm) of a chick embryo of 2 days at the thoracic level. Stained by HCl-toluidin blue. A = dorsal aorta; D = dermo-mytome; F = floor plate; N = notochord; NT = neural tube; S = sclerotome. x 166.

resulted in dislodgement of the myotome into the ventrolateral direction. The morphology of the neural tube was changed when the distance between implant and neural tube was less than 80 µm. The most pronounced effect was an increase in size of the ipsilateral half of the neural tube (Fig. 2), which was maximal when the neural tube and implant were close to each other, and the distance between both notochords exceeded 200 µm. Sections treated to demonstrate the presence of AChE (AChE is a marker for neuroblasts, Miki et al., 1981a) revealed that this increase for the larger part was due to an increase in the area of neuroblasts (Fig. 4b). Quantitative studies demonstrated that on the experimental side, more cells had achieved a high level of differentiation than at the right side, and that the number of neuroblasts as well as the total number of cells in the neural tube half were increased (preliminary results). Moreover, the ventral root had its origin over an extended area of the outer surface now (Fig. 4b). These results indicate that the proliferation of the matrix layer must be enhanced and that more cells are differentiated into neuroblasts as a reaction to the presence of an additional notochord.

With respect to the proliferation pattern of the neural tube, the following data are relevant. At the thoracic level, a high proliferation rate was measured in the basal plate at about 3 days; it gradually shifts in the dorsal direction, where the maximal proliferation rate was determined at about day 6 (Hamburger, 1948; Corliss and Robertson, 1963; Mitolo, 1967). Thus, the period of maximal proliferation of the basal plate coincides with the period of formation of the larger part of the neuroblasts. In our experiments, this high level of proliferation must even be enhanced in order to deliver the supernumerary quantity of neuroblasts.

In fact, it is likely that we are dealing with a synergistic action of the two noto-

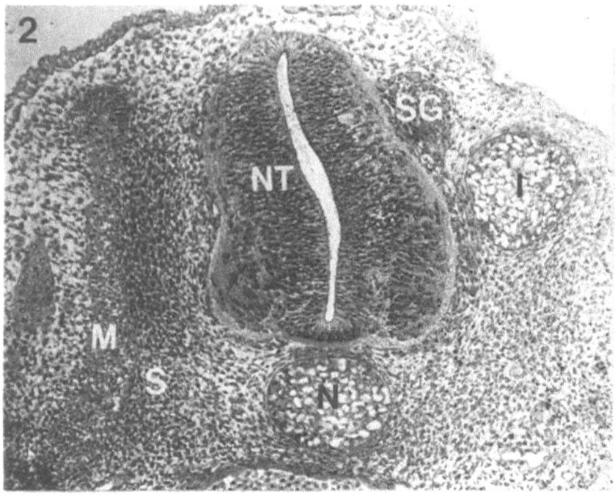

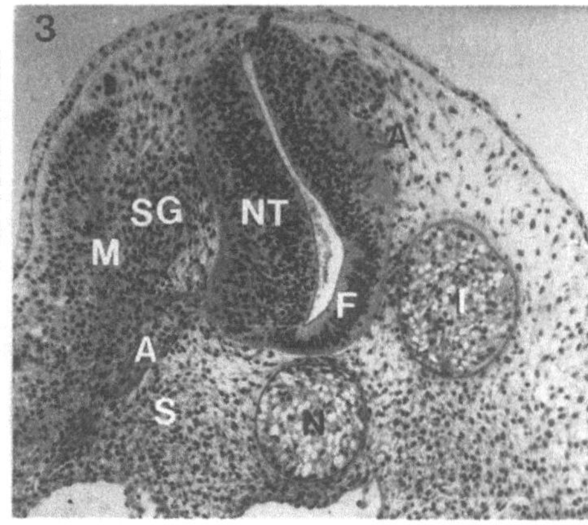

Figure 2: Transverse section (2 µm) of a chick embryo of 4 days at the thoracic level. Stained with toluidin blue. The additional notochord (I) was implanted at two days. Note the enlargement of the right ventral horn area of the neural tube. M = myotome; N = notochord; NT = neural tube; S = sclerotome; SG = spinal ganglion. x 70.

Figure 3: Transverse section (5 µm) of a chick embryo of 4 days at the thoracic level. Stained with HCl-toluidin blue. The notochord (I) was implanted at 1.5 days. An extended floor plate-like structure (F) is present. Axons (A) pass the right outer surface at a relatively dorsal position. M = myotome; N = notochord; NT = neural tube; S = sclerotome; SG = spinal ganglion. x 85.

chords, which results in the observed enhancement of neuroblast development. It implies that the one natural notochord also will promote these actions, though to a lesser extent. Thus, the notochord would enhance the proliferation of the basal matrix cells and possibly promote the differentiation of matrix cells into neuroblasts.

If we indeed deal with a synergistic action of both notochords, it has to be questioned why a second "floor plate" did not develop as Watterson (1965) suggested. The most likely explanation is that the second notochord is present from two days onwards, which might be beyond the period of induction. Therefore, a series of implantation experiments was performed on 1.5 days (van Straaten et al., 1985b). As a result, an extended floor plate area was present (Fig. 3), when the implant was less than 25 µm apart from the neural tube. When the notochords where at a larger distance from each other, an additional "floor plate" was seen (Fig. 4c) and the area in between demonstrated an increased number of neuroblasts, comparable to the results of the foregoing experiments. Relatively few mitotic figures were present in the additional floor plate (preliminary results), and this fits well with Watterson's statement, that contact of the notochord with the neural tube causes a floor plate to arise, probably by inhibition of proliferation.

Absence of the Notochord

If the notochord enhances proliferatin and differentiation of neuroblasts and is necessary for floor plate development, absence of the notochord should result in the opposite effect. However, normal neural tubes were described after removal of the notochord at the neural plate stage in amphibians (Kitchin, 1949), and removal of Hensen's node in chick embryos (Grabowski, 1956). On the contrary, in notochord-defective amphibian embryos, caused by X-radiation, no floor plate was observed in the neural tube (Youn and Malacinsky, 1981), and chorda resection experiments of Bergquist (1959) and Burda (1968) point to a proliferation-inhibiting effect of the notochord on the neural tube as a whole.

We have performed surgical extirpation of a notochordal fragment in the chick embryo of 1.5 days in the prospective area of somite 20-25. At 4 days, in the longitudinal direction, the neural tube was reduced in diameter over a restricted area, while the notochord was absent overe a considerable length (up to more than 10 somites). In the thinned portions of the neural tube (Fig. 4d), a floor plate was absent; neuroblasts were located in a narrow peripheral band, their number was small and the degree of differentiation low. Axons crossing the limiting membrane, if present, were concentrated in the median plane as one ventral root. Again, these results indicate that the notochord is necessary for the development of a floor plate. The reduced size of the neural tube indicates that the proliferation must be reduced, and the reduced area of neuroblasts indicates a low rate of differentiation. These results fit well with the foregoing results and conclusion, namely that the notochord a) is necessary for the development of the floor plate, b) would enhance proliferation of matrix cells and possibly promotes differentiation of matrix cells into neuroblasts, and c) plays a role in the determination of the location of axonal exits.

Effects of the Notochord on Proliferation and Differentiation

In our experiments, the involvement of the notochord on proliferation appears to be twofold. Firstly, close contact between the additional notochord and neural tube form 1.5 days onwards results in a thin-walled area with a low proliferation rate. Secondly, the presence of the additional notochord at a longer distance, or in later ontogenetic stages (from 2 days onwards), results in the opposite, namely an enhancement of proliferation of matrix cells and their differentiation into neuroblasts. Absence of the notochord from 1.5 onwards results in the reverse effects. If these findings are extrapolated to the normal situation, it is likely that between 1.5 and 2 days the notochord determines the adjacent neural cells into floor plate cells (probably by inhibition of proliferation), while proliferation is stimulated in those cells not blocked by this action, i.e., the matrix cells next to the now determined floor plate cells.

Differentiation of matrix cells into neuroblasts may be just a consequence of the

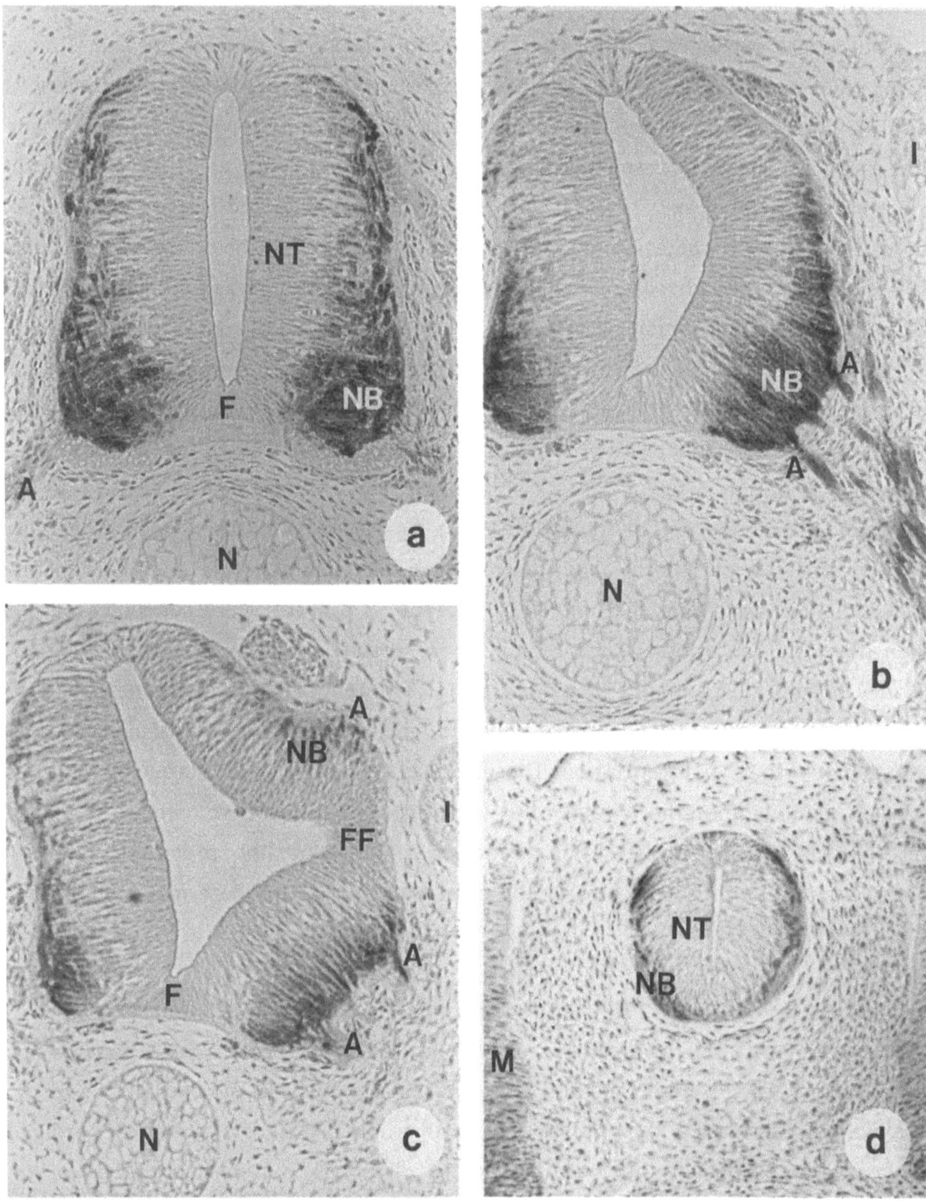

Figure 4: Transverse sections (10 µm) of chick embryos of 4 days at the thoracic level, treated for the presence of AChE (van Straaten et al. (1986) Acta Histochem., Suppl. **32**: 185-190). Magn. 192x. Neuroblasts (NB) can clearly be distinguished. A = axons; F = floor plate; I = implant; N = notochord; NT = neural tube. x 192. a: control embryo. b: Notochord (I) implanted at 2 days. Enlargement of the area of neuroblasts and the area where axons pass the outer surface. c: Notochord (I) implanted at 1.5 days. A second floor plate-like structure (FF) is present. An extended area of neuroblasts and axons can be seen in between both notochords. d: Notochord extirpated at 1.5 days. The neural tube size is markedly reduced and only a few neuroblasts are present. M = myotome.

magnitude and time course of proliferation, which implies that the first neuroblasts have to arise in the basal plate, but the increased number of differentiated neuroblasts, in the implantation experiments, may also point to an involvement of the notochord in the process of differentiation.

Effect of the Notochord on the Location of Ventral Roots

The matrix cells which differentiate first grow their axons out of the neural tube in the ventrolateral region (Cajal, 1960). It is likely that this event determines the location of the ventral root and that matrix cells which differentiate later will use the same passage for their axonal outgrowth. With this assumption some of the implantation results can be explained. By the synergistic action of both notochords, not only the matrix cells next to the floor plate, but the whole matrix zone, enclosed by both notochords, will be involved in the enhanced rate of proliferation and differentiation. As a consequence several matrix cells will develop simultaneously in this zone and thus several "first" axons will grow out simultaneously at several locations.

Interaction Between Notochord and Neuroblasts

Our experiments have proven that the early development of the basal matrix of the neural tube is influenced by the notochord. The way by which this influence takes place still has to be elucidated. The floor plate may arise as a result of a direct contact between notochord and neural tube but, with respect to the other events, it is likely that other cells, and/or ECM, play an important role in transmission of the information. Cells around the neural tube, which may play such a role, are the sclerotomal cells, neural crest cells and other mesenchymal cells. The notochord may exert its effect by interfering with either the arrangement and/or the functional state of these cells, or ECM around the neural tube, which in turn will influence the adjacent neural tissue. Several data underly this concept. Initially, the sclerotome faces the basal plate of the neural tube (Watterson, 1965; see Fig. 1) and possibly determines the sequence of axonal roots in longitudinal direction (Keynes and Stern, 1984). In vitro, the notochord attracts sclerotomal cells (Cheney and Lash, 1978) and in vivo an orderly arrangement of these cells with respect to the notochord and to the ventrolateral part of the neural tube can be seen (Jacob et al., 1975; see Fig. 4b). Thus, the notochord may indirectly be involved in the determination of the ventral roots by influencing the arrangement of sclerotomal cells, and the extension of the exit of ventral roots after addition of an extra notochord could thus result from the rearrangement of sclerotomal cells. Neural crest cells may also be involved. These cells migrate from their origin along the outer side of the neural tube through a fibronectin-rich ECM (Rogers et al., 1986). From in vitro experiments, it is known that neural crest cells on fibronectin sheaths avoid chondroitin-sulphate

(Newgreen, 1982), which is secreted by the notochord in large amounts. The cells will deviate from their course along the neural tube surface somewhere. Thiery et al. (1985) found that neural crest cells interface with the first neuroblasts in the neural tube, probably at the site of their lowest point of migration along the neural tube. In this way, the notochord indirectly might determine the location where the first axons leave the neural tube.

The influence of the notochord on the proliferation of the neural tube cells may be mediated by the extracellular matrix. Hyaluronic acid is related to proliferation and cell migration (Stern, 1984; Toole, 1981), and appears to be present around the neural tube in large amounts (Pintar, 1978; Delpech et al., this volume).

The influence of the surrounding mesenchyme on the development on the CNS has been studied with respect to the role of the meningeal cells on the cerebellum. Destruction of this layer by 6-OHDA during ontogenesis results in a disturbed foliation and lamination pattern, and proliferation is reduced (Pehlemann et al., 1985; Sievers et al., 1985, and this volume). It was suggested that meningeal cells are necessary for the development and maintenance of the glia limitans superficialis. It is known that in the development and maintenance of epithelia an intricate interaction between the epithelial cells and the surrounding mesenchyme exists. The basal lamina, in between both tissues, plays an important role in transmission of information (Bernfield et al., 1984). When this model is applied to the neural tube, it is possible that cells and ECM surrounding the neural tube play a decisive role in the devlopment of this part of the CNS.

Conclusions

From our experiments, it is concluded that the notochord a) is necessary for the development of the floor plate, b) enhances proliferation of matrix cells and possibly promotes differentiation of matrix cells into neuroblasts, and c) plays a role in the determination of the location of axonal exits. The notochord may exert its effects by direct contact, as will be the case with respect to floor plate determination, or may be involved in the arrangement and/or functional state of cells and ECM which surround the neural tube. These surroundings structures, in turn, will interact with the neural tissue, resulting in the specific development of the matrix cells into neuroblasts.

References

Bergquist, H. (1959) Experiments on the overgrowth phenomenon in the brain of chick embryos. Anat. Anz. **106:** 429-441.

Bernfield, M.R., S.D. Banerjee, J.E. Koda, and A.C. Rapraeger (1984) Remodelling of the basement membrane as a mechanism of morphogenetic tissue interaction. In Role of Extracellular Matrix in Development, R. Trelstad, ed., pp. 545-572. Alan R. Liss, New York.

Burda, D.J. (1968) Studies on the experimental induction of overgrowth in chick embryos.

Anat. Rec. **161**: 419-425.
Cajal, S. Ramon y (1960) *Studies on Vertebrate Neurogenesis*. Translated by L. Guth. Charles C. Thomas, Publisher.
Cheney, C.M., and J.W. Lash (1978) Chick somites: Heterogeneity in their response to notochord. J. Cell Biol. **79**: 21a.
Corliss, C.E., and G.G. Robertson (1963) The pattern of mitotic density in the early chick neural epithelium. J. Exp. Zool. **153**: 125-140.
Fuyita, S. (1963) The matrix cell and cytogenesis in the developing central nervous system. J. Comp. Neurol. **120**: 37-42.
Fuyita, S. (1964) Analysis of neuron differentiation in the central nervous system by tritiated thymidine autoradiography. J. Comp. Neurol. **122**: 311-327.
Grabowski, C.T. (1956) The effects of excision of Hensen's node on the early development of the chick embryo. J. Exp. Zool. **133**: 301-343.
Hamburger, V. (1948) The mitotic pattern in the spinal cord of the chick embryo in relation to histogenetic processes. J. Comp. Neurol. **88**: 221-284.
Hamburger, V., and H.L. Hamilton (1951) A seriers of normal stages in the development of the chick embryo. J. Morphol. **88**: 49-92.
Hamburger, V. (1980) Trophic interactions in neurogenesis: A personal historical account. Ann. Rev. Neurosc. **3**: 269-278.
Hollyday, M., and V. Hamburger (1977) An autoradiographic study on the formation of the lateral motor column in the chick embryo. Brain Res. **132**: 197-208.
Jacob, H.J., B. Christ, M. Jacob, and P. Ahlström (1976) Differenzierung des Neuralrohres. Experimentelle Untersuchungen an Hühner- und Wachtelembryonen. Acta Anat. **94**: 204-220.
Jacob, M., H.J. Jacob, and B. Christ (1975) Die frühe Differenzierung des chordanahen Bindegewebes. Raster- und transmissionselektronenmikroscopische Untersuchungen an Hühnerembryonen. Experientia **31**: 1083-1086.
Jacobson, M. (1978) *Developmental Neurobiology*. 2nd ed. Plenum Press, New York.
Källen, B. (1965) Early morphogenesis and pattern formation in the central nervous system. In *Organogenesis*, R. de Haan and H. Ursprung, eds., pp. 107-128. Holt, Rhinehart and Winston, New York.
Kanemitsu, A. (1977) Development of spinal motor neurons. Adv. Neurol. Sci. **21**: 204-217.
Keynes, J., and C.D. Stern (1984) Segmentation in the vertebrate nervous system. Nature **310**: 786-789.
Kitchin, I. (1949) The effects of notochordectomy in Amblystoma mexicanum. J. Exp. Zool. **112**: 393-415.
Landmesser, L.T. (1980) The generation of neuromuscular specificity. Ann. Rev. Neurosci. **3**: 279-302.
Limborgh, J. van (1956) The influence of adjacent mesodermal structures upon the shape of the neural tube and plate in bird embryos. Acta Morph. Neerl.-Scand. **1**: 155-166.
Lyser, K.M. (1964) Early differentiation of motor neuroblasts in the chick embryo as studied by electron microscopy. I. General aspects. Dev. Biol. **10**: 433-466.
Martin, A.H. (1971) Mitotic patterns in the neural tube of the chick after unilateral removal of 1-3 somites. Acta Embryol. Exp.: 9-16.
Martin, A.H. (1977) Neural tube - somite interaction, effect of neural tube reversal upon neuron development. Acta Embryol. Exp. **3**: 305-313.
Miki, A., Y. Atoji, and H. Mizoguti (1981a) The relationship between the proliferating ability and acetylcholinesterase activity of the neural tube cells in early chick embryos. Acta Histochem. Cytochem. **14**: 461-475.
Miki, S., Y. Atoji, and H. Mizoguti (1981b) Time of origin of the neuroblasts in the neural tube of the chick embryo determined by histochemical observation of acetylcholinesterase activity. Acta Histochem. Cytochem. **14**: 641-653.
Mitolo, V. (1967) Quantitative data on the mitotic activity in the brachial and thoracic segments of the chick embryo neural tube from the 4th to the 8th day. Acta

Embryol. Exp. **10:** 62-74.

Newgreen, D. (1982) The role of the extracellular matrix in the control of neural crest cell migration. In *Extracellular Matrix*, S. Hawkes and J.L. Wang, eds., pp. 141-145. Acad. Press, New York.

Pehleman, F.-W., J. Sievers, and M. Berry (1985) Meningeal cells are involved in foliation, lamination and neurogenesis of the cerbellum: Evidence from 6-hydroxydopamine-induced destruction of meningeal cells. Dev. Biol. **110:** 136-146.

Pintar, J.E. (1978) Distribution and synthesis of glycosaminoglycans during quail neural crest morphogenesis. Dev. Biol. **67:** 444-464.

Rogers, S.L., K.J. Edson, P.C. Letourneau, and S.C. McLoon (1986) Distribution of laminin in the developing peripheral nervous system of the chick. Dev. Biol. **113:** 429-435.

Sievers, J., F.-W. Pehlemann, H.-G. Baumgarten, and M. Berry (1985) Selective destruction of meningeal cells by 6-hydroxydopamine: A tool to study Meningeal-neuroepithelial interaction in brain development. Dev. Biol. **110:** 127-135.

Steding, G. (1962) Experimente zur Morphogenese des Rückenmarkes. Acta Anat. **49:** 199-231.

Stern, C.D. (1984) Mini-review: Hyaluronidases in early embryonic development. Cell Biol. Intern. Reports **8:** 703-717.

Straaten, H.W.M. van, F. Thors, E.L. Hoessels, J.W.M. Hekking, and J. Drukker (1985a) Effect of a notochordal implant on the early morphogenesis of the neural tube and neuroblasts. Histometrical and histological results. Dev. Biol. **110:** 247-254.

Straaten, H.W.M. van, J.W.M. Hekking, F. Thors, E.L.J.M. Wiertz, and J. Drukker (1985b) Induction of an additional floor plate in the neural tube. Acta Morphol. Neerl.-Scand. **23:** 91-97.

Takaya, H. (1977) Types of neural tissue induced through the presumptive notochord of newt embryo. Differentiation **7:** 187-192.

Thiery, J.P., A. Delouvée, M. Grumet, and G.M. Edelman (1985) Initial appearance and regional distribution of the neuron-glia cell adhesion molecule in the chick embryo. J. Cell Biol. **100:** 442-456.

Toole, B.P. (1981) Glycosaminoglycans in morphogenesis. In *Cell Biology of Extracellular Matrix*, E.D.Hay, ed., pp. 259-294. Plenum Press, New York.

Thors, F., E.M.J. de Kort, and R. Nieuwenhuys (1982) On the development of the spinal cord of the clawed frog, Xenopus laevis. II Experimental analysis of differentiation and migration. Anat. Embryol. **164:** 443-454.

Watterson, R.L., C. Goodheart, and G. Lindberg (1955) The influence of adjacent structures upon the shape of the neural tube and neural plate of chick embryos. Anat. Rec. **122:** 539-559.

Watterson, R.L. (1965) Structure and mitotic behavior of the early neural tube. In *Organogenesis*, R.L. de Haan and H. Ursprung, eds., pp. 129-159. Holt, Rhinehart and Winston, New York and Chicago.

Weiss, P. (1955) Nervous system. In *Analysis of Development*, B.H. Willier, P. Weiss, and V. Hamburger, eds., pp. 346-401. Saunders, Philadelphia.

Wiertz-Hoessels, E.L.M.J., K. Hara, J.W.M. Hekking, H.W.M. van Straaten, F. Thors, and J. Drukker (1986) Differentiation of gut entoderm in dependence of the notochord, submitted for publication.

Youn, B.W., and G.M. Malacinsky (1981) Axial structure development in ultraviolet-irradiated (notochord-defective) amphibian embryos. Dev. Biol. **83:** 339-352.

IS THERE A SPECIFIC INTERACTION BETWEEN DONOR EMBRYONIC ENTODERM (CHICK OR QUAIL) AND HOST EMBRYONIC MESODERM (CHICK)?

E. Wiertz-Hoessels and J. Drukker
Dept. of Anatomy and Embryology
Medical Faculty, University of Limburg
Maastricht
The Netherlands

Introduction

The notochord plays a well known role in neural induction. There are various grounds for considering the possibility that it also exerts an influence in endoderm differentiation.

Firstly, in normal development a direct contact is seen between endoderm and notochord in early stages. A similar direct contact is seen between notochord and neural tube. Van Straaten et al. (1985) have made plausible that this contact has implications for the development of the neural tube, namely the formation of the floor plate.

Further, the occurrence of entero-endocrine cells throughout the whole digestive tract may be an indication that there exists a relation between notochord and these aminergic cells. They belong, together with undeniably neural crest derived cells (C-cells of the thyroid gland, type I cells of the glomus caroticum, glomus aorticum, adrenal medulla), to the peripheral APUD-cells (Pearse, 1969). The work of Nieuwkoop et al. (1952) and Hoessels (1957) indicates that neural crest develops only when presumptive neuroderm has been under the influence of the notochord.

Earlier experiments showed that presumptive intra-embryonic endoderm of the chick in stage 4^- to 5 (Hamburger and Hamilton, 1951) has a low differentiation capacity if grafted to the coelom of a 2-day-old host (Fig. 1). These grafts seldom survive 5 days or longer (Fig. 2). If a fragment of notochord of a 2-day-old donor is added to this endoderm, the developing grafts nearly always survive for longer times (Wiertz-Hoessels et al., 1986). Entero-endocrine cell development could invariably be demonstrated (Fig. 3, 4). So the notochord appeared to enhance the differentiation of the presumptive intra-embryonic endoderm of the chick.

Experiments

Excision of endoderm (Fig. 5) is carried out according to Rosenquist's (1971) map which

NATO ASI Series, Vol. H5
Mesenchymal-Epithelial Interactions in Neural Development
Edited by J. R. Wolff et al.

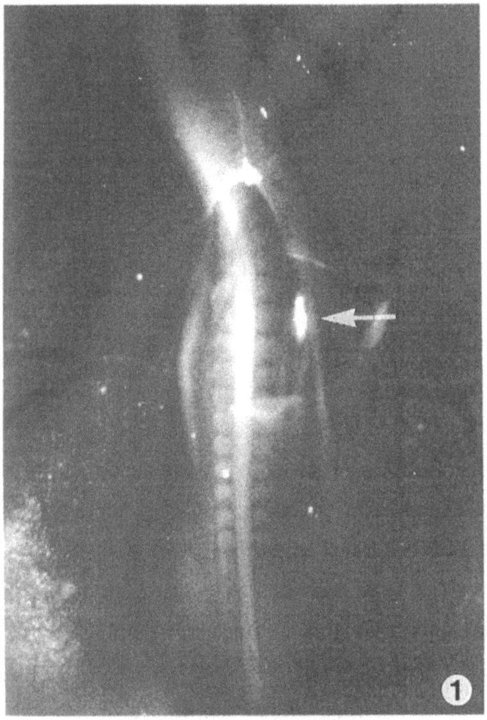
1

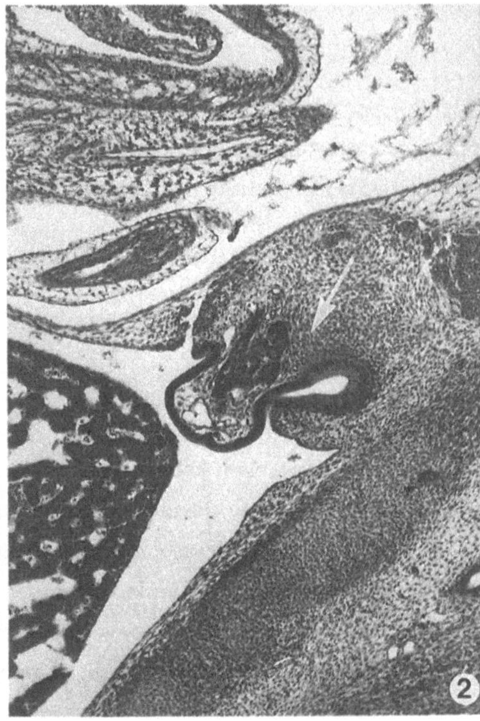
2

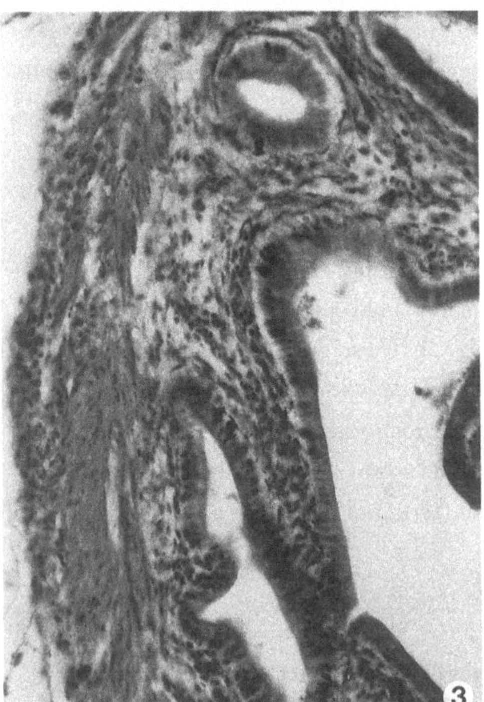
3

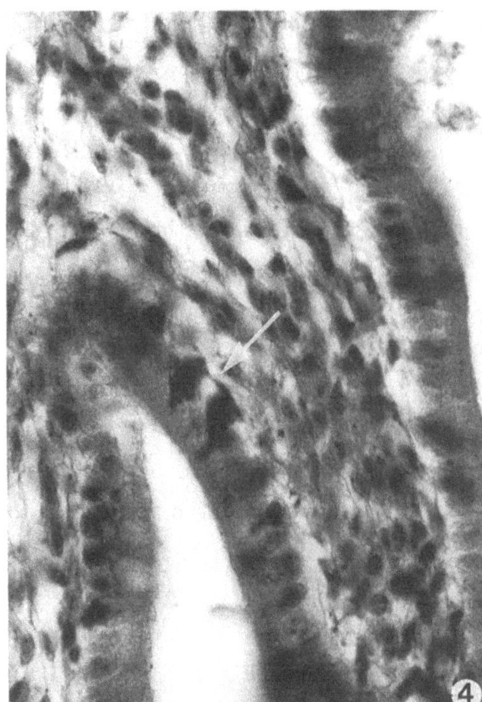
4

Figure 1: The graft (arrow) is implanted in the coelom on the right-hand side of a two days chick host at the level of somite 16-18.

Figure 2: Graft (arrow) developing from chick intra-embryonic entoderm without notochordal fragment, 5 days post operation. Haematoxylin-eosin. Magn.: 40 x.

Figure 3: Graft developing from chick intra-embryonic entoderm together with a fragment of notochord, 14 days post operation. Ungewitter's silver staining. Magn.: 160 x.

Figure 4: Entero-endocrine cells (arrow) appear to occur in the epithelium in this graft. Ungewitter's silver staining. Magn.: 570 x.

corresponds to stage 4- to 5 according to Hamburger and Hamilton (1951). No endodermal determination has taken place at this stage. Notochord is excised from a 2-day-old donor. This age is chosen because the notochord contains mono-amines at this moment (Kirby and Gilmore, 1960; Allen and Newgreen, 1977). These mono-amines might be related to induction of aminergic cells.

Implantation in the host's coelom is performed at the level of somite 16-18 on the right-hand side.

Because of our interest in entero-endocrine cells, the hosts are cultivated usually 14 days post operation. In normal development, entero-endocrine cells are observed from day 11 onwards. The positive notochordal influence led us to the question whether the notochord might have an enhancing influence on presumptive extra-embryonic entoderm as well, leading to the formation of enteric structures. It should be noted that intra-embryonic endoderm develops from the primitive streak (Vakaet, 1970; Rosenquist, 1971; Veini and Hara, 1975), while extra-embryonic endoderm is derived from the hypoblast. The grafts were prepared from the peripheral area (Fig. 5).

In a third series, intra-embryonic quail endoderm was taken, instead of chick endoderm, in order to elucidate whether entero-endocrine cells develop from the entodermal graft or from the host.

Analysis of the Grafts

a) A number of grafts developing from presumptive intra-embryonic entoderm of chick and quail in combination with 2-day-old fragments of notochord have a spherical shape without a lumen. The epithelium is present at the surface. In most grafts lumina occur. Sometimes tube-like lumina with septa can be observed.

Most of the grafts develop into more or less complex structures (Fig. 6). Regularly well developed structures resembling esophagus, bronchius, lung, air sac, proventriculus, gizzard, small intestine and pancreas are found. They may occur in combinations which conform to normal development (esophagus- and lung-like, proventriculus- and gizzard-like), or in combinations which do not occur in normal development (lung- and pancreas-

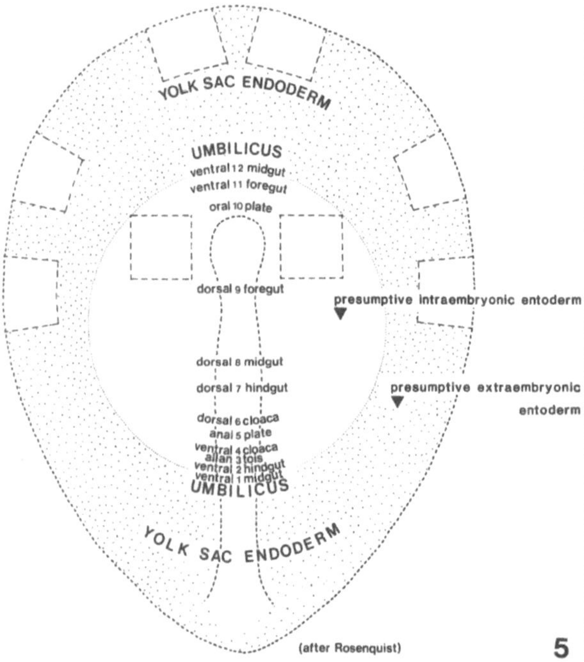
YOLK SAC ENDODERM
UMBILICUS
ventral 12 midgut
ventral 11 foregut
oral 10 plate
dorsal 9 foregut
presumptive intraembryonic entoderm
dorsal 8 midgut
dorsal 7 hindgut
presumptive extraembryonic entoderm
dorsal 6 cloaca
anal 5 plate
ventral 4 cloaca
allan 3 tois
ventral 2 hindgut
ventral 1 midgut
UMBILICUS
YOLK SAC ENDODERM
(after Rosenquist)

5

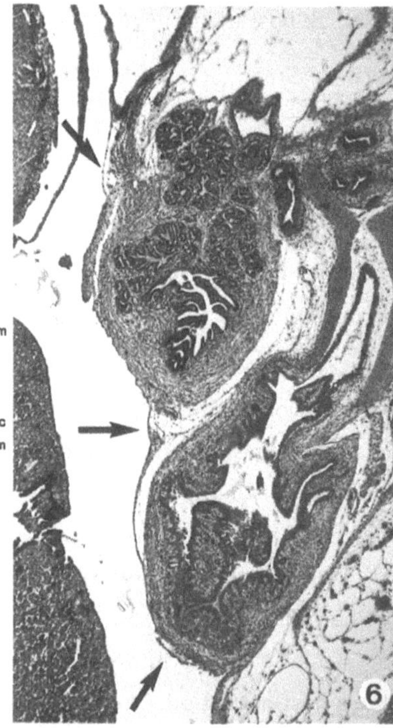

6

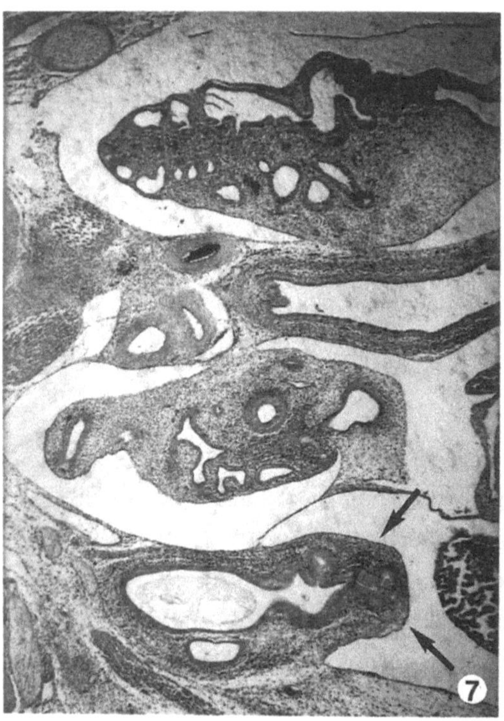

7

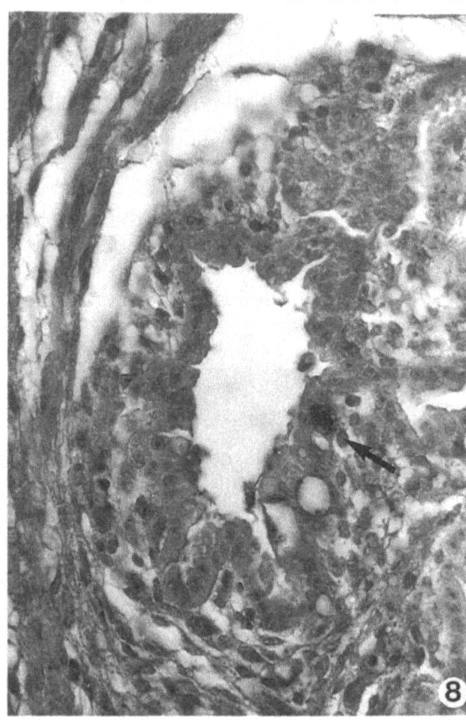

8

Figure 5: Rosenquist's map indicating the location of presumptive intra- and extra-embryonic entoderm at stage 4^- to 5 (Hamburger and Hamilton, 1951). The dashed rectangles mark the areas excised.

Figure 6: Graft developing from quail presumptive intra-embryonic entoderm, together with a fragment of notochord, 14 days post operation. In the area of the host's somatopleura-heart-liver, a complex graft (arrowed) containing respiratory, proventriculus, pancreas and small intestinal structures is found. Haematoxylin eosin. Magn.: 40 x.

Figure 7: Graft developing from chick presumptive extra-embryonic entoderm together with a fragment of notochord, 6 days post operation. In the host's somatopleura the graft is developing into proventriculus. The proventriculus-like tip of the graft (arrowed) is directed toward the host's liver. The graft occurs at the level of the host's lungs. Haematoxylin-eosin. Magn.: 40 x.

Figure 8: Glucagon-containing entero-endocrine cell (arrow) in a proventriculus-structure, developing from a quail intra-embryonic graft combined with a fragment of notochord, 14 days post operation. PAP technique. Magn.: 570 x.

like). Whether the identified parts of the gut or its derivatives in grafts really are homologous to the organs in normal development (as to branching patterns of glands and lung, presence of enzymes) is questionable. Besides, structures are found which do have well differentiated epithelium but which do not resemble any structure in normal development.

The notochord is still recognizable in grafts after 7 days of cultivation. After 14 days, generally no notochordal rudiment is observed. In a number of cases, however, cartilaginous structures occur. They may have developed from sklerenchymal cells which had not been removed from the notochord. Since no trypsine is used in preparing the notochord, it is difficult to obtain a completely clean notochord fragment.

b) Grafts developing from presumptive extra-embryonic endoderm of chick in combination with a 2-day-old fragment of notochord preliminarily show that presumptive extra-embryonic endoderm, in the presence of notochord, develops into several enteric structures, comparable to those observed in grafts from intra-embryonic endoderm (Fig. 7).

This development from presumptive extra-embryonic endoderm is even more puzzling than that of presumptive intra-embryonic endoderm. The former has a different origin and does not give rise to enteric entoderm in normal development.

Entero-endocrine Cells

In the chimaera experiments it was clear that many entero-endocrine cells in the grafts had the heterochromatin structure specific for the quail nucleus. It cannot absolutely be ruled out, however, that there are entero-endocrine cells with the chick nuclear structure.

Rawdon and Andrew (1981) made an inventory of the occurrence of various peptides in different areas in the chick at hatching stage. In structures resembling specific regions of the digestive tract and its derivatives, we investigated immunocytochemically the possibility that entero-endocrine cells develop which contain specific peptides (aside from amines) which are present during normal development. In accordance with their data the presence of somatostatin, pancreatic polypeptide and glucagon in the proventriculus (Fig. 8) and the small intestine was detected. Neurotensin was observed in the proventriculus. Somatostatin was occasionally observed in esophagus in our grafts. In pancreas, we were able to detect insulin, glucagon and somatostatin. Serotonin was present in nearly all develping structures.

We interpret these findings as support for the morphological identification of the structures observed in the grafts.

Neutrality of the Coelomic Environment

The coelom as well as the eye chamber are regarded as a neutral surrounding. Yet, there will be some influences from the host. It is feasible that these influences increase during further development. Whether the host's influence is restricted to the contribution of muscle and connective tissue is questionable. The graft is able to differentiate because of ingrowth of the host's blood vessels; substances within the blood may affect the graft. Nerve fibers usually accompany blood vessels. Sometimes we observed autonomic ganglia, which might have migrated from the host into the graft.

From the fact that parts of a graft, protruding into the coelom, have the tendency to degenerate, it can be concluded that the coelom does not exert an enhancing influence. This is in accordance with Hara's (1961) conclusion that the coelom can be considered as a neutral environment in our experiments.

Role and Infleunce of the Host's Mesoderm

Though the graft is inserted in the host's coelom at the level of somite 16-18, the implant may attach at different sites. In our experiments, the endodermal graft may attach to splanchnic and also to somatic and nephrogenic mesoderm, and to more than one type of mesoderm. The host's mesoderm is more than one day older than the grafted endoderm. The next question is whether the mesoderm of the host to which the graft is attached, determines what enteric structure will develop from the endoderm. According to Rosenquist (1971), the presumptive <u>intra</u>-embryonic endoderm is not determined at this stage; the fate of <u>extra</u>-embryonic endoderm is to cover the yolk sac. Thus, it is likely that the structures observed in the grafts are induced by the host's mesoderm.

It is generally accepted that the local mesenchyme determines the specificity of the enteric epithelium (Sigot and Marin, 1970; Goldin, 1980; Haffen et al., 1982). Goldin (1980) described the mesenchymal influence in terms of different growth rates (e.g., different types of mesenchyme induce different branching patterns in lung development) on the one hand, and in the production of growth stimulating factors (e.g., the so-called Mesodermal Factor, MF) on the other hand (e.g., pancreas development can take place without mesenchyme if the culture medium contains MF). It is most remarkable therefore that a number of enteric structures occurs regularly regardless of the graft's localization and independent of the origin of the mesoderm to which it is attached and which provides the mesodermal component of the gut part of the tissue.

Not only splanchnic, but also somatic and nephrogenic host mesoderm contributes to gut formation in the grafts. No relation has been observed between the development of the graft and the implantation site in the host (lung, heart, body wall, septum transversum, liver, mesonephros). Not only the regional mesenchymal cells should be taken into consideration when looking for factors influencing the endoderm, but both regional growth factors and the regional extracellular matrix cannot be excluded.

Concluding Remarks

Presumptive intra-embryonic endoderm of chick and quail embryos, and presumptive extra-embryonic endoderm of the chick develop into various enteric structures when grafted together with a fragment of notochord into the coelom of a 2-day-old host. Peptide-like immunoreactivity in the entero-endocrine cells corresponds to that in similar divisions of the gut during normal development. It appears that both presumptive intra- and extra-embryonic entoderm have a number of common differentiation tendencies which are not all expressed during normal development.

Entodermal differentiation takes place irrespective of the site of attachment, be it somatopleura, nephrogenic mesoderm or splanchnopleura. Conversely, the local mesoderm differentiates into structures like the deeper layers of the lmucosa, submucosa and muscularis similar to those normally surrounding gut epithelium. The nature of an epithelio-mesenchymal interaction remains obscure. The role of the notochord should be further elucidated. It may exert an influence directly on the entoderm, on the local mesoderm, or on both. In our model, the notochord may promote the expression of the competence of entoderm and/or mesoderm to interact mutually, as well as with other neighboring structures.

Summary

Presumptive intra-embryonic endoderm has a low differentiation tendency when grafted to the coelom of a 2-day-old host. Addition of a fragment of notochord enhances development considerably. A similar influence is found if presumptive extra-embryonic endoderm is used. Various enteric structures and derivatives develop, irrespective of the site of attachement in the host. Mesoderm for the graft is provided by the host's splanchnopleura or somatopleura, nephrogenic mesoderm or by more than one type of mesoderm. Recognizable gut divisions in grafts present APUD cells containing the same peptides as comparable gut structures in normal development.

References

Allen, I.J., and D.F. Newgreen (1977) Catecholamine accumulation in neural crest cells and the primary sympathetic chain. Am. J. Anat. **149:** 413-422.

Goldin, G.V. (1980) Towards a mechanism for morphogenesis in epithelio-mesenchymal organs. Quart. Rev. Biol. **55:** 251-265.

Haffen, K., B. Lacroix, M. Kedinger, and P.M. Simon-Assmann (1983) Inductive properties of fibroblastic cell cultures derives from rat intestinal mucosa on eptihelial differentiation. Differentiation **23:** 226-233.

Hamburger, V., and H.L. Hamilton (1951) A series of normal stages in the development of the chick embryo. J. Morphol. **88:** 49-92.

Hara, K. (1961) Regional neural differentiation induced by prechordal and presumptive chordal mesoderm in the chick embryo. Thesis, Utrecht.

Hoessels, E.L.M.J. (1957) Evolution de la plaque préchordale d'Ambystoma mexicanum; sa différenciation propre et sa puissance inductrice pendant la gastrulation. Thesis, Maastricht.

Kirby, M.L., and S.A. Gilmore (1960) A fluorescence study on the ability of the notochord to synthesise and store catecholamines in early chick embryos. Anat. Rec. **173:** 469-478.

Nieuwkoop, P.D., E.C. Boterenbrood, A. Kremer, F.F.S.N. Bloemsma, E.L.M.J. Hoessels, G. Meyer, and F.J. Verheyen (1952) Activation and organization of the central nervous system in Amphibians. J. Exp. Zool. **120:** 1-107.

Pearse, A.G.E. (1969) The cytochemistry and ultrastructure of polypeptide hormone producing cells of the APUD series and the embryologic, physiologic and pathologic implications of the concept. J. Histochem. Cytochem. **17:** 303-313.

Rawdon, B.B., and A. Andrew (1981) An immunocytochemical survey of endocrine cells in the gastro-intestinal tract of chicks at hatching. Cell Tissue Res. **220:** 279-292.

Rosenquist, G.C. (1971) The location of the pregut entoderm in the chick embryo at the primitive streak stage as determined by radio-autographic mapping. Dev. Biol. **26:** 323-335.

Sigot, M., and L. Marin (1970) Organogenèse de l'estomac de l'embryon de Poulet. Evolution de l'épithélium du proventricule au contact de surfaces conditionnées par des mésenchymes. J. Embryol. Exp. Morph. **24:** 43-63.

Straaten, H.W.M. van, J.W.M. Hekking, F. Thors, E.L.M.J. Wiertz-Hoessels, and J. Drukker (1985) Induction of an additional floor plate in the neural tube. Acta Morphol. Neerl.-Scand. **23:** 91-97.

Vakaet, L. (1970) Cinephotomicrographic investigations of gastrulation in the chick blastoderm. Arch. Biol. **81:** 387-426.

Veini, M., and K. Hara (1975) Changes in the differentiation tendencies of the hypoblast-free Hensen's node during "gastrulation" in the chick embryos. Roux's Archives **177:** 89-100.

Wiertz-Hoessels, E.L.M.J., K. Hara, J.W.M. Hekking, H.W.M. van Straaten, F. Thors, and J. Drukker (1986) Differentiation of gut entoderm in dependence of the notochord. Submitted for publication.

INFLUENCES OF MENINGEAL CELLS ON THE DEVELOPMENT OF THE BRAIN

J. Sievers, D. Hartmann, S. Gude, F.W. Pehlemann, and M. Berry*
Department of Anatomy
University of Kiel
Federal Republic of Germany

* Department of Anatomy
Guy's Hospital Medical School
London
U.K.

Most epithelia are unable to develop in the absence of their surroundings mesenchyme, and the latter influences such fundamental events in epithelial development as cell proliferation, branching morphogenesis and functional differentiation (reviews in Grobstein, 1955; Wessells, 1977; Bernfield and Banerjee, 1978; Hay, 1981, 1984; Cunha et al., 1983; Bernfield et al., 1984). Already the earliest stages of neural development, i.e., the formation of the neural plate in the dorsal ectoderm of the gastrula, are induced by the underlying mesoderm (reviews in Spemann, 1938; Jacobson, 1984). In the spinal cord this neural induction is followed by a series of tissue interactions between neural plate and neural tube on the one hand and the adjacent chorda dorsalis and the somites on the other hand which result in the characteristic shape of the developing spinal cord (see van Straaten and Drukker, this volume). Little is known about similar influences on the development of the brain, yet the mesenchymal partner in such tissue interactions would be the meninges, and the occurrence and significance of a postulated mesenchymal-epithelial interaction during brain development can be assessed if the meningeal cells are eliminated in different stages of CNS ontogeny. In this article we will summarize the evidence collected in our laboratories over the last few years on the influences of meningeal cells on brain development, which we have obtained utilizing a pharmocological method for the selective destruction of meningeal cells. These findings show that in addition to secreting as yet unidentified, probably diffusible molecules stimulating cell proliferation (see Gensburger et al., and Pehlemann et al. this volume), the meningeal cells stabilize the external surface of the cerebellum and hippocampus by producing components of both the interstitial matrix and the basal lamina and thus form the foun-

NATO ASI Series, Vol. H5
Mesenchymal-Epithelial Interactions in Neural Development
Edited by J. R. Wolff et al.

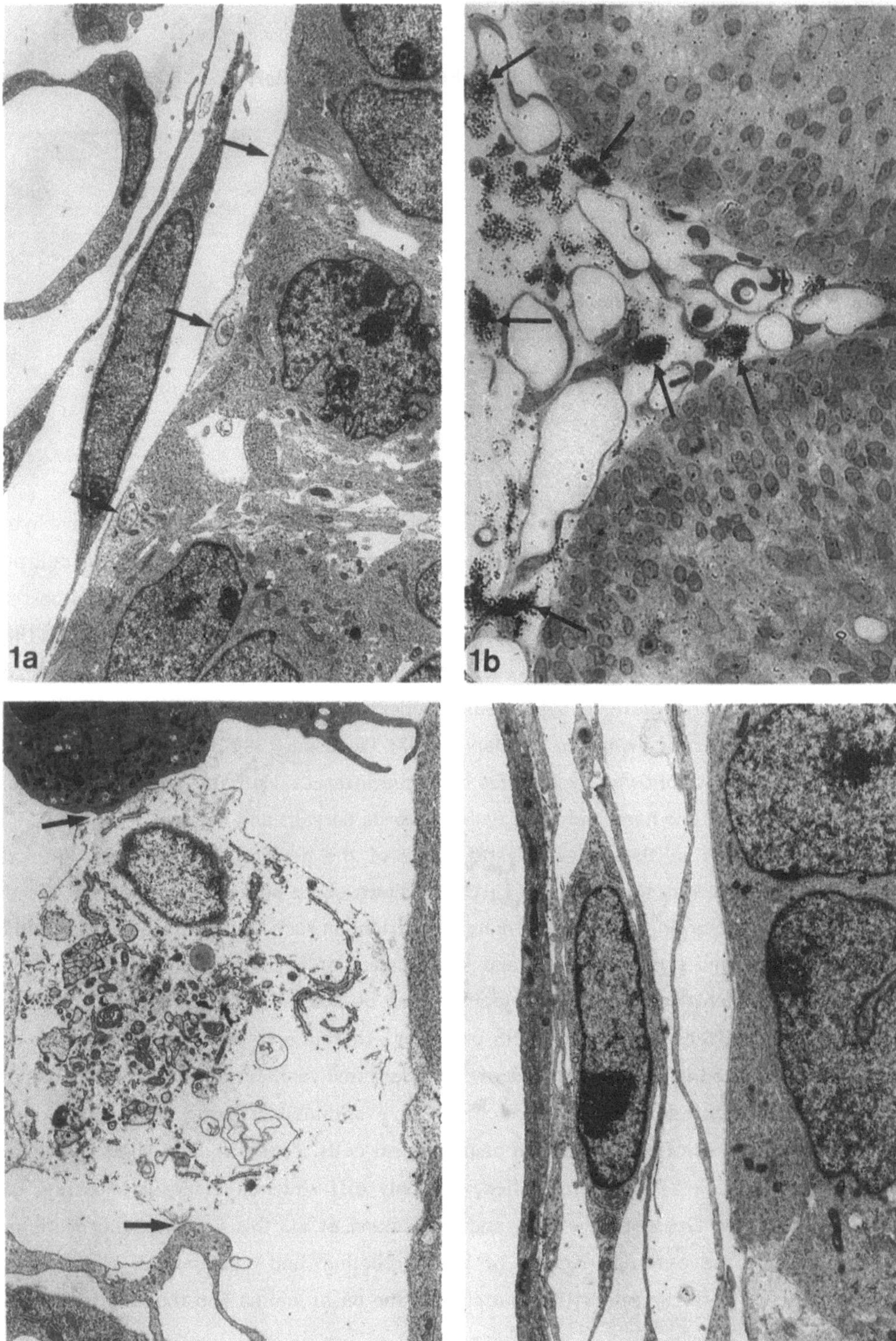
1a
1b
1c
1d

Figure 1: a) Cerebellar surface of a newborn rat. The cerebellar neuroepithelial surface which consists of glial end-feet (arrows) and cell bodies and processes of external granule cells, is covered by several layers of flat meningeal cells. x 5,600. b) Cerebellar surface of a newborn rat 1 hr p.i. of 25 µg ^{3}H-6-OHDA into the IVth ventricle. Note heavy labeling of meningeal cells (arrows), but absence of label over endothelial cells of blood vessels and external granule cells and glial cells in the cerbellum. x 560. c) Meningeal cells 24 hr p.i. of 100 µg 6-OHDA into the IVth ventricle. The plasmalemma is ruptured, the intracellular membrane systems are disintegrated and spilling out of the cell, macrophages surround the cell (arrows). x 6,800. d) Cerebellar surface of a neonate rate 24 hr p.i. of 25 µg 6-OHDA after pretreatment with normetanephrine to block the extraneuronal uptake 2. The meningeal cells do not show any morphological alterations. x 6,800. (Figs. 1a and b from Sievers et al., 1983b; Fig. 1c from Sievers et al., 1985).

dation of the radial glial scaffold present in these regions which, in turn, is involved in establishing the precise pattern of folia and laminae by guiding neuronal migration.

The Selective Destruction of Meningeal Cells with 6-hydroxydopamine and Its Correlation with the Occurrence of Abnormal Brain Development

We have collected 6 independent lines of evidence which show 1) that meningeal cells can be selectively destroyed with the neurotoxin 6-hydroxydopamine (6-OHDA), and 2) that the ensuing absence of meningeal cells is correlated with the appearance of developmental defects in the cerebelum and hippocampus. (i) As shown by autoradiography, ^{3}H-6-OHDA is accumulated intracellularly - as prerequisite for its toxic action (Jonsson, 1980) - only in catecholaminergic neurons and axons and in some extraneuronal cells, most conspicuously, meningeal cells (Fig. 1b), but not into glial cells or undifferentiated neuronal stem cells, e.g., the external granule cells of the cerebellar cortex (Fig. 1b, Sievers et al., 1983a). (ii) The uptake of 6-OHDA into meningeal cells is mediated by the extraneuronal uptake 2 (according to Iversen, 1975), because it can be blocked with inhibitors of uptake 2 (for example, normetanephrine, Fib. 1c, d) but not of the neuronal uptake 1 (for example, nomifensine) (Sievers et al., 1985). In support of the pharmacological demonstration of uptake 2 capacity in meningeal cells, the enzyme catechol-O-methyl-transferase which forms a protein complex with the uptake 2 membrane pump (Bönisch, 1980), has also been localized in the meninges (Kaplan et al., 1979, 1981). Using the differential blockade paradigm, we have shown that inhibition of uptake 2 prevents both the destruction of meningeal cells and the occurrence of developmental malformations, while blockade of uptake 1 has no effect on either event (Fig. 2, Sievers et al., 1985; Pehlemann et al., 1985). (iii) The 6-OHDA-induced alterations of brain development do not show a linear dose-response curve. They are completely absent after doses of 6-OHDA which are not sufficient to destroy the meningeal cells, and present, but always of the same magnitude, at any dose above the threshold to destroy all meningeal cells (Sievers et al., 1985; Pehlemann et al., 1985). (iv) Transplantation of fetal meningeal

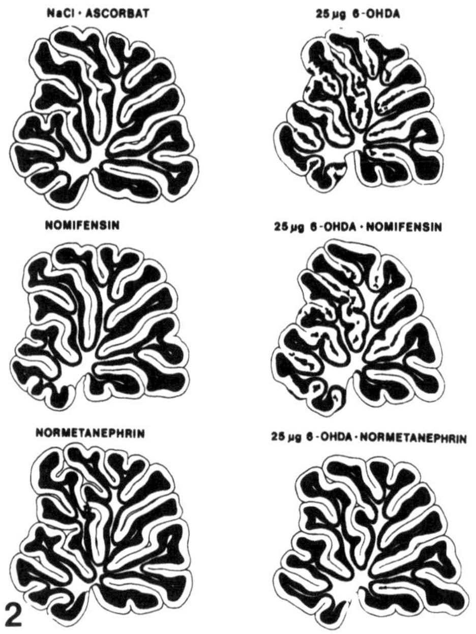

2

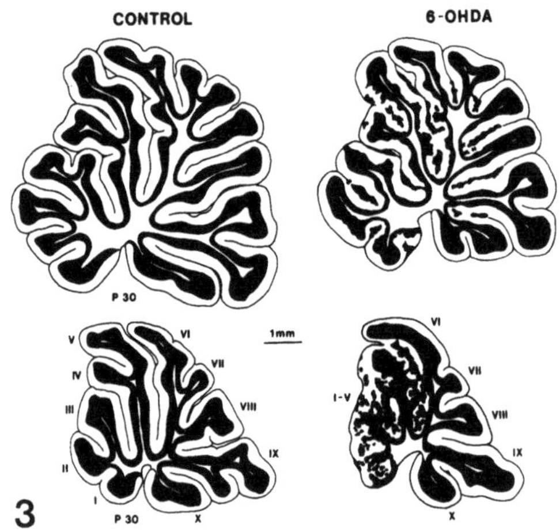

3

Figure 2: Camera lucida drawings of paramedian sagittal sections through the cerebellar vermis of rats 30 days after treatment with either vehicle solution (NaCl plus 0.1% ascorbate), or 25 µg 6-OHDA, and after treatments with either nomifensine (to block the extraneuronal uptake 2) before the injection of 25 µg 6-OHDA, with the respective control experiments. Note that the changes in cerebellar morphology induced by 6-OHDA-treatment are similar in localization and magnitude after blockade of the neuronal uptake 1 with nomifensine, but completely absent after blockade of the extraneuronal uptake 2 with normetanephrine. (From Pehlemann et al., 1985)

Figure 3: Camera lucida drawings of parasagittal sections of the cerebellum of rats (above) and hamsters (below) 30 days after injection of either ascorbate saline vehicle (control) or 6-OHDA on the day of birth. In the controls, folia I to X are separated by fissures, and lamination is regular (the solid black internal granular layer separates the superficial molecular layer from the deep medullary tree, the Purkinje cell layer is at the interface between molecular layer and internal granular layer). The most conspicuous alterations in treated animals are a loss of folial integrity (in the hamster folia I to V are not recognizable) and a disruption of lamination by extensive granule cell ectopia (indicated by the black areas separated from the internal granular layer). (From Sievers et al., 1986a).

cells onto the cerebellum of newborn rats whose endogenous meningeal cells had been destryoed with 6-OHDA 12 hours before, alleviates the 6-OHDA-induced cerebellar defects (Pehlemann and Sievers, 1986). (v) Treatment of newborn rats with DSP-4 which is accumulated by the uptake 1 for catecholamines, and thus destroys the catecholaminergic innervation of the cerebellum (Jonsson et al., 1982), but not by the extraneuronal uptake 2 (Böke and Sievers, in preparation), does neither destroy meningeal cells, nor cause any alterations in cerebellar development comparable to those after 6-OHDA-treatment (Böke and Sievers, in preparation). (vi) The effects of destruction of meningeal cells on brain development exhibit a critical period that is different from that seen after various other experimentally used cell toxins or x-rays (Sievers et al., 1986b, and see below).

Destruction of Meningeal Cells Disturbs Foliation and Lamination in the Cerebellar Cortex

The alterations of cerebellar development after destruction of meningeal cells in new-born rats or hamsters are so severe that they can be recognized even on a gross macro-scopical level (Fig. 3). In addition to an overall decrease in size (Allen et al., 1981; Sievers et al., 1980, 1981, 1983b; Pehlemann et al., 1985; v. Knebel Doeberitz et al., 1986) - the basis of which is a reduction in cell proliferation (see Pehlemann et al., this volume) - the shape is affected due to the fusion of neighboring folia, and the regular lamination of the cerebellar cortex is disturbed as a result of the displacement of granule cells and Purkinje cells as well as the disruption of the normal glial scaffold of Bergmann glia (Fig. 4). The magnitude of all these defects is the greater the earlier the ontogenetic stage at which the meningeal cells are destroyed (v. Knebel Doeberitz et al., 1986; Sievers et al., 1986b, see below), in the most severe cases resulting in the complete ab-sence of foliation and lamination in the anterior cerebellar vermis (Fig. 4).

These massive alterations of cerebellar structure are at the end of a chain of events that is triggered by the destruciton of meningeal cells within the first 6 hours after injection of 6-OHDA. The death and removal of meningeal cells is followed by a thinning and break up of the basal lamina at the cerebellar surface. Both external granule cells and Bergmann glia end-feet which are normally in contact with the basal lamina (Fig. 1a) start to form blebs in areas where the overlying basal lamina is lost (Fig. 5a), and to grow into the subarachnoid space (Fig. 5b). With time, the external granule cells migrating from the cerebellar cortex into the subarachnoid space rapidly increase in number and accumulate predominantly in the depths of fissures. They are accompanied by Bergmann glia fibers that sprout from their end-feet, and elongate and branch among the steadily increasing numbers of fissural ectopic granule cells (Fig. 6). This outgrowth of both external granule cells and Bergmann glia fibers results in a dissolution of the normal glia limitans superficialis, a disruption of the cortical glial scaffold, and a loss of normal meningeal-glial relationships which is most pronounced in the anterior vermis of the hamster (Fig. 7). Adjacent folia thus become fused through the intervening ectopic neurons and glia. The egress of external granule cells and the outgrowth and branching of Bergmann glia fibers rapidly fill the subarachnoid space between the basal parts of the folia, and effectively prevent the reestablishment of a meningeal cell population and,

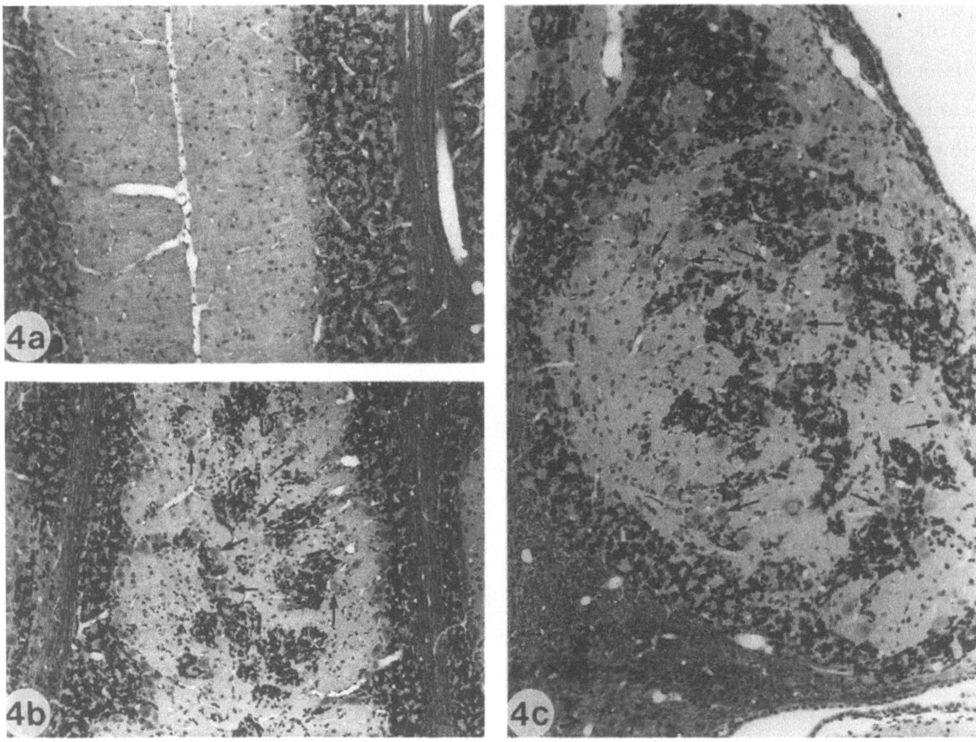

Figure 4: Hamster cerebellar cortex 20 days p.i. of either vehice (a) or 6-OHDA (b and c) on the day of birth. In the primary fissure (b) many Purkinje cells (arrows) and large groups of granule cells are located ectopically within the fused molecular zone of adjacent folia; in the anterior part of the vermis (c) lamination is completely lost, and granule cells and Purkinje cells are irregularly scattered throughout fused folia I to III. a and b x 260, c x 340 (From v. Knebel Doeberitz et al., 1986).

consequently, normal glia-meningeal relationships over these parts of the cerebellar surface during the recovery phase. By contrast, the superfical parts of the affected folia, and those formed in later ontogenetic stages, become recovered with meningeal cells, regenerating from the dura, and thereafter develop normally. The maldevelopment of the Purkinje cell layer in hamsters is not sufficiently understood, but the displacement of Purkinje cells is preceded by a corresponding displacement of the cell bodies of the Bergmann glia to which Purkinje cells normally have contact. A similar correlation between the ectopic position of Purkinje cells and Bergmann glia cell bodies has also been observed in the human nodulus (Braak, 1974).

Thus, destruction of meningeal cells destabilizes the superficial glia limitans and the glial scaffold of the cerebellum. These changes in the organization and differentiation of glia, in turn, cause the defects in segregation and lamination of neurons as a result of the disturbance of the glial-neuronal interaction mandatory for normal neuronal migration (Rakic, 1971, 1982) during cerebellar development.

Meningeal Cells Influence Cerebellar Development Over a Critical Period

Destruction of meningeal cells at different stages in ontogeny has different effects on the development of the cerebellum. The most evident correlation is between the state of foliation at the time of treatment and the later distribution of the 6-OHDA-induced defects in that these predominantly occur in those fissures present or forming at the time of destruction of meningeal cells (Sievers et al., 1986b, Fig. 8a). The clefts between epithelial lobules are predilection sites for the occurrence of morphogenetic defects in many other epithelia after various experimental manipulations to alter the extracellular matrix at the epithelial surface (Grobstein and Cohen, 1965; Wessells and Cohen, 1968; Spooner and Faubion, 1980). Thus, we have interpreted the preferential occurrence of 6-OHDA-induced defects in cerebellar fissures to indicate (i) that there is either a differential composition and/or turnover of the extracellular matrix over the cerebellum analogous to that over other epithelia with a branching morphogenesis in which the basal lamina in interlobular clefts is stabilized by mesenchyme-derived interstitial fibrillary collagen (Bernfield et al., 1984), or (ii) that this distribution of defects is caused by different rates of regeneration of meningeal cells from the overlying dura mater, which is fastest over the folial apices and slowest into the fissures (Sievers et al., 1986b).

The end of the critical period of meningeal cell influence on the cerebellum (in the hamster or postnatal day 5) correlates with the maturation of the glia limitans superficialis. In the cerebellum, both Bergmann glia end-feet and processes or cell bodies of external granule cells are in contact with the basal lamina (Fig. 1a, Sievers et al., 1981; Hausmann and Sievers, 1985); however, their proportions change with age. In the hamster, for example, the lowest proportion of glial end-feet is present on postnatal day 1, amounting to about 70% of the whole internal surface (Sievers et al., 1986b). This percentage increases rapidly within the next few days and on postnatal day 5 reaches a plateau of about 90% (Fig. 8b) which remains at this level up to the end of the proliferative period of the external granular layer. Thus, the end of the inducibility of cerebellar defects after destruction of meningeal cells coincides with the beginning of the plateau-phase of the development of the glia limitans superficialis which at this stage may have reached a threshold end-feet density to maintain the integrity of the cerebellar surface alone (see below).

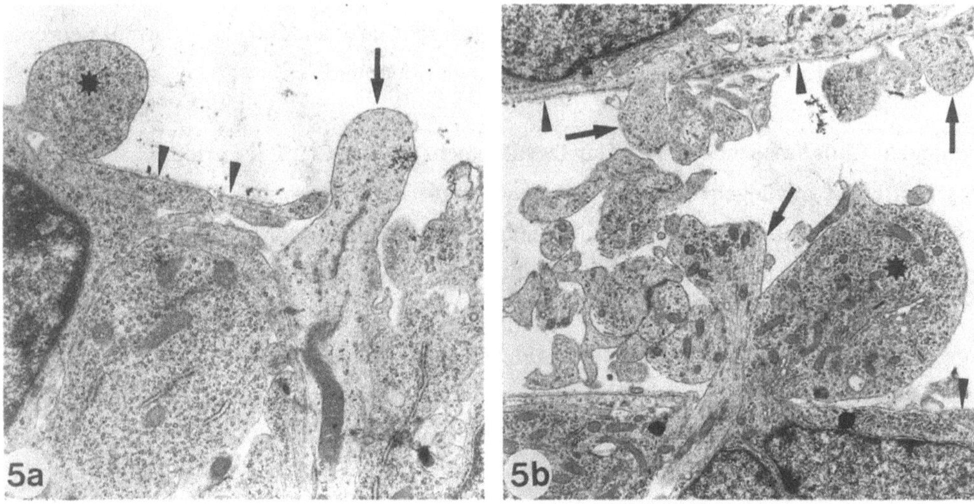
5a
5b

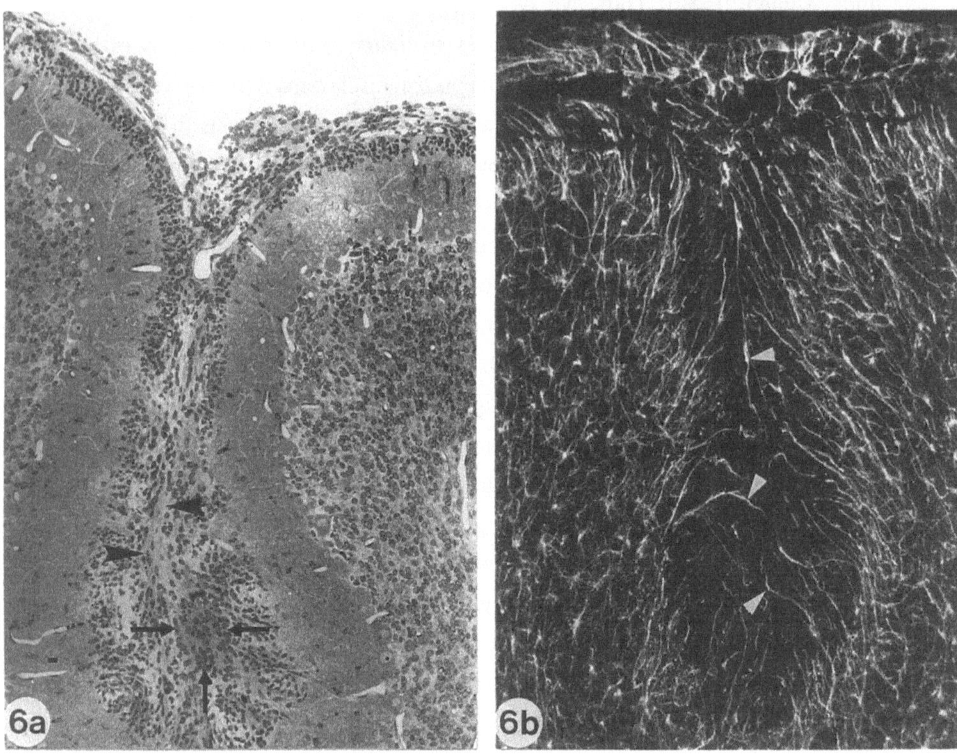
6a
6b

Figure 5 (facing page): Cerebellar surface 5 days p.i. of 6-OHDA. The basal lamina is thinned (arrowheads) and ruptured. External granule cells (asterisks) and Bergmann glia fibers (arrows) protrude into the subarachnoid space of the fissures that are devoid of meningeal cells. a x 12,360, b x 7,820. (From Sievers et al., 1981 and 1983b).

Figure 6 (facing page): a) Fusion of cerbellar fissure between folia VI and VII 15 days p.i. of 6-OHDA. Emigrating external granule cells from adjacent folia have obliterated the fissure; some of them have formed colonies of differentiated granule cells (between the arrows), whose axons have grown towards the cerebellar surface (large arrowheads). x 140 (From Sievers et al., 1983b). b) Immunohistochemical demonstration of glial fibrillar acidic protein (GFAP) to show the glial scaffold in the region of an obliterated fissure. The radial alignment of Bermgann glia fibers towards the folial surface (compare with Fig. 7a) is lost, instead, most fibers are directed towards the cerebellar surface below the anterior medullar velum (above). Many of the Bergmann glia fibers have grown out of their original position and invaded the region of fusion, where they branch among the ectopic granule cells (arrowheads). x 270.

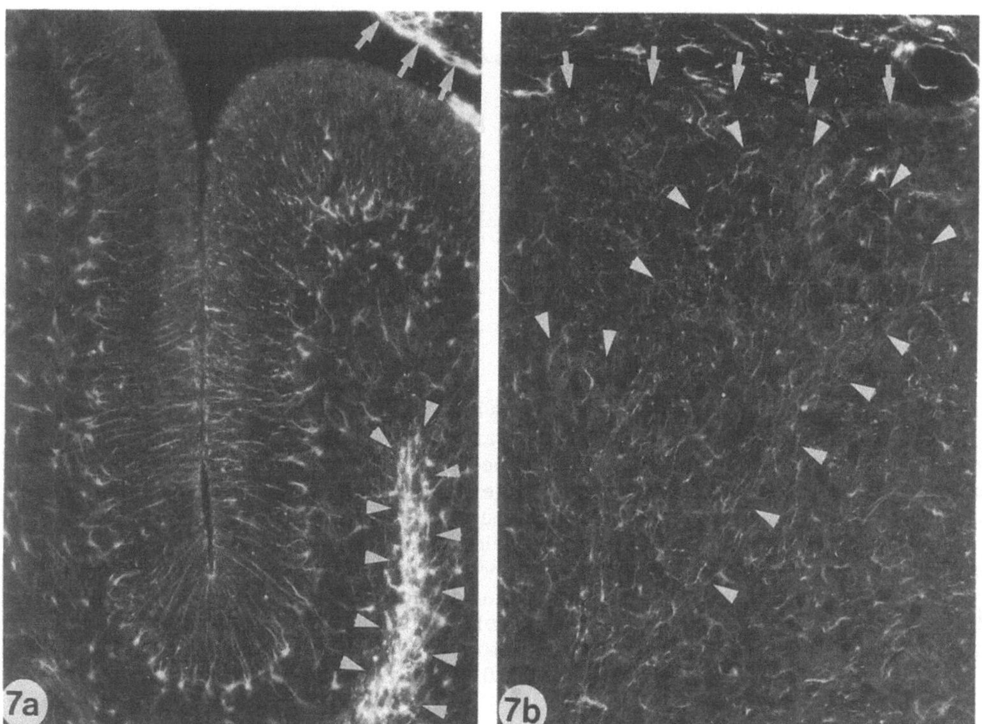

Figure 7: Immunohistochemical demonstration of the glial scaffold (using antibodies to GFAP) in control (a) and 6-OHDA-treated (b) hamsters 10 days p.i. at birth. Identical regions of the rostral cerebellar vermis are shown, the surface of the mesencephalon is at the top (arrows). For orientation refer to Fig. 4c. The typical radially orientated glial scaffold consisting of Bergmann glia is present in the control animal. The intense immunoreactivity at the right corner is due to the fibrous astrocytes of the medullary layer (arrowheads). In contrast, in the treated animal, the Bergmann glia scaffold is completely absent, and the dense accumulation of fibrous astrocytes is split up into widely spaced parallel bundles (arrowhead). a and b x 170.

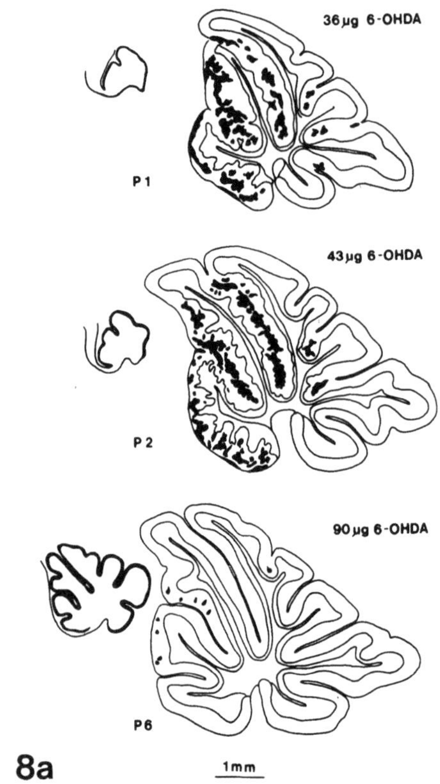

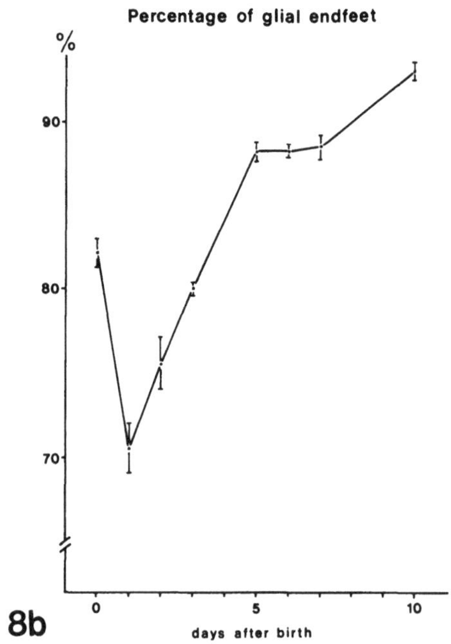

Figure 8: a) Composite camera lucida drawings of cerebellar vermes (3 animals per group) of 30-day-old hamsters that were treated with 6-OHDA on postnatal (P) day 1, 2 or 6 (the outlines of the cerebella at the time of treatment are shown to the left of the adult vermes). The ectopic groups of granule cells are shown in solid black. Note that granule cell ectopia spreads caudally (compare the animals treated on P1 or P2 with those treated on P0, Fig. 3) as fissures develop caudally at the time of treatment, and that animals treated on P6 are relatively normal (the ectopias shown are the sum of 3 animals). Note also that on P6 the primordia of all folia are present. b) Percentage of glial end-feet covering the internal surface of the basal lamina in hamster cerebella of different ages. The rapid increase from 70% to about 90% between postnatal days 1 and 5 is followed by a plateau between days 5 and 10. (From Sievers et al., 1986a).

Destruction of Meningeal Cells Prevents the Formation of the Infrapyramdal Blade of the Dentate Gyrus

After destruction of meningeal cells over the medial cerebral hemispheres in newborn hamsters, the dentate gyrus of adult hamsters lacks the lateral three-fourths of its infrapyramidal blade (Figs. 9a, b, Hartmann et al., 1986a). These alterations are most severe in the septal portion of the dentate gyrus and less pronounced in the temporal part. They are prevented by blocking the extraneuronal uptake 2 with normetanephrine (Fig. 9a,

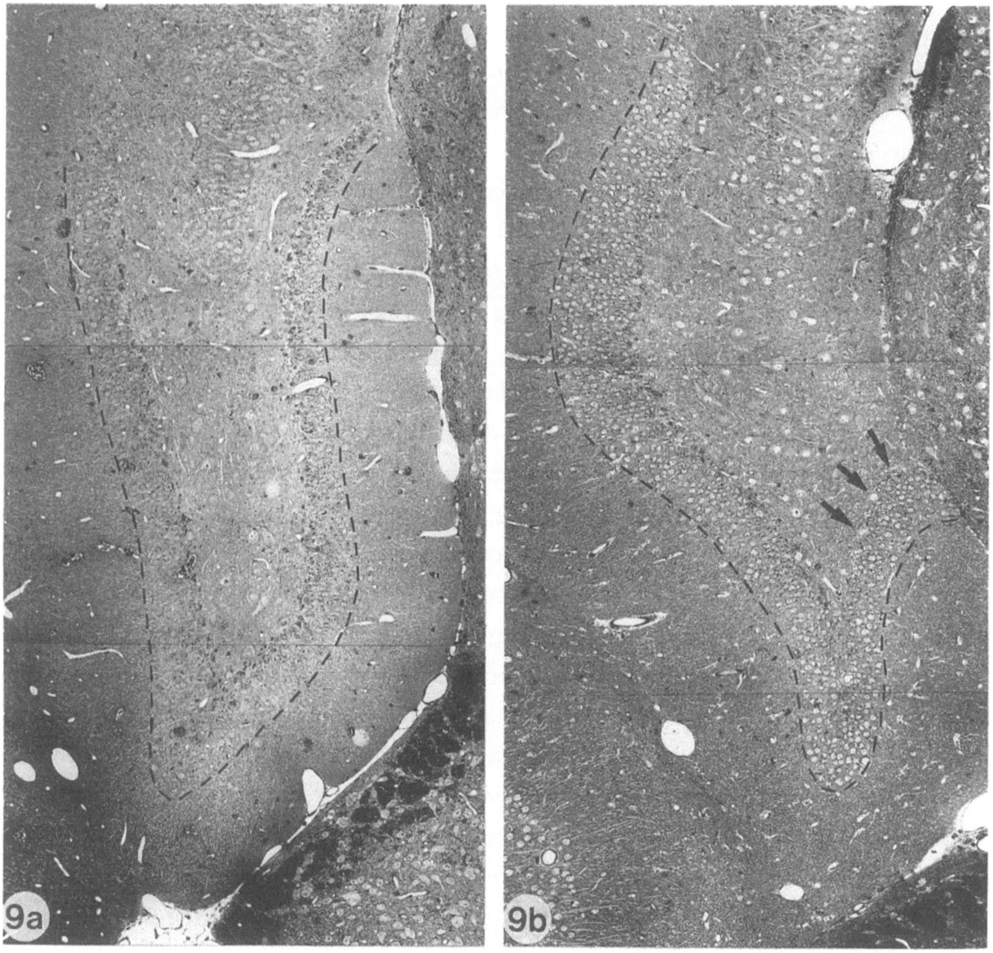

Figure 9: Coronal sections through the hippocampus at the level of the optic chiasm in hamsters 30 days after interhemisphaeric injection of normetanephrine plus 6-OHDA (a) or 6-OHDA alone (b). The outer contours of the dentate granule cell layers are delimited by the broken black line. The infrapyramidal blade of the dentate gyrus is almost completely lacking in animals treated with 6-OHDA alone, while it is present in animals that had been pretreated with normetanephrine to block the extraneuronal uptake of 6-OHDA into meningeal cells. a and b x 100.

Hartmann et al., 1986a) showing that they are selectively caused by the effects of 6-OHDA on meningeal cells, and also exhibit a critical period that ends on postnatal day 4. The radial glial scaffold of the dentate gyrus, with cell bodies located in the subgranular zone and radial processes running through the stratum granulosum and the molecular layer, is absent in the missing lateral parts of the prospective infrapyramidal blade, and irregularly orientated in the rudiment of the infrapyramidal blade.

The rearrangement of the dentate gyrus starts within the first few days after destruction of meningeal cells. In these early postinjection stages the growing suprapyramidal blade is elongated by the addition of granule and glial cells (apparently destined for the infrapyramidal blade under normal conditions) and extends up to the surface of the medial cerebral hemisphere (Figs. 10a, b). Lateral to the tip of the suprapyramidal blade, in the medialmost part of the prospective infrapyramidal blade, is a cell-sparse zone that is followed laterally by an accumulation of irregularly orientated undifferentiated cells adjacent to the fimbria of the fornix (Fig. 10b). Many of these undifferentiated cells which seem to be arrested along the normal path of migrating dentate precursor cells (across the fimbria, tangentially along the surface of the dentate gyrus and into the hilus; see Hausmann et al., this volume), can be identified as glial cells both in controls and in treated animals (Figs. 10c, d; Hartmann et al., 1986b). Thus, the topographical distribution of undifferentiated cells in the prospective infrapyramidal blade (Fig. 10b), and their colocalization with GFAP immunoreactivity (Fig. 10d) indicates firstly, that many of these cells are astroglial cells, and, secondly, that the destruction of meningeal cells interfered with their translocation from the fimbria into the secondary proliferative zone of the dentate gyrus (Hartmann et al., 1986b). The resultant reduction of glial cells migrating into the secondary proliferative zone of the dentate gyrus probably forces the granule cells generated within this region to aggregate with the granule cells and glial cells of the suprapyramidal blade which is thereby elongated (Fig. 10b, d). In later stages after destruction of meningeal cells (i.e., 15 and 30 days post injection) the accumulation of undifferentiated cells in the lateral portion of the prospective infrapyramidal blade disappears and a short rudiment of an infrapyramidal blade is formed.

Although many details of these alterations of the development of the dentate gyrus remain to be elucidated, we propose that glial cells move continuously from the ventricular zone along (or out of) the radial glial scaffold of the fimbria fornicis and the surface of the dentate gyrus into the secondary proliferative zone, where they aggregate with the dentate stem cells and the postmitotic granule cells to form the radial glial scaffold of the dentate granular layer. Either the movement and/or the radial alignment of these glial cells is disturbed by the destruction of meningeal cells and the subsequent changes in the composition of the extracellular matrix at the dentate surface.

Meningeal Cells Synthesize at Least 3 Types of Fibrillary Collagens and the Major Constituents of the Basal Lamina

Immunohistochemical studies with antibodies against various extracellular matrix components show that in vitro meningeal cells synthesize the fibrillary collagens type I, III, and VI, and the basal lamina constituents collagen type IV, laminin, nidogen, and heparan sulfate proteoglycan, as well as fibronectin that is associated with the basal lamina

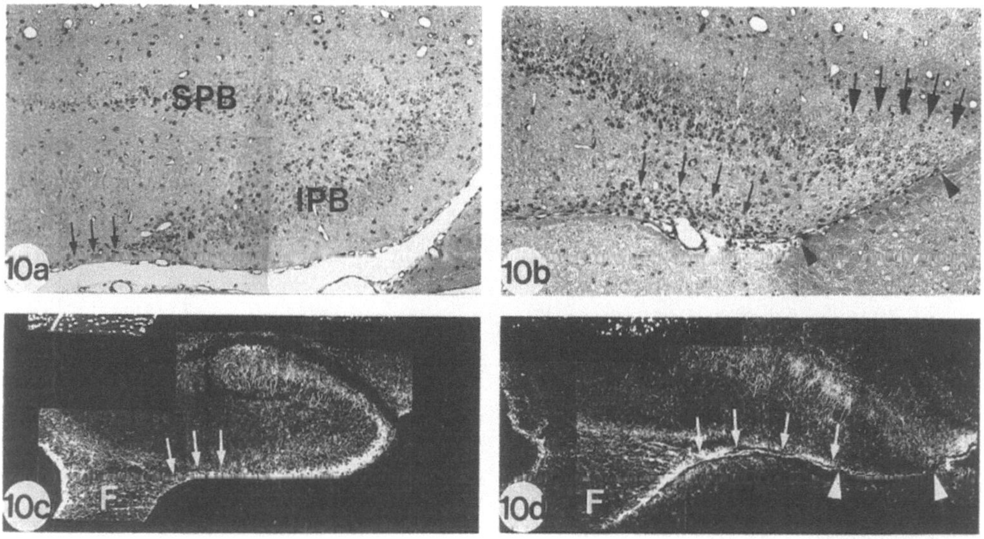

Figure 10: Coronal sections through the septal hippocampus in hamsters 7 days after interhemispheric injection of either ascorbate/saline vehicle (a and c) or 6-OHDA (b and d) on the day of birth. In control animals (a) the suprapyramidal blade (SPB) the tip and the infrapyramidal blade (IPB) of the dentate gyrus are almost completely developed, only few undifferentiated cells move into the lateralmost part of the infrapyramidal blade from the fimbria tangentially along the surface of the hippocampus (small arrows). In treated animals (b) the suprapyramidal blade is present, appears thicker as in controls and does not bend into the infrapyramidal blade but is elongated up to the medial surface (large arrows). In the medial part of the prospective infrapyramidal blade (between the arrowheads) very few cells are present, while further laterally many undifferentiated cells have accumulated (small arows) which lie in the tangential migratory path between fimbria and hilus. The immunohistochemical demonstration of GFAP shows that the radial glial cells of the dentate gyrus are completely developed in controls (c) and that some immunoreactive cells are present between fimbria (F) and the lateral parts of the infrapyramidal blade (small arrows). In treated animals (d) radial glia are only present in the suprapyramidal blade. In the cell sparse zone of the medial part of the prospective infrapyramidal blade (between the arrowheads) immunoreactivity is completely absent, while in the lateral part where the undifferentiated cells have accumulated strong immunoreactivity is found (small arrows). a and b x 84, c and d x 50.

(Gude et al., 1986; Fig. 11). In sections, meningeal cells over the cerebellum of normal rats and hamsters are stained with antibodies against the fibrillary collagens I, III, and VI and fibronection, collagen type VI, nidogen and laminin (Fig. 11). After destruction of meningeal cells, the intensity of immunoreactivity against the fibrillary collagens, fibronectin, collagen type IV and, most conspicuously, laminin, at the surface of the cerebellum is decreased (Gude et al, 1986). These findings correlate well with the ultrastructural observation of a thinning and rupture of the basal lamina after destruction of meningeal cells (Fig. 5). Thus, meningeal cells are not only capable of producing fibrillary

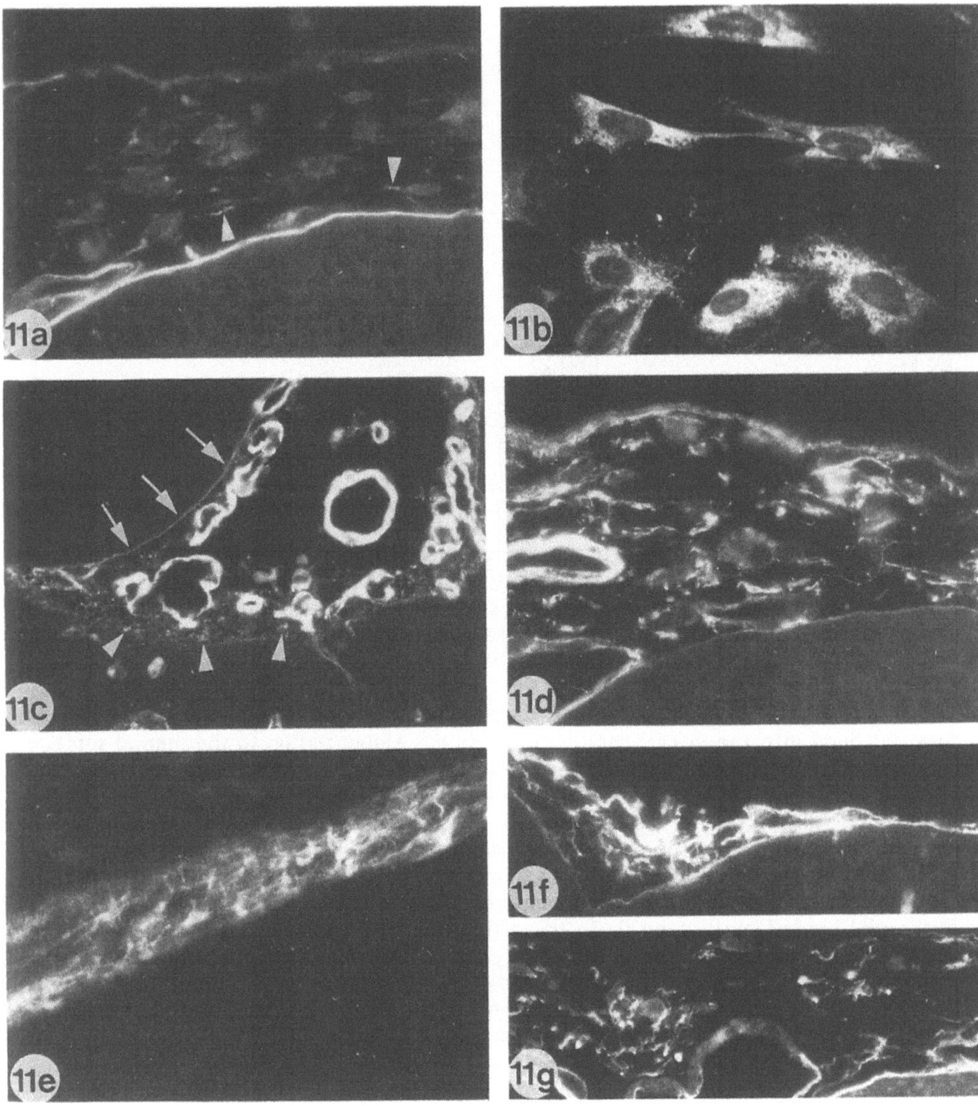

Figure 11: a) immunohistochemical demonstration of punctate laminin reactivity on meningeal cells (arrowheads) over the cerebellum of a newborn rat. Note strong reactivity of the cerebellar and vascular basal lamina. b) Laminin positive meningeal cells in vitro, 24 hours after incubation with ascorbate. c) Punctate collagen type IV immunoreactivity (arrowheads on meningeal cells over the cerebellum of newborn hamsters. Note the low intensity of the cerebellar basal lamina (arrows) in comparison to that surrounding blood vessels. d) Fibronectin-immunoreactivity on meningeal cells over the cerebellum of a 5 day old hamster. e) Collagen type I immunoreactivity in the fissure between cerebellum and mesencephalon of a newborn rat. Collagen type II (f) and collagen type VI (g) immunoreactivity on the cerebellar surface of a 4 day old hamster. a x 544, b x 384, c x 340, d x 544, e x 544, f x 340, g x 340.

collagens and fibronectin for cell attachment, but also all structural constituents of a compete basal lamina, which most likely is deposited onto the surface of the brain, supplementing the basal lamina produced by the glial end-feet (see below).

Possible Mechanisms of the Influences of Meningeal Cells on the Development of the Brain

There are several common denominators that characterize the circumstances which operate in disturbing neural development after destruction of meningeal cells with 6-OHDA. (i) Alterations in development appear only in regions and in ontogenetic stages in which the glia limitans superficialis is not complete, but interrupted by intercalated neuroepithelial stem cells. So far, they have only been observed in the cerebellum and the dentate gyrus and in early stages of development of the cerebral cortex (Lidov and Molliver, 1982). (ii) In the case of the cerebellum, the defects are localized in regions, which are known to be sensitive to alterations of the extracellular matrix, i.e., cerebellar fissures (equivalent to the interlobular clefts, see Bernfield et al., 1984). (iii) The destruction of meningeal cells affects primarily the amount and composition of the extracellular matrix at the brain surface and the development and structure of the glial scaffold, while the defects in neuronal development seem to be secondary to these events. (iv) Meningeal cells are capable of producing not only several components of the mesenchymal interstitial matrix at the brain surface, but also all structural components of the basal lamina which usually is thought to be produced by glial end-feet alone. The synthesis of basal lamina constituents and their deposition into the basal lamina by mesenchymal cells has also been reported for muscle fibroblasts (Kühl et al., 1982, 1984, Sanderson et al., 1986) and for Wi 38 fibroblasts (Alitalo, 1980) in vitro.

The foregoing list raises some hypotheses for the explanation of the developmental defects arising after destruciton of meningeal cells. The exclusive occurrence in brain regions with an incomplete glia limitans superficialis could be explained by postulating that the non-glial neuroepithelial stem cells (e.g., external granule cells) are incapable of producing basal lamina, and that the basal lamina over these cells is either produced or stabilized (see below) by the meningeal cells and therefore becomes defective after their destruction. If this should be the case, one can predict that meningeal cells would become necessary for the stabilization of the outer surface of the developing brain as soon as neuronal stem cell lines start to separate from the undifferentiated neuroepithelial cells and/or the glial stem cell lines, probably immediately after closure of the neural tube.

The preferential occurrence of developmental defects in the cerebellar fissures could either be explained by transferring the basement membrane remodeling hypothesis of Bernfield et al. (1984) to the cerebellum (Sievers et al., 1986a), which places the emphasis on the production of fibrillary collagens by the mesenchymal cells in order to stabilize morphogenetically less active regions of a folding epithelium, or by postulating that the regional differences of maldevelopment are the result of different rates of repopulation of the cerebellar surface with meningeal cells which produce integral parts of the basal lamina. In this case, meningeal cells regenerating from the dura mater would reach the folial apices much earlier than the interfolial fissures and, thus, prevent the occurrence of cerebellar defects (see above). In both instances, meningeal cells would be required to stabilize the basal lamina of the brain by producing components of the extracellular matrix which either are directly incorporated into the basal lmaina or indirectly alter the metabolism of the basal lamina.

In addition to the loss of extracellular matrix components, destruction of meningeal cells also causes changes in the development of the glial cells in affected brain regions. These appear as (i) disruption of the glia limitans superficialis, (ii) disorganisation of the glial scaffold by outgrowth of processes and displacement of cell bodies, (iii) cessation of basal lamina production, (iv) loss of S-100 immunoreactivity (Gude et al., 1986) and (v) arrest of glial cell migration in the hippocampal anlage. Thus, meningeal cells appear to act (i) as spatial cues for glial cells to form end-feet and a scaffold of defined three dimensional orientation, and (ii) as a necessary prerequisite to induce basal lamina production by glial cells. It remains to be elucidated whether the extracellular matrix produced by the meningeal cells is suffient to mediate these influences, or whether additional molecules or cell-cell contacts are involved in these mesenchymal-epithelial interactions.

Acknowledgements: The antibodies against collagens type I, IV and laminin were kindly supplied by Dr. K. v.d. Mark, antibodies against collagen type IV and heparan sulfate proteoglycan by Dr. M. Paulsson, MPI für Biochemie, Martinsried; and those against collagens type III and VI by Dr. D. Schuppan, Medizinische Klinik, Klinikum Steglitz, FU Berlin. We thank Waltraud Müller, Sibille Piontek, Marlis Rall, Ingrid Riettiens and Rosemarie Sprang for excellent technical assistance, and Jutta Schlahn for typing the manuscript.

References

Alitalo, K. (1980) Production of both interstitial and basement membrane procollagens by fibroblastic Wi 38 cells from human embryonic lung. Biochem. Biophys. Res. Commun. **93:** 873-880.

Allen, C., J. Sievers, M. Berry, and S. Jenner (1981) Experimental studies on cerebellar foliation. II. A morphometric analysis of cerebellar fissuration defects and growth retardation after neonatal treatment with 6-OHDA in the rat. J. Comp. Neurol. **202:** 771-784.

Bernfield, M.R., and S.D. Banerjee (1978) The basal lamina in epithelial-mesenchymal morphogenetic interactions. In *Biology and Chemistry of Basement Membranes*, N.A. Kefalides, ed., pp. 137-148. Academic Press, New York.

Bernfield, M.R., S.D. Banerjee, J.E. Koda, and A.C. Rapraeger (1984) Remodeling of the basement membrane as a mechanism of morphogenetic tissue interaction. In *The role of extracellular matrix in development*, R. Trelstad, ed., pp. 545-572. Alan R.

Liss, New York.
Böke, R., and J. Sievers (1986) Effects of the noradrenergic neurotoxic DSP4 on the development of the locus coeruleus and the cerebellum. In preparation.
Bönisch, H. (1980) Extraneuronal transport of catecholamines. Pharmacology. **21:** 93-108.
Braak, H. (1974) On the intermediate cells of Lugaro within the cerebellar cortex of man. Cell Tissue Res. **149:** 399-411.
Cunha, G.R., L.W.K. Chung, J.M. Shannon, O. Taguchi, and H. Fujii (1983) Hormone-induced morphogenesis and growth: Role of mesenchymal-epithelial interactions. Rec. Prog. Hormone Res. **39:** 559-598.
Grobstein, C. (1955) Tissue interaction in the morphogenesis of mouse embryonic rudiments in vitro. In *Aspects of Synthesis and Order in Growth*, D. Rudnick, et., pp. 233-256. Princeton University Press, Princeton.
Grobstein, C., and J. Cohen (1965) Collagenase: Effect on the morphosgenesis of embryonic salivary epithelium in vitro. Science **150:** 626-628.
Gude, S., J. Burmester, J. Sievers, and F.W. Pehlemann (1986) Meningeal cells synthesize components of both the interstitial matrix and the basal lamina at the surface of the brain. In preparation.
Hartmann, D., J. Sievers, and F.W. Pehlemann (1986a) Die Zerstörung der Meningealzellen über dem medialen cerebralen Cortex neugeborener Hamster verhindert die Bildung des infrapyramidalen Blattes des Gyrus dentatus. Anat. Anz. **161:** 136.
Hartmann, D., J. Sievers, and F.W. Pehlemann (1986b) Die Entwicklung des Gyrus dentatus nach Zerstörung der Meningealzellen: Analyse der Zellwanderung und des Gliagerüstes. Anat. Anz (Erg.-H.), in press.
Hausmann, B., and J. Sievers (1985) Cerebellar external granule cells are attached to the basal lamina from the onset of migration up to the end of their proliferative activity. J. Comp. Neurol. **241:** 50-62.
Hay, E.D. (1981) Collagen and embryonic development. In *Cell Biology of Extracellular Matrix*, E.D. Hay, ed., pp. 379-409. Plenum, New York.
Hay, E.D. (1984) Cell-matrix interaction in the embryo: Cell shape, cell surface, cell skeletons, and their role in differentiation. In *The Role of Extracellular Matrix in Development*, R. Trelstad, ed., pp. 1-31. Alan R. Liss, New York.
Iversen, L.L. (1975) Uptake processes for biogenic amines. In *Handbook of Psychopharmacology*, L.L. Iversen, S.D. Iversen, and S.H. Snyder, eds., pp. 96-117, vol. 3. Plenum, New York.
Jacobson, M. (1984) Cell lineage analysis of neural induction: Origins of cells forming the induced nervous system. Dev. Biol. **102:** 122-129.
Jonsson, G. (1980) Chemical neurotoxins as denervation tools in neurobiology. Ann. Rev. Neurosci. **3:** 169-187.
Jonsson, G., H. Hallmann, and E. Sundström (1982) Effects of the noradrenaline neurotoxin DSP4 on the postnatal development of central noradrenaline neurons in the rat. Neuroscience **7:** 2895-2907.
Kaplan, G.P., B.K. Hartmann, and C.R. Creveling (1979) Immunhistochemical demonstration of catechol-O-methyltransferase in mammalian brain. Brain Res. **167:** 41-52.
Kaplan, G.P., B.K. Hartmann, and C.R. Creveling (1981) Localization of catechol-O-methyltransferase in the leptomeninges, choroid plexus and ciliary epithelium: Implications for the separation of central and peripheral catechols. Brain Res. **204:** 353-360.
von Knebel Doeberitz, Ch., J. Sievers, M. Sadler, F.-W. Pehlemann, M. Berry, and P. Halliwell (1986) Destruction of meningeal cells over the newborn hamster cerebellum with 6-hydroxydopamine prevents foliation and lamination in the rostral cerebellum. Neuroscience **17:** 409-426.
Kühl, U., R. Timpl, and K. v.d. Mark (1982) Synthesis of type IV collagen and laminin in cultures of skeletal muscle cells and their assembly on the surface of myotubes. Dev. Biol. **93:** 344-354.
Kühl, U., M. Öchalan, R. Timpl, R. Mayne, E. Hay, and K. v.d. Mark (1984) Role of

muscle fibroblasts in the deposition of type IV collagen in the basal lamina of myotubes. Differentiation **28:** 164-172.
Lidov, H.G.W., and M.E. Molliver (1982) The structure of cerebral cortex in the rat following prenatal administration of 6-hydroxydopamine. Dev. Brain Res. **3:** 81-108.
Pehlemann, F.W., J. Sievers, and M. Berry (1985) Meningeal cells are involved in foliation, lamination and neurogenesis of the cerebellum: Evidence from 6-hydroxydopamine-induced destruction of meningeal cells. Dev. Biol. **110:** 136-146.
Pehlemann, F.W., and J. Sievers (1986) Die Transplantation fetaler Meningealzellen auf das Kleinhirn neugeborener Ratten verringert das Auftreten 6-OHDA-induzierter Veränderungen der Schichtenbildung. Anat. Anz. **161:** 154-155.
Rakic, P. (1971) Neuron-glia relationship during granule cell migration in developing cerebellar cortex: A Golgi and electron microscopic study in Macacus rhesus. J. Comp. Neurol. **141:** 283-312.
Rakic, P. (1982) The role of neuronal-glial cell interaction during brain development. In *Neuronal-glial Cell Interrelationships*, T.A. Sears, ed., pp. 25-38. Springer, Heidelberg and New York.
Sanderson, R.D., J.M. Fitch, T.R. Linsenmayer, and R. Mayne (1986) Fibroblasts promote the formation of a continuous basal lamina during myogenesis in vitro. J. Cell Biol. **102:** 740-747.
Sievers, J., H.P. Klemm, S. Jenner, H.G. Baumgarten, and M. Berry (1980) Neuronal and extraneuronal effects of intracisternally administered 6-hydroxydopamine on the developing rat brain. J. Neurochem. **34:** 765-771.
Sievers, J., U. Mangold, M. Berry, C. Allen, and H.G. Schlossberger (1981) Experimental studies on cerebellar foliation. I. A qualitative morphological analysis of cerebellar fissuration defects after neonatal treatment with 6-OHDA in the rat. J. Comp. Neurol. **203:** 751-769.
Sievers, H., J. Sievers, H.G. Baumgarten, N. König, and H.G. Schlossberger (1983a) Distribution of tritium label in the neonate rat brain following intracisternal or subcutaneous administration of ^{3}H-6-OHDA. An autoradiographic study. Brain Res. **275:** 23-45.
Sievers, J., U. Mangold, and M. Berry (1983b) 6-OHDA-induced ecotopia of external granule cells in the subarachnoid space covering the cerbellum. Genesis and topography. Cell Tissue Res. **230:** 309-336.
Sievers, J., F.-W. Pehlemann, H.G. Baumgarten, and M. Berry (1985) Selective destruction of meningeal cells by 6-hydroxydopamine: A tool to study meningeal-neuroepithelial interaction in brain development. Dev. Biol. **110:** 127-135.
Sievers, J., F.-W. Pehlemann, and M. Berry (1986a) Influences of meningeal cells on brain development: Findings and hypothesis. Naturwissenschaften **73:** 8 8-194.
Sievers, J., Ch. von Knebel Doeberitz, F.W. Pehlemann, and M. Berry (1986b) Meningeal cells influence cerebellar development over a critical period. Anat. Embryol., in press,.
Spemann, H. (1938) *Experimentelle Beiträge zu einer Theorie der Entwicklung*. Springer, Berlin.
Spooner, B.R., and J.M. Faubion (1980) Collagen involvement in branching morphogenesis of embryonic lung and salivary gland. Dev. Biol. **77:** 84-102.
Wessells, N.K. (1977) *Tissue Interactions and Development*. Benjamin, New York.
Wessells, N.K., and J.A. Cohen (1968) Effects of collagenase on developing epithelia in vitro: Lung, ureteric bud, and pancreas. Dev. Biol. **18:** 294-309.

Section 3

CELL PROLIFERATION, GROWTH AND DIFFERENTIATION

FIBROBLAST GROWTH FACTOR: MOLECULAR AND BIOLOGICAL PROPERTIES

D. Gospodarowicz and G. Neufeld
Cancer Research Institute
University of California, Medical Center
San Francisco, CA
U.S.A.

Introduction

Many aspects of gene replication and expression have been solved by modern molecular biology and the study of viruses, bacteria and yeasts. The fundamental problem of cell proliferation and differentiation in eukaryotic cells still remains in most parts unresolved. The recently developed field of growth factors could provide answers at least in part as to the nature of the factors involved in cell proliferation and differentiation in developing organisms. In the following review we outline our present knowledge on one specific growth factor, Fibroblast growth factor (FGF). This factor which exists under two closely related forms (one basic, the other acidic) has been shown to trigger the proliferation and differentiation of a wide variety of mesoderm and neuroectoderm derived cells. Its mode of action could be intimately linked to that of extracellular matrix (ECM), which is known to regulate cell proliferation and gene expression, as well as morphogenetic tissue interactions.

Biochemistry of FGF

Bovine brain and pituitary basic and acidic FGF's have been purified to homogeneity (Gospodarowicz et al., 1984a, 1985c; Thomas et al., 1984) and subsequently characterized (Esch et al., 1985a, b; Gimenez Gallego, 1985). The basic form of FGF has a pI of 9.6 (Gospodarowicz et al., 1984a, 1985c) while the acidic form has a pI of 5.6 to 6.0 (Thomas et al., 1984). Both basic and acidic FGF are heat and acid labile (Gospodarowicz et al., 1985c; Gospodarowicz and Cheng, 1986). This precluded the use of reverse phase HPLC for their isolation in their native form. Both basic and acidic FGF can easily be purified using the extremely selective method of heparin sepharose affinity chromatography (Gospodarowicz et al., 1984a; Lobb and Fett, 1984).

Basic FGF has by now been purified from most mesoderm derived tissues, which have been shown to be FGF sensitive either in vitro or in vivo. These include brain

NATO ASI Series, Vol. H5
Mesenchymal-Epithelial Interactions in Neural Development
Edited by J. R. Wolff et al.

(Gospodarowicz et al., 1984a), pituitary (Gospodarowicz et al., 1984c, 1984a; Bohlen et al., 1984), retina (Baird et al., 1985a), corpus luteum (Gospodarowicz et al., 1985a), adrenal gland (Gospodarowicz et al., 1986c), kidney (Baird et al., 1985b), placenta (Gospodarowicz et al., 1985b), macrophages (Baird et al., 1985c) prostate (Nishi et al., 1985) and thymus (Table 1). SDS gel electrophoresis resolves the purest preparations of basic FGF into one or two very homogeneous bands with an apparent MW of 16 and 15 Kda respectively (Gospodarowicz et al., 1985c, 1986c). Structural studies have shown that basic FGF is a single chain polypeptide composed of 146 amino acids which can also exist in an amino terminal truncated form missing the first 15 amino acids: des. 1-15 FGF (Esch et al., 1985a; Gospodarowicz et al., 1985a). Depending on the organ from which it is purified, either the complete or truncated forms of basic FGF can be present. In pituitary, brain and retina, the complete form predominates, while in kidney and corpus luteum, only the truncated form can be detected. In adrenal gland and placenta, both forms co-exist. The truncated form of FGF is as potent as native FGF. Thus, the NH_2 terminal FGF region is neither involved in biological activity nor in binding to FGF cell surface receptors (Gospodarowicz et al., 1985a; Neufeld and Gospodarowicz, 1985).

So far acidic FGF has only been found in brain and retina (Thomas et al., 1984; Lobb and Fett, 1984; Baird et al., 1985a; Bohlen et al., 1985). Since acidic FGF is 30- to 100-fold less potent than basic FGF, it contributes to only 8% and 0.15%, respectively, of the total mitogenic activity present in crude brain or retinal extracts (Baird et al., 1985a; Bohlen et al., 1985). SDS gel electrophoresis of acidic FGF resolved the purest preparation of acidic FGF into two closely associated bands with MW in the range of 15 to 15.5 Kda (Thomas et al., 1984; Bohlen et al., 1985). Structural studies indicate that they represent a pair of microheterogeneous forms differing in the presence of absence of an amino terminal hexapeptide (Esch et al., 1985b; Gimenez Gallego et al., 1985). The larger form of acidic FGF is a single chain polypeptide composed of 140 amino acids (Esch et al., 1985b; Gimenez Gallego, 1985).

Comparison of the primary structure of basic and acidic FGF have shown that basic and acidic FGF are two very closely related mitogens with a 55% absolute homology (Esch et al., 1985b). Up to 64% of the remaining sequence may involve nucleotide substitution, where a single base change could result in amino acid replacement and concomitant homology. Both mitogens possess two potential binding domains for heparin (Esch et al., 1985a, b) one being located either near the NH_2 terminal region, while the other is near the carboxy terminal (residues 18 to 22 and 107 to 110, for basic FGF, and residue 9 to 12 and 100 to 102, for acidic FGF). Basic FGF also contains the inverse minimal cellular recognition site present in fibronectin and secondary structural analysis of the mitogen predicts that residues 37 to 40 and 78 to 81, which are the exact inverse of the Arg-Gly-Asp-Xaa fibronectin sequence, may exist on the exterior of the protein.

Thus, these residues might enable the protein to attach to cell surface recognition sites (Esch et al., 1985b). A weak homology exists between basic or acidic FGF and interleukin 1 (25 and 27% respectively) (Esch et al., 1985b; Gimenez Gallego et al., 1985). A weak homology has also been reported between the region covering residues 102 to 111 of acidic FGF and several neuropeptides (Gimenez Gallego et al., 1985).

The FGF's Gene

The homology between the primary structure of basic and acidic FGF suggests that they are derived from a single ancestral gene (Esch et al., 1985a, b; Gimenez Gallego et al., 1985). The cloning of the FGF genes and complementary DNA sequences of both basic and acidic FGF, which have recently been reported, suggest that through a process of gene duplications and evolutionary divergence, they have become separate gene products (Abraham et al., 1986). In view of the high degree of homology between those 2 mitogens, the strategy for isolating cloned cDNA coding for acidic and basic FGF is of interest.

Two unique sequence oligonucleotide hybridization probes, corresponding to amino acids 1-16 and 18-34 of bovine acidic FGF, were first synthesized (Abraham et al., 1986). The sequences of these oligonucleotides were based on considerations of the codons used on sequenced bovine genes and on the underrepresentation of the dinucleotide CpG in mammalian DNA (Derynck et al., 1984). A bovine genomic library was screened in duplicate with these probes, and phages that hybridized to both probes were rescreened with a short, degenerate oligonucleotide probe corresponding to acidic FGF residues 7 to 12. Two phages, λBA2 and λBA3, that hybridized to all three probes, were purified and shown by restriction enzyme analysis to contain overlapping genomic sequences. A 250 base pair AluI fragment, containing an acidic FGF exon, was subcloned from both λBA2 and λBA3 and its nucleotide sequence was determined. This exon coding for residues 1 to 41 of acidic FGF was then used as a highly specific probe for subsequent cDNA cloning.

To clone basic FGF, use was made of the homology between the basic and acidic FGF amino acid sequences (residues 18 to 31 for basic FGF and residues 9 to 22 for acidic FGF) to design a new 40 base long oligonucleotide probe. At positions where the two amino acid sequences of acidic and basic FGF were identical, the probe contained the same codon choice found in the acidic FGF genomic fragment, while at positions where the sequences differ, the probe incorporated the minimum possible number of nucleotide changes between the two sequences. It was expected that this oligonucleotide would hybridize to both acidic and basic FGF sequences, but that at conditions of higher stringency, it would be specific for basic FGF. The appropriate hybridization conditions for this specificity were established through a series of Southern blots of bovine genomic DNA that were washed at increasing stringencies. At both a 50° and 55°C wash temperature, a series of hybridizing fragments were detected, including those known to correspond to acidic FGF, from an analysis of clones λBA2 and λBA3 (Abraham et al., 1986). At a 65°C wash temperature, these acidic FGF fragments are no longer detectable and the remaining 3.5 kb PstI and 10.0 kb EcoRI fragments presumably did correspond to basic FGF. A bovine pituitary cDNA library of approximately 10^6 independent recombinants was screened with this probe using a 65°C wash temperature. A single hybridizing recombinant, λBB2, was isolated and shown, by DNA sequence analysis, to be a 2.1 Kb cDNA clone encoding bovine basic FGF. The nucleotide sequence corresponding to the published 146 amino acid sequence of bovine basic FGF was identified in λBB2. Basic FGF does not appear to be synthesized initially as a C-terminally extended form, since the codon for amino acid 146 is followed immediately by a TGA termination codon. The 3'-untranslated region of λBB2 is 1554 neucleotides long, and contains neither a polyadenyalation recognition signal (Proudfoot and Brownlee, 1976) nor a poly-A tail, suggesting that the cDNA clone is incomplete at the 3'-end.

An open reading frame extends upstream from the proline at position +1 of mature basic FGF, to the end of the cDNA clone. It is therefore possible that basic FGF is synthesized initially in a precursor form with an N-terminal extension. Features present in this upstream region argue against it being entirely a coding sequence. An ATG (methionine) codon was found nine amino acids upstream from the proline at position +1 and was in the same translational reading frame. The sequence surrounding this potential initiator codon (GGGCCATGG) is a good match to the consensus sequence (Kozak, 1983) proposed for the start of eucaryotic translation (CC$^{A}_{G}$CCATGG). When a comparison of this region was made with the corresponding region of bovine acidic FGF, the amino acid sequence homology that had been observed previously between the two FGF's (Esch et al., 1985; Gimenez Gallego et al., 1985) was seen to extend upstream. The proposed initiator ATG for acidic FGF aligned exactly with that for basic FGF and is also in the context of a good match to the consensus sequence (GAGCCATGG).

With both basic and acidic FGF, use of this initiator codon would result in a primary translation product of 155 amino acids. Cleavage from that precursor protein of the first 9 (basic FGF) or 15 amino acids (acidic FGF) would result in the generation of the mature proteins which can then be cleaved further in homologous position to give the NH_2 truncated form of basic des. 1-15 FGF or acidic des. 1-6 FGF.

FGF Receptors

FGF-specific receptors have been identified using cross-linking techniques on all cell types which have been shown to respond to basic or acidic FGF. The density of cell surface receptor in FGF sensitive normal diploid cells is low, (10^3 to 10^4 receptors/cell) (Neufeld and Gospodarowicz, 1985). In the case of established cell lines such as baby hamster kidney (BHK-21) cells, higher density of FGF cell surface receptors can readily be detected, and it is with that cell type that it has been characterized (Neufeld and Gospodarowicz, 1985, 1986). Binding studies have indicated that the apparent dissociation constant for the interaction of acidic FGF and BHK-21 cells is 2.5 x 10^{-10} M. Approximately 87,000 binding sites are found per cell (Neufeld and Gospodarowicz, 1985). This compares with an apparent dissociation constant for basic FGF of 2.7 x 10^{-10} M and 120,000 binding sites per cell (Neufeld and Gospodarowicz, 1985). As expected from the high degree of structural homology between basic and acidic FGF, both mitogens can displace each other from their binding sites. This suggests that they bind to similar receptors (Neufeld and Gospodarowicz, 1986). The results of cross-linking studies where iodinated, basic or acidic FGF were cross-linked, using disuccinimidyl suberate, to their cell surface receptor indicate that qualitatively, both mitogens interact with the same membrane components (M_r 145,000 and 125,000 species) (Neufeld and Gospodarowicz, 1986). Quantitatively, basic FGF appears to display a higher affinity than acidic FGF for the M_r 145,000 receptor species, while acidic FGF displays a higher affinity than basic FGF for the M_r 125,000 MW receptor species (Neufeld and Gospodarowicz, 1986). This could help in explaining the different biological potencies of basic vs. acidic FGF. Both classes of receptors are present on all cell types responsive to FGF's but not on several cell types that are not FGF responsive. The fact that lectin inhibits binding suggests that the FGF receptor contains N-acetylglucosamine and galactose residues.

In contrast to other mitogens, FGF does not appear to be actively degraded by the cells (Neufeld and Gospodarowicz, 1985). Also, in contrast to other mitogens, tyrosine phosphorylation of the FGF receptor can not be detected even when reagents as sensitive as antibodies directed against phosphorylated tyrosine are used (Neufeld and Gospodarowicz, 1985). Threonine and serine phosphorylation of the FGF receptor have not yet been examined. These phosporylation sites could be more relevant to the biological activity of FGF than tyrosine phosporylation, since FGF has been shown to activate protein C kinase (Tsuda et al., 1985; Kaibuchi et al., 1986). This raises the possibility that the FGF receptor, by analogy with the EGF receptor (Cochet et al., 1984), could be a substrate for the Ca^{2+}/phospholipid-dependent, threonine-specific protein C kinase.

The apparent lack of tyrosine phosphorylation of the FGF receptor could explain its lack of intracellular degradation. Recent evidence has strongly suggested that receptor phosphorylation may play a crucial role in inducing receptor entry into the cell and modulating the intracellular routing of receptor mitogen complexes (Hanover et al., 1985). Receptor phosphorylation on tyrosine residues seems to lead ultimately to lysosomal delivery (Beguinot et al., 1985). In contrast, when only serine or threonine residues are phosphorylated, the receptors are not delivered to lysosomes, where they would be degraded, but, instead, appear to be recycled rapidly back to the cell surface (Hanover et al., 1985; Beguinot et al., 1985).

The Mitogenic Response to FGF

So far, the mitogenic response to FGF has been studied best with basic FGF. Addition of the mitogen to quiescent cultures of Swiss 3T3 cells rapidly induced diacylglycerol formation, protein kinase C activation and Ca^{2+} mobilization (Tsuda et al., 1985; Kaibuchi et al., 1986). This, as pointed out above, could correlate with threonine phosphorylation of the FGF receptor.

Using cultures of vascular endothelial cells, the early effect of FGF on membrane motility has been studied (Gospodarowicz et al., 1979a). It was observed that both the short- and the long-range types on concanavalin A receptor mobility were increased in bovine vascular endothelial cells preincubated with FGF. A 6-9-hr preincubation with FGF gave an early maximal increase (two- to four-fold) in the binding of cells to concanavalin A molecules and already had a significant effect at 3 hr. A longer preincubation period (12-24 hr) with FGF is required to induce, in a large proportion (60-80%) of the cells, the ability to form caps upon addition of concanavalin A. The early effects of FGF on increased membrane fluidity and increased ruffling activity correlate, as has been shown in the case of EGF (Schlessinger and Geiger, 1981) and PDGF (Westermark et al., 1982), with rapid changes in the dynamic structure of the actin cytoskeleton (Gospodarowicz, 1985).

The induction of the "pleiotypic" and mitogenic response by FGF has been analyzed using confluent cultures of BALB/c 3T3 cells. The addition of FGF to confluent and resting cultures leads to an increased uptake of uridine, total amino acids, and thymidine (Rudland et al., 1974). The time course of the increase in net incorporation into the cell is almost identical for both FGF and serum additions. Amino acid uptake increased immediately, followed by a 15 min lag time for uridine and a 3-5 hr lag time for thymidine. The rate of protein synthesis increased (by threefold) within 5 hr, that of mRNA (about 15-20%) within 5 hr, and DNA (50- to 100-fold) within 24 hr. These increases are almost identical for cultures activated with FGF or serum, both in the kinetics and extent of the increases. Finally, cell division occurs in both cases 25-30 hr later. The abilities of FGF and serum to induce polysome re-formation measured 6 hr after additions are also virtually identical. Approximately 28% of the mRNA in resting BALB/c 3T3 cells is located in polysomes, and 72% in the 30S-80S messenger ribonucleoprotein complex. Addition of either FGF or fresh serum results in greater than 85% of the mRNA being located in the polysomes. This increase in polysome formation correlates with an increase in protein synthesis.

Based on data indicating that proteins need to be made for the rate of DNA synthesis to increase in response to growth stimulation, (Nilsen-Hamilton et al., 1980, 1981;

Hamilton et al., 1985) proteins that are specifically synthesized in growth arrested BALB/c 3T3 in response to FGF and to serum have been identified. The most dramatic, specific increases in labeling with (^{35}S) methionine is of secreted rather than intracellular proteins. The secreted levels of these proteins begin to rise within 2 hr after the addition of FGF to the growth arrested 3T3 cells (Hamilton et al., 1985). The secreted levels of one of these proteins with a molecular weight of 39,000, which is also called a "major excreted protein" (MEP; Gottesman, 1978), increases about eightfold in response to FGF (Nilsen-Hamilton et al., 1980). FGF also increases the secreted level of a set of five proteins that share the property of being "superinduced" by cycloheximide. Cycloheximide acts in synergism with FGF. These cellular proteins have been named "superinducible proteins" or SIPs. They have molecular weights ranging from 12,000 (SIP 12) to 62,000 (SIP 62). Their function is presently unknown.

FGF regulates the expression of cellular oncogenes, in particular that of c-fos and c-myc which have been shown to be involved in cell differentiation (Muller and Wagner, 1984) and cell proliferation (Armelin et al., 1984) respectively. The very rapid appearance of c-fos mRNA and protein after FGF stimulation argues that c-fos activation affects gene expression by a direct mechanism, whereas activation of the c-myc may occur via indirect mechanisms (Muller et al., 1984). The sudden disappearance of c-fos mRNA also contrasts strikingly with the gradual decline of c-myc mRNA (Muller et al., 1984).

In Vitro Biological Effect of FGF

Most of the biological studies with FGF have been done with the basic form. Only recently has the biological activity of acidic FGF started to be investigated (Bohlen et al., 1985). As expected from the homologies between the acidic and basic form, acidic FGF has a biological effect identical to that of basic FGF, although it is, depending on the cell type, 30 to 100 fold less potent (Bohlen et al., 1985). The lesser potency of acidic FGF makes it a weaker agonist than the basic form, and may reflect the qualitatively different interactions of acidic vs. basic FGF with the 145 and 125 Kda cell surface receptor species (Neufeld and Gospodarowicz, 1986).

Effect on cell morphology and cell transformations

Basic FGF has both acute and long-term effects on the morphology and growth pattern of responsive cells. Human skin fibroblasts maintained with FGF in serum-supplemented medium become extremely elongated with long slender projections which are retraction fibrils resulting from an increase in locomotory activity (Gospodarowicz and Moran, 1975). The same is true for vascular endothelial and smooth muscle cells which, in the presence of FGF, become bipolar (Gospodarowicz, 1975). Even more striking is the effect of FGF on cell morphology when it is added to a confluent and resting monolayer of

BALB/c 3T3 cells. Such cells, which show a regular cobblestone pattern in confluent culture exposed to serum-supplemented medium, become spindle shaped and grow in an irregular crisscross pattern when exposed to FGF for 6-12 hr (Gospodarowicz and Moran, 1974). Large membrane ruffles, often associated with macropinocytotic vesicles, can be seen on most of the cells. Morphologically, the FGF-treated cells look "transformed" since reduced cell-substratum adhesion, crisscross growth patterns, and increased membrane ruffling are all features typical of transformation. The similarity in phenotype of transformed and FGF-treated cells is consistent with the hypothesis that transformation and FGF action might at least partially share a common metabolic pathway. This effect is reversible since, upon removal of FGF, the cells revert to their normal phenotype. FGF has also been reported to induce the growth, in soft agar, of non-transformed cells, and to potentiate the effect of ß-transforming growth factor (ßTGF) (Rizzino et al., 1985). Comparison of the locomotor activity of arterial smooth muscle cells and endothelial cells in media supplemented with serum and FGF, respectively, has given circumstantial evidence that FGF stimulates cell migration (Gospodarowicz et al., 1985a, b).

Effect on cell proliferation

In its native form, basic FGF is a potent mitogen for mesoderm-derived cells (Gospodarowicz 1985a, b) (Table 1). When tested on vascular endothelial cells derived from either large vessels or capillaries, it triggers cell proliferation at concentrations as low as 5 pg/ml, with half-maximal effect occurring at 22.5-39 pg/ml (1.5-2.6 pM) and saturation at 140-280 pg/ml (9.3-18.6 pM) (Gospodarowicz et al., 1985c). Those concentrations are 20-60-fold lower than those required to see comparable effects on cell growth or initiation of DNA synthesis with α-transforming growth factor (αTGF), epidermal growth factor (EGF), or platelet-derived growth factor (PDGF). Similar potency has been reported when basic FGF is tested on other cell types ranging from established cell lines such as BALB/c-3T3 to normal diploid cell strains such as bovine granulosa and adrenal cortex cells, myoblasts, corneal endothelial cells, vascular smooth muscle cells, glial cells and human umbilical endothelial cells (Gospodarowicz et al., 1985c; Gospodarowicz, 1985a, b).

Until now, basic FGF has mostly been used in vitro to develop new cell strains (Gospodarowicz, 1979; Gospodarowicz and Zetter, 1977), particularly from the vascular and corneal endothelia (Gospodarowicz et al., 1976a, 1978b, c, 1979c). FGF is mitogenic both for cells seeded at clonal density and for low-density cultures (Gospodarowicz and Zetter, 1977). Although the response of the cells can vary widely, depending on the types of sera and media used (Gospodarowicz and Lui, 1981), addition of FGF to the cultures of most mesoderm-derived cells results in a greatly reduced average doubling time; in the case of vascular endothelial cells, for example, it can decrease from 72 to 18 hr. This results primarily from a shortening of the G_1 phase of the cell cycle (Gospodarowicz and Savion, 1980b).

Effect on cell differentiation

FGF stabilizes the phenotypic expression of cultured cells. This is a particularly interesting characteristic of FGF, since it has made possible the long-term culturing of cell types that otherwise would lose their normal phenotypes in culture when passaged repeatedly at low cell density. This biological effect of FGF has been best studied using vascular or corneal endothelial cell cultures cloned and maintained in the presence of FGF and then deprived of it for various time periods (Gospodarowicz et al., 1979, 1981; Greenburg et al., 1980; Tseng et al., 1982; Vlodavsky and Gospodarowicz, 1979). It has been related to the ability of FGF to control the synthesis of various extracellular matrix components that would in turn affect cell polarity and gene expression (Gospodarowicz et al., 1979c, 1982a). A similar effect of FGF on the differentiation of costal chondrocytes has been reported (Kato and Gospodarowicz, 1985). In the case of sheep preadipocyte fibroblasts, it has been reported that FGF promotes their differentiation into adipocytes (Broad and Ham, 1983). A similar observation has been made by Serrero and Khoo (1982) using the 1246 cell line. In the case of a rat pheochromocytoma (PC12 cell line) still able to respond to nerve growth factor (NGF) by initiation of neurite outgrowth, basic FGF acts as a differentiation factor, inducing both neurite outgrowth and ornithine decarboxylase activity (Togari et al., 1983, 1985).

Effect on cell senescence

FGF can also significantly delay the ultimate senescence of cultured cells. In the case of granulosa cells, addition of FGF to either clonal or mass cultures of granulosa cells can extend their lifespan in culture from 10 generations to 60 generations (Gospodarowicz and Bialecki, 1978). Adrenal cortex cell lines cloned in the presence of FGF show a similar dependence on this growth factor during their limited in vitro lifespan (Simonian et al., 1979). Its removal from the culture medium results not only in a greatly extended doubling time but also in rapid cell senescence. In the case of vascular and corneal endothelial cells, FGF has been shown to extend the lifespan of the cultures (Gospodarowicz et al., 1978b, 1981). Corneal endothelial cells maintained in the absence of FGF have a lifespan of 20 to 30 generations, whereas in its presence, they can proliferate for 200 generations (Gospodarowicz et al., 1981). In summary, the following effects of FGF have been observed with various cell types: it stimulates migration, it increases the rate of proliferation, it stabilizes phenotypic expression, and it extends cultural lifespan. As pointed out above, these effects correlate with the ability of FGF to activate cellular oncogenes which have been implicated in cell proliferation, differentiation and longevity.

Table I: Organ Distribution and Microheterogeneity of Basic FGF

Tissue	Species	Biological Activity[a]	Radio-Immunoassay[b]	Purified to Homogeneity	AA Residue	Yield of Basic FGF (μg/kg)
Brain	Bovine	+	+	+	1-146	33-50
	Human	+	+	+	1-146	ND[c]
Pituitary	Bovine	+	+	+	1-146	350-600
Kidney	Bovine	+	-	+	16-146	90-150
Adrenal glands	Bovine	+	+	+	1-146 and 16-146	200
Corpus Luteum	Bovine	+	-	+	16-146	50-70
Placenta	Human	+	+	+	1-146 and 16-146	30-40
Retina	Bovine	+	+	+	1-146	ND
Macrophage	Murine	+	+	-	?	ND
Chondrosarcoma	Rodent	+	+	?	?	ND
Thymus	bovine	+	+	+	1-146 and 16-146	35-70

a Biological activity detected on the various cell types shown in Table II.

b Radioimmunoassay as described in Bohlen et al., 1984.

c ND: not determined

Table II: Cell types for which basic or acidic FGF is mitogenic.

Normal Diploid Cells	Basic FGP	Acidic FGF
Glial and astroglial cells	+ (D)	+
Trabecular meshwork cells	+	?
Capillary, large vessel and endocardium endothelial cells	+ (D)	+ (D)
Corneal endothelial cells	+ (D)	+ (D)
Fibroblast	+	+
Myoblast	+ (D)	+ (D)
Vascular smooth muscle	+	+
Chondrocytes	+ (D)	+ (D)
Blastema cells	+	?
Adrenal cortex cells	+	+
Granulosa cells	+	+
Established Cell Lines		
Rat fibroblast-1	+	?
Balb/c 3T3	+	+
Swiss 3T3	+	+
BHK-21	+	+
PC-12	(D)	(D)

(D) = induce differentiation

Table III: **Pseudonyms under which Basic and Acidic FGF are Recorded in Literature**

ACIDIC FGF	BASIC FGF
1 Embryonic kidney derived angiogenesis factor I	1 Leukemia growth factor (Thymus
2 Glial growth factor	2 Macrophage growth factor
3 Astroglial growth factor 1	3 Brain derived neurotrophic factor
4 Endothelial cell growth factor (ECGF)	4 Embryonic kidney derived angiogenesis factor 2
5 Retina derived growth factor	5 Prostatic growth factor
6 Heparin binding growth factor class I	6 Astroglial growth factor 2
7 Endothelial growth factor	7 Endothelial growth factor
8 Eye derived growth factor II	8 Tumor Angiogenesis factor
	9 Hepatoma growth factor
	10 Chondrosarcoma growth factor
	11 Cartilage derived growth factor I
	12 Eye-derived growth factor I
	13 Heparin binding growth factors class II
	14 Myogenic growth factor
	15 Human placenta purified factor (Moscatelli)

Structurally related

acidic FGF = basic FGF

same receptors

acidic FGF is an analog of basic FGF

Modulation of the bioactivity of FGF by heparin

Heparin and its closely related glycosaminoglycan heparan sulfate have both negative and positive effects on cell growth, depending on the cell type considered. With smooth muscle cells, heparin is a potent growth inhibitor (Castellot et al., 1985). Similar effects on the part of heparan sulfate glycosaminoglycans either obtained from exponentially growing smooth muscle cells (Fritze et al., 1985) or produced by vascular endothelial cells (Castellot et al., 1981) have been reported. In other cell types, such as Schwann cells (Ratner et al., 1985), or human umbilical endothelial cells (Thornton et al., 1983), heparan sulfate proteoglycans or heparin have been shown to positively regulate cells. In the case of human vascular endothelial cells where heparin can potentiate the mitogenic activity of endothelial cell growth factor (ECGF), an acidic FGF-like peptide, it has been suggested that the association between heparin and mitogen induces a conformational change that increases or stabilizes its biological activity (Schreiber et al., 1985). In recent studies, heparin has been shown to protect both acidic and basic FGF from acid or heat inactivation (Gospodarowicz and Cheng, 1986). It has been proposed that the inherent liability of basic or acidic FGF to denaturing conditions could reside in the fact that none of the cystine residues, identified in their primary sequence, participate in sulfhydril bridging (Esch et al., 1985a, b). Since basic, as well as acidic, FGF have extremely high affinity for heparin, one could speculate that in low salt concentrations, FGF molecules, which have two heparin binding sites, could bind one molecule of glycosaminoglycan to 2 heparin binding domains. This would result in a cyclization of the FGF molecule which could stabilize the tertiary structure of FGF and would make it both heat and acid resistant. The protective effect of heparin on the biological effect of FGF could have physiological significance. In previous studies, it has been shown that even at 37°C and at neutral pH, the in vitro half life of FGF is 24 hr (Westall et al., 1983). Interaction of heparin, or heparan sulfate, with FGF will likely result in greatly extending its half life, making low concentrations of FGF appear more active than they would have otherwise. In addition, with certain cell types, for which heparin or heparan sulfate is not growth inhibitory, and which respond to FGF, heparin, by acting at the cell level, could make them responsive to concentrations of the mitogens to which the cells would otherwise not respond.

FGF and the Extracellular Matrix

In the early stage of embryonic development, when different tissues composing a given organ are formed as a result of strictly timed and spatially interrelated proliferative and differentiative events, interaction of cells with newly formed ECM has been shown to promote cell proliferation and to stabilize new phenotypic expression (Wessel, 1977; Hay, 1981; Saxen et al., 1981). ECM components have been indicated as playing a role in inductive tissue interaction in somite chondrogenesis (Kratchowil, 1972), the differentiation of corneal epithelium (Hay, 1980, 1984), salivary gland morphogenesis (Bernfield et al., 1984), tooth germ development (Thesleff et al., 1981), and kidney nephron formation (Eckblom, 1984).

FGF has been shown to have profound effects on ECM production by various cell types for which it is mitogenic. These include corneal and vascular endothelial cells, as well as chondrocytes. For these 3 cell types, the ability of FGF to promote cell proliferation and stabilization of their phenotypic expression correlates with an ability to modulate the synthesis and distribution of various ECM components such as collagen, fibronectin, and proteoglycans (Gospodarowicz et al., 1981; Greenburg et al., 1980; Tseng et al., 1982; Vlodavsky et al., 1979; Kato and Gospodarowicz, 1985). It raises the possibility that one of the ways FGF could promote cell proliferation and stabilization of phenotypic expression is through its effect on ECM production. This possibility has been explored using dishes coated with an ECM produced by cultured corneal endothelial cells (Gospodarowicz et al., 1982b; Gospodarowicz and Tauber, 1980). This matrix is composed of elastin, basement-membrane collagen types IV, V and interstitial collagen type III (Gospodarowicz, 1984). Also present are proteoglycans, composed mostly of heparan sulfate and dermatan sulfate proteoglycans and glycoproteins such as fibronectin and laminin (Gospodarowicz, 1984). When the ability of the ECM to support cell proliferation was analyzed with various cell types, it proved capable of replacing FGF (Gospodarowicz et al., 1982b; Gospodarowicz and Giguiere, 1982; Gospodarowicz and Greenburg, 1981; Gospodarowicz and Tauber, 1980; Gospodarowicz, 1984). As expected from its composition, it exhibits cell specificity and can only support the growth and differentiation of cell types which have been shown to be FGF sensitive and does not support the growth of cell types that are not responsive to FGF. For example, in order to induce kidney epithelial cells to divide and to express their morphogenesis properly when confluent (tubule formation), one has to maintain the cells on an ECM with a composition similar to that of kidney tubules in vivo (Gospodarowicz et al., 1984b). Such matrices are produced by the mouse-derived teratocarcinoma PF-HR-9 cell line and are composed exclusively of collagen type IV, heparan sulfate proteoglycans, laminin, and entactin, as well as a 185 kDa glycoprotein whose function is at present unknown (Gospodarowicz, 1984).

These observations raise the possibility that growth factors such as FGF can replace the ECM and make the cells fully responsive to progression factors present in serum and plasma when cells are maintained on plastic. This could result from the ability of FGF to induce the synthesis and secretion of the cell's ECM. Such induction could in turn make the cells sensitive to progression factors present in serum or plasma, which in subsequent studies were shown to be transferrin, high density lipoproteins, and insulin (review in Gospodarowicz et al., 1982a).

Mechanisms Through Which ECM Components Could Replace or Mimic the Effect of FGF

Little is known about the ECM factors that have a permissive effect on cell proliferation and could restore the cell response to plasma factors. These factors could be intrinsic components of the ECM. They could consist of as-yet-unidentified serum or plasma factors as well as such growth factors as FGF that have become part of the ECM during its formation. Also, cellular growth-promoting factors released during cell lysis could be absorbed onto it and be responsible for its growth-promoting properties.

The possibility that the growth promoting ability of ECM results from serum or plasma factors adsorbed on it is not likely, since the ECM produced by corneal endothelial cells, growing in defined medium supplemented with HDL, transferrin, and insulin are as potent in supporting cell growth as ECM produced by cells growing in serum- or plasma-supplemented medium (Gospodarowicz et al., 1983b). Likewise, the possibility that the factors involved in supporting cell growth could be intrinsic components of the ECM is not supported by the fact that of the various components that have been analyzed for their ability to support cell growth, neither collagen type I, II, III, IV, V, or fibronectin alone could replace the ECM (Gospodarowicz et al., 1980a). Proteoglycans and laminin were not analyzed in that study. The possibility therefore exists that, although no single ECM component is active in supporting cell growth, a cooperative effect between its components could result in such an effect. The intricate nature of the ECM as exemplified by the complex interaction of its known components (collagen, glycoproteins, proteoglycans, and glycosaminoglycans), which form a highly stable scaffolding upon which cells rest in vivo, makes its reconstitution from its purified constituents a formidable task and the testing of that hypothesis difficult.

Another possibility is that growth factors themselves could be adsorbed on the ECM. This is supported by the study of Shing et al. (1984). These authors have purified a basic FGF-like capillary endothelial cell growth factor from chondrosarcoma that displays a very high affinity for heparin. Since heparan sulfate is structurally (Wasteson et al., 1977) and, at high concentrations, biologically similar to heparin (Kraemer, 1971), it is conceivable that the capillary growth factor could be closely associated in the ECM with heparan sulfate proteoglycans.

This possibility was analyzed by exposing denuded ECM to inactivating treatment such as heat or acid exposure (Gospodarowicz et al., 1983; Greenburg and Gospodarowicz, 1982), which have previously been shown to inactivate FGF. Although none of these treatments were effective in destroying the ability of the ECM to promote cell growth, thereby suggesting that FGF was not involved, such results should be interpreted with caution since it has recently been shown that heparin binding to FGF makes it heat and acid resistant (Gospodarowicz and Cheng, 1986). It is therefore likely that if FGF was to bind tightly to heparan sulfate proteoglycans present in the ECM, heat or acid treatment

(Gospodarowicz et al., 1983b; Greenburg and Gospodarowicz, 1982) would not totally destroy the biological activity of bound FGF.

If FGF did bind to heparan sulfate proteoglycans, this could represent a storage form of the mitogen. However, it is unlikely that under this form it would be active since although it would be capable of binding to cell surface FGF receptor, present in the basal cell surface, the immobilization of FGF on the heparan sulfate proteoglycans would prevent microaggregation of the FGF receptor complexes. This has been shown by others (Schlessinger, 1980) to be required for mitogenesis. Hydrolysis of the ECM, however, could result in the liberation of heparan sulfate glycosaminoglycans-FGF complexes which would be biologically active. In this context, it is interesting to note that during morphogenesis, the areas of greatest mitotic activity are located where ECM is hydrolyzed (Bernfield et al., 1984). Similarly, in the case of the embryonic kidney, angiogenesis correlates with the hydrolysis of the kidney mesenchymal stroma (Ekblom et al., 1981). In the adult phase of development, the heparan sulfate present in ECM can be degraded by heparitinase, an inducible enzyme which has been shown to be released either by platelets, when they attach to the subendothelium, or by macrophages once they are activated (Yahalom et al., 1984; Matzner et al., 1985; Savion et al., 1984). This could ultimately lead to the solubilization of heparan-sulfate-FGF complexes which would be biologically active, and could participate in various repair processes, including wound healing (see below).

In Vivo Effects of FGF

Limb and lens regeneration

Certain amphibian species have a well-developed capacity to regenerate lost appendages. The dependence of limb regeneration on nerves was reported as early as 1823 by Todd, and studies by Singer have clarified and extended the concept of neurotrophic control in this process (Singer, 1952, 1954, 1974; Singer and Craven, 1948). Denervation of an adult newt limb at the time of amputation or prior to the formation of the blastema effectively blocks the regeneration process. If denervation is delayed until after a blastema has formed, however, differentiation of the blastema cells and morphogenesis of new limb parts occur despite the lack of nerves (Singer, 1974; Powell, 1969). This finding implies that the neurotrophic effect is needed for the cell proliferation involved in blastema formation but not for the later events of regeneration, which include differentiation of blastema cells into muscle, cartilage, and other tissues of the new limb. The molecular basis of the neurotrophic phenomena is poorly understood, but the available evidence suggests that such control is mediated by means of neurosecretory products unrelated to nerve impulses or neurotransmitters (Dresden, 1969; Guttman, 1976).

Blastema cells are thought to originate from either dedifferentiated myoblasts or chondrocytes (Hay, 1974). These are two cell types for which basic FGF has been shown to be a potent mitogen (Gospodarowicz et al., 1976b; Gospodarowicz and Mescher, 1977). Infused FGF promotes the resumption of mitotic activity in denervated newt forelimb blastemas in vivo (Gospodarowicz and Mescher, 1981; Gospodarowicz et al., 1978d; Mescher and Gospodarowicz, 1979), and when added to cultured blastemas, basic FGF, at a concentration of 0.1 ng/ml, is as effective as an optimal dose of brain extract in promoting (^{3}H)thymidine incorporation (Mescher and Loh, 1980), ^{14}C labeled amino acid incorporation, and mitotic activity (Carlone and Foret, 1979). Similar concentrations of basic FGF have also been shown to maintain total acetylcholinesterase activity in cultured newt triceps muscle above that in untreated contralateral controls after 1 week (Carlone et al., 1981). It has also been reported (Gospodarowicz and Mescher, 1981; Gospodarowicz et al., 1975) that administration of FGF promotes heterotrophic growth of amputated adult frog (Rana pipiens) for limbs comparable to that observed by Singer (1974) after surgical augmentation of the nerve supply. These observations lend support to the possibility that basic FGF is involved in one of the earlier steps of regeneration, namely recruitment of primitive cells leading to blastema formation.

It has been well-documented for several decades that lens regeneration from the dorsal iris in vivo is dependent upon a stimulus coming from the neural retina (for reviews, see Yamada, 1977; Reyer, 1977). In organ culture, the presence of retina (Yamada and McDevitt, 1976) or pituitary (Connelly et al., 1973) is required for consistent lens regeneration. Yamada has suggested that the factor involved in lens regeneration acts indirectly to shorten preferentially the cell cycle time in the dorsal iris epithelial cells in vivo as compared to the ventral iris, thus permitting the dorsal iris epithelial cells to go through the requisite number of cell cycles for conversion into lens cells before mitosis ceases once again in the iris epithelium (Yamada, 1982). In both retina (Baird et al., 1985a) and pituitary (Gospodarowicz et al., 1985a), the major form of growth factor identified is the basic form of FGF, together with smaller quantities of acidic FGF in the case of retina (Baird et al., 1985a; Courty et al., 1985). Basic FGF and its agonist, acidic FGF, could therefore play a role in lens regeneration in vivo. Evidence that it could be so has been presented by Yamada, who has reported that crude preparation of retinal derived FGF could induce lens regeneration in vivo (Yamada, 1982) and by the fact that both retina derived or pituitary derived FGF are potent mitogens for cultured lens epithelial cells, as well as iris cells (Courty et al., 1985; Gospodarowicz et al., 1977b).

Angiogenic activity of FGF

Basic FGF has been shown to be a potent angiogenic factor in vivo when tested in assays as different as the rabbit cornea, the chick chorioallantoic membrane (CAM) or the hamster cheek pouch (Gospodarowicz et al., 1979a, 1984a, 1985a, 1986c). Acidic FGF, although less potent than the basic form, has also been shown to be angiogenic in the CAM and rabbit cornea assay (Thomas et al., 1985; Lobb et al., 1985). The presence of basic FGF as the main mitogen present in corpus luteum (Gospodarowicz et al., 1985a), adrenal gland (Gospodarowicz et al., 1986c), kidney (Baird et al., 1985b) and retina (Baird et al., 1985a) correlates extremely well with the strong angiogenic properties of these organs.

In the case of corpus luteum, angiogenesis plays an important role in its development (Bassett, 1943). This is manifested by the extremely rapid and radical vascular changes that take place in the capillary wreath surrounding the follicle at the time of ovulation. From the endothelium of these vessels, capillary sprouts begin to grow into the granulosa cell layer and develop by 48 hr into a totally new complex network of sinusoidal vessels, which invade the previously avascular granulosa cell layers, induce luteinization, and later nourish the parenchyma of the corpus luteum (Bassett, 1943). In previous studies (Gospodarowicz and Thakral, 1978), it was demonstrated that rabbit luteinizing granulosa and luteal cells produced a diffusible substance that could trigger the early vascular changes that occur during the development of the corpus luteum (Gospodarowicz and Thakral, 1978) and elicit a strong angiogenic response on the part of the host. Basic FGF, which accounts for the full mitogenic potential of corpora lutea crude extract Gospodarowicz et al., 1985a), and is mitogenic for corpus luteum derived capillary endothelial cells, as well as angiogenic in vivo, could be responsible in part or in toto for the angiogenic activity of the developing corpus luteum.

In the case of adrenal cortex, after incomplete ablation, it will readily regenerate. When adrenal cortex tissue fragments are implanted under the kidney capsule, the grafts rapidly become vascularized by the host capillaries (Turner and Bagnara, 1981), and within a short time period, an adrenal cortex forms again and becomes functional. Kiss and Krompecher (1962) have reported that adrenal extracts are most effective in stimulating vascularization (Kiss and Krompecher, 1962; Kiss and Szabo, 1969). The isolation of basic FGF from adrenal glands extracts, which accounts for the full mitogenic potential of the extract, could also account for its angiogenic potential. Interestingly, FGF has also been shown to be mitogenic for adrenal cortex cells in vitro (Kato and Gospodarowicz, 1984). Therefore, the release of FGF from injured adrenal glands could auto-stimulate the regeneration of that tissue, acting both by triggering proliferation of adrenal cells as well as by triggering capillary invasion which would bring in nutrients. Also of interest is the fact that previous reports have described enhanced in vivo chondrossification by crude adrenal gland extract (Kiss and Krompecher, 1962; Kiss and Szabo, 1969). This is in agreement with the fact that basic FGF has been shown to stimulate in vitro chondrocyte proliferation (Gospodarowicz and Mescher, 1977; Kato and Gospodarowicz, 1984) and differentiation (Kato and Gospodarowicz, 1985) and to induce in vivo chondrossification

(Jentsch et al., 1980).

The case of the kidney, where basic FGF has been identified as the mitogen which would account for the full mitogenic as well as angiogenic potential of kidney crude extracts, is perhaps one of the most interesting since it would suggest that the appearance of FGF could be developmentally regulated. Previous studies have shown that embryonic kidney implants (Preminger er al., 1980; Sariola et al., 1983) can elicit a strong angiogenic response once implanted on the CAM. Likewise, kidney extract contains chemoattractants for vascular endothelial cells (Glaser et al., 1980). When embryonic kidney rudiments at different stages of development are grafted on the CAM it was observed that both whole kidney anlagen and experimentally induced isolated kidney mesenchyme became richly vascularized on CAM, whereas uninduced mesenchyme grafted under identical conditions remained avascular. Similarly, in explants where induction was incomplete and only a portion of the mesenchyme was converted into epithelial structures, the undifferentiated part remained avascular while the vessels were seeking the differentiated portion (Sariola et al., 1983). Second, it was shown that while the formation of an endothelial glomerular loop was a rule in the grafted whole kidneys, it was a rare exception in the experimentally induced, similarly grafted mesenchyme, which otherwise showed good differentiation of the various epithelial components (Sariola et al., 1983).

The first set of observations suggests that vascularization of the nephric blastema was linked to its induction and early differentiation. Morever, since the invasion of the host capillaries was both temporally and spatially associated with the formation of pretubular mesenchymal aggregates, it would be reasonable to search for angiogenic growth factor in these condensates. Within this context, it is likely that basic FGF, which has been shown to be the preponderant form of angiogenic factor in adult bovine or murine kidney, could be responsible for the developmentally regulated invasion by capillaries into the avascular metanephric mesenchyme. It is also possible that the biosynthesis of FGF within the epithelial structure could be controlled by factors from the epithelial ureter bud which induce differentiation of the mesenchym (Grobstein, 1956). This in turn would lead to a stimulation of blood vessel invasion (Preminger et al., 1980; Sariola et al., 1983).

The retina also provides a good example where the presence of basic FGF and the related agonist, acidic FGF, could explain the angiogenic potential of that tissue, as well as a few proliferative retinal disorders. Over 30 years ago, Michaelson (1948) proposed the existence of a vasculogenic factor, produced by the ischemic retina, that could induce neovascularization. Although 20 years were required for direct in vitro and in vivo evidence demonstrating angiogenic and endothelial cell-stimulating activity of the retina (D'Amore et al., 1981; Barritault et al., 1981), the original work was pursued by Ashton et al. (1954) who predicted the involvement of a putative vasculogenic factor in the etiology of diabetic retinopathy, retinal fibroplasia, and retinal vein occlusion.

With the identification of FGF as the vasculogenic factor postulated by Michaelson (1948), it is interesting to speculate on the physiological and pathophysiological implications of this finding. The effects of FGF on fibroblasts as well as on endothelial cells (Gospodarowicz and Moran, 1975; Gospodarowicz et al., 1986a, b) may explain the association of fibrosis with the neovascularization that is seen in conditions of retrolental fibroplasia and diabetic retinopathy. The possibility that either or both forms of FGF may be involved in the etiology of vasoproliferative diseases of the eye is an attractive hypothesis that emphasizes the importance of expanding these results.

Wound healing

In the case of wound healing, basic FGF, which has been detected in macrophages (Baird et al., 1985c) and could be released from the injured tissues, could play a crucial role since, in contrast to other growth factors, such as PDGF or ßTGF, it has been shown to stimulate in vitro as well as in vivo the proliferation of all the cell types involved in wound healing. These include capillary endothelial cells, vascular smooth muscle cells, and fibroblasts, notwithstanding other cell types which are involved in the wound healing of specialized territory such as chondrocytes, myoblasts, etc. Indeed, it has been reported by Buntrock et al. (1982a, b) that basic FGF increases the formation of granulation tissue in vivo. This increase is associated with stimulation of the synthetic function of fibroblast and myofibroblast (Buntrock et al., 1982a, b, c). Similar results were reported by Davidson et al. (1985) using polyvinyl alcohol sponges implanted subcutaneously in rats and injected with basic FGF. Although the name given in that study to the growth factor was that of cartilage derived growth factor (Davidson et al., 1985), this factor was later shown to be identical to basic FGF (Lobb et al., 1986).

Tumor angiogenesis

Solid tumors are angiogenesis dependent and once "tumor-take" has occurred, an increase in the tumor cell population must be preceded by an increase in new convergent capillaries (Algire et al., 1945; Tannock, 1968). This has led Folkman and his group to propose that tumors can release growth factors which would induce capillary endothelial cell proliferation and new capillary formation (Gimbrone et al., 1974; Folkman and Cotran, 1976). The isolation of such factors from cultured tumor cells or solid tumors has been the object of intense and unrewarding investigations for the past ten years. The realization that endothelial cell growth factors have a high affinity for heparin and the use by Shing and his colleagues of Heparin-Sepharose affinity chromatography (Shing et al., 1984) for the isolation of these endothelial cell growth factors has finally led to the purification of the angiogenic factor from either solid tumor (chondrosarcoma, melanoma) or from cultured tumor cells (hepatoma). Once characterized, it was found to be identical to basic FGF (Lobb et al., 1986). Interestingly enough, the hepatoma growth factor contained two of the 9 amino acid extension peptides (Klagsbrun et al., 1986) that the cloning of basic FGF cDNA would have predicted to exist in the unprocessed molecule (Abraham et al., 1986). Therefore, basic FGF is likely to be involved in the angiogenic process triggered by tumor cells in a way probably similar to that induced by normal tissue, where angiogenesis takes place (corpus luteum, kidney, wound healing). Since basic FGF was isolated in 1974 (Gospodarowicz, 1974), shown to be mitogenic for endothelial cells as early as 1976 (Gospodarowicz et al., 1976a; Gospodarowicz et al., 1978a), and angiogenic in vivo in 1979 (Gospodarowicz et al., 1979a), one cannot help but wonder why it has taken until 1986 to realize that basic FGF is involved in tumor angiogenesis. Three factors probably contributed to the early failure to recognize this. First of all, it was believed that the angiogenic process was unique to tumors (Ausprunk et al., 1975), although ample evidence existed that it also occurred in normal tissue (Gospodarowicz and Thakral, 1978; Gospodarowicz et al., 1978e). The process of neovascularization of normal tissue was then proposed to be due to a grafting process of pre-existing capillaries within grafted tissues with those of the host (Ausprunk et al., 1975). Since basic FGF was isolated from normal tissue, it was thought not to be involved in the tumor angiogenic process. Secondly, it was believed that the growth and substrate requirement of endothelial cells derived from large vessels was different from those of capillaries (Zetter, 1981). This was later shown to be incorrect, both cell types having exactly the same growth and substrate requirements, irrespective of the large vessel or capillary territories from which they were derived (Gospodarowicz et al., 1986a, b). Thirdly, the unexpected skepticism that greeted the report on the existence of basic FGF in brain tissue (Gospodarowicz, 1974) or the finding that basic FGF was mitogenic for vascular endothelial cells in vitro (Gospodarowicz et al., 1976a, 1978a) and angiogenic in vivo (Gospodarowicz et al., 1979a), played a decisive role in the failure, early on, to recognize basic FGF as the tumor angiogenic factor. The reports by Gajdusek et al. (1980) and Schwartz et al. (1981) that basic FGF was not mitogenic for vascular endothelial cells (reviewed in Schwartz and Gajdusek, 1982), that of Thomas et al. (1980) that all the mitogenic activity in bovine brain could be accounted for by acidic FGF, and the repeated reports of the group of Maciag that acidic FGF is the main mitogen for vascular endothelial cells (Maciag et al., 1984), while basic FGF has no effect whatsoever on that cell type (Maciag et al., 1978, 1982; Maciag, 1984; Lemmon et al., 1982) seem rather difficult to believe in retrospect (reviewed by Gospodarowicz et al., 1982c, 1983a).

The available data, although they strongly support the fact that basic FGF and various tumor derived endothelial cell mitogens are identical, do not demonstrate that

FGF could account for the full angiogenic activity of tumor cells. In addition to producing endothelial cell growth factors, tumor cells also produce chemotatic factors which could play an important role in tumor angiogenesis. One of these, named angiogenin, was recently isolated from colon carcinoma cells and characterized (Fett et al., 1985).

Conclusion

The widespread distribution of basic FGF as well as the wide range of cell types on which it acts has given rise to a plurality of names for basic FGF and at least 15 pseudonyms exist for it. Even in the case of acidic FGF, which seems to be restricted to brain and retina, 8 pseudonyms exist. If one considers that basic and acidic FGF, despite their physicochemical differences, are two closely related mitogens acting through common cell surface receptors, and if one accepts the concept that it is the receptor which dictates the specificity of function, 23 apparently different growth factors will, once their structure is established by the various groups working with them, likely be compressed into a single one: FGF. Of all the growth factors which have been identified to date, none have such a wide range of effects on so many cell types. This is in fact what has led many scientists to be skeptical as to the existence of FGF and to propose that many growth factors, each responsible for one specific effect of FGF, were present in the original preparation. Once FGF was structurally characterized and cloned, this was shown to be incorrect.

Because of its unique relationship to components of the ECM such as heparan sulfate proteoglycans, FGF could probably be of value in explaining morphogenetic events related to hydrolysis of specific territory of basement membranes. Although a more popularized aspect of FGF is that of a mitogen, one should emphasize that it also acts as a differentiation factor for cell types as different as vascular or corneal endothelial cells, or chondrocytes. Here again a common thread becomes apparent, since FGF modulates in all these cell types the production of ECM components which could influence phenotypic expression. Both its mitogenic and differentiation effect on various cell types of the nervous system would be worth investigating from the point of view of developmental biology, since FGF has been shown to be mitogenic for glial cells, oligodendrocytes, and astroglial cells, as well as to trigger the differentiation of nerve cells.

In vivo, the aspects of FGF in developmentally regulated processes are just starting to be investigated. Although its effects on capillary endothelial cells in vitro, and its distribution in various organs or tumors known to have an angiogenic potential, provide a common denominator for its angiogenic activity in vivo, one would have a limited view of its potential targets in vivo, if one was to limit oneself to the angiogenic properties of FGF. The cell types listed in Table 2 clearly indicate in which tissue an effect of FGF should be investigated on either tissue formation or repair processes. In this respect, the development of cDNA probes for both basic and acidic FGF should provide useful tools for the study of the FGF expression in these various tissues.

Evidence that a viral oncogene may code for a growth factor or part of a growth factor receptor has recently emerged from studies on the PDGF structure and that of the EGF receptor structure (Waterfield et al., 1983; Downward et al., 1984). In the case of FGF, no sequence homology between this mitogen and the 20 or so oncogene products so far identified has been found. In view of the similarities of mode of action between PDGF and FGF, there is little doubt that FGF is the product of an oncogene yet unidentified. The way to identify its transforming activity could be by recombination of the transduced cellular FGF gene with the genome of a non-transforming virus. Such recombinants could acquire the characteristics of a rapidly transforming retrovirus.

Acknowledgements. We thank the U.S. Public Health Service for financial support (grants HL 20197 and EY 02186). The group of Dr. R. Guillemin (Salk Institute for Biological Studies, La Jolla, CA) was primarily responsible for determining the primary structure of basic and acidic FGF, while that of Dr. J. Fiddes (California Biotechnology, Mountain View, CA) was primarily responsible for the cloning of both basic and acidic FGF.

References

Abraham, J.A., A. Mergia, J.L. Whang, A. Tumolo, J. Friedman, K.A. Hyerild, D. Gospodarowicz, and J.C. Fiddes (1986) Nucleotide sequence of a bovine cDNA clone encoding the angiogenic protein, basic fibroblast growth factor. Science, in press.

Algire, G.H., H.W. Chalkley, F.Y. Legallais, and H.D. Park (1945) Vascular reactions of normal and malignant tumors in vivo. I. Vascular reactions of mice to wounds and to normal and neoplastic transplants. J. Natl. Cancer Inst. **6:** 73-85.

Armelin, H.A., M.C.S. Armelin, K. Kelly, T. Stewart, P. Leder, B.H. Cochran, and C.D. Stiles (1984) Functional role of c myc in mitogenic response to platelet derived growth factors precedes activation of c myc. Nature **310:** 655-660.

Ashton, N., B. Ward, and G. Serpell (1954) Effect of oxygen on developing retinal vessels with particular reference to the problem of retrolental fibroplasia. Br. J. Ophthalmol. **38:** 397-433.

Ausprunk, D.H., D.R. Knighton, and J. Folkman (1975) Vascularization of normal and neoplastic tissues grafted to the chick chorioallantois: Role of host and preexisting graft blood vessels. Am. J. Pathol. **79:** 597-618.

Baird, A., F. Esch, D. Gospodarowicz, and R. Guillemin (1985a) Retina derived endothelial cell growth factors: Partial molecular characterization identity with acidic and basic fibroblast growth factor. Biochemistry **24:** 7855-7859.

Baird, A., F. Esch, N. Ling, and D. Gospodarowicz (1985b) Isolation and partial characterization of an endothelial cell growth factor from the bovine kidney: Homology with basic fibroblast growth factor. Regulatory Peptides **12:** 201-213.

Baird, A., P. Mormede, and P. Bohlen (1985c) Immunoreactive fibroblast growth factor in cells of peritoneal exudate suggest its identity with macrophage derived growth factor. Biochem. Biophys. Res. Comm. **126:** 358-364.

Barritault, D., C. Arruti, and Y. Courtois (1981) Is there a ubiquitous growth factor in the eye? Differentiation (Berlin) **18:** 29-42.

Bassett, D.L. (1943) The changes in the vascular pattern of the ovary of the albino rat during the estrous cycle. Am. J. Anat. **73:** 251-262.

Beguinot, L., J.A. Hanover, S. Ito, N. Richert, M.C. Willingham, and I. Pastan (1985)

Phorbolesters induced transient internalization without degradation of unoccupied epidermal growth factor receptors. Proc. Natl. Acad. Sci. (USA) **82:** 2774-2778.

Bernfield, M., S.D. Banerjee, J.E. Koda, and A.C. Rapraeger (1984) Remodelling of the basement membrane as a mechanism of morphogenetic interaction. In *The Role of Extracellular Matrix in Development*, R.L. Trelstad, ed., pp. 545-572. Allan R. Liss, Inc., New York.

Bohlen, P., A. Baird, F. Esch, N. Ling, and D. Gospodarowicz (1984) Partial molecular characterization of pituitary fibroblast growth factor. Proc. Natl. Acad. Sci. (USA) **81:** 5364-5368.

Bohlen, P., F. Esch, A. Baird, and D. Gospodarowicz (1985) Amino terminal sequence and comparison to basic fibroblast growth factor. EMBO (Europ. Molec. Biol. Org.) Journal **4:** 1951-1956.

Broad, T.E., and R.G. Ham (1983) Growth and adipose differentiation of sheep preadipocytes fibroblasts in serum free medium. Eur. J. Biochem. **135:** 33-39.

Buntrock, P., K.D. Jentzsch, and G. Heder (1982a) Stimulation of wound healing using brain extract with FGF activity. Quantitative and biochemical studies into formation of granulation tissue. Exp. Pathol. **21:** 46-53.

Buntrock, P., K.D. Jentzsch, and G. Heder (1982b) Stimulation of wound healing using brain extract with FGF activity. II. Histological and morphometric examination of cells and capillaries. Exp. Pathol. **21:** 62-67.

Buntrock, P., M. Buntrock, I. Marx, D. Kranz, K.D. Jentzsch, and G. Heder (1984) Stimulation to wound healing using brain extract with Fibroblast growth factor (FGF) activity. Exp. Pathol. **26:** 247-254.

Carlone, R.L., and J.E. Foret (1979) Stimulation of mitosis in cultured limb blastemata of the newt Notophthalmus viridescens. J. Exp. Zool. **21:** 245-252.

Carlone, R.L., M Ganagarajah, M.P. Rathbone (1981) Bovine pituitary fibroblast growth factor has neurotrophic activity for newt limb regenerates and skeletal muscle in vitro. Exp. Cell Res. **132:** 15-21.

Castellot, J.J., M.L. Addinizio, R.D. Rosenberg, and M.J. Karnovsky (1981) Cultured endothelial cells produce a heparin like inhibitor of smooth muscle cell growth. J. Cell Biol. **90:** 372-279.

Castellot, J., D. Cochran, and M. Karnovsky (1985) Effect of heparin on vascular smooth muscle cells. J. Cell. Physiol. **124:** 21-28.

Cochet, C., G.N. Gill, J. Meisenhelder, J.A. Cooper, and T. Hunter (1984) C-Kinase phosphorylates the epidermal growth factor receptor and reduces its EGF-stimulated tyrosine protein kinase activity. J. Biol. Chem. **259:** 2553-2558.

Connelly, T.G., J.R. Ortiz, and T. Yamada (1972) Influence of the pituitary on Wolffian lens regeneration. Dev. Biol. **31:** 310.

Courty, J., C. Loret, M. Mooenner, B. Chevallier, O. Lagente, Y. Courtois, and D. Barritault (1985) Bovine retina contains 3 growth factor affinities with different affinities for heparin: Eye derived growth factor I, II, III. Biochimie **67:** 265-269.

D'Amore, P.A., B.M. Glaser, S.K. Brunson, and A.H. Fenselau (1981) Angiogenic activity from bovine retina: Partial purification and characterization. Proc. Natl. Acad. Sci. (USA) **78:** 3068-3072.

Davidson, J., M. Klagsbrun, K. Hill, A. Buckley, R. Sullivan, P.Brewer, and S. Woodward (1985) Accelerated wound repair, cell proliferation and collagen accumulation are produced by cartilage derived growth factor. J. Cell Biol. **100:** 1219-1227.

Derynck, R., A.B. Roberts, M.E. Winkler, E.Y. Chen, and D.G. Goeddel (1984) Human transforming growth factor and precursor structure and expression in E Coli. Cell **38:** 287-294.

Dresden, M.H. (1969) Denervation effects on newt limb regeneration: DNA, RNA and protein synthesis. Dev. Biol. **19:** 311-320.

Downward, J., Y. Yarden, E. Mayes, G. Scarce, N. Totly, P. Stockwell, A. Ullrich, J. Schlessinger, and M.D. Waterfield (1984) Close similarity of epidermal growth factor receptor and v-er-B oncogenes protein sequences. Nature **307:** 521-527.

Ekblom, B., E. Lehtonen, L. Saxen, and R. Timpl (1981) Shift in collagen types as an early response to induction of the metanephric mesenchyme. J. Cell Biol. **89:** 276-283.

Ekblom, P. (1984) Basement membrane proteins and growth factors in kidney differentiation. In *The Role of Extracellular Matrix in Development*, R.L. Treistad, ed., pp. 173-206. Alan R. Liss, Inc., New York.

Esch, F., A. Baird, N. Ling, N. Ueno, F. Hill, L. Deneroy, R. Klepper, D. Gospodarowicz, P. Bohlen, and R. Guillemin (1985a) Primary structure of bovine pituitary basic fibroblast growth factor (FGF) and comparison with the amino terminal sequence of bovine brain acidic FGF. Proc. Natl. Acad. Sci. (USA) **85:** 6507-6511.

Esch, F., N. Ueno, A. Baird, F. Hill, L. Denoroy, N. Ling, D. Gospodarowicz, and R. Guillemin (1985b) Primary structure of bovine brain acidic fibroblast growth factor (FGF) Biochem. Biophys. Res. Comm. **133:** 554-562.

Fett, J.W., D.Y. Styrdom, R.R. Lobb, E.M. Alderman, J. Lemuel-Bethune, J.F. Riordan, and B.L. Vallee (1985) Isolation and characterization of angiogenin, an angiogenic protein, from human carcinoma cells. Biochemistry **24:** 5480-5486.

Fritze, M.S., C.F. Reilly, and R.D. Rosenberg (1985) An antiproliferative heparan sulfate species produced by post confluent smooth muscle cells. J. Cell. Biol. **100:** 1041-1049.

Folkman, J., and R.S. Cotran (1976) Relation of vascular proliferation to tumor growth. Int. Rev. Expt. Pathol. **16:** 207-248.

Gajdusek, C.M., P. Di Corleto, R. Ross, and S.M. Schwartz (1980) An endothelial cell derived growth factor. J. Cell Biol. **85:** 467-472.

Glaser, B.M., H. D'Amore, S. Seppa, H. Seppa, and E. Schiffman (1980) Adult tissues contain chemoattractants for vascular endothelial cells. Nature **288:** 438-484.

Gimbrone, M.A., Jr., R.S. Cotran, S.B. Leapman, and J. Folkman (1974) Tumor growth and neovascularization: An experimental model using the rabbit cornea. J. Natl. Cancer Inst. **52:** 413-427.

Gimenez-Gallego, G., J. Rodkey, C. Bennett, M. Rios Candelore, J. Disalvo, and K. Thomas (1985) Brain-derived acidic fibroblast growth factor: Complete amino acid sequence and homologies. Science **230:** 1385-1388.

Gospodarowicz, D. (1974) Localization of a fibroblast growth factor and its effect alone and with hydrocortisone on 3T3 cell growth. Nature **249:** 123-127.

Gospodarowicz, D. (1979) Fibroblast and epidermal growth factors: Their uses in vivo and in vitro in studies on cell functions and cell transplantation. Mol. Cell Biochem. **25:** 79-110.

Gospodarowicz, D. (1984) Preparation of extracellular matrices produced by cultured bovine corneal endothelial cells and PF HR-9 teratocarcinoma cells: Their use in the study of cell proliferation and differentation. In *Methods in Molecular and Cell Biology*, vol. 1, D. Barnes, D. Sirbasdu, and G. Sato, eds., pp. 275-294. A.R. Liss, New York.

Gospodarowicz, D. (1985a) Fibroblast growth factor. In *Hormonal proteins and Peptides*, vol. 12, Li Ch, ed., pp. 205-230. Academic Press, New York.

Gospodarowicz, D. (1985b) Biological activity in vivo and in vitro of pituitary and brain fibroblast growth factor. In *Mediators in Cell Growth and Differentiation*, R.J. Ford and A.L. Maizel, eds., pp. 109-134. Raven Press, New York.

Gospodarowicz, D., and H. Bailecki (1978) The effects of the epidermal and fibroblast growth factor on the replicative lifespan of bovine granulosa cells in culture. Endocrinology **103:** 854-858.

Gospodarowicz, D., and J. Cheng (1986) Heparin protects basic and acidic FGF from inactivation. J. Cell Physiol., in press.

Gospodarowicz, D., and L. Giguere (1982) Growth factors, effects on corneal tissue. In *Cell Biology of the Eye*, D.S. McDevitt, ed., pp. 97-135. Academic Press, New York.

Gospodarowicz, D., and G. Greenburg (1981) Growth control of mammalian cells. Growth factors and extracellular matrix. In *The Biology of Normal Human Growth*, M. Ritzen, A. Aperia, K. Hall, A. Larsson, A. Zetterberg, and R. Zetterstrom, eds., pp. 1-21. Raven Press, New York.

Gospodarowicz, D., and G.-M. Lui (1981) Effect of substrata and fibroblast growth factor on the proliferation in vitro of bovine aortic endothelial cells. J. Cell. Physiol. **109:** 385-392.

Gospodarowicz, D., and A.L. Mescher (1977) A comparison of the response of cultured myoblasts and chondrocytes to fibroblast and epidermal growth factors. J. Cell Physiol. **93:** 117-128.

Gospodarowicz, D., and A.L. Mescher (1981) Fibroblast growth factor and vertebrate regeneration. In *Advances in Neurology: Neurofibromatosis*, V.M. Riccardi and J.J. Mulvihill, eds., vol. 29, pp. 149-171. Raven Press, New York.

Gospodarowicz, D., and J.S. Moran (1974) Effect of fibroblast growth factor, insulin, Dexamethasone and serum on the morphology of BALB/c 3T3 cells. Proc. Natl. Acad. Sci. (USA) **71:** 46-48.

Gospodarowicz, D., and J.S. Moran (1975) Mitogenic effect of fibroblast growth factor on early passage cultures of human and murine fibroblasts. J. Cell. Biol. **66:** 451-556.

Gospodarowicz, D., and J.-P. Tauber (1980) Growth factors and extracellular matrix. Endocr. Rev. **1:** 201-227.

Gospodarowicz, D., and T.K. Thakral (1978) Production of a corpus luteum angiogenic factor responsible for proliferation of capillaries and neovascularization of the corpus luteum. Proc. Natl. Acad. Sci. (USA) **75**, 847.

Gospodarowicz, D., and Zetter, B. (1977) In *Joint WHO/IABS Symposium on the Standardization of Cell Substrates for the Production of Virus Vaccine*, pp. 109-130. S. Karger, Basel, New York.

Gospodarowicz, D., H. Bialecki, T.K. Thakral (1979a) The angiogenic activity of the fibroblast and epidermal growth factor. Exp. Eye Res. **28:** 501-504.

Gospodarowicz, D., K.S. Brown, and C.R. Birdwell, (1978a) Control of proliferation of human vascular endothelial cells of human origin. I. Characterization of the response of human umbilical vein endothelial cells to fibroblast growth factor, epidermal growth factor, and thrombin. J. Cell. Biol. **77:** 774-788.

Gospodarowicz, D., J. Cheng, and M. Lirette (1983a) Bovine brain and pituitary fibroblast growth factor: Comparison of their abilities to support the proliferation of human and bovine vascular endothelial cells. J. Cell Biol. **97:** 1677-1685.

Gospodarowicz, D., D.C. Cohen, and D.K. Fujii (1982a) Regulation of cell growth by the basal lamina and plasma factors: Relevance to embryonic control of cell proliferation and differentiation. In *Cold Spring Harbor Conference on Cell Proliferation, vol. 9, Growth of Cells in Hormonally Defined Media*, G. Sato, A. Pardee, and D. Sirbasku, eds., pp. 95-124. Cold Spring Harbor.

Gospodarowicz, D., D. Delgado, and I. Vlodavsky (1980a) Permissive effect of the extracellular matrix on cell proliferation in vitro. Proc. Natl. Acad. Sci. (USA) **77:** 4094-4098.

Gospodarowicz, D., D.K. Fujii, I. Vlodavsky (1982b) Basal lamina and the control of proliferation of malignant and normal cells. In *Expression of Differentiated Functions in Cancer Cells*, R.P. Revoltella, G.M. Ponteri, C. Basilico, G. Rovera, R.C. Gallo, and J.H. Subak-Sharpe, eds., pp. 121-139. Raven Press, New York.

Gospodarowicz, D., R. Gonzalez, and D.K. Fujii (1983b) Are factors originating from serum plasma, or cultured cells involved in the growth-promoting effect of the extracellular matrix produced by cultured bovine corneal endothelial cells? J. Cell Physiol. **114:** 191-202.

Gospodarowicz, D., G. Greenburg, H. Bialecki (1978b) Factors involved in the modulation of cell proliferation in vivo and in vitro: The role of fibroblast and epidermal growth factors in the proliferative response of mammalian cells. In Vitro **14:** 85-118.

Gospodarowicz, D., G.-M. Lui, and J. Cheng (1982c) Purification in high yield of brain fibroblast growth factor by preparitive isoelectric focusing at pH 9.6. J. Biol. Chem. **257:** 12266-12278.

Gospodarowicz, D., S. Massoglia, and J. Cheng (1986a) Effect of retina derived acidic and basic fibroblast growth factor and lipoproteins on the proliferation of bovine retina derived capillary endothelial cells. Expt. Eye Res., in press.

Gospodarowicz, D., A.L. Mescher, and C.R. Birdwell (1978c) Determination of cellular shape by the extracellular matrix and its correlation with the control of cellular growth. National Cancer Inst. Monogr. **48:** 109-130.

Gospodarowicz, D., A.L. Mescher, and J.S. Moran (1978d) Cellular specificities of fibroblast growth factor and epidermal growth factor. Symp. Soc. Dev. Biol. **35:** 33-61.

Gospodarowicz, D., I. Vlodavsky, and N. Savion (1980b) The extracellular matrix and the control of proliferation of vascular endothelial and vascular smooth muscle cells. J. Supramol. Struct. **13:** 339-372.

Gospodarowicz, D., I. Vlodavsky, and N. Savion (1981) The role of fibroblast growth factor and the extracellular matrix in the control of proliferation and differentiation of corneal endothelial cells. Vision Res. **21:** 87-103.

Gospodarowicz, D., C.R. Ill, P.J. Hornsby, and G.N. Gill (1977a) Control of bovine adrenal cortical cell proliferation by fibroblast growth factor. Lack of effect of epidermal growth factor. Endocrinology **100:** 1080-1089.

Gospodarowicz, D., J. Lepine, S. Massoglia, and I. Wood (1984b) Comparison of the ability of basement membranes produced by corneal endothelial and mouse-derived endodermal PF-HR-9 cells to support the proliferation and differentiation of normal diploid bovine kidney tubule epithelial cells in vitro. J. Cell. Biol. **99:** 947-961.

Gospodarowicz, D., S. Massoglia, J. Cheng, and D.K. Fujii (1986b) Effect of fibroblast growth factor and lipoproteins on the proliferation of endothelial cells derived from bovine adrenal and brain cortex, bovine corpus luteum, and brain capillaries. J. Cell Physiol., in press.

Gospodarowicz, D., A.L. Mescher, K. Brown, and C.R. Birdwell (1977b) The role of fibroblastic growth factor and epidermal growth factor in the proliferative reponse of the corneal and lens epithelium. Exp. Eye Res. **25:** 631-649.

Gospodarowicz, D., J.S. Moran, D. Braun, and C.R. Birdwell (1976a) Clonal growth of bovine endothelial cells in tissue culture: Fibroblast growth factor as a survival agent. Proc. Natl. Acad. Sci. (USA) **73:** 4120-4124.

Gospodarowicz, D., P. Rudland, J. Lindstrom, and K. Benirschke (1975) Fibroblast growth factor: Localization, purification, mode of action, and physiological significance. In *Advances in Metabolic Disorders*, H. Luft, ed., vol 8, pp. 301, 335. Stockholm.

Gospodarowicz, D., I. Vlodavsky, P. Fielding, and C.R. Birdwell (1978e) The effect of epidermal and fibroblast growth factor upon cell proliferation using vascular and corneal endothelial cells as a model. In *Birth Defects*, J.W. Littlefield and J. de Grouchy, eds., pp. 233-271. Excerpta Medica, Amsterdam, The Netherlands.

Gospodarowicz, D., J. Weseman, J. Moran, and J. Lindstrom (1976b) Effect of fibroblast growth factor on the division and fusion of bovine myoblasts. J. Cell Biol. **70:** 395-407.

Gospodarowicz, D., J. Cheng, G.-M. Lui, A. Baird, and P. Bohlen (1984a) Isolation by heparin separose affinity chromatography of brain fibroblast growth factor: Identity with pituitary fibroblast growth factor. Proc. Natl. Acad. Sci. (USA) **81:** 6963-6967.

Gospodarowicz, D., S. Massoglia, J. Cheng, G.-M. Lui, and P. Bohlen (1985c) Isolation of bovine pituitary fibroblast growth factor purified by fast protein liquid chromatography (FPLC). Partial chemical and biological characterization. J. Cell Physiol. **122:** 323-393.

Gospodarowicz, D., J. Cheng, G.-M. Lui, A. Baird, F. Esch, and P.Bohlen (1985a) Corpus luteum angiogenic factor is related to fibroblast growth factor. Endocrinology **117:** 2283-2291.

Gospodarowicz, D., J. Cheng, G.-M. Lui, D.K. Fujii, A. Baird, and P. Bohlen (1985b) Fibroblast growth factor in the human placenta. Biochem. Biophys. Res. Comm. **30:** 554-562.

Gospodarowicz, D., I. Vlodavsky, G. Greenberg, J. Alvarado, L.K. Johnson, and J. Moran

(1979b) Cellular Shapes is determined by the extracellular matrix and is responsible for the control of cellular growth and function. In *Cold Spring Harbor Conferences on Cell Proliferation, vol. 6, Hormones and Cell Culture*, R. Ross and G. Sato, eds., pp. 561-592. Cold Spring Harbor.

Gospodarowicz, D., I. Vlodavsky, G. Greenburg, J. Alvarado, L.K. Johnson, and J. Moran (1979c) Studies in atherogenesis and corneal transplantation using cultured vascular and corneal endothelia. Recent Prog. Horm. Res. **35**: 375-448.

Gottesman, M.M. (1978) Transformation dependent secretion of low molecular weight protein by murine fibroblasts. Proc. Natl. Acad. Sci. (USA) **75**: 2767-2771.

Greenburg, G., and D. Gospodarowicz (1982) Inactivation of a basement membrane component responsible for cell proliferation but not cell attachment. Exp. Cell Res. **140**: 1-14.

Greenburg, G., I. Vlodavsky, J.M. Foidart, and D. Gospodarowicz (1980) Conditioned medium from endothelial cell cultures can restore the normal phenotypic expression of vascular endothelium maintained in vitro in the absence of fibroblast growth factor. J. Cell. Physiol. **103**: 333-347.

Grobstein, C. (1956) Trans-filter induction of tubules in mouse metanephrogenic mesenchyme. Exp. Cell Res. **10**: 424-440.

Guttman, E. (1976) Neurotrophic relations. Annu. Rev. Physiol. **38**: 177-216.

Hamilton, R.T., M. Nilsen-Hamilton, and G.A. Adams (1985) Superinduction by cycloheximide of mitogen induced secreted proteins produced by Balb/c 3T3 cells. J. Cell Physiol. **123**: 201-208.

Hanover, J.A., and R.B. Dickson (1985) The possible link between receptor and phosphorylation and internalization. Trends Pharmaco. Sci. (Dec.): 457-459.

Hay, E.D. (1974) Cellular basis of regeneration. In *Concepts of Development*, J. Lash and J.R. Whittaker, eds. pp. 404-428. Sinauer, Stanford, Conn.

Hay, E.D. (1980) Development of the vertebrate cornea. Int. Rev. Cytol. **63**: 263-322.

Hay, E.D. (1981) Extracellular matrix. J. Cell Biol. **91**: 201-223.

Hay, E.D. (1984) Cell matrix interaction in the embryo: Cell shape, cell surface, cell skeleton, and their role in differentiation. In *The Role of Extra Cellular Matrix in Development*, R.L. Trelstad, ed., pp. 1-3. Alan R. Liss, Inc., New York.

Jentzsch, K.D., G. Wellmitz, G. Heder, P. Petzold, P. Buntrock, and P.A. Oehme (1980) Bovine brain fraction with fibroblast growth factor actively inducing articular cartilage. Acta biol. med. germ. **39**: 97-981.

Kato, Y., and D. Gospodarowicz (1984) Growth requirements of low density rabbit costal chondrocyte cultures maintained in serum-free medium. J. Cell Physiol. **120**: 354-363.

Kato, Y., and D. Gospodarowicz (1985) Sulfated proteoglycan synthesis by rabbit costal chondrocytes grown in the presence and absence of fibroblast growth factor. J. Cell Biol. **100**: 477-485.

Kaibuchi, K., T. Tsuda, A. Kikuchi, T. Tanimoto, T. Yamashita, and Y. Takai (1986) Possible involvement of protein kinase C and calcium ion in growth factor induced expression of c myc oncogene in Swiss 3T3 cells. J. Biol. Chem. **261**: 1187-1192.

Kiss, F.A., and I. Krompecher (1962) Promotion of vascularisation with adrenal extracts explanted to chorioallantoic membrane. Acta Biol. Acad. Sci. Hung. (Suppl) **4**: 39-57.

Kiss, F.A., and S. Szabo (1969) Total adrenal extract enhancing vascularization and ossification. Acta Morphol. Acad. Sci. Hung. **17**: 351-355.

Klagsbrun, et al. (1986). Proc. Natl. Acad. Sci., in press.

Kozak, M. (1984). Nucleic Acids Res. **12**: 857.

Kraemer, P.M. (1971) Heparan sulfates of cultured cells II acid soluble and precipitable species of different cell lines. Biochemistry **10**: 1445-1451.

Kratochwil K. (1972) Tissue interactions during embryonic development. In *General Properties in Tissue Interaction in Carcinogenesis*, D. Train, ed, pp. 1-48. Academic Press, New York.

Lemmon, S.K., M.C. Riely, K.A. Thomas, G.A. Hoover, T. Maciag, and R. Bradshaw

(1982) Bovine fibroblast growth factor: Comparison of brain and pituitary preparations. J. Cell Biol. **95:** 162-169.

Lobb, R.R., and J.W. Fett (1984) Purification of 2 distinct growth factors from bovine neural tissue by heparin affinity chromatography. Biochemistry **23:** 6925-6929.

Lobb, R.R., E.M. Alderman, and J.W. Fett (1985) Induction of angiogenesis by bovine brain derived class I heparin binding growth factor. Biochemistry **24:** 4969-4973.

Lobb, R., J. Sasse, Y. Shing, P. D'Amore, J. Fett, R. Sullivan, J. Jacobs, and M. Klagsbrun (1986) Purification and characterization of heparin binding endothelial cell growth factors. J. Biol. Chem. **261:** 1924-1928.

Maciag, T. (1984) Angiogenesis. Progress in Hemostasis and Thrombosis **7:** 167-182.

Maciag, T., R. Friesel, L. Mehlman, and A.B. Schreiber (1984) Heparin binds endothelial cell growth factor, the principal endothelial cell mitogen in bovine brain. Science **225:** 932-934.

Maciag, T., J. Cerundolo, S. Isley, P.R. Kelley, and R. Forand (1979) An endothelial cell growth factor from bovine hypothalamus: Identification and partial characterization. Proc. Natl. Acad. Sci (USA) **76:** 5674-5678.

Maciag, T., G.A. Hoover, J.Van der Spek, M.B. Stermerman, and R. Weinstein (1982) Growth and differentiation of human umbilical vein endothelial cells in culture. In *Cold Spring Harbor Conference on Cell Proliferation*, vol. 9, pp. 525-538.

Matzner, Y., M. Bar Ner, J. Yahalom, R. Ishai-Michaels, Z. Fuks, and I. Vlodavsky (1985) Degradation of heparan sulfate in the sub-endothelial matrix by readily released heparanese from human neutrophils. J. Clin. Inv. **76:** 1306-1313.

Mescher, A.L., and D. Gospodarowicz (1979) Mitogenic effect of a growth factor derived from myelin on denervated regenerates of newt forelimbs. J. Exp. Zool. **207:** 497-503.

Mescher, A.L., and J.J. Loh (1980) Newt forelimb regeneration blastemas in explanation and effects of various growth-promoting substances. J. Exp. Zool. **216:** 235-245.

Michaelson, J.C. (1948) The mode of development of the vascular system of the retina, with some observations on its significance for certain retinal diseases. Trans. Opthalmol. Soc. U.K. **68:** 137-180.

Muller, R., R. Bravo, J. Burkshardt, and T. Curran (1984) Induction of c fos gene and protein by growth factors precedes activation of c myc. Nature **312:** 716-720.

Muller, R., and E.F. Wagner (1984) Differentiation of F9 teratocarcinoma stem cells after transfer of c-fos proto-oncogenes. Nature **311:** 438-442.

Neufeld, G., and D. Gospodarowicz (1985) The identification and partial characterization of the fibroblast growth factor receptor of baby hamster kidney cells. J. Biol. Chem. **260:** 13860-13868.

Neufeld, G., and D. Gospodarowicz (1986) Basic and acidic fibroblast growth factor interact with the same cell surface receptor. J. Biol. Chem., in press.

Nilsen-Hamilton, M., R.T. Hamilton, W.R. Allen, and S.L. Massoglia (1981) Stimulation of the release of two glycoproteins from mouse 3T3 cells by growth factors and by agents that increase intralysosomal pH. Biochem. Biophys. Res. Commun. **101:** 411-417.

Nilsen-Hamilton, M., J.M. Shapiro, S.L. Massoglia, and R.T. Hamilton (1980) Selective stimulation by mitogens of incorporation of ^{35}S-methionine into a family of proteins released into the medium by 3T3 cells. Cell **20:** 19-28.

Nishi, N., Y. Matuo, Y. Mugurama, Y. Yoshitake, K. Nishikawa, and F. Wada (1985) A human prostatic growth factor (hPGF): Partial purification and characterization. Biochem. Biophys. Res. Com. **132:** 1103-1109.

Powell, J.A. (1969) Analysis of histogenesis and regenerative ability of denervated forelimb regenerates of Triturus viridescens. J. Exp. Zool. **170:** 125-148.

Preminger. G.M., W.E. Koch, F.A. Fried. and J. Mandell (1980) Utilization of the chick chorioallantoic membrane for in vitro growth of the embryonic murine kidney. Amer. J. Anat. **159:** 17-24.

Proudfoot, N.J., and G.G. Brownlee (1976) 3' non-coding region sequences in eukaryotic messenger RNA. Nature **263:** 211-213.

Ratner, N., R.P. Bunge, and L. Glaser (1985) A neuronal cell surface heparan sulfate proteoglycan is required for dorsal root ganglion neuron stimulation of Schwann cell proliferation. J. Cell Biol. **101**: 744-754.
Reyer, R.W. (1977) The amphibian eye: Development and regeneration. In *Handbook of Sensory Physiology* VII/5, ed. F. Crescitelli, pp. 309-390. Springer-Verlag, Berlin and New York.
Rudland, P.S., W. Seifert, and D. Gospodarowicz (1974) Growth control in cultured mouse fibroblasts: Induction of the pleiotypic and mitogenic responses by a purified growth factor. Proc. Natl. Acad. Sci. (USA) **71**: 2600-2604.
Rizzino, A., E. Ruff, and H. Rizzino (1985a) PDGF and FGF induce and/or potentiate the soft agar growth of non-transformed cells. J. Cell Biol. **101**: Abst. 435, p. 115.
Sariola, H., P. Ekblom, E. Lehtonen, and L. Saxen (1983) Differentiation and vascularization of the metanephric kidney grafted on the chorioallantoic membrane. Dev. Biol. **96**: 926-927.
Savion, N., I. Vlodavsky, and Z. Fuks (1984) Interaction of T-lymphocytes and macrophages with cultured vascular endothelial cells. J. Cell. Physiol. **118**: 169-178.
Saxen, L., P. Ekblom, and E. Lehtonen (1982) The kidney as a model system for determination and differentation. In *The Biology of Normal Human Growth*, M. Ritzen, A. Aperia, K. Hall, A. Larsson, A. Zettersberg and R. Zetterstrom, eds., pp. 177-127. Raven Press, New York.
Schlessinger, J., and B. Geiger. Epidermal growth factor induces redistribution of actin and actinin in human epidermal carcinoma cells. Exp. Cell Res. **134**: 273-279.
Schlessinger, J. (1980) The mechanism and role of hormone induced clustering of membrane receptor. Trends Biol. Sci. **5**: 210-214.
Schreiber, A.B., J. Kenney, W.J. Kowalski, R. Friesel, T. Mehlaman, and T. Maciag (1985) Interaction of endothelial cell growth factor with heparin: Characterization by receptor and antibody recognition. Proc. Natl. Acad. Sci. (USA) **82**: 6138-6142.
Schwartz, S.M., and C.M. Gajdusek (1982) Growth factors and the vessel wall. Prog. Hemost. Thromb. **5**: 85-112.
Schwartz, S.M., C.M. Gajdusek, and S.C. Selden, III. (1981) Vascular wall growth control: The role of the endothelium. Arteriosclerosis **I**: 107-126.
Serrero, G., and J.C. Khoo (1982) An in vitro model to study adipocyte differentiation in serum free medium. Anal. Biochem. **120**: 351-359.
Simonian, M.H., P.J. Hornsby, C.R. Ill, M.J. O'Hare, and G.N. Gill (1979) Characterization of cultured bovine adrenal cortical and derived clonal lines: Regulation of steroidogenesis and culture life span. Endocrinology **105**: 99-108.
Shing, Y., J. Folkman, R. Sullivan, C. Butterfield, J. Curray, and M. Klagsbrun (1984) Heparin affinity: Purification of a tumor-derived capillary endothelial cell growth factor. Science **223**: 1296-1298.
Singer, M. (1954) Induction of regeneration of the forelimb of the post-metamorphic frog by augmentation of the nerve supply. J. Exp. Zool. **126**: 419-471.
Singer, M. (1952) The influence of the nerve in regeneration of the amphibian extremity. Q. Rev. Biol. **27**: 169-200.
Singer, M. (1974) Neurotrophic control of limb regeneration in the newt. Ann. N.Y. Acad. Sci. **228**: 308-322.
Singer, M., and L. Craven (1948) The growth and morphogenesis of the regenerating forelimb of adult triturus following denervation at various stages of development. J. Exp. Zool. **108**: 279-308.
Tannock, I.F. (1945) The relation between cell proliferation and the vascular system in a transplanted mouse mammary tumor in vivo. I. Vascular reactions of mice to wounds and to normal and neoplastic transplants. J. Natl. Cancer Inst. **6**: 73-85.
Thesleff, I., H.J. Barrach, J.M. Foidart, A. Vaheri, R.M. Pratt, and G.R. Martin (1981) Tissue interactions in tooth development. A review. Dev. Biol. **81**: 182-192.
Thomas, K.A., M.C. Riley, S.K. Lemmon, N.C. Baglan, and R.A. Bradshaw (1980) Brain fibroblast growth factor: Nonidentity with myelin basic protein fragments. J. Biol. Chem. **255**: 5517-5520.

Thomas, K.A., M. Rios-Candelore, and S. Fitzpatrick (1984) Purificaton and characterization of acidic fibroblast growth factor from bovine brain. Proc. Natl. Acad. Sci. (USA) **81:** 357-361.
Thomas, K.A., M. Rios-Candelore, G. Gimenez-Gallego, J. DiSalvo, C. Bennett. J. Rodkey, and S. Fitzpatrick (1985) Pure brain derived fibroblast growth factor is a potent angiogenic vascular endothelial cell mitogen with sequence homology to interleukin. I. Proc. Natl. Acad. Sci (USA) **82:** 6409-6413.
Thornton, S.C., S.N. Mueller, and E.M. Levine (1983) Human endothelial cells: Use of heparin in cloning and long term serial cultivation. Science **222:** 623-625.
Todd, J.T. (1823) On the process of reproduction of the members of the aquatic salamander. Q. J. Mic. Sci. Lit. Arts **16:** 84-96.
Togari, A., D. Baker, G. Dickens, and G. Guroff (1983) The neurite-promoting effect of fibroblast growth factor on PC-12 cells. Biochem. Biophys. Res. Commun. **114:** 1189-1193.
Togari, A., G. Dickens, H. Huzuya, and G. Guroff (1985) The effect of fibroblast growth factor on PC-12 cells. J. Neurosciences **5:** 307-316.
Turner, D.C., and J.T. Bagnara (1981) *General Endocrinology*, 5th ed., pp. 349-386. Saunder, Philadelphia.
Tseng, S., N. Savion, R. Stern, and D. Gospodarowicz (1982) Fibroblast growth factor modulates synthesis of collagen in cultured vascular endothelial cells. Eur. J. Biochem. **122:** 355-360.
Tsuda, T., K. Kaibuchi, Y. Kawahara, H. Fukuzaki, and Y. Takai (1985) Induction of protein kinase C and calcium ion mobilization by fibroblast growth factor in Swiss 3T3 cells. FEBS letter, in press.
Vlodavsky, I., and D. Gospodarowicz (1979) Structural and functional alterations in the surface of vascular endothelial cells associated with the formation of a confluent cell monolayer and with the withdrawal of fibroblast growth factor. J. Supramol. Struct. **12:** 73-14.
Vlodavsky, I., L.K. Johnson, G. Greenburg, and D. Gospodarowicz (1979) Vascular endothelial cells maintained in the absence of fibroblast growth factor undergo structural and functional alterations that are incompatible with their in vivo differentiated properties. J. Cell Biol. **83:** 468-486.
Wasteson, A., B. Glimelius, C. Busch, B. Westermark, C.-H. Heldlin, and B. Norling (1977) Effect of a platelet endoglycosidase on cell surface heparin sulfate of human cultured endothelial and glial cells. Thromb. Res. **11:** 309-321.
Waterfield, M.D., G.T. Scrace, N. Whittle, P. Stroobant, A. Johnson, A. Wasteson, B. Westermark, C.H. Heldlin, J.S. Huang, and T.F. Deuel. Platelet-derived growth factor is structurally related to the putative transforming protein p 28^{sis} of simian sarcoma virus. Nature **304:** 35-39.
Westall, F.C., R. Rubin, and D. Gospodarowicz (1983) Brain-derived growth factor: A study of its inactivation. Life Sciences **33:** 2425-2429.
Westermark, B., C.H. Heldin, B. Ek, A. Johnson, K. Mellstrom, M. Nister, and A. Wasteson (1982) Biochemistry and biology of platelet-derived growth factor. In *Growth and Maturatoin Factors*, G. Guroff, ed., vol. I, pp. 75-112. Wiley, New York.
Wessels, N.K. (1977) Extracellular materials and tissue interaction, In *Tissue Interaction and Development*, pp. 213-229. W.A. Benjamin, Inc., New York.
Yamada, T. (1977) Control mechanisms in cell type conversion in newt lens regeneration. In Monogr. Dev. Biol., vol. 13, A. Wolsky, ed., pp. 1-126. Karger, Basel, New York.
Yamada, T. (1982) Transdifferentiation of lens cells and its regulation. In *Cell Biology of the Eye*, McDevitt, ed., pp. 193-234. Acad. Press, New York.
Yamada, T. and D.S. McDevitt (1974) Direct evidence for transformation of differentiated iris epithelial cells into lens cells. Dev. Biol. **38:** 104-124.
Yahalom, J., A. Eldor, Z. Fuks, and I. Vlodavsky (1984) Degradation of sulfated proteoglycans in the subendothelial basement membrane by human platelet heparitinase. J. Clin. Inv. **74:** 1842-1849.
Zetter, B.R. (1981) The endothelial cells of large and small blood vessels. Diabetes **30:** 24-28.

HORMONAL REGULATION OF EPITHELIAL MORPHOGENESIS, GROWTH AND CYTODIFFERENTIATION IN FETAL AND ADULT UROGENITAL TRACTS: ROLES OF MESENCHYMAL-EPITHELIAL INTERACTIONS

G.R. Cunha[1,2], R.M. Bigsby[1], A.A. Donjacour[1], and P.S. Cooke[1]

[1] Department of Anatomy
University of California
San Francisco, CA

[2] Cancer Research Laboratory
University of California
Berkeley, CA
U.S.A.

The importance of mesenchymal-epithelial interactions for normal development is well established for a variety of structures and organs within the integumental, skeletal, gastro-intestinal, and neural systems (Kratochwil, 1972; Bernfield et al., 1973; Deuchar, 1975; Sengel, 1976; Saxen et al., 1976; Wessells, 1977; Fallon, 1983). Likewise, in the urogenital tract epithelial morphogenesis, growth, cytodifferentiation and function are induced and specified regionally by mesenchyme (Cunha, 1976a; Cunha et al., 1980a, 1983a). A special feature of the genital tract is that sex steroids and their analogs have profound developmental effects (Takasugi, 1976; McLachlan et al., 1980; Newbold and McLachlan, 1985; Bern and Talamantes, 1981; Clark and McCormack, 1977, 1980; Forsberg and Kalland, 1981; Taguchi and Nishizuka, 1985; Burns, 1961). These hormones may act in part by affecting mesenchymal-epithelial interactions in developing accessory sexual organs; such hormonal regulation provides unique opportunities for detailed analyses of the mechanism and control of these tissue interactions.

Mesenchymal Induction of Epithelial Morphogenesis and Cytodifferentiation

Mesenchyme in the developing genital tract elicits the formation of a variety of epithelial morphologies: branched ductal architecture (prostate); elaborately folded and branched mucosa (seminal vesicle); simple tubular glands (uterus); or simple tubular organs (ductus deferens and vagina). Through analysis of tissue recombinants composed of epithelial and mesenchymal elements from the same organ (homotypic tissue recombinants), it has been shown that mesenchyme is required for epithelial morphogenesis

NATO ASI Series, Vol. H5
Mesenchymal-Epithelial Interactions in Neural Development
Edited by J. R. Wolff et al.

(Cunha, 1972; 1976a, b). In heterotypic tissue recombinants, in which epithelium and mesenchyme are derived from different organ rudiments, mesenchyme can induce and specify new patterns of epithelial architecture (Table 1). In certain inductions the expression of new histological patterns is associated with radical changes in epithelial cytodifferentiation and concomitant alterations in functional or biochemical properties (Cunha, 1976b; Cunha et al., 1980a, 1983a, b, c; Neubauer et al., 1983). For example, under endogenous androgenic stimulation afforded as a result of grafting tissue recombinants into male hosts, epithelium of the fetal or adult urinary bladder, a highly specialized stratified epithelium, is induced by urogenital sinus mesenchyme (UGM, an embryonic prostatic inductor) to differentiate into a simple columnar epithelium that possesses an array of functional properties typical of the prostate (Table 2) (Cunha and Lung, 1978a; Cunha et al., 1980a, b, 1983b; Neubauer et al., 1983). Analogously, when simple columnar uterine epithelium (UE) is combined with vaginal mesenchyme (VM), it is

Table 1: Instructive inductions in heterotypic recombinations of epithelium and mesenchyme of the urogenital tract.

Tissue Recombinant		Host Sex		Morphological Reference
Mesenchyme	Epithelium		Endpoint*	
Urogenital Sinus	Bladder	Male	Prostate	Cunha and Lung (1978) Cunha et al. (1980b, 1983b)
Urogenital Sinus	Vagina	Male	Prostate	Cunha (1975)
Uterus	Vagina	Female	Uterus	Cunha (1976b)
Vagina	Uterus	Female	Vagina	Cunha (1976b)
Bulbourethral Gland	Bladder	Male	Bulbo-urethral gland	Cooke (unpublished)
Seminal Vesicle	Upper Wolffian duct	Male	Seminal Vesicle	Cunha (unpublished)
Cervix	Uterus	Female	Cervix	Cunha (unpublished)

* Epithelial and mesenchymal tissue were separated following tryptic digestion, recombined, and grown overnight in vitro to permit readhesion of the tissues. Subsequently, the tissue recombinants were grafted under the kidney capsule of either male or female hosts. Developmental endpoints have been assessed morphologically as well as with other criteria (see Table 2).

Table 2: Epithelial characteristics in prostate, urinary bladder and heterotypic tissue recombinants*.

Type of Analysis or Feature	Specimen		
	Bladder	Prostate	UGM + BFE
Histology	Transitional	Glandular	Glandular
Electron Microscopy	Non-Secretory Asymmetric Membrane	Secretory Symmetric Membrane	Secretory Symmetric Membrane
Histochemistry			
Alkaline Phosphatase	+	-	-
Alcian Blue	-	+	+
Nonspecific Esterase	- or ±	+	+
Prostate Antigens	-	+	+
Androgen Receptors	-	+	+
Androgen-dependent DNA Synthesis	-	+	+
Protein Synthesis (2-D gels)	bladder	prostate	prostate-like

* From Cunha et al., (1983b); Neubauer et al., (1983)

induced to form a stratified squamous vaginal epithelium (Cunha, 1976b). Moreover, the superficial layer of the newly formed stratified epithelium undergoes cyclic changes in cytodifferentiation, forming alternating layers of cornified and mucified cells in response to cyclic variations of ovarian steroids (Cunha, 1976b). These cyclic changes in cytodifferentiation are unique to vaginal epithelium. Patterns of protein synthesis in VM + UE recombinants (2-dimensional gels) are also distinctly vaginal (Cunha et al., 1983c). Finally, the epithelium of VM + UE recombinants expresses a uniquely vaginal pattern of a cell surface heparan sulfate proteoglycan (Cunha and Bernfield, unpublished). Thus, inductive interactions that result in changes in epithelial cytodifferentiation (simple columnar to stratified squamous; transitional to simple columnar) are associated with profound changes in functional and biochemical properties. In contrast, minor mesenchyme-induced changes in glandular or ductal branching morphology are not associated with changes in functional activity (Sakakura et al., 1976; Yasugi et al., 1985).

Mesenchyme-induced changes in epithelial morphogenesis are frequently accompanied by increased epithelial proliferative activity. Such proliferative effects of mesenchyme are most dramatic in tissue recombinants composed of urogenital sinus mesenchyme (UGM) and bladder epithelium (BLE), in which prostate is induced. Following grafting of UGM + BLE recombinants under the renal capsule of intact male hosts, rapid vascularization of the tissue recombinants insures that adequate levels of testosterone will reach the graft from the host. As early as 3 days after grafting, the basal cells of the BLE appear hyperplastic while apical cell layers begin to slough (Cunha et al., 1983b). At 3 to 4 days after grafting, solid epithelial buds begin to grow into the underlying UGM; these buds elongate, branch, and canalize during subsequent growth (Cunha et al., 1983b). At 4 and 7 days after grafting epithelial [^{3}H]-thymidine labelling index is markedly elevated (Shannon and Cunha, 1984; Cunha, unpublished). Under conditions of androgen deprivation (castrated hosts) both the morphogenetic and proliferative processes are absent in UGM + BLE recombinants (Shannon and Cunha, 1984).

Examination of UGM + BLE recombinants at 10 days after grafting demonstrates that the change in epithelial differentiation from the bladder to the prostatic phenotype is accompanied by changes in the expression of the heparan sulfate proteogylcan described by Bernfield and colleagues. Fig. 1 shows epithelial cells in various stages of the induction process. In some areas bladder epithelial cells retain their urothelial phenotype. In other areas, solid epithelial cords, some of which are undergoing canalization, emerge from the basal surface of the BLE. Fully differentiated prostatic ducts lined by a simple columnar epithelium are also present. Bladder epithelial cells exhibit intense immunostaining for the heparan sulfate proteoglycan while the simple columnar epithelium of mature prostatic ducts exhibits reduced immunostaining that is largely confined to the basal aspect of the cell. As the epithelial cells progress through the developmental stages from the urothelial to the prostatic phenotype there is a concomitant reduction in staining. This reduction of a specific gene product demonstrates modification in cell surface properties during mesenchyme-induced changes in epithelial cytodifferentiation.

Another significant alteration in the bladder epithelium as it is induced to form prostate is the appearance of androgen receptors. Normal bladder epithelium is devoid of nuclear androgen receptors. However, BLE is induced to express androgen receptors following induction by UGM (Cunha et al., 1980c; Neubauer et al., 1983). Other changes in functional characteristics, which occur as a result of prostatic induction, are listed in Table 2.

Mesenchyme as a Mediator of Hormonal Effects on Epithelium

In the developing and mature prostate gland, androgenic effects upon the epithelium

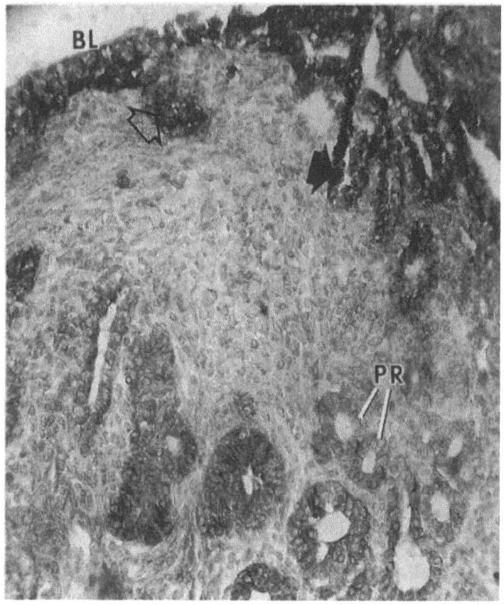

Figure 1: A tissue recombinant composed of 16-day fetal urogenital sinus mesenchyme associated with adult urinary bladder epithelium. Following 10 days of growth under the kidney capsule of an intact male host, the graft contains uninduced bladder epithelium (BL), solid epithelial downgrowths (open arrow), canalizing epithelial down growths (closed arrows) and maturing prostatic ducts (PR). Immunoperoxidase staining with monoclonal antibody 281-2 (see Bernfield, this volume) directed against a heparan sulfate proteoglycan demonstrates intense staining of the bladder epithelium and solid epithelial downgrowths. Canalizing downgrowths and maturing prostatic ducts show a progressive reduction in immunostaining. (200 X).

appear to be mediated via the mesenchyme. This conclusion is based in part upon the fact that in the fetal and early neonatal prostate the epithelium lacks nuclear androgen binding sites (androgen receptors), while the mesenchyme exhibits nuclear androgen binding (Shannon and Cunha, 1983; Takeda et al., 1985). More conclusive evidence has been obtained from recombination studies utilizing components derived from androgen-receptor-positive, wild-type and androgen-receptor-deficient, Tfm testicular feminization) mice. For example, prostatic ductal morphogenesis, epithelial proliferation, and secretory cytodifferentiation occur in wild-type UGM + Tfm BLE recombinants grafted into intact male hosts, indicating that androgenic effects are expressed in an androgen-insensitive, androgen-receptor-deficient Tfm epithelium (Cunha and Lung, 1978; Lasnitzki and Mizuno, 1980; Cunha et al., 1980a; Shannon and Cunha, 1984; Cunha and Chung, 1981). Also, once these tissue recombinants have been allowed to mature into prostate-like tissues, an experimental regimen of androgen deprivation followed by androgen stimulation induces proliferation of these Tfm epithelial cells (Cunha and Chung, 1981; Sugimura et al., 1986). Thus, intra-epithelial androgen receptors are not required for expression of many androgenic effects within developing and mature prostatic epithelial cells. Similarly, androgen-induced regression of fetal mammary epithelial rudiments in the male is mediated by the mesenchyme (Kratochwil and Schwartz, 1976; Drews and Drews, 1977; Wasner et al., 1983).

The premise that androgens elicit their effects upon prostatic and mammary epithelium via the mesenchyme raises the possibility that other classes of sex steroids may elicit effects on epithelia of other genital organs via similar mesenchyme-mediated processes. Estrogens have profound effects upon the developing female genital tract (Bern and Talamantes, 1981; McLachlan et al., 1980; Newbold and McLachlan, 1985; Forsberg and Kalland, 1981; Takasugi, 1976; Cunha et al., 1977). While an estrogen-receptor-deficient mutation is not available, striking parallels have been observed between the developing prostate and the developing female genital tract relative to the expression of hormonal effects versus the presence of hormone receptors. Autoradiographic studies have shown that uterine and vaginal epithelial cells of neonatal mice do not concentrate radiolabelled estrogens in their nuclei and, therefore, lack estrogen receptors (Fig. 2) (Cunha et al., 1982; Bigsby and Cunha, 1986). However, DNA synthesis

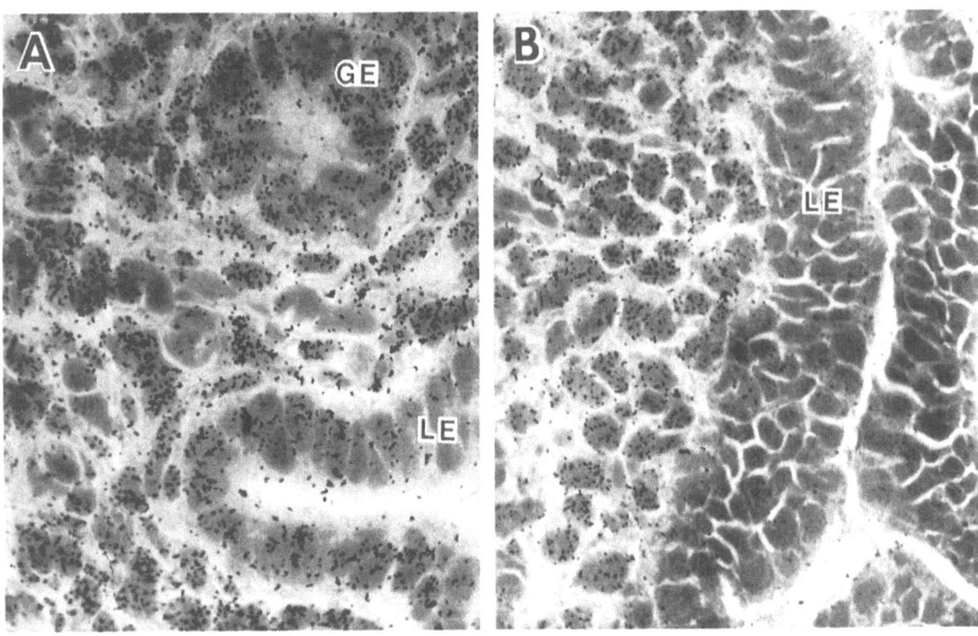

Figure 2: Estradiol autoradiography of mouse uterine tissues. Pieces of uterus from immature (21-day-old) (A) or neonatal (4-day-old) (B) mice were incubated with 10 μM [^{3}H]estradiol, washed in isotonic saline, frozen and sectioned at 4 μm for thaw-mount autoradiography. Note the lack of nuclear localization of silver grains over the luminal epithelium (LE) of the neonatal uterus (B); while in the 21-day-old uterus (A) both luminal epithelium and glandular epithelium (GE) exhibit nuclear concentration of steroid. At both ages, mesenchymally derived cells concentrate the steroid with their nuclei. (800 X) (from Bigsby and Cunha, 1986).

and proliferation in neonatal uterine epithelial cells are responsive to estrogen (Fig. 3), a situation analagous to androgen-induced proliferation in androgen-receptor-negative prostatic epithelial cells as discussed above. At later stages, nuclear estrogen binding sites are detectable in uterine epithelial cells of immature (21-day) and adult mice by identical autoradiographic techniques (Cunha et al., 1982; Stumpf and Sar, 1976; Bigsby and Cunha, 1986). Recently these autoradiographic observations have been corroborated by whole cell estrogen uptake studies performed on isolated uterine epithelial and stromal cells (Bigsby and Cunha, 1986). In agreement with the autoradiographic observations, high affinity estrogen binding activity, which was both specific and saturable, was measured in epithelial and stromal cells from uteri of 21-day-old mice (Fig. 4). Only estrogens (estradiol-17B, diethylstilbestrol, and moxestrol) competed effectively for this specific uptake; progesterone and dihydrotesterone had no effect. In contrast, uterine epithelial cells from 4- and 5-day-old mice did not exhibit saturable or specific uptake, while the mesenchymal cells from theseuteri contained a full complement of specific

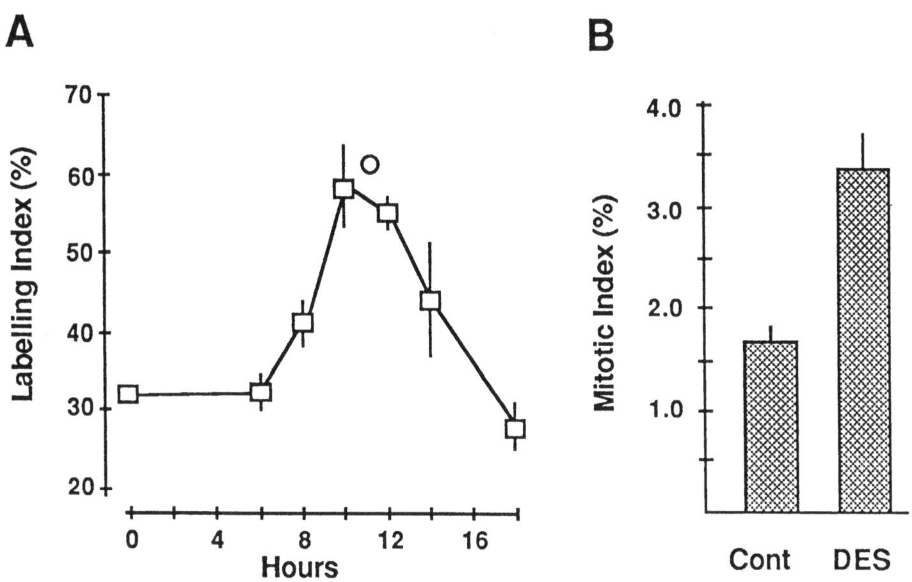

Figure 3: Thymidine labelling index (A) and mitotic index (B) of uterine epithelium of 5-day-old mice following estrogen treatment. A: [^{3}H]TdR was injected into each animal at various times following a single 5 µg injection of diethylstilbestrol (DES) [□], or at 11h after 17ß-estrdiol (O). One hour later the animal was killed and the uterine horns were processed for autoradiography. Labelling index (LI) is expressed as the percent of epithelial cells labelled. Androgens, progestins, and glucocorticoids did not stimulate epithelial DNA synthesis (not shown). B: 16 to 18 h after DES treatment animals were treated with colchicine for 2 h and then their uteri were processed for histological examination. The number of cells arrested in metaphase were counted and expressed as a percent. The mean index ±SEM of 3-5 animals at each time is presented. (modified from Bigsby and Cunha, 1986).

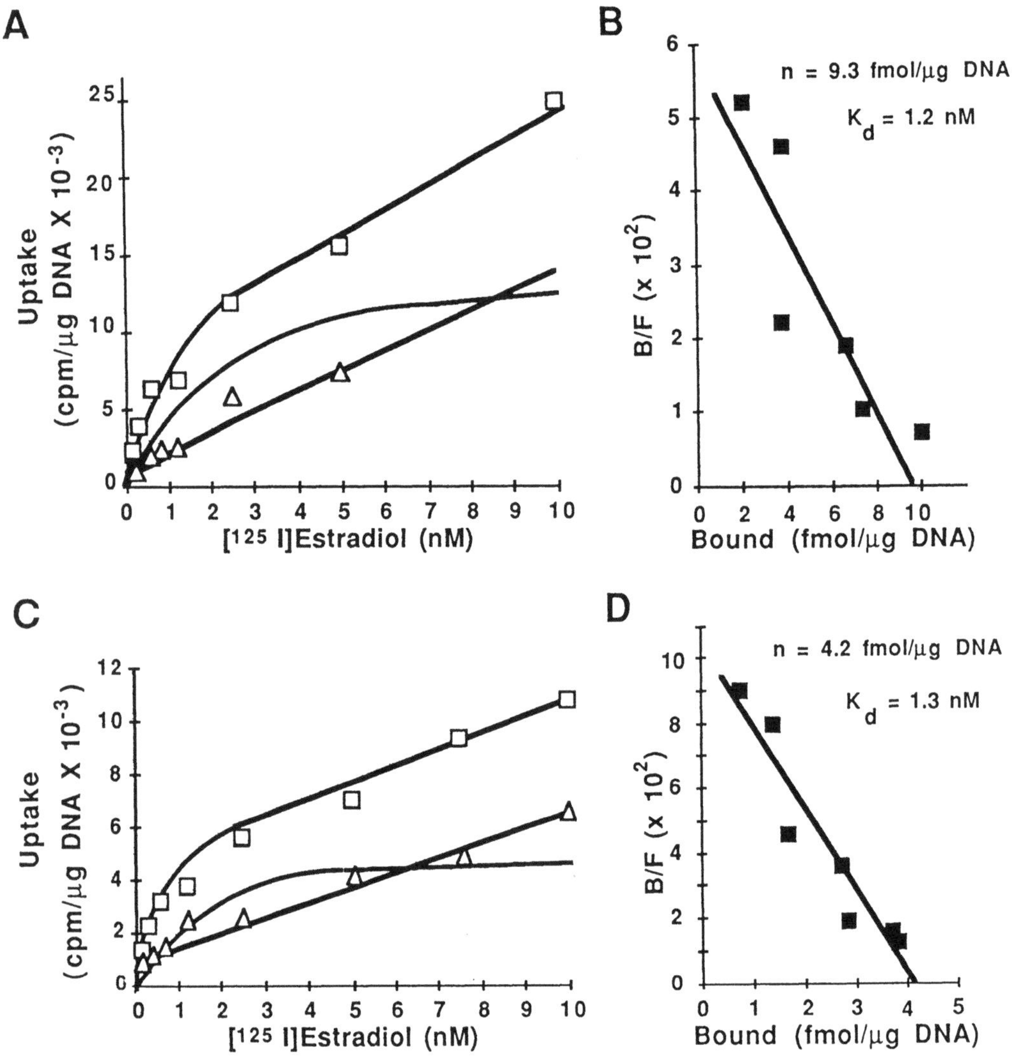

Figure 4: Estradiol uptake by cells from uteri of 21-day-old mice. Epithelial and mesenchymal cells were isolated following enzymatic digestion and incubated at 37°C with [^{125}I] estradiol at 0.15 to 10 μM in the absence or presence at 200-fold concentration of the radio inert estrogen analog, moxestrol. After extensive washing of the cells at 4°C the uptake of radiolabel was measured and related to the DNA content. Specific uptake was calculated by subtracting the non-specific uptake, uptake in the presence of moxestrol, from the total uptake. Total uptake (□—□), nonspecific uptake (△—△), and specific uptake (———) by epithelial cells (A) and mesenchymal cells (C) are presented as saturation curves. Corresponding Scatchard plots of specific uptakes are presented in (B) and (D). (from Bigsby and Cunha, 1986).

estrogen binding sites (Fig. 5). Even at 12 hours after estrogen injection, when the uterine epithelium is responding to estrogen stimulation, estrogen receptors are undetectable in that tissue. These observations indicate that intra-epithelial estrogen receptors are not necessary for estrogen stimulation of uterine epithelial proliferation in the neonatal mouse and suggest that the estrogen-receptor-positive mesenchyme mediates this stimulation.

The time course of the proliferative response and the maxiumum labelling index achieved in uterine epithelium of the neonatal mouse are nearly identical to those which have been reported in adult mice following a single injection of estrogen (Martin et al., 1973). These similarities between the neonatal and adult uteri suggests that the mechanism of stimulation may also be similar. Although adult uterine epithelial cells have estrogen receptors (Stumpf and Sar, 1976; McCormack and Glasser, 1980), they too may be proliferating in response to estrogen-induced regulators from stromal cells rather than through direct estrogenic action. This possibility is supported by the inability of estrogen to stimulate proliferation of uterine epithelial cells in culture (Casimiri et al., 1981; Tomooka et al., 1986). While other hypotheses have been put forward which also explain these results (Sirbasku and Benson, 1979; Soto and Sonnenshein, 1984), the strength of the

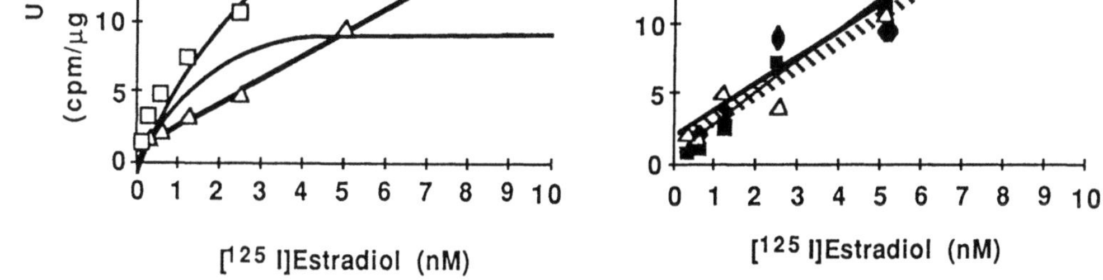

Figure 5: Estradiol uptake by uterine cells by 4-day-old mice. Saturation plots of the uptake by mesenchymal cells (A), symbols as in Fig. 4, and epithelial cells (B), symbols according to legend. Epithelial uptake is presented for cells taken from control 4-day-old animals and from 5-day-old animals that had been treated with DES 12 h earlier. Note the lack of difference between total and nonspecific uptake in the epithelium indicating in a lack of estrogen receptors. (from Bigsby and Cunha, 1986).

analogy between the observations in the developing male and female organs weigh in favor of the concept of a stromally mediated effect.

Stromal-Epithelial Interactions in Adult Genital Organs

In the genital tract significant changes in tissue architecture and epithelial cytodifferentiation are not restricted to fetal or neonatal periods. In adulthood alterations in the animal's hormonal status affect the structure and function of the reproductive tract and mammary gland. These changes during adulthood may well occur via the same mechanisms that operate during fetal and neonatal development, i.e., stromal induction and epithelial competence.

One of the fundamental tenets of developmental biology is that as embryonic development proceeds there is a gradual restriction in developmental potential from a totipotent to a pluripotent state and then ultimately an irreversible commitment to a single developmental endpoint (Blau et al., 1986). In epithelial tissue the eventual commitment to a particular phenotype is thought to be due to a series of inductive interactions (Nieuwkoop, 1985). In virtually all systems analyzed, commitment is attained during embryonic, fetal, neonatal or larval stages (Wessells, 1977; Kratochwil, 1972; Cunha, 1976a; Deuchar, 1975; Nieuwkoop. 1985). Based upon the stability of the differentiated state, it has been presumed that in adulthood cellular commitment (determination) has been achieved and is irreversible under normal (non-neoplastic) conditions. Exceptions to this universally held concept have recently emerged; these include those results of heterotypic tissue recombinants described above in which a differentiated epithelium can change its phenotype and biochemical characteristics in response to induction by heterologous mesenchyme (Tables 1 and 2). There are but two interpretations of these findings: either neonatal and adult ephithelia retain a subpopulation of undifferentiated, uncommitted cells which are capable of differentiating into a novel phenotype, or, alternatively, the induced cells can undergo dedifferentiation and subsequent reprogramming under the influence of heterologous mesenchyme. While it is not yet possible to distinguish between these two alternatives, it is important to recognize that determination and commitment are in fact operationally defined terms and, thus, potentially artificial concepts. For instance, a tissue is considered to be irreversibly determined or committed if its gene expression is restricted so that it continues to exhibit its normal phenotype even after being experimentally associated with a variety of heterotypic inductors. While the observed commitment may indeed reflect actual intrinsic properties of the tissue, "developmental commitment" (determination) may merely be a function of the competence, ingenuity and thoroughness of the investigator or of the power of the technical approaches utilized. For this reason, the progressive loss of developmental plasticity and the expression of determination during progressive developmental stages must be considered with some skepticism and caution. Clearly, with appropriate techniques nuclei of cells derived from highly differentiated tissues of adult frogs are multipotent and able to participate in embryonic development when transplanted into enucleated frog eggs (DiBerardino et al., 1984). Remarkable changes in cellular commitment have also been reported in amphibians during limb (Hay, 1968) and lens (Reyer, 1954) regeneration. In both systems, highly specialized cells revert to an undifferentiated phenotype and then subsequently may redifferentiate into completely different cell types. In Drosophila, culture of imaginal disc cells can lead to transdetermination, which represents a heritable switch from one phenotype, such as antenna, to another, such as wing (Hadorn, 1968). Utilizing somatic cell hybrids composed of mouse muscle and human non-muscle cells, Blau et al. (1986) have elegantly demonstrated the activation of human muscle-specific genes in the heterokaryons. These examples of experimentally altered gene regulation attest to the plasticity of the differentiated state.

Further evidence of the retention of developmental properties in adulthood is the expression of inductive activities by stroma of the adult vagina and mammary gland (Cunha, 1976b; Daniel and DeOme, 1965; Sakakura et al., 1979a, b). Moreover, uterine and vaginal stromal cells, grown as long-term primary or passaged cultures, continue to be able to function as both instructive and permissive inductors and to promote proliferative and differentiative hormonal responses in epithelial cells with which they are experimentally reassociated (Cooke et al., 1986). Finally, adult prostatic epithelial cells exhibit remarkable morphogenetic and growth responses when associated with embryonic urogenital sinus mesenchyme (Norman et al., 1986). These observations on both the developing and mature genital tract emphasize the importance of connective tissue as the primary regulator of epithelial proliferation, morphogenesis, hormonal responsiveness, and cytodifferentiation, and affirm the continued importance of stromal-epithelial interactions in adulthood.

Acknowledgements. The authors would like to acknowledge the assistance of Simona Ikeda for preparation of the figures and Gary Lee Myers for typing the manuscript. This study was supported in part by the following NIH grants: HD 17491, CA 05388, AM 32157, HD 06701, HD 06580, and HD 06520.

References

Bern, H.A., and F.J. Talamantes (1981) Neonatal mouse models and their relation to disease in the human female. In *Developmental Effects of Diethylstilbestrol (DES) in Pregnancy*, A.L. Herbst and H.A. Bern, eds., pp. 129-147. Thieme-Stratton, New York.

Bernfield, M.R., R.H. Cohn, and S.D. Banerjee (1973) Glycosaminoglycans and epithelial organ formation. Am. Zool. **13**: 1067-1083.

Bernfield, M.R., S.D. Banerjee, J.E. Koda, and A.C. Rapraeger (1984) Remodeling of the basement membrane as a mechanism of morphogenetic tissue interactions. In *The Role of Extracellular Matrix in Development*, R.L. Trelstad, ed., pp. 545-596. A.R.

Liss, New York.
Bigsby, R.M., and G.R. Cunha (1986) Estrogen stimulation of DNA synthesis in epithelium lacking estrogen receptors. Endocrinology, in press.
Blau, H.M., G.K. Pavlath, E.C. Hardeman, C.-P. Chin, L. Silberstein, S.G. Webster, S.C. Miller, and C. Webster (1986) Plasticity of the differentiated state. Science **230:** 758-766.
Burns, R.K. (1961) Role of hormones in the differentiation of sex. In *Sex and Internal Secretions*, 3rd Edition, W.C. Young, ed., pp. 76-158. Williams and Wilkins, Baltimore.
Casimiri, V., N.C. Tath, N.C. Parvey, and A. Psychoyos (1980) Effect of sex steroids on rat endometrial epithelium and stroma cultured separately. J. Steroid Biochem. **12:** 293-298.
Clark, J.H., and S.A. McCormack (1977) Clomid or nafuxidine administered to neonatal rats causes reproductive tract abnormalities. Science **197:** 164-165.
Clark, J.H., and S.A. McCormack (1980) The effect of Clomid and other triphenylethylene derivatives during pregnancy and the neonatal period. J. Steroid Biochem. **12:** 47-53.
Cooke, P.S., D.K. Fujii, and G.R. Cunha (1986) Vaginal and uterine stroma maintain their inductive properties following primary culture. In Vitro, submitted.
Cunha, G.R. (1972) Epithelio-mesenchymal interactions in primordial gland structures which become responsive to androgenic stimulation. Anat. Rec. **172:** 179-196.
Cunha, G.R. (1975) Age-dependent loss of sensitivity of female urogenital sinus to androgenic conditions as a function of the epithelial-stromal interaction in mice. Endocrinology **95:** 665-673.
Cunha, G.R. (1976a) Epithelial-stromal interactions in development of the urogenital tract. Int. Rev. Cytol. **47:** 137-194.
Cunha, G.R. (1976b) Stromal induction and specification of morphogenesis and cytodifferentiation of the epithelia of the Müllerian ducts and urogenital sinus during development of the uterus and vagina in mice. J. Exp. Zool. **196:** 361-370.
Cunha, G.R., B. Lung, and K. Kato (1977) Role of the epithelial-stromal interaction during the development and expression of ovary-independent vaginal hyperplasia. Dev. Biol. **56:** 52-67.
Cunha, G.R., and B. Lung (1978) The possible influences of temporal factors in androgenic responsiveness of urogenital tissue recombinants from wild-type and androgen-insensitive (Tfm) mice. J. Exp. Zool. **205:** 343-342.
Cunha, G.R., L.W.K. Chung, J.M. Shannon, and B.A. Reese (1980a) Stromal-epithelial interactions in sex differentiation. Biol. Reprod. **22:** 19-42.
Cunha, G.R., B. Lung, and B. Reese (1980b) Glandular epithelial induction by embryonic mesenchyme in adult bladder epithelium of balb/c mice. Invest. Urol. **17:** 302-304.
Cunha, G.R., B.A. Reese, and M. Sekkingstad (1980c) Induction of nuclear androgen-binding sites in epithelium of the embryonic urinary bladder by mesenchyme of the urogenical sinus of embryonic mice. Endocrinology **107:** 1767-1770.
Cunha, G.R., and L.W.K. Chung (1981) Stromal-epithelial interactions: Induction of prostatic phenotype in urothelium of testicular feminized (Tfm/y) mice. J. Steroid Biochem. **14:** 1317-1321.
Cunha, G.R., J.M. Shannon, K.D. Vanderslice, M. Sekkingstad, and S.J. Robboy (1982) Autoradiographic analysis of nuclear estrogen binding sites during postnatal development of the genital tract of female mice. J. Steroid Biochem. **17:** 281-286.
Cunha, G.R., L.W.K. Chung, J.M. Shannon, O. Taguchi, and H. Fujii (1983a) Hormone-induced morphogenesis and growth: Role of mesenchymal-epithelial interactions. Rec. Prog. Horm. Res. **39:** 559-598.
Cunha, G.R., H. Fujii, B.L. Neubauer, J.M. Shannon, L.W. Sayer, and B.A. Reese (1983b) Epithelial-mesenchymal interactions in prostatic development. I. Morphological observations of prostatic induction by urogenital sinus mesenchyme in epithelium of the adult rodent urinary bladder. J. Cell Biol. **96:** 1662-1670.
Cunha, G.R., J.M. Shannon, O. Taguchi, H. Fujii, and B.A. Meloy (1983c) Epithelial-mesenchymal interactions in hormone-induced development. In *Epithelial-*

Mesenchymal Interactions in Development, R.H. Sawyer and J.F. Fallon, eds., pp. 51-74. Praeger Scientific Press, New York.

Daniel, C.W., and K.B. DeOme (1965) Growth of mouse mammary glands in vivo after monolayer culture. Science **149:** 634-636.

Deuchar, E.M. (1975) In *Cellular Interactions in Animal Development*, pp. 119-147. Wiley and Sons, New York.

DiBerardino, M.A., N.J. Hoffner, and L.D. Etkin (1984) Activation of dormant genes in specialized cells. Science **224:** 946-952.

Drews, U., and U. Drews (1977) Regression of mouse mammary gland anlagen in recombinants of Tfm and wild-type tissues: Testosterone acts via the mesenchyme. Cell **10:** 401-404.

Fallon, J.F., D.A. Rowe, J.M. Frederick, and B.K. Simandl (1983) Studies on epithelial-mesenchymal interactions during limb development. In *Epithelial-Mesenchymal Interactions in Development*, R.H. Sawyer and J.F. Fallon, eds., pp. 3-26. Praeger Press, New York.

Forsberg, J.-G., and J. Kalland (1981) Neonatal estrogen treatment and epithelial abnormalities in the cervicovaginal epithelium of adult mice. Cancer Res. **41:** 721-734.

Hadorn, E. (1968) Transdetermination in cells. Sci. Am. **219:** 110-123.

Hay, E.D. (1968) Dedifferentiation and metaplasia in vertebrate and invertebrate regeneration. In *The Stability of the Differentiated State*, H Ersprung, ed., pp. 85-108. Springer, New York.

Kratochwil, K. (1972) Tissue interaction during embryonic development: General properties. In *Tissue Interactions in Carcinogenesis*, D. Tarin, ed., pp. 1-47. Academic Press, New York.

Kratochwil, K., and P. Schwartz (1976) Tissue interaction in androgen response of the embryonic mammary rudiment of mouse: Identification of target tissue of testosterone. Proc. Natl. Acad. Sci. (USA) **73:** 4041-4044.

Lasnitzki, I., and T. Mizuno (1980) Prostatic induction and interaction of epithelium and mesenchyme from normal wild-type and androgen-insensitive mice with testicular feminization. J. Endocrinol. **85:** 423-428.

Martin, L., C.A. Finn, and G. Trinder (1973) Hypertrophy and hyperplasia in the mouse uterus after estrogen treatment: An autoradiographic study. J. Endocrinol. **56:** 133-144.

McCormack, S.A., and S.R. Glasser (1980) Differential response of individual uterine cell types from immature rats treated with estradiol. Endocrinology **106:** 1034-1049.

McLachlan, J.A., R.R. Newbold, and B.L. Bullock (1980) Prenatal exposure to diethylstilbestrol in mice: Long term effects on the female genital tract. Cancer Res. **40:** 3988-3999.

Neubauer, B.L., L.W.K. Chung, K. McCormick, O. Taguchi, T.C. Thompson, and G.R. Cunha (1983) Epithelial-mesenchymal interactions in prostatic development. II. Biochemical observations of prostatic induction by urogenital sinus mesenchyme in epithelium of the adult rodent urinary bladder. J. Cell Biol. **96:** 1671-1676.

Newbold, R.R., and J.A. McLachlan (1985) Diethylstilbestrol-associated defects in murine genital tract development. In *Estrogens in the Environment* II, J.A. McLachlan, ed., pp. 288-318. Elsevier, New York.

Nieuwkoop, P.O. (1985) Inductive interactions in amphibian development and their general nature. J. Embryol. Exp. Morph. (Supplement) **89:** 333-347.

Norman, J.T., G.R. Cunha, and Y. Sugimura (1986) The induction of new ductal growth in adult prostatic epithelium in response to an embryonic prostatic inductor. Prostate, in press.

Reyer, R.W. (1954) Regeneration of the lens in the amphibian eye. Quart. Rev. Biol. **29:** 1-46.

Sakakura, T., Y. Nishizuka, and C.J. Dawe (1976) Mesenchyme-dependent morphogenesis and epithelium-specific cytodifferentiation in mouse mammary gland. Science **194:** 1439-1441.

Sakakura, T., Y. Nishizuka, and C.J. Dawe (1979a) Capacity of mammary fat pads of adult C3H/HeMs mice to interact morphogenetically with fetal mammary epithelium. J. Natl. Cancer Inst. **63:** 733-736.

Sakakura, T., Y. Nishizuka, and Y. Nishizuka (1979b) Persistence of responsiveness of adult mouse mammary gland to induction by embryonic mesenchyme. Dev. Biol. **72:** 201-210.

Saxen, L., M. Karkinen-Jaaskelainen, E. Lehtonen, S. Nordling, and J. Wartiovaara (1976) Inductive tissue interactions. In *The Cell Surface in Animal Embryogenesis and Development*, G. Poste, and G.L. Nicolson, eds., pp. 331-408. Elsevier/North-Holland, New York.

Sengel, P. (1976) *Morphogenesis of Skin*. Cambridge University Press, New York.

Shannon, J.M., and G.R. Cunha (1983) Autoradiographic localization of androgen binding in the developing mouse prostate. Prostate **4:** 367-373.

Shannon, J.M., and G.R. Cunha (1984) Characterization of androgen binding and deoxyribonucleic acid synthesis in prostate-like structures induced in testicular feminized (Tfm/Y) mice. Biol. Reprod. **31:** 175-183.

Sirbasku, D.A., and R.H. Benson (1979) Estrogen-inducible growth factors that may act as mediators (Estromedins) of estrogen-promoted tumor cell growth. In *Hormones and Cell Culture*, G.H. Sato and R. Ross, eds., pp. 477-498. Cold Spring Harbor Laboratory, Cold Spring Harbor, NY.

Soto, A.M., and C. Sonnenschein (1984) Mechanism of estrogen action on cellular proliferation: Evidence for indirect and negative control of cloned breast tumor cells. Biochem. Biophys. Res. Commun. **122:** 1097-1103.

Stumpf, W.E., and M. Sar (1976) Autoradiographic localization of estrogen, androgen, progestin, and glucocorticosteroid in "target tissues" and "non-target tissues." In *Receptors and Mechanism of Action of Steroid Hormones*, J.R. Pasqualini, ed., pp. 41-84. Marcel Dekker, New York.

Sugimura, Y., and G.R. Cunha (1986) Androgenic induction of deoxyribonucleic acid synthesis in prostate-like glands in urothelium of testicular feminized (Tfm/y) mice. Prostate, submitted.

Taguchi, O., and Y. Nishizuka (1985) Reproductive tract abnormalities in female mice treated neonatally with tamoxifen. Am. J. Obstet. Gynecol. **151:** 675-678.

Takeda, H., T. Miguno, and I. Lasnitzki (1985) Autoradiographic studies of androgen-binding sites in the rat urogenital sinus and postnatal prostate. J. Endocrinol. **104:** 87-92.

Takasugi, N. (1976) Cytological basis for permanent vaginal changes in mice treated neonatally with steroid hormones. Int. J. Cytology **44:** 193-224.

Tomooka, Y., R.P. DiAugustine, and J.A. McLachlan (1986) Proliferation of mouse uterine epithelial cells in vitro. Endocrinology **118:** 1011-1018.

Wasner, G., I. Hennermann, and K. Kratochwil (1983) Ontogeny of mesenchymal androgen receptors in the embryonic mouse mammary gland. Endocrinology **113:** 1771-1780.

Wessells, N.K. (1977) *Tissue Interactions and Development*. 276 p. W.A. Benjamin, Inc., Menlo Park, CA.

Yasugi, S., Matsushita, S., and Mizuno, T. (1985) Gland formation induced in the allantoic and small-intestinal endoderm by the preventricular or mesenchyme is not coupled with pepsinogen expression. Differentiation **30:** 47-52.

INFLUENCE OF MENINGEAL CELLS ON THE DEVELOPMENT OF NEURONAL CELLS IN PRIMARY CULTURE

C. Gensburger, I. Barakat, G. Labourdette and M. Sensenbrenner
Centre de Neurochimie du CNRS and INSERM U.44
5, rue Blaise Pascal
Strasbourg
France

Interactions between neural and mesodermal tissues have been demonstrated to occur at early developmental stages (Landesman, 1967; Toivenen and Saxen, 1968). Recent studies have shown that meninges, which are of mesodermal origin and surround the central nervous system, play an important role during cerebellar development, even postnatally (Allen et al., 1981; Pehlemann et al., 1985).

The development of culture systems of a nearly homogeneous population of precursor cells of neurons has allowed us to test the effects of meningeal cells on neuronal development. We have shown that the addition of meningeal cells, or extracts, enhances the proliferation (Barakat et al., 1981, 1982) and the morphological maturation (Cam et al., 1975) of neuroblasts from chick embryo cerebral hemispheres in primary culture. More recently, we have described stimulating effects of meningeal cells on rat cerebral neuronal cells (Gensburger et al., 1986). Treatment of the cultures with meningeal cell extracts, or conditioned media, has indicated that the observed stimulatory effect may be mediated by diffusible factors.

The work from our laboratory on the influence of meningeal cells on the proliferation and maturation of cerebral neuronal cells in primary culture will be reviewed.

Culture Systems Enriched in Primitive Neuroblasts

To investigate the effect of meningeal cells on the proliferation and maturation of cultured neuronal cells, the use of cultures highly enriched in neuronal precursors (neuroblasts) appeared to be necessary.

Chick neuroblast cultures

Neurons from the cerebral hemispheres of chick embryos can be cultured as a nearly pure population when grown on a polylysine substrate (Sensenbrenner et al., 1978; Pettmann et

NATO ASI Series, Vol. H5
Mesenchymal-Epithelial Interactions in Neural Development
Edited by J. R. Wolff et al.

al., 1979). However, under such conditions the replication rate of the cells is low, probably due to the fact that neuronal differentiation begins precociously (Baraket et al., 1982). In contrast, when neuroblasts are cultured on a collagen substrate instead of polylysine they multiply more actively (Barakat et al., 1982). Indeed, we have shown that cell density and cell-cell interactions are of importance in favoring neuroblast proliferation. Dissociated cells seeded at high cell density on a collagen substrate yield cell aggregates, while on polylysine the cells remain more dispersed. Moreover, we have demonstrated that the proliferation activity of chick neuroblasts is greater in a medium containing 5% fetal calf serum (FCS) compared to a medium with 20% FCS (Barakat et al., 1984). We have shown that FCS contains an inhibitory factor of neuroblast proliferation (Barakat et al., 1983). The medium containing 5% FCS favoured the maintenance of neuronal cells in a neuroblast stage.

In several reports, we have provided evidence that neuroblasts dissociated from cerebral hemispheres of young chick embryos (3- to 7-day-old) proliferate in culture (Sensenbrenner et al., 1980; Barakat and Sensenbrenner, 1981; Barakat et al., 1981). The highest proliferative activity was obtained with brain cells derived from 6-day-old chick embryos.

All these observations indicate that cells from 6-day-old embryos cultured on a collagen substrate in Eagle's minimal essential medium (MEM) supplemented with 5% FCS can be considered as an adequate model system to investigate the proliferation of chick neuroblasts from cerebral hemispheres (Barakat et al., 1984). Until day 5, these cultures are essentially constituted of round or elongated cells, some of them possessing long processes. Most of these cells have been characterized as neuronal cells by histochemical demonstration of the presence of acetyl-cholinesterase (AChE) activity (Fig. 1). In brain cell cultures from chick embryos, all neuronal cells stain for AChE, while glial cells never contain this enzyme (Sensenbrenner et al., 1973).

Rat neuroblast cultures

Rat brain cultures highly enriched in neuronal cells were generally obtained by seeding the cells on a polylysine or polyornithine substrate combined with a treatment with antimitotics (Yavin and Yavin, 1974; Rüegg and Hefti, 1984; Keller et al., 1985). Chemically defined media have also been used to diminish the non-neuronal cell contamination and resulted in virtually pure cultures of neuronal cells (Ahmed et al., 1983; Kumar, 1983; Romijn et al., 1984; Löffler et al., 1986).

In our laboratory, we have established a culture system, highly enriched in primitive neuroblasts, by seeding cells partially cleaned of their meningeal membranes from cerebral hemispheres of 13-day-old rat embryos on a polylysine substrate, and by growing them in a serum-free chemically defined medium (Gensburger et al., 1986). Under these conditions, the culture consists essentially of process-bearing cells between 2 and 8 days. Most of these cells have been identified as neurons by immunoreactivity against neurofilament (NF) proteins (Fig. 2). Many of these cells were capable of multiplying during the first 3 days of culturing, and thus, such an in vitro system, offers advantages for investigating the proliferation and the subsequent maturation of rat brain neuroblasts.

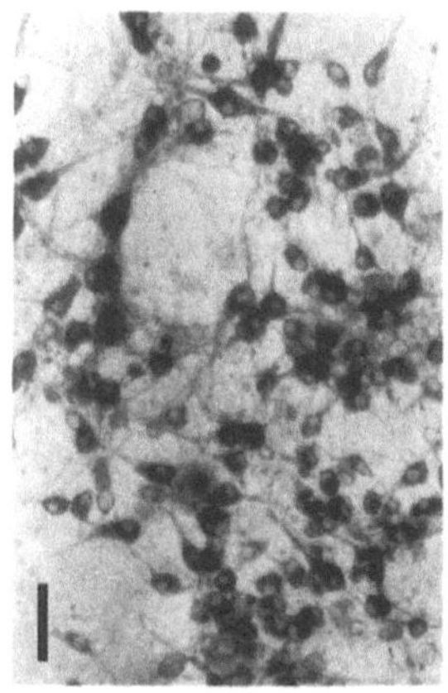

Figure 1: Brain cells from cerebral hemispheres of 6-day-old chick embryos cultured during 5 days. Neurons are visualized by histochemical staining for AChE activity. Bar = 50 μm.

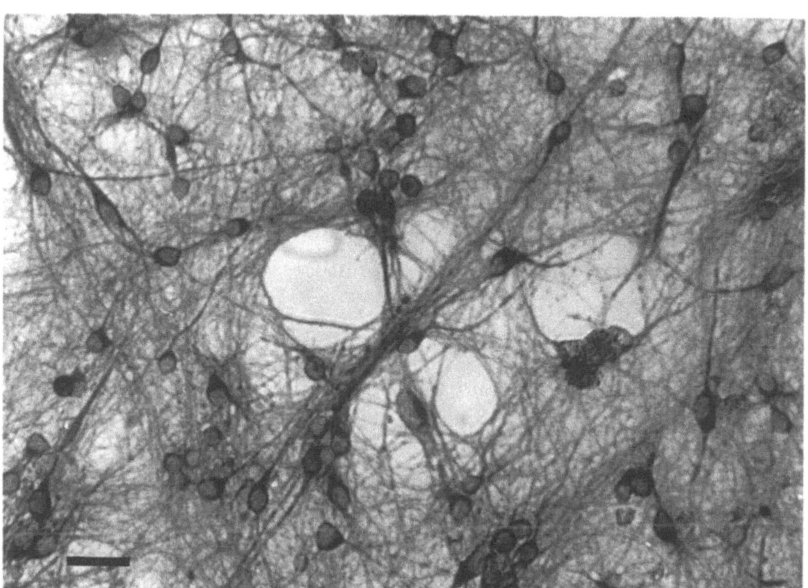

Figure 2: Brain cells from cerebral hemispheres of 13-day-old rat embryos cultured during 7 days. Neurons are visualized by immunostaining for neurofilament proteins. Bar = 50 μm.

Influence of Meningeal Cells on the Proliferation of Chick Cerebral Neuroblasts

To investigate the proliferation of neuroblasts and the influence of meningeal cells, four culture groups have been set up: the first where grain cells from 5-6-day-old chick embryo were cultured alone; the second, where brain cells were cultured in the presence of meningeal extract; the third, where brain cells were mixed with meningeal cells, and the fourth, where brain cells were grown on a preformed layer of meningeal cells (Barakat et al., 1981). Meningeal cells and extracts, as well as meningeal cell layers, were obtained from meninges of 8-day-old chick embryo cerebral hemispheres.

The proliferative activity of the cultured neuronal cells was evaluated by cell counts, determinations of DNA and measurements of tritiated thymidine (^{3}H-TdR) incorporation combined with autoradiography (Barakat et al., 1981, 1982).

In a first study, we have shown that in control cultures, i.e., those without meningeal cells, mainly small and medium aggregates of neuronal cells are observed after 7 days in culture (Fig. 3a). In contrast, in the presence of meningeal cells, very large clusters of neuronal cells predominate (Fig. 3b). The addition of meningeal extract to the culture medium also enhances the formation of large clumps of neuronal cells. This stimulatory effect seems to be limited to meningeal cells, since a monolayer of glial cells or fibroblasts never stimulated the formation of large cell aggregates (Barakat et al., 1981). A quantitative evaluation of neuronal aggregate formations (Fig. 4) and number of AChE stained neuronal cells has confirmed the morphological observations.

Autoradiographic studies using ^{3}H-thymidine (^{3}H-TdR) have shown the presence of labelled neuronal cells in control cultures, indicating that neuroblasts from young chick embryos are still able to proliferate under some culture conditions. In the presence of meningeal cells, many neurons are labelled (Fig. 5a), suggesting that the proliferation of the neuroblasts is stimulated. In subsequent work, where the brain cells are plated at high or low density, we have demonstrated that the addition of meningeal extract stimulates the multiplication of neuroblasts only in cultures where the cells were in closer contact with each other (Barakat et al., 1982).

The results of our studies indicate that cell density and cell-cell interactions are of importance in favoring chick neuroblast proliferation. Furthermore, meningeal cells were shown to influence the mitogenic activity of these neuroblasts.

Influence of Meningeal Cells on the Proliferation of Rat Neuroblasts

To study the effect of meningeal cells on neuroblast proliferation, cerebral hemispheres from 13-day-old rat embryos were cleaned either entirely or only partially of their meningeal membranes, then mechanically dissociated and the cells cultured for 2 hr in a serum-containing medium (Dulbecco's modified Eagle's medium (DMEM) plus 20% FCS), then switched to a serum-free defined medium. The medium consisted of DMEM plus

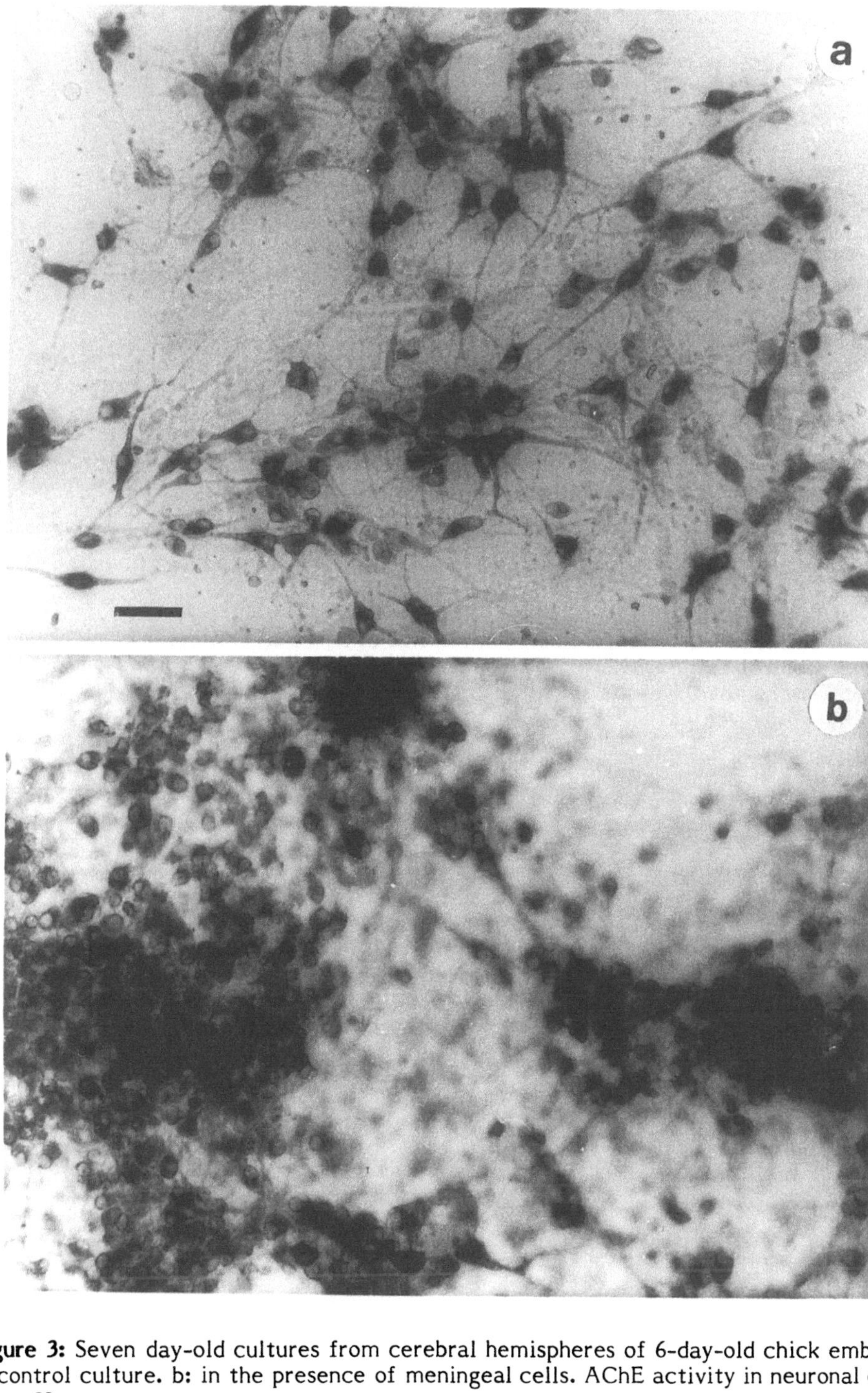

Figure 3: Seven day-old cultures from cerebral hemispheres of 6-day-old chick embryos. a: control culture. b: in the presence of meningeal cells. AChE activity in neuronal cells. Bar = 50 µm.

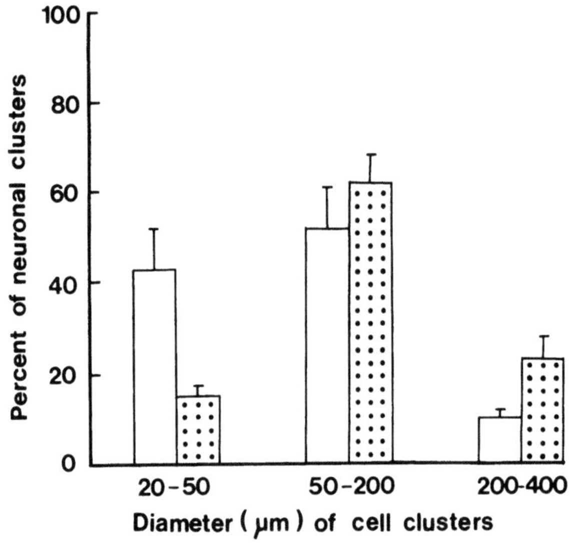

Figure 4: Influence of meningeal cells on the size of neuronal clusters in chick brain cell cultures. Control culture: ....; in the presence of meningeal cells: .... 500 to 1,000 clusters were counted. Results are means .. SD.

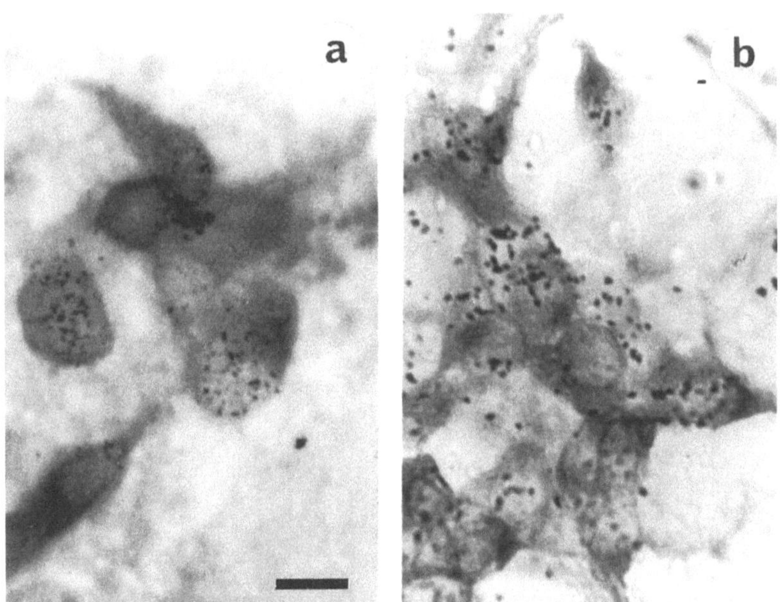

Figure 5: Autoradiographs showing labelled neuronal cells exposed to ^{3}H-TdR from 24 to 48 hr and grown for 8 days in the presence of meningeal cells. a: chick neurons stained with toluidine blue. b: rat neurons immunostained for NF proteins. Bar = 10 µm.

insulin (5 µg/ml), transferrin (50 µg/ml), progesterone (20 mM), putrescine (100 µM) and sodium selenite (30 nM) (Gensburger et al., 1986).

Autoradiography after ^{3}H-TdR incorporation, combined with immunocytochemical staining for NF proteins has permitted us to identify proliferating neuroblasts. When the cells were treated with ^{3}H-TdR during the first 3 days of culture, radioactive neuronal cells (Fig. 5b) were seen in both culture conditions, i.e., with or without meningeal cells. A quantitative evaluation of labelled cells is given in Table 1. In the presence of meningeal cells, the proliferative activity of the neuroblasts is always higher (about 2 times) compared to control cultures.

These observations indicate that neuroblasts from cerebral hemispheres of 13-day-old rat embryos are capable of dividing under our culture conditions. Other authors have reported that neuronal cells from rat brain divide in dispersed cell cultures (Kriegstein and Dichter, 1984; Asou et al., 1985). Our results demonstrate further that cells from meningeal membranes influence the proliferation of rat neuroblasts in culture.

Since the presence of meningeal cells enhances the proliferation of neuroblasts, the question arises whether cellular contacts or trophic substances released from meningeal cells are responsible for the observed stimulatory effect. Therefore, we tested the effects of crude extracts from meningeal membranes or from meningeal cell cultures as well as of menigeal conditioned medium (CM). Meninges from brains of 13-day-old rat embryos and newborn rats have been used. We found that only the CM exerted an influence on rat neuronal cell proliferation. Large cellular aggregates were present in the treated cultures and more labelled neuronal cells were seen. A quantitative analysis of ^{3}H-TdR incorporation indicated a two-fold increase under the influence of meningeal cell CM (Table 1).

The fact that the presence of meningeal cells and CM from these cells give rise to similar effects strongly suggests that the stimulatory effect is mediated by soluble factors.

Table 1: Proliferation of neuroblasts from 13-day-old rat embryos in culture.

Exposure to ^{3}H-TdR (0.1 µCi/Ml	Culture Conditions Without meningeal cells (control)	With meningeal cells	With meningeal cell conditioned medium
2-24 h	12.5	29.0	
24-48 h	24.0	50.0	47.0
48-72 h	14.5	33.5	

Results are expressed as percent of labelled NF^{+} neuronal cells to total NF^{+} cells.

Influence of Meningeal Cells on the Maturation of Neuroblasts

In 1975, we showed that meningeal membrane extract influences the morphological maturation of young chick neuroblasts (Cam et al., 1975). Under the effect of this ex-extract about 2 to 3 times more large multipolar neurons and thick fiber bundles develop. More recently, we have investigated the influence of meningeal cells on the maturation of young rat neuroblasts. In cultures derived from cerebral hemispheres of 13-day-old rat embryos most cells develop long processes which form a dense fiber network. These fibers remain relatively thin while, in the presence of meningeal cells, both fine and thick processes are found. The cell somata, as well as the processes, are stained for NF proteins. In the presence of meningeal cells, or of meningeal CM, the immunostaining is stronger in most neurons.

The ultrastructure of the neurons examined after 2 weeks in vitro revealed an immature appearance in control cells. The processes contain many microtubules, but few neurofilaments. Synaptic contacts are seldom and contain very few small clear vesicles. In contrast, in the presence of meningeal cells or of CM the synapses are more abundant, with characteristic membrane thickening and clusters of clear vesicles. However, most cells show the features characteristic of young neurons.

After 3 weeks, most neurons have degenerated in the absence of meningeal cells, while in their presence the neuronal cells survive for longer periods (3 to 4 weeks).

Conclusion

Neuroblasts, derived from either cerebral hemispheres of 5- to 6-day-old chick embryos or 13-day-old rat embryos, under certain culture conditions, are capable of proliferation during the 3 first days of culturing. In the presence of meningeal cells, or conditioned medium, the proliferative activity of chick and rat neuroblasts is enhanced, while meningeal cells extract has a stimulatory effect on chick neuroblasts.

Neuronal-meningeal cell interactions are not necessary to obtain a stimulatory effect. However, the contact with meningeal cells produces a higher stimulatory effect on chick neuronal cells. This stimulatory effect on rat neuroblasts as well as on chick neuroblasts seems to be mediated by soluble substances released from meningeal cells.

With time in culture, most of the neuronal cells reach a limited stage of maturation. The meningeal cells stimulate the morphological maturation and especially the development of synaptic formations.

Acknowledgements: The authors wish to thank Mrs. M. Miehe for helpful assistance in the electron microscopical work, Dr. J.F. Leterrier for the generous gift of NF antiserum and Mrs. S. Ott for typing the manuscript. This work was supported by CNRS and INSERM, and by a grant from the French "Foundation pour la Recherche Médicale."

References

Ahmed, Z., P.S. Walker, and R.E. Fellows (1983) Properties of neurons from dissociated fetal rat brain in serum-free culture. J. Neurosci. **3**: 2448-2462.

Allen, C., J. Sievers, M. Berry, and S. Jenner (1981) Experimental studies on cerebellar foliation. II. A morphometric analysis of cerebellar fissuration defects and growth retardation after neonatal treatment with 6-OHDA in the rat. J. Comp. Neurol. **203**: 771-783.

Asou, H., N. Iwasaki, S. Hirano, and D. Dahl (1985) Mitotic neuroblasts in dissociated cell cultures from embryonic rat cerebral hemispheres express neurofilament protein. Brain Res. **332**: 355-357.

Barakat, I., J. Courageot, G. Devilliers, and M. Sensenbrenner (1984) Effects of chick brain extract on the proliferation of chick neuroblasts cultured in media supplemented with low and high serum concentrations. Neurochem. Res. **9**: 263-272.

Barakat, I., G. Labourdette, and M. Sensenbrenner (1983) Inhibitory effects of fetal calf serum on proliferation of chick neuroblasts in culture. Dev. Neurosci. **6**: 169-183.

Barakat, I., and M. Sensenbrenner (1981) Brain extracts that promote the proliferation of neuroblasts from chick embryo in culture. Dev. Brain Res. **1**: 355-368.

Barakat, I., M. Sensenbrenner, and G. Vincendon (1982) The importance of cell contact for the proliferation of neuroblasts in culture and its stimulation by meningeal extract. Neurochem. Res. **7**: 287-300.

Barakat, I., E. Wittendorp-Rechenmann, R.V. Rechenmann, and M. Sensenbrenner (1981) Influence of meningeal cells on the proliferation of neuroblasts in culture. Dev. Neurosci. **4**: 363-372.

Cam, Y., M. Sensenbrenner, and P. Mandel (1975) A comparative study of the effects of brain extracts and mesodermal membrane extracts on nerve cell differentiation. Experientia **31**: 1430-1431.

Gensburger, C., G. Labourdette, and M. Sensenbrenner (1986) Influence of meningeal cells on the proliferation and maturation of rat neuroblasts in culture. Exptl. Brain Res., in press.

Keller, M., R. Jackisch, A. Seregi, and G. Herting (1985) Comparison of prostanoid forming capacity of neuronal and astroglial cells in primary cultures. Neurochem. Int. **7**: 655-665.

Kriegstein, A., and M.A. Dichter (1984) Neuron generation in dissociated cell cultures from fetal rat cerebral cortex. Brain Res. **295**: 184-189.

Kumar, S. (1983) Growth of neural cell cultures in a chemically defined, serum-free culture medium. Neurochem. Res. **8**: 847-852.

Landesman, R. (1967) Neural-mesodermal interactions subsequent to neural induction in Ambystoma. Devl. Biol. **16**: 341-367.

Löffler, F., S.M. Lohmann, B. Walckhoff, U. Walter, and B. Hamprecht (1986) Immunocytochemical characterization of neuron-rich primary cultures of embryonic rat brain cells by established neuronal and glial markers and by monospecific antisera against cyclic nucleotide-dependent protein kinases and the synaptic vesicle protein synapsis I. Brain Res. **363**: 205-221.

Pehlemann, F.W., J. Sievers, and M. Berry (1985) Meningeal cells are involved in foliation, lamination, and neurogenesis of the cerebellum: evidence from 6-hydroxydopamine-induced destruction of meningeal cells. Devl. Biol. **110**: 136-146.

Pettmann, B., J.C. Louis, and M. Sensenbrenner (1979) Morphological and biochemical maturation of neurones cultured in the absence of glial cells. Nature **281**: 378-380.

Romijn, H.J., F. van Huizen, and P.S. Wolters (1984) Towards an improved serum-free, chemically defined medium for long-term culturing of cerebral cortex tissue. Neurosci. & Biobehavioral Rev. **8**: 301-334.

Rüegg, U.T., and F. Hefti (1984) Growth of dissociated neurons in culture dishes coated with synthetic polymeric amines. Neurosci. Letters **49**: 319-324.

Sensenbrenner, M., J. Booher, and P. Mandel (1973) Histochemical study of dissociated nerve cells from embryonic chick cerebral hemispheres in flask cultures. Experientia **29**: 699-701.

Sensenbrenner, M., K. Maderspach, L. Latzkovits, and G.G. Jaros (1978) Neuronal cells from chick embryo cerebral hemispheres cultivated on polylysine-coated surfaces. Dev. Neurosci. **1**: 90-101.

Sensenbrenner, M., E. Wittendorp, I. Barakat, and R.V. Rechenmann (1980) Autoradiographic study of proliferating brain cells in culture. Devl. Biol. **75**: 268-277.

Toivenen, S., and L. Saxen (1968) Morphogenetic interaction of presumptive neural and mesodermal cells mixed in different ratio. Science **159**: 539-540.

Yavin, E., and Z. Yavin (1974) Attachment and culture of dissociated cells from rat embryo cerebral hemispheres on polylysine-coated surface. J. Cell. Biol. **62**: 540-546.

INFLUENCE OF MENINGEAL CELLS ON CELL PROLIFERATION IN THE CEREBELLUM

F.W. Pehlemann[1], S. Mohr[1], H. Korr[2], J. Sievers[1], and M. Berry[3]

[1] **Department of Anatomy, University of Kiel**
Kiel
Federal Republic of Germany

[2] **Department of Anatomy**
RWTH Aachen
Federal Republic of Germany

[3] **Department of Anatomy**
Guy's Hospital Medical School
London
U.K.

Epithelial morphogenesis is dependent on the surrounding mesenchyme which, in addition to other influences, stimulates epithelial cell proliferation (review in Cunha et al., 1983, and this volume). We have previously deduced a similar role for the meninges, the mesenchyme surrounding the brain, of the cerebellum, because after destruction of meningeal cells in newborn rats and hamsters, the number of cerebellar granule cells becomes reduced by almost 20% in adult animals (Sievers et al., 1980, 1981a, b; Allen et al., 1981; Pehlemann et al., 1985a, b; v. Knebel Doeberitz et al., 1986). This proposal is supported by findings of Sensenbrenner and colleague (Barakat et al., 1981, 1982; Gensburger et al., this volume) who showed that, in vitro, addition of either meningeal cells or meningeal extracts increased and prolonged the proliferation of neuronal stem cells. In order to specify which brain cells are influenced by meningeal cells, we performed a cell-kinetic study on the proliferation and migration of cells in the cerebellar cortex with tritiated thymidine after destroying the meningeal cells in newborn rats.

Material and Methods

In order to selectively destroy meningeal cells (Sievers et al., 1985), newborn rats were given a single intracisternal injection of 50 µg 6-hydroxydopamine (6-OHDA, free base) within 12 hr after birth. Controls received only vehicle injections (0.9% NaCl plus 1 mg/ml ascorbate). At 3, 5 or 10 days post injection (p.i.) control and treated animals (groups of 10) received a single subcutaneous injection of 5 µCi ^{3}H-thymidine/g body weight. 1 or 24 hr p.i. Five animals from each group were killed and their cerebella pre-

NATO ASI Series, Vol. H5
Mesenchymal-Epithelial Interactions in Neural Development
Edited by J. R. Wolff et al.

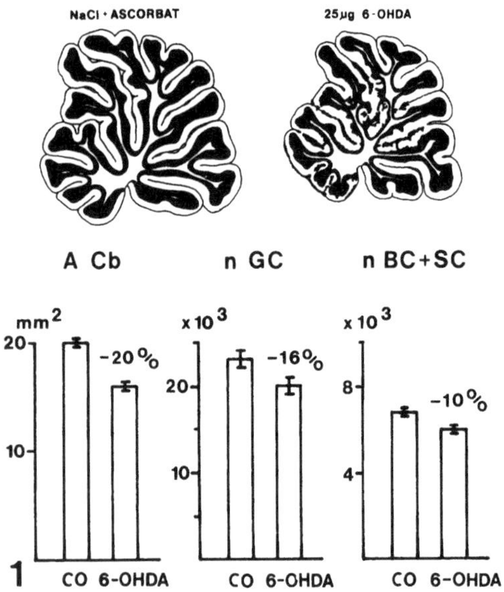

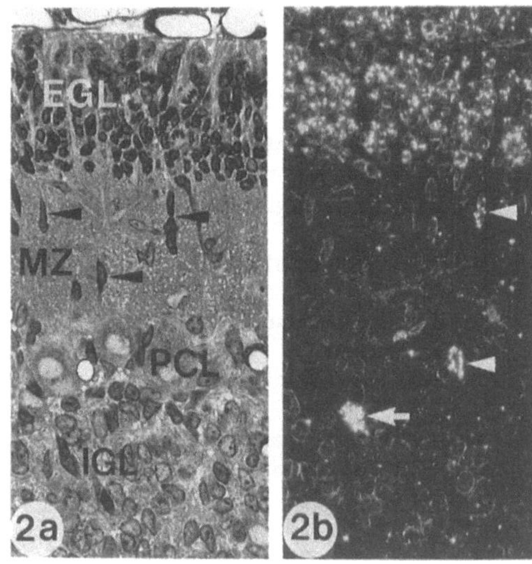

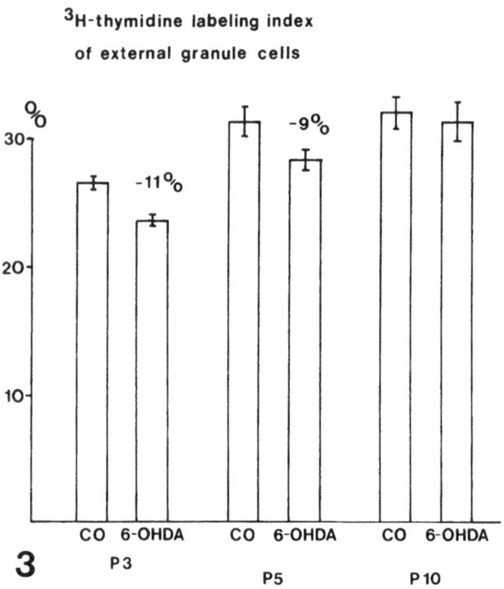

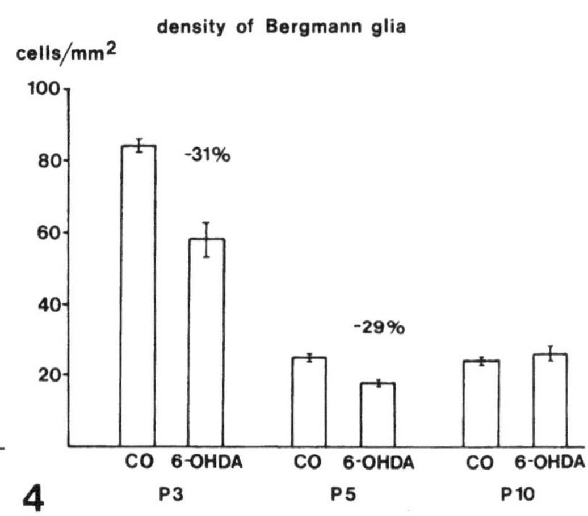

Figure 1: Above: Camera lucida drawings of paramedian sagittal sections of the cerebellar vermis of rats 30 days after injection of either ascorbate (left) or 6-OHDA (right) on the day of birth. In the controls, folia I to X (Larsell, 1952) are separated by fissures, and lamination is regular (the solid black internal granular layer separates the superficial molecular layer from the deep medullary tree. The Purkinje cell layer is at the interface between molecular layer and internal granular layer). The most conspicuous alterations in treated animals (right) are a decrease of the area of the vermis, a reduction of fissuration, and a disruption of lamination by extensive granule cell ectopia (indicated by the black areas separated from the internal granular layer).

Below: Quantitative data documenting the qualitative changes in cerebellar development shown above. Left: total area of paramedian sagittal vermal sections; middle:

total number of granule cells (GC) on a standardized vermal section of 1 µm thickness; right: total number of basket (BC) and stellate (SC) cells, on a standardized vermal section. CO = control injected animals, 6-OHDA = animals treated with 25 µg 6-OHDA on the day of birth. The percent reduction of treated groups in comparison to the control groups is given only when statistically significant.

Figure 2: Cerebellar cortex of 11 days old rats in normal light microscopy (a), and with dark field optics (b), 24 hours after injection of tritiated thymidine. The silver grains are seen as bright spots. EGL = external granular layer; PCL = Purkinje cell layer; IGL = internal granular layer; arrow = labelled Bergmann glial cell in the Purkinje cell layer; arrowheads = migrating granule cells. x 560

Figure 3: ^{3}H-thymidine labeling index of external granule cells; P = postnatal day; CO = control injected animals; 6-OHDA = animals which received 25 µg 6-OHDA on PO; vertical bar = standard error of the mean. Percent reduction of treated groups in comparison to control groups is given only when statistically significant.

Figure 4: Density (number of cells/mm^2 of paramedian sagittal vermal sections of 1 µm thickness) of Bermann glia cells labeled with tritiated thymidine; for further explanation see Fig. 3.

pared for autoradiography. Additionally, results are reported from experiments which were designed to delineate a critical period of meningeal cell influence on cerebellar development. In these, hamsters of different postnatal ontogenetic stages (P0 to P10) were treated with 6-OHDA to destroy their meningeal cells, and the effects of this treatment on cell proliferation were evaluated by measuring the number of granule cells at the age of 30 days (see Sievers et al., 1986).

Results and Discussion

Fig. 1 summarizes the findings from previous experiments (Pehlemann et al., 1985 a, b; Ruhnke et al., in preparation) which show that after treatment of newborn rats with 25 µg 6-OHDA, the total area of the cerebellum is reduced by 20%, the number of granule cells by 16% and the number of basket and stellate cells by 10%. Moreover, many of the granule cells lie ectopically either in obliterated fissures between adjacent folia, or on the folial surface. In the hamster, the loss of granule cells after destruction of meningeal cells is 20% (v. Knebel Doeberitz et al., 1986).

This reduction in the number of granule cells, the majority of which arise from the external granular layer during the postnatal ontogenetic period (review in Altman, 1982), is preceded by a small and short lived reduction of the ^{3}H-thymidine labeling index of external granule cells (Figs. 2 and 3). At 3 days after destruction of meningeal cells by 6-OHDA the labeling index of external granule cells is decreased by 11%, at 5 days p.i. by 9% and at 10 days p.i. the labeling index does not differ from controls.

The discrepancy between the small reduction in proliferation of external granule cells and the large reduction in granule cell number at the end of development is most

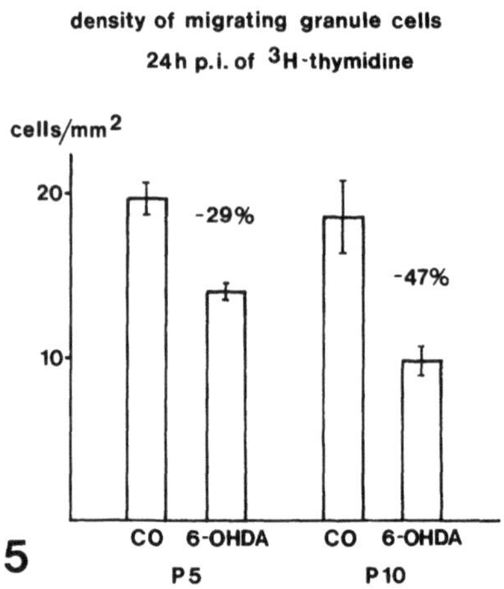

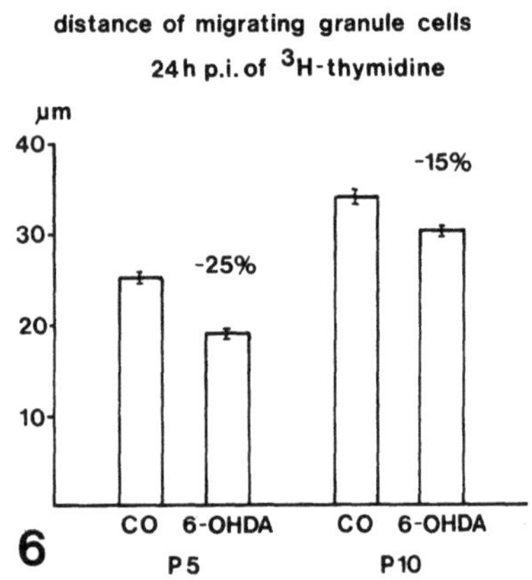

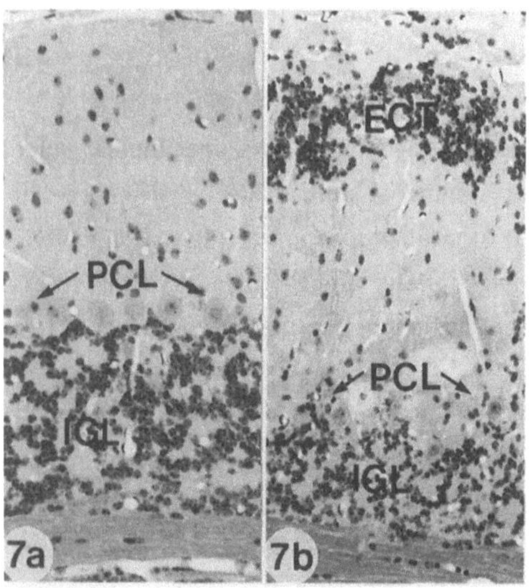

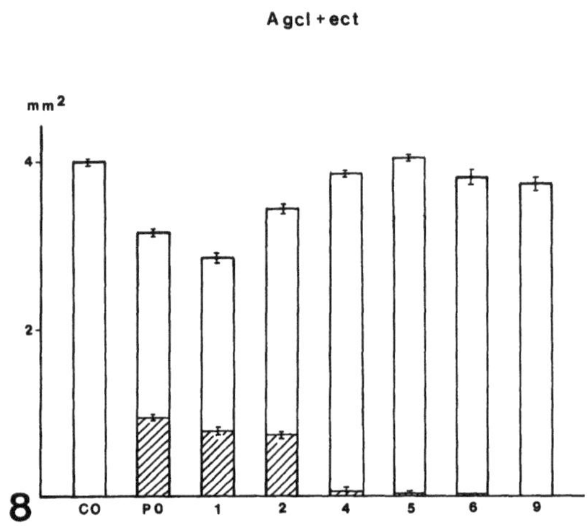

Figure 5: Density (number of cells/mm^2 of paramedian sagittal vermal sections of 1 µm thickness) of labeled migrating granule cells 24 h p.i. of ^{3}H-thymidine; for further explanation see Fig. 3.

Figure 6: Distance covered by migrating labeled granule cells (µm) from the internal border of the external granular layer 24 h p.i. of ^{3}H-thymidine; for further explanation see Fig. 3.

Figure 7: Cerebellar cortex of 30 days old hamsters. a) Control injected animal, b) animal which received a single injection of 60 µg 6-OHDA on postnatal day 4. In the treated animal the inner granular layer (IGL) is reduced in width. A large number of granule cells lie ectopically (ECT) arrested on their way from the cerebellar surface to the inner granular layer. (PCL = Purkinje cell layer). a and b x 350.

Figure 8: Total area (A) of granule cells (gcl) in paramedian sagittal vermal sections of

30 days old hamsters after injection of 6-OHDA at different ages, beginning on the day of birth (PO) up to the 9th postnatal day (P9). The hatched parts of the columns represent the amount of granule cells situation in ectopias (ect). Values of the animals of the P0, P1 and P2 groups differ significantly from the controls, in later stages (P4 and beyond) there is no difference. Note that the granule cell ectopias disappear at the same stage (P4 and later). From Sievers et al., 1986.

likely explained by the fact that, during the early postnatal period, proliferation of the external granular layer produces mainly external granule cells and few postmitotic granule cells (Altman, 1969, 1982). The ensuing reduction in the stem cell population inevitably gives rise to a complementary deficit in the number of postmitotic granule cells in the later stages of cerebellar development. By contrast, the proliferation of Bergmann glia is more profoundly reduced, by 31% at 3 days p.i. and by 29% at 5 days p.i. (Fig. 4). At 10 days p.i. Bergmann glia proliferation is slightly elevated above control values, but this increase is not statistically significant.

The large reduction in the rate of proliferation of Bergmann glia relative to external granule cells has repercussions on cerebellar cortical growth. According to presently accepted concepts of cerebellar development, granule cell migration from the external into the internal granular layer is guided and controlled by the Bergmann glial fibers (review in Rakic, 1982; but see Sotelo and Rio, 1980; Altman, 1975) and, as a prerequisite, the granule cell and Bergmann glia populations must develop together. Any interference with the harmonious development of the two cell populations is likely to perturb granule cell migration, especially if congestion develops in the migratory path (Bergmann glial fibers) because of an excess of granule cells relative to Bergmann glia.

This prediction has been borne out by the results of the present study. (i) The density of labeled granule cells migrating from the external into the internal granular layer 24 hr p.i. of ^{3}H-thymidine is reduced by 29% at 5 days p.i. of 6-OHDA and by 47% at 10 days p.i. (Fig. 5). (ii) The average distance covered by migrating labeled granule cells 24 hr p.i. of ^{3}H-thymidine is reduced by 25% at 5 days p.i. of 6-OHDA and by 15% at 10 days p.i. (Fig. 6). (iii) Ectopia develops along the granule cell migratory path within the molecular zone (Fig. 7), testifying to a disturbed relationship between granule cells and Bergmann glia. Thus, taken together, this indirect experimental evidence supports the concept of Rakic (1971, 1982) that granule cell guidance is provided by Bergmann glial fibers.

Like the influence on the stability of the cerbellar surface and glial scaffold (see Sievers et al., this volume), the influence of meningeal cells on cell proliferation in the cerebellum is confined to a critical period. Thus, after destruction of meningeal cells over the hamster cerebellum at different ontogenetic stages, the total number of granule cells in adult animals (as parameter for cell production by the external granular layer) is significantly reduced only in the groups treated on postnatal day 0, 1 and 2 (Fig. 8, Sievers et al., 1986). Destruction of meningeal cells on postnatal day 4 and later does not significantly reduce granule cell number (Fig. 8), although a small amount of intracortical granule cell ectopia still signals a slight disturbance of granule cell migration after treatment on postnatal days 4, 5 and 6 which might be due to interference with Bergmann glia proliferation as outlined above.

Conclusion

Destruction of meningeal cells over the cerebellum of neonatal rats and hamsters reduces the proliferation of Bergmann glia cells and external granule cells. The glial cell population is much more severely affected than the neuronal stem cells and this numerical mismatch results in defects in intracortical granule cell migration. At present it is unclear whether the effects of meningeal cells on cell proliferation are mediated by the extracellular matrix at the cerbellar surface to which they contribute (see Sievers et al., this volume), since substrate attachment is one prerequisite for epithelial cell proliferation (Weiss et al., 1975). On the other hand, meningeal cells could secrete mitogens for the neuroepithelium (as suggested for other mesenchymal cells stimulating epithelial cell proliferation, see Cunha et al., 1983) which diffuse across the basement membrane and stimulate cell proliferation in the cerebellum. Such diffusible molecules would explain the stimulatory effects of meningeal extracts and conditioned media on neuronal cell proliferation in rat and chicken embryos in vitro (Barakat et al., 1981, 1982; Gensburger et al., this volume). Such mitogens, or course, could also be incorporated into the basal lamina (Greenburg and Gospodarowicz, 1982), and in this case, loss of the basal lamina, as seen after destruction of meningeal cells, would deprive the neuroepithelial cells of both a substrate for attachment and of mitogenic molecules.

References

Allen, C., J. Sievers, M. Berry, and S. Jenner (1981) Experimental studies on cerebellar foliation. II. A morphometric analysis of cerebellar fissuration defects and growth retardation after neonatal treatment with 6-OHDA in the rat. J. Comp. Neurol **202**: 771-784.

Altman, J. (1969) Autoradigraphic and histological studies of postnatal neurogenesis. III. Dating the time of production and onset of differentiation of cerebellar microneurons in rats. J. Comp. Neurol. **136**: 269-294.

Altman, J. (1975) Postnatal development of the the cerebellar cortex in the rats. IV. Spatial organization of bipolar cells, parallel fibers and glial palisades. J. Comp. Neurol. **163**: 427-448.

Altman, J. (1982) Morphological development of the rat cerebellum and some of its mechanisms. Exp. Brain Res. (Suppl.) **6**: 8-49.

Barakat, I., E. Wittendorp-Rechenmann, R.V. Rechenmann, and M. Sensenbrenner (1981) Influence of meningeal cells on the proliferation of neuroblasts in culture. Dev. Neurosci. **4**: 363-372.

Barakat, I., M. Sensenbrenner, and G. Vincendon (1982) The importance of cell contact for the proliferation of neuroblasts in culture and its stimulation by meningeal extract. Neurochem. Res. **7**: 287-300.

Cunha, G.R., L.W.K. Chung, J.M. Shannon, O. Taguchi, and H. Fujii (1983) Hormone-induced morphogenesis and growth: Role of mesenchymal-epithelial interactions. Rec. Prog. Hormone Res. **39**: 559-598.

Greenburg, G., and D. Gospodarowicz (1982) Inactivation of a basement membrane component responsible for cell proliferation but not for cell attachment. Exp. Cell Res. **140**: 1-14.

von Knebel Doeberitz, Ch., J. Sievers, M. Sadler, F.-W. Pehlemann, M. Berry, and P. Halliwell (1986) Destruction of meningeal cells over the newborn hamster cerebellum with 6-hydroxydopamine prevents foliation and lamination in the rostral cerebellum. Neuroscience **17**: 409-426.

Larsell, D. (1952) The morphogenesis and adult pattern of the lobules and fissures of the cerbellum of the white rat. J. Comp. Neurol. **79**: 231-356.

Pehlemann, F.-W. and J. Sievers (1985a) Meningealzellen sind beteiligt an Formbildung, Lamination und Neurogenese des Rattenkleinhirns. Verh. Anat. Ges. **79**: 451-452.

Pehlemann, F.-W., J. Sievers, and M. Berry (1985b) Meningeal cells are involved in foliation, lamination and neurogenesis of the cerbellum: Evidence from 6-hydroxydopamine-induced destruction of meningeal cells. Dev. Biol. **110**: 136-146.

Rakic, P. (1971) Neuron-glia relationship during granule cell migration in developing cerebellar cortex. A Golgi and electron microscopic study in Macacus rhesus. J. Comp. Neurol. **141**: 283-312.

Rakic, P. (1982) The role of neuronal-glial cell interaction during brain development. In *Neuronal-glial Cell Interrelationships*, T.A. Sears, ed., pp. 25-38. Springer, Heidelberg and New York.

Sievers, J., H.P. Klemm, S. Jenner, H.G. Baumgarten, and M. Berry (1980) Neuronal and extraneuronal effects of intracisternally administered 6-hydroxydopamine on the developing rat brain. J. Neurochem. **34**: 765-771.

Sievers, J., M. Berry, and H.G. Baumgarten (1981a) The role of noradrenergic fibers in the control of postnatal cerebellar development. Brain Res. **207**: 200-208.

Sievers, J., U. Mangold, M. Berry, C. Allen, and H.G. Schlossberger (1981b) Experimental studies on cerebellar foliation. I. A qualitative morphological analysis of cerebellar fissuration defects after neonatal treatment with 6-OHDA in the rat. J. Comp. Neurol. **203**: 751-769.

Sievers, J., F.-W. Pehlemann, H.G. Baumgarten, and M. Berry (1985) Selective destruction of meningeal cells by 6-hydroxydopamine: A tool to study meningeal-neuroepithelial interaction in brain development. Dev. Biol. **110**: 127-135.

Sievers, J., Ch. v. Knebel Doeberitz, F.-W. Pehlemann, and M. Berry (1986) Meningeal cells influence cerebellar development over a critical period. Anat. Embryol., in press.

Sotelo, C., and J.P. Rio (1980) Cerebellar malformations obtained in rats by early postnatal treatment with 6-aminonicotinamide. Role of neuron-glia interactions in cerebellar development. Neuroscience **5**: 1737-1759.

Weiss, L., G. Poste, A. MacKearnin, and K. Willett (1975) Growth of mammalian cells on substrates coated with cellular microexudates. J. Cell Biol. **64**: 135-145.

ROLE OF MEDIUM AND SUBSTRATUM ON PROLIFERATION OF ASTROGLIA IN VITRO

A. Michler-Stuke
Zentrum Anatomie
Universität Göttingen
Göttingen
Federal Republic of Germany

Introduction

In the adult brain astrocytes represent a multifunctional cell population. Uptake and/or release of putative neurotransmitters like GABA, glutamate and taurine (Schousboe et al., 1976; Sellström and Hamberger, 1977; Balcar et al., 1977) are among the various functions in which they are involved. By taking up substances like potassium (Somjen, 1975) they may also play a role in the regulation of the extracellular ionic environment in the nervous system.

To study some of their functions in more detail, primary cultures of astrocytes derived from neonatal rat or mice brain have proven to be useful. Grown in a medium devoid of serum those cultures represent a fairly well defined system where trophic requirements for growth and differentiation of astrocytes can be studied.

Several serum-free media for the growth of primary astroglial cell cultures have been developed (Morrisson and DeVellis, 1981; Fischer et al., 1982; Weibel et al., 1984). However, they either require a preselection procedure for certain cell types or the intermediate presence of serum.

Despite the important influence of soluble factors for growth and differentiation the question has to be addressed of what the influence might be of the cellular substratum on either proliferation or differentiation of astrocytes. In vivo, cells interact with an extracellular matrix (ECM) on which they grow and differentiate (Grobstein, 1975). Using cell lines as a model, it has been shown that proliferation of human neuroblastoma cells can be enhanced by homotypic ECM (Michler-Stuke and Bottenstein, 1981). Further investigations have shown that some constituents of the ECM change in amount and distribution during brain development (Toole, 1976; Aquino et al., 1984; Delpech and Delpech, 1984), suggesting an important influence of these molecules on maturation of the brain.

NATO ASI Series, Vol. H5
Mesenchymal-Epithelial Interactions in Neural Development
Edited by J. R. Wolff et al.

The aim of the present study was to devise a serum-free, defined medium for astroglial cells which is able to support cell growth and differentiation without preceding serum incubation. This medium is based on a previously described one, developed to support proliferation of rat and human glioblastoma cell lines (Michler-Stuke and Bottenstein, 1982). One further question was whether constituents of the ECM like hyaluronic acid, chondroitin sulfate A and C and heparin can influence proliferation of astroglial cells.

Material and Methods

Primary cultures of astrocytes were derived from the cerebral hemispheres of neonatal rats. The tissue was minced into small pieces in nutrient medium consisting of Dulbecco's modified Eagle's medium (DME) supplemented with 10% fetal calf serum (FCS) and containing 1.2g/l sodium bicarbonate and 15 mM N-2-hydroxylethylpiperazine-N-2-ethanesulfonic acid (HEPES). Tissue fragments were incubated with 0.25% trypsin in phosphate buffered saline (PBS) for 10 min at 37°C. To inactivate the trypsin, an appropriate amount of nutrient medium was added followed by centrifugation of the tissue for 2 min at 150 g. After replacing the supernatant with serum-free DME, tissue fragments were triturated 10-15 times with a 10 ml pipette and subsequently filtered through a 200 µm metal mesh. Cells were plated at a final density of 10^6 cells/35 mm plastic petri dish.

The culture dishes were either coated with poly-D-lysine (PLYS) at a concentration of 0.05 mg/ml, poly-L-ornithine (PORN; 0.5 mg/ml) or collagen type III (COL; 1 mg/ml). Culture media were either DME + 10% FCS, G2 medium or variations thereof called G4 and G5 medium. The composition of the media is listed in Table 1. To serum-free media, fibronectin (FN) was added separately into each dish at a concentration of 1.5-2.0 µg/ml. Modifications of the culture substratum were achieved by adding hyaluronic acid (HA), chondroitin sulfate A or C (ChSA, ChSC) or heparin (Hep) in various concentrations one day after plating the cells.

Proliferation rates were determined by either counting the number of cells in a defined visual field using an inverted Zeiss microscope with phase contrast optics or by determining the number of cells employing a Coulter counter. Cell counts were done either 5 or 7 days after plating.

The expression of glial fibrillary acidic protein (GFAP) was demonstrated by labeling prefixed cells with 1) anti-GFAP-antiserum and 2) rhodamine conjugated goat-antirabbit-antibody. For these experiments, astrocytes were plated on PLYS-coated coverslips and, after fixation and staining, mounted on slides in a glycerol/PBS mixture of 4:1. Cultures were viewed with a Zeiss flourescence microscope with the appropriate filters.

Table 1: Composition of G2, G4, and G5 medium.

Supplements	G2	G4	G5
Sodium selenite (nM)	30	30	30
Hydrocortisone (nM)	10	10	10
Insulin (ug/ml)	-	5	5
Biotin (ng/ml)	10	10	10
Transferrin (ug/ml)	50	50	50
FGF (ng/ml)	5	5	5
EGF (ng/ml)	-	-	10
Fibronectin (ug/ml)	1.5-2.0	1.5-2.0	1.5-2.0

Supplements were added to serum-free DME

Results

Growth of astroglial cells in defined medium.

Astrocytes which are grown of PLYS-coated dishes in G2 medium proliferate slowly during the first 5 days and decline in cell number thereafter (Fig. 1). When insulin is added to the medium (G4 medium) cell growth is comparable to control cultures grown in serum supplemented medium. The final cell density reached after 11 days in vitro is about 77% of that reached in serum grown cultures. The addition of epidermal growth factor (EGF) significantly enhances cell proliferation above values obtained in serum supplemented cultures. Growth reponse to insulin and EGF is concentration dependent, with the optimal concentration for insulin being 5 µg/ml, and for EGF 10 µg/ml (see also Table 1). Higher concentrations of EGF resulted in a decline of cell number.

Morphology of astroglial cells in different media

Depending on the type of medium used, cell morphology varied significantly after 7 days in cultures as was demonstrated by immunocytochemical staining with anti-GFAP-antibody (Fig. 2). Cells which were grown in serum supplemented medium had a flat epithelial- like cell shape with only a few cells bearing short processes (Fig. 2a). In G4 and G5 medium, long branched processes could be observed and perikarya were small and slightly spherical (Fig. 2c, d). Flat epithelial-like cells were rarely seen in these media. The overall appearance was that of a more advanced morphological differentiation of astrocytes which could be due to the presence of insulin and EGF, since no such significant changes in morphology were observed in G2 medium (Fig. 2b). In all media tested, GFAP was expressed after 5 days in vitro and was still present after more than 2 weeks.

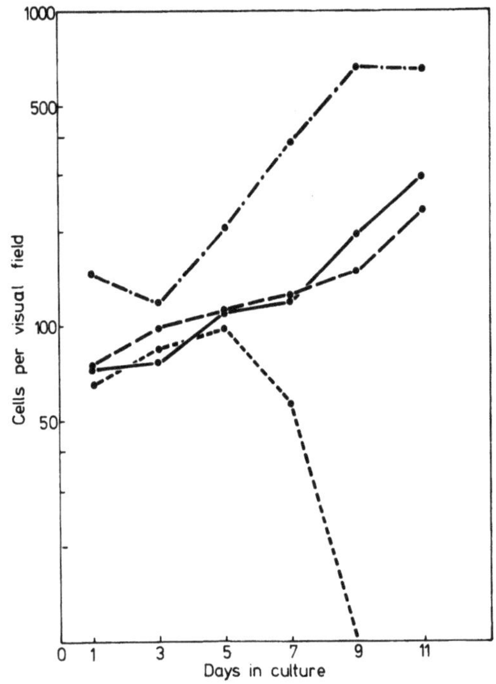

Figure 1: Proliferation of astroglial cells from primary cultures of rat forebrain in different media. Cell inoculum was 10^6 cells/35 mm petri dish. S.E.M. of triplicate cultures was less than 10% of the mean value. (——) DME + 10% FCS, (-----) G2 medium, (– – –) G4 medium, (·–··–·) G5 medium (from Michler-Stuke et al., 1984)

Table 2: Growth of astroglial cells on different surface coatings of petri dishes

	G4 medium	G5 medium
PLYS	100 ± 5	100 ± 3
PORN	96 ± 6	104 ± 6
COL	74 ± 5	85 ± 7

Cell numbers were counted 5 days after plating using an inverted microscope and are expressed as relative cell number (PLYS = 100%) over the mean ± S.E.M. of triplicate cultures. COL = collagen; PLYS = polylysine; PORN = polyornithine.

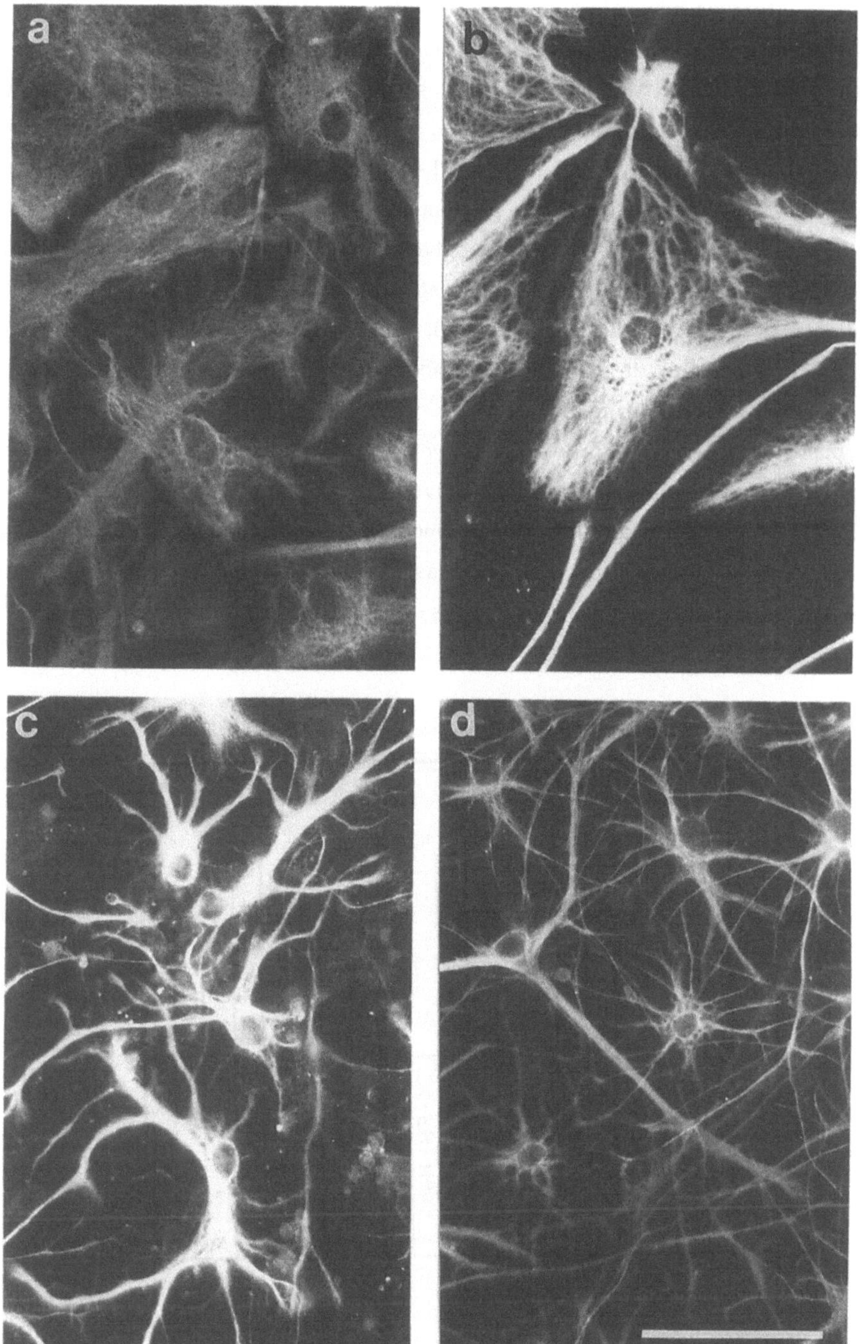

Figure 2: Astroglial cells from primary cultures of rat forebrain in different media as revealed by immunocytochemical staining of GFAP. Cultures were fixed after 7 days in vitro. a) DME + 10% FCS, b) G2 medium, c) G4 medium, d) G5 medium. Bar represents µm.

Influence of different substrata on cell proliferation

Different surface coatings of petri dishes as well as various constituents of the ECM were tested for their ability to influence astroglial proliferation. In G4 medium, no significant difference in cell growth was obtained when cells were plated either in PLYS- or PORN-coated petri dishes, whereas in COL-coated dishes proliferation was reduced (Table 2). These differences were less pronounced when G5 medium - containing EGF - was employed. When different glycosaminoglycans (GAGs) were added to G4 medium in the presence of FN, a marked increase in cell proliferation was observed with most of the supplements tested. Hyaluronic acid (HA), chondroitin sulfate A (ChSA), and heparin (Hep) stimulated cell growth by a factor of about 1.5, chondroitin sulfate (ChSC) turned out to be slightly inhibitory (Fig. 3a).

A positive effect of GAGs was seen even when FN was omitted from G4 medium. Without FN, astrocytes were not able to survive in this medium, while addition of HA and ChSA resulted in proliferation rates being comparable to or even higher than those measured in G4 medium with FN added. In contrast, heparin seemed more dependent on the presence of FN, since its stimulatory effect is not as pronounced in the absence of FN.

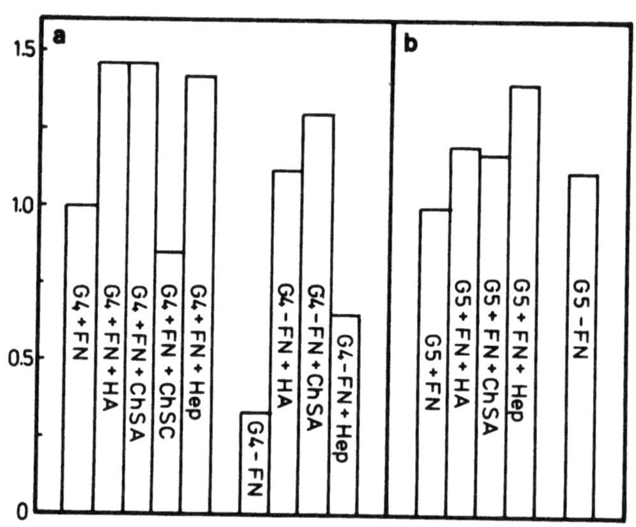

Figure 3: Growth of astrocytes in defined medium after addition of hyaluronic acid (HA, 10 µg/ml), chondroitin sulfate A and C (ChSA, ChSC, 100 µg/ml), and heparin (Hep, 1 µg/ml). The effects of these glycosaminoglycans are compared between asrocytes grown (a) in G4 medium or (b) in G5 medium and in the presence (+FN) or absence (-FN) of fibronectin. Cell numbers were counted after 7 days in vitro and expressed as relative numbers of triplicate cultures. S.E.M. was less than 10%.

Cells grown in G5 medium did not respond by a significant increase in cell proliferation upon addition of GAGs. However in G5 medium, in contrast to the G4 medium, GN, otherwise a stringent plating requirement, was no longer needed in the presence of EGF (Fig. 3b).

Discussion

The defined media described in the present paper were developed on the basis of a previously described one (G2 medium) for the growth of rat and human glioma cell lines (Michler-Stuke and Bottenstein, 1982). Significant improvement in survival and proliferation of astrocytes derived from primary cultures was obtained by addition of insulin and EGF to G2 medium. These two components - and insulin in particular - seemed to be vital substances for these cells although they were not stringently required by glioma cells (Michler-Stuke and Bottenstein, 1982).

The action of insulin on brain cells still remains unclear. There is no known response to insulin in brain tissue although insulin receptors are widely distributed in the CNS as demonstrated by radio ligand binding (Havrankova et al., 1978). Recently, specific insulin binding sites were found on neurons of primary cultures of rat brain (Raizada et al., 1982) and astroglial cell-enriched fractions showed a very high density of these receptors (Albrecht et al., 1982).

The influence of EGF on proliferation has been shown for a variety of cell types (Carpenter and Cohen, 1979; Gospodarowicz et al., 1978). In a serum-free medium, it can stimulate cell growth of normal human glial cells (Heldin et al., 1980) as well as proliferation of cerebellar astrocytes from primary cultures grown either in serum supplemented medium (Leutz and Shachner, 1981) or in serum-free defined medium (Fischer et al., 1982). This is in correspondence to recent findings, where binding of EGF to purified populations of astrocytes was demonstrated (Simpson et al., 1982). Since it apparently is not able to pass the blood-brain barrier (Nave et al., 1985), it still remains to be investigated whether or not EGF exerts comparable effects in vivo as obtained in cell culture. However, astrocytes are influenced by insulin and EGF in a way that may be described as a change of cell morphology towards a higher degree of maturation.

Several studies have dealt with the important influence of the ECM on a variety of cell functions (for review see Hynes, 1981; Gospodarowicz et al., 1980; Hay, 1980). Glycosaminoglycans are ubiquitous and characteristic components of extracellular matrices (Toole, 1976) and are found in greater amounts in the developing than in the adult brain (Margolis et al., 1975; Singh et al., 1969).

Detailed studies on the distribution and localization of hyaluronic acid and chondroitin sulfate proteoglycans in immature postnatal rat cerebellum revealed that both components are primarily localized extracellularly (Ripellino et al., 1985; Aquino et al.,

1984). At day 7 postnatally, only faint intracellular staining of astrocytes was present (Aquino et al., 1984). Assusming a comparable situation in rat forebrain, it may suggest an important role of hyaluronic acid and chondroitin sulfate for growth and/or maturation of astrocytes. This may explain the stimulatory effect of the two substances when added to G4 medium. In G5 medium, e.g., in the presence of EGF, the situation is somewhat different: in this medium no significant increase of cell proliferation is observed after addition of glycosaminoglycans. This effect might be explained, if EGF changes membrane properties by influencing the incorporation of glycosamines into glycosaminoglycans as was shown by Lembach (1976) for cultured human fibroblasts. This would also be in line with the observation that in the presence of EGF astrocytes do not required FN as a plating substratum. Under these conditions they may be capable to express FN, as was recently shown by Liesi et al. (1986) for early primary cultures from newborn rat brain.

Summary

A serum-free, defined medium was developed to support survival and growth of astroglial cells from primary cultures of newborn rat forebrain. Substantial increase in proliferation was obtained by adding insulin and EGF to the medium. These two components also influenced the morphology of astrocytes towards a higher degree of maturation. In defined medium astrocytes express glial fibrillary acidic protein which could be demonstrated 5 days after plating. Modifications of the culture dish surface revealed that a poly-D-lysine coating was most suitable for cell growth in contrast to collagen. Upon addition of glycosaminoglycans like hyaluronic acid, chondroitin sulfate A and C and heparin an increase in proliferation by a factor of about 1:5 was obtained in defined medium devoid of EGF. In the presence of EGF no significant stimulation of growth by glycosaminoglycans was observed. Interestingly, in EGF containing medium astrocytes were no longer dependent on fibronectin, otherwise a stringent plating requirement.

References

Albrecht, J., B. Wroblewska, and M.J. Mossakowski (1982) The binding of insulin to cerebral capillaries and astrocytes of the rat. Neurochem. Res. **7:** 489-494.

Aquino, D.A., R.U. Margolis, and R.K. Margolis (1984) Immunocytochemical localization of a chondroitin sulfate proteoglycan in nervous tissue. II. Studies in developing brain. J. Cell. Biol. **99:** 1130-1139.

Balcar, V.J., J. Borg, and P. Mandel (1977) High affinity uptake of L-glutamate and L-aspartate by glial cells. J. Neurochem. **28:** 87-93.

Carpenter, G., and S. Cohen (1979) Epidermal growth factor. Ann. Rev. Biochem. **48:** 193-216.

Delpech, A., and B. Delpech (1984) Expression of hyaluronic acid-ginding glycoprotein, hyaluronectin, in the developing rat embryo. Dev. Biol. **101:** 391-400.

Fischer, G., A. Leutz, and M. Schachner (1982) Cultivation of immature astrocytes of mouse cerebellum in a serum-free, hormonally defined medium. Appearance of the

mature astrocytic phenotype after addition of serum. Neurosci. Lett. **29:** 279-302.
Gospodarowicz, D., D. Delgado, and I. Vlodavsky (1980) Permissive effect of the extracellular matrix on cell proliferation. Proc. Natl. Acad. Sci. (USA) **77:** 4094-4098.
Gospodarowicz, D., G. Greenburg, H. Bialecki, and B.R. Zetter (1978) Factors involved in the modulation of cell proliferation in vivo and in vitro: The role of fibroblast and epidermal growth factors in the proliferative response of mammalian cells. In Vitro **14:** 85-118.
Grobstein, C. (1975) Developmental role of intracellular matrix: Retrospective and prospective. In *Extracellular Matrix Inifluences on Gene Expression*, H. Slavkin and R. Greulich, eds., pp. 6-16. Academic Press, New York.
Havrankova, J., J. Roth, and M. Brownstein (1978) Insulin receptors are widely distributed in the central nervous system of the rat. Nature, Lond. **272:** 827-829.
Hay, E.D. (1980) Development of the vertebrate cornea. Int. Rev. Cytol. **633:** 263-322.
Heldin, C., A. Wasteson, and B. Westermark (1980) Growth of normal human glial cells in a defined medium containing platelet-derived growth factor. Proc. Natl. Acad. Sci. (USA) **77:** 6611-6615.
Hynes, R.O. (1981) Relationships between fibronectin and the cytoskeleton. Cell Surface Rev. **7:** 97-136.
Lembach, K.J. (1976) Enhanced synthesis and extracellular accumulation of hyaluronic acid during stimulation of quiescent human fibroblasts by mouse epidermal growth factor. J. Cell Physiol. **89:** 277-288.
Leutz, A., and M. Schachner (1981) Epidermal growth factor stimulates DNA-synthesis of astrocytes in primary cerebellar cultures. Cell Tissue Res. **220:** 393-404.
Liesi, P., T. Kirkwood, and A. Vaheri (1986) Fibronectin is expressed by astrocytes cultured from embryonic and early postnatal rat brain. Exp. Cell. Res. **163:** 175-185.
Margolis, R.U., R.K. Maragolis, L.B. Chang, and C. Pret (1975) Glycosaminoglycans of brain during development. Biochemistry **14:** 85-88.
Michler-Stuke, A., J.R. Wolff, and J. Bottenstein (1984) Factors influencing astrocyte growth and development in defined media. Int. J. Devl. Neuroscience **2:** 575-584.
Michler-Stuke, A., and J. Bottenstein (1982) Proliferation of glial-derived cells in defined media. J. Neurosci. Res. **7:** 215-228.
Michler-Stuke, A., and J. Bottenstein (1981) Homotypic extracellular matrices enhance the proliferation of human neuroblastoma cells. Soc. Neurosci. Abstr. **1:** 149.
Morrisson, R., and J. De Vellis (1981) Growth of purified astrocytes in a chemically defined medium. Proc. Natl. Acad. Sci. (USA) **78:** 7205-7209.
Nave, K.A., R. Probstmeier, and M. Schachner (1985) Epidermal growth factor does not cross the blood-brain barrier. Cell Tissue Res. **241:** 453-457.
Raizada, M.K., J.F. Stamler, J. Quinlan, S. Landas, and M.I. Phillips (1982) Identification of insulin receptor containing cells in primary cultures of rat brain. Cell. Mol. Neurobiol. **2:** 47-52.
Ripellino, J.A., M.M. Klinger, R.U. Margolis, and R.K. Margolis (1985) The hyaluronic acid binding region as a specific probe for the localization of hyaluronic acid in tissue sections. J. Histochem. Cytochem. **33:** 1060-1066.
Schousboe, A., H. Fosmark, and G. Svenneby (1976) Taurine uptake in astrocytes cultured from dissociated mouse brain hemispheres. Brain Res. **116:** 158-164.
Sellström, A., and A. Hamberger (1977) Potassium-stimulated-aminobutyric acid release from neurons and glia. Brain Res. **119:** 189-198.
Simpson, D.L., R. Morrisson, J. De Vellis, and H.R. Herschman (1982) Epidermal growth factor binding and mitogenic activity on purified populations of cells from the central nervous system. J. Neurosci. Res. **8:** 453-462.
Singh, M., E.V. Chandrasekaran, E. Cherian, and B.K. Bacchawat (1969) Isolation and characterization of glycosaminoglycans in brain and different species. J. Neurochem. **16:** 1157-1162.
Somjen, G.G. (1975) Electrophysiology of neuroglia. Ann. Rev. Physiol. **37:** 163-190.
Toole, B.P. (1976) Morphogenetic role of glycosaminoglycans (acid mucopolysaccharides) in brain and other tissue. In *Neuronal Recognition*, S.H. Barondes, ed., pp. 275-329.

Chapman and Hall, London.
Weibel, M., B. Pettmann, G. Daune, G. Labourdette, and M. Sensenbrenner (1984) Chemically defined medium for rat astroglial cells in primary culture. Int. J. Devl. Neuroscience **2**: 355-366.

Section 4

CELL MOVEMENT

HYALURONATE-CELL INTERACTIONS IN MORPHOGENESIS AND TUMORIGENESIS

B.P. Toole[1], C.B. Knudson[2], W. Knudson[2], R.L. Goldberg[3], G. Chi-Rosso[1], and C. Biswas[1]

[1] Department of Anatomy and Cellular Biology
Tufts University Schools of Medicine
Dental Medicine and Veterinary Medicine
Boston, MA

[2] Department of Biochemistry
Chicago, IL

[3] Ciba-Geigy
Ardsley, NY

U.S.A.

Introduction

Hyaluronate is a ubiquitous component of the extracellular matrices in which embryonic mesenchymal cells migrate and proliferate. It is also present in tissues at early stages of healing, remodelling and regeneration (for reviews, see Toole, 1976, 1981; Toole et al., 1984). Hyaluronate is a high molecular weight, negatively charged glycosaminoglycan consisting of repeating disaccharide units of N-acetylglucosamine and glucuronate. Its physical and chemical properties contribute to an extracellular milieu which is important both to the structural integrity of the above tissues and to the morphogenetic processes that take place within them. The contribution of hyaluronate is due to at least two factors: i) its remarkable hydrodynamic properties; ii) its modes of interaction with the cell surface. It is the latter upon which this brief review will mainly focus.

Hyaluronate in Embryonic Development, Regeneration and Tumorigenesis

Many studies have been published indicating that hyaluronate synthesis accompanies cell movement in the embryo. For example, hyaluronate accumulates in the pathways of migration of corneal mesenchyme (Toole and Trelstad, 1971), neural crest cells (Pratt et al., 1975; Derby, 1978; Pintar, 1978), endocardial cushion cells (Markwald et al., 1978) and sclerotomal mesenchyme (Kvist and Finnegan, 1970; Toole, 1973; Solursh et al., 1979). Typically, increased hyaluronidase activity and decreased hyaluronate synthesis occur as the cells cease migration and commence to differentiate in these systems. Simi-

NATO ASI Series, Vol. H5
Mesenchymal-Epithelial Interactions in Neural Development
Edited by J. R. Wolff et al.

lar but much less precise correlations have been made in relation to development of the brain (Polansky et al., 1974; Margolis et al., 1975; Werz et al., 1985, Normand et al., 1985). Recent morphological studies suggest, however, that hyaluronate is localized in specific regions undergoing morphogenesis in the developing cerebral cortex and cerebellum (Delpech and Delpech, this volume; Margolis, this volume).

Fibroblast invasion is a characterization of the early stages of tissue repair and remodelling in the adult and also takes place within extracellular matrices enriched in hyaluronate, e.g., in tendon repair (Reid and Flint, 1974), bone fracture callus formation (Maurer and Hudack, 1952), and skin wound healing (Bertolami and Donoff, 1978). Particularly striking is the relationship of hyaluronate synthesis to blastema formation in the regenerating newt limb and its removal during subsequent differentiation (Toole and Gross, 1971; Smith et al., 1975; Mescher and Munaim, 1986).

Hyaluronate is also a major component of the extracellular matrix of many types of tumors, e.g., mammary (Takeuchi et al., 1976), hepatic (Kojima et al., 1975), kidney (Hopwood and Dorfman, 1978) and lung (Horai et al., 1981; Kawai et al., 1985) in the human. We have compared the level of hyaluronate in the rabbit V2 carcinoma under conditions where the tumor exhibits invasive and non-invasive phenotpye (Toole et al., 1979). High concentrations were found to be present in the invasive tumors especially in the peripheral zones of invasion. However, the benign form of the tumor contained similar or lesser amounts of hyaluronate than that of the surrounding normal tissues. Since these results suggested the possibility that hyaluronate may facilitate tumor cell invasion, we have begun to investigate the role of tumor cell-fibroblast interactions in the regulation of hyaluronate synthesis in a similar manner to that carried out prevoiusly (Biswas, 1982, 1985a) in studies of tumor collagenase (against type I collagen).

Some types of tumor and transformed cells actively synthesize hyaluronate in cell culture (Hopwood and Dorfman, 1977; Mikuni-Takagaki and Toole, 1979; Angello et al., 1982; Ullrich and Hawkes, 1983; Kimata et al., 1983) but others do not, even when the tumors from which they were derived contain high concentrations of hyaluronate (Bhavanandan, 1981; Underhill and Toole, 1982; W. Knudson et al., 1984a, b). In recent studies, however, we have shown that cocultures of several tpyes of human tumor cells and normal fibroblasts synthesize much higher amounts of hyaluronate than either cell type alone and that cell contact is necessary for the effect (W. Knudson et al., 1984b). We have also shown that tumor cell membranes and detergent extracts thereof mimic the stimulatory effect of the intact tumor cells (W. Knudson and Toole, 1985). These studies suggest strongly that hyaluronate stimulatory factor resides in the plasma membrane of the tumor cells. In other studies, marked stimulation of fibroblast collagenase (Biswas, 1982, 1984), hyaluronate (Merrilees and Finlay, 1985) and proteoglycan (Iozzo, 1985) synthesis has been obtained using conditioned medium from tumor cell cultures. In the case of collagenase, the stimulatory factor has also been shown to be present in the tumor cell plasma membrane (Biswas, 1985b; Biswas and Nugent, 1986).

The results of the above studies indicate that interactions between normal fibroblasts and invading tumor cells may be important in the regulation of synthesis of both hyaluronate and collagenase in the invasion zone of tumors. It is obvious that tumor cell invasion is the result of numerous facts, e.g., changes in cytoskeletal and cell surface properties, interactions with cells of the immune and vascular systems, and penetration of extracellular matrices. With respect to the last of these parameters, two important requirements would be: i) destruction of pre-existing matrix barriers; and ii) reconstruction of a suitable extracellular milieu for cell movement and proliferation. Presumably, production of collagenase would be important in the former, and production of hyaluronate in the latter.

It is of interest to note that collagen removal and hyaluronate production may go hand in hand in many processes involving tissue remodelling. For example, interactions of several types of epithelia with cells of mesenchymal origin lead to increases in production of hyaluronate (Merrilees and Scott, 1980, 1981; Merrilees et al., 1983) and collagenase (Johnson-Muller and Gross, 1978; Johnson-Wint and Bauer, 1985). In embryonic development of the kidney, the first event observed to occur after interaction of metanephric mesenchyme with the ingrowing ureteric bud epithelium is removal of collagen from the mesenchyme (Ekblom, 1984, and this volume). High levels of hyaluronate are also associated with these early stages of development of the kidney (Belsky and Toole, 1983). Early events in newt limb regeneration, occurring at the crucial stage of blastema formation, are increases in collagenase activity (Grillo et al., 1968; Mailman and Dresden, 1979) and hyaluronate production (Toole and Gross, 1971; Smith et al., 1975; Mescher and Munaim, 1986). As in tumorigenesis, regeneration appears to require removal of pre-existing structural barriers which may interfere with cell proliferation and migration as well as reconstruction of an embryonic-like extracellular matrix prior to redifferentiation.

Hyaluronate and Cell Behavior

Hyaluronate has been shown to influence several types of cell behavior, but its effect depends greatly on at least three parameters. These are the size and concentration of the hyaluronate and the type of cell in question. For example, hyaluronate bonded to the cell culture substratum inhibits differentiation of myoblasts (Kujawa et al., 1986a) but promotes the differentiation of chondroblasts (Kujawa and Caplan, 1986). The latter phenomenon, however, is caused efficiently only by hyaluronate of molecular weight 2-4 x 10^5 and is not at all influenced by hyaluronate of greater than 10^6 (Kujawa et al., 1986b). Also, in other chondrogenic systems, various sizes of hyaluronate in solution are inhibitory (Toole, 1973; Wiebkin and Muir, 1973; Solursh et al., 1980). Hyaluronate has been shown to inhibit migration of endothelial cells (Feinberg and Beebe, 1983) but to promote the movement of several other cell types (Turley and Roth, 1979; Bernanke and Markwald, 1979; Abatangelo et al., 1983; Turley et al., 1985; Doillon and Silver, 1986). In the case of leukocytes, hyaluronate inhibits movement, adherence and phagocytosis at high concentrations (Brandt, 1970, 1974; Forrester and Balazs, 1980; Forrester and Wilkinson, 1981; Forrester and Lackie, 1981) but at low concentrations it has a marked stimulatory effect both in vitro and in vivo (Hakansson et al., 1980; Hakansson and Venge, 1985). Hyaluronate also exhibits a concentration dependent effect on cell aggregation. At high concentrations it blocks aggregation of several cell types, whereas at low concentrations it can mediate aggregation (Fraser and Clarris, 1970; Pessac and Defendi, 1972; Wasteson et al., 1973; Underhill and Dorfman, 1978; Love et al., 1979; Underhill and Toole, 1981; Forrester and Lackie, 1981; Wright et al., 1981). In studies of certain virally transformed cells it has been shown that aggregation is due to cross bridging by hyaluronate of receptors present in the plasma membrane of the cells. However, at high concentrations of hyaluronate these receptors become saturated, thus inhibiting cross bridging (Underhill and Toole, 1981; Underhill, 1982) (also see below).

It is not all clear which of the hyaluronate-mediated phenomena described above and elsewhere are relevant to cell behavior in vivo. However, there is a significant body of convincing evidence relating to mesenchymal cell movement. It is thought that hyaluronate promotes mesenchymal cell migration in three ways. First, high concentrations of hyaluronate form meshworks which exert considerable osmotic pressures (Meyer, 1983; Meyer et al., 1983). These swelling pressures may lead to the formation of hydrated pathways separating barriers to cell movment such as collagen fibers and cell layers. Examples of this are seen in the accumulation of hyaluronate and the concomitant tissue hydration and separation of tissue structures at the time of migration of corneal mesenchyme (Toole and Trelstad, 1971) and neural crest cells (Pratt et al., 1975). Second, hyaluronate is essential to the structure of hydrated pericellular coats which in several cases have been shown to inhibit close cell interactions (Fraser and Clarris, 1970; McBride and Bard, 1979; Underhill and Toole, 1982). Prevention of tight intercellular adhesion is necessary for continued cell movement or proliferation. For example, embryonic myoblasts exhibit a prominent hyaluronate-dependent coat but this coat is lost on fusion of the cells to form differentiated myotubes (Orkin et al., 1985). Exposure of myoblasts to a hyaluronate-conjugated substratum prevents their fusion and permits them to continue to proliferate (Kujawa and Tepperman, 1983; Kujawa et al., 1986a). Third, hyaluronate is associated with reduced adhesion to substrata (Culp et al., 1979; Barnhart et al., 1979; Fisher and Solursh, 1979; Mikuni-Takagaki and Toole, 1980; Abatangelo et al., 1982; Turley et al., 1985). Of particular interest are the observations of Turley and associates

(1985) showing that addition of a hyaluronate-binding protein and hyaluronate together to fibroblasts causes an increase in their motility, reduces cell spreading and increases cell underlapping. The hyaluronate-binding protein was localized to the leading lamellae and retraction processes of motile cells leading to the conclusion that these phenomena are due to reduced adhesion resulting from hyaluronate-substratum interactions (Turley and Torrance, 1984; Turley et al., 1985).

Hyaluronate Receptors and Hyaluronate-Cell Surface Interactions in Morphogenesis and Differentiation

Methods for the analysis of hyaluronate-cell surface interactions were first developed in this laboratory using 3T3 and simian virus-transformed 3T3 (SV-3T3) cells. Underhill and Dorfman (1978) had shown that divalent cation independent aggregation of SV 3T3 and other virally transformed cells was mediated by cell surface hyaluronate, since it was inhibited by Streptomyces hyaluronidase which is specific for hyaluronate. They also showed that aggregation was inhibited by high concentrations of exogenous hyaluronate, suggesting the possibility that the mechanism of aggregation was via hyaluronate cross-bridging of hyaluronate-binding sites on the surface of the cells. Subsequent studies led to the detailed characterization of a specific, high affinity (Kd of approximately 1 nM), hyaluronate receptor present in the plasma membrane of the SV-3T3 cells (Underhill and Toole, 1979, 1980; Underhill et al., 1983, 1985; Goldberg et al., 1984; Chi-Rosso and Toole, 1986; reviewed in Toole et al., 1984).

SV-3T3 and 3T3 cells differ radically in their mode of interaction of hyaluronate with the cell surface. First, 3T3 cells possess hyaluronate dependent pericellular coats whereas SV-3T3 cells do not (Underhill and Toole, 1982). Second, 3T3 cells do not exhibit divalent cation independent aggregation (Underhill and Dorfman, 1978; Underhill, 1982) and bind hyaluronate to their surface with considerably less avidity than SV-3T3 cells (Underhill and Toole, 1981; Goldberg et al., 1984). Detailed analysis of the mechanism of retention of endogenous cell surface-bound hyaluronate at the cell surface (Goldberg et al., 1984) has led to the following conclusions: i) that SV-3T3 cells possess both occupied and unoccupied hyaluronate receptors of the same type; ii) that all of the hyaluronate retained at the surface of SV-3T3 cells is bound to this class of receptors; iii) that the number of these receptors on 3T3 cells is much lower than on SV-3T3 cells; iv) that the hyaluronate retained at the surface of 3T3 cells, which is much greater in amount than for SV-3T3 cells (Underhill and Toole, 1982), is retained by a different mechanism than that for SV-3T3 cells, one which does not involve hyaluronate receptors.

The methods used to characterize the relationship of hyaluronate to the surface of 3T3 and SV-3T3 cells have recently been applied to embryonic cells, particularly those of the chick embryo limb bud (C. Knudson and Toole, 1984, 1985a, b). Mesodermal cells in the early limb bud (3 to 3.5 days of embryonic development; Hamburger and Hamilton stages 18 to 21) appear morphologically homogeneous and are separated by extensive hayaluronate-filled spaces (Toole, 1973; Singley and Solursh, 1981). The prechondrogenic and premyogenic areas of the limb mesoderm then begin to condense (3.5 to 4 days; stages 22 to 24) and subsequently differentiate. Using cells isolated from each of these stages of limb bud development we have examined: i) the expression of hyaluronate-dependent pericellular coats; ii) the synthesis of glycosaminoglycans; iii) the presence of membrane-bound hyaluronate receptors; and iv) the mode of retention of endogenous hyaluronate at the cell surface.

We have observed a close correlation between the ability of cells from limb buds of the various developmental stages to produce hyaluronate-dependent coats in culture and the volume of intercellular space separating them in vivo (C. Knudson and Toole, 1985a). Cells of the early mesoderm, which are separated by hyaluronate-rich spaces in vivo (Singley and Solursh, 1981), exhibit pronounced coats in vitro. At the stages when, in vivo, the prechondrogenic and premyogenic areas of the limb bud are becoming condensed, i.e., separated by less matrix (Thorogood and Hinchliffe, 1975; Singley and Solursh, 1981), the isolated cells do not exhibit visible coats. During differentiation of chondrocytes, which are separated by large spaces in vivo, large pericellular coats are re-expressed (Goldberg and Toole, 1984; C. Knudson and Toole, 1985a). These coats are still dependent on hyaluronate even though their major components are now proteoglycan and collagen (Goldberg and Toole, 1984). Fusion of myoblasts to form differentiated myotubes, however, is accompanied by loss of pericellular coats (Orkin et al., 1985).

A crucial stage in the differentiation of limb mesoderm is the condensation of cells which precedes actual differentation. At this stage, at least four major events take place. These are: i) the failure to express hyaluronate-dependent coats in cell culture, as discussed above; ii) a dramatic decrease in the ratio of hyaluronate to chondroitin sulfate-proteoglycan produced by the mesodermal cells (C. Knudson and Toole, 1985a); iii) expression of membrane-bound receptors for hyaluronate (C. Knudson and Toole, 1985b); and, iv) a radical change in the mode of retention of hyaluronate at the cell surface from a non-receptor mediated to a receptor mediated mechanism (C. Knudson and Toole, 1984). The significance of expression of the hyaluronate receptor at this stage is not yet known but may relate to receptor-mediated endocytosis and degradation of hyaluronate from the matrix (Orkin et al., 1982) or to cell aggregation during condensation (Underhill and Dorfman, 1978; Underhill and Toole, 1981; Underhill, 1982). The hyaluronate receptors persist on the surface of differentiated chondrocytes and participate in the assembly of their pericellular matrix (C. Knudson and Toole, 1985b; McCarthy and Toole, 1985) but are lost from the surface of differentiated myotubes (C. Knudson and Toole, unpublished).

Interaction of the ectoderm and mesoderm may be important in determining the above events. For example, it has been observed that, at the time of premyogenic and prechondrogenic condensation, the mesoderm underlying the ectoderm in vivo retains mesenchymal morphology and the large hyaluronate-rich spaces between cells (Singley and Solursh, 1981). This regionalization of condensed and non-condensed mesoderm appears to be under the control of the ectoderm in that it prevents differentiation of nearby mesoderm to cartilage and causes the retention of mesenchymal characteristics including hyaluronate-rich intercellular spaces (Solursh et al., 1981). The effect of this epithelial-mesenchymal interaction on the expression of hyaluronate receptors is not yet known.

Of considerable interest is the discovery of at least two different tpyes of hyaluronate-binding proteins in brain (Delpech and Halavent, 1981; Marks and Toole, 1985). Although the role of these proteins is not known, it is likely that they also are involved in important morphogenetic processes in similar fashion to those discussed above.

Conclusion

Hyaluronate is a major component of the extracellular matrices which surround proliferating and migrating mesenchymal cells at early stages of morphogenesis of embryonic tissues and organs, as well as in adult tissue regeneration and tumor cell invasion. During tumorigenesis, and most probably during morphogenesis, the synthesis of hyaluronate is regulated by cell-cell interactions. Using various types of cells in culture, we have demonstrated two distinct modes of interaction of hyaluronate with the cell surface. The first of these is characteristic of normal fibroblast cell lines and of early limb mesoderm, but is of unknown mechanism. The second involves specific, non-covalent, high affinity interaction of the polysaccharide with multiple receptors in the plasma membrane. Cell differentiation, e.g., in the embryonic limb, involves transitions in these two types of hyaluronate-cell interactions. During chick embryo limb development, mesodermal cells enter the limb, accumulate by proliferation, condense and then differentiate to cartilage and muscle cells. The mesodermal cells initially exhibit prominent hyaluronate-dependent pericellular coats, but these are lost at the time of condensation. Also, at the time of condensation, hyaluronate receptors appear on the surface of the mesodermal cells. During subsequent cartilage formation, pericellular coats are regained. These coats are now enriched in proteoglycan which is held to the surface of the chondrocytes via interaction with hyaluronate. During myogenesis, large hyaluronate-dependent coats which surround myoblasts are lost on fusion of these cells to form multinucleated myotubes. During this differentiation of the mesoderm, the hyaluronate receptors are retained on the surface of the chondrocytes but lost from the myotubes. Since hyaluronate and hyaluronate-binding proteins are also present in and undergo large changes in amount during brain development, it will be of interest to determine their role in cell-cell interactions, migration and differentiation of neural cells.

The recent work described in this review was supported by grants from the National Institutes of Health, DE05838 (to B.P.T.), CA 38817 and CA 41701 (to C.B.).

References

Abatangelo, G., R. Cortivo, M. Martelli, and P. Vecchia (1982) Cell detachment mediated by hyaluronic acid. Exp. Cell Res. **137**: 73-78.

Abatangelo, G., M. Martelli, and P. Vecchia (1983) Healing of hyaluronic acid-enriched wounds: Histological observations. J. Surg. Res. **35**: 410-416.

Angello, J.C., K.G. Danielson, L.W. Anderson, and H.L. Hosick (1982) Glycosaminoglycan synthesis by subpopulations of epithelial cells from a mammary adenocarcinoma. Cancer Res. **42**: 2207-2210.

Barnhart, B.J., S.H. Cox, and P.M. Kraemer (1979) Detachment variants of Chinese hamster cells. Hyaluroinic acid as a modulator of cell detachment. Exp. Cell Res. **119**: 327-332.

Belsky, E., and B.P. Toole (1983) Hyaluronate and hyaluronidase in the developing chick embryo kidney. Cell Diff. **12**: 61-66.

Bernanke, D.H., and R.R. Markwald (1979) Effects of hyaluronic acid on cardiac cushion tissue cells in collagen matrix cultures. Texas Rep. Biol. Med. **39**: 271-285.

Bertolami, C.N., and R.B. Donoff (1978) Hyaluronidase activity during open wound healing in rabbits: A preliminary report. J. Surg. Res. **25**: 256-259.

Bhavanandan, V.P. (1981) Glycosaminoglycans of cultured human fetal uveal melanocytes and comparison with those produced by cultured human melanoma cells. Biochemistry **20**: 5595-5602.

Biswas, C. (1982) Tumor cell stimulation of collagenase production by fibroblasts. Biochem. Biophys. Res. Commun. **109**: 1026-1034.

Biswas, C. (1984) Collagenase stimulation in cocultures of human fibroblasts and human tumor cells. Cancer Lett. **24**: 201-207.

Biswas, C. (1985a) Matrix influence on the tumor cell stimulation of fibroblast collagenase production. J. Cell Biochem. **28**: 39-45.

Biswas, C. (1985b) Tumor cell stimulation of fibroblast collagenase production: Membrane association of the tumor cell stimulator. Fed. Proc. **44**: 1337.

Biswas, C., and M.A. Nugent (1986) Membrane association of the collagenase stimulatory factor form B-16 melanoma cells. Submitted for publication.

Brandt, K. (1970) Modification of chemotaxis by synovial fluid hyaluronate. Arth. Rheum. Rheum. **13**: 308-309.

Brandt, K. (1974) The effect of synovial hyaluronate on the ingestion of monosodium urate crystals by leukocytes. Clin. Chim. Acta **55**: 307-315.

Chi-Rosso, G., and B.P. Toole (1986) Hyaluronate-binding protein of simian virus 40-transformed 3T3 cells: Membrane distribution and reconstitution into lipid vesicles. Submitted for publication.

Culp, L.A., B.A. Murray, and B.J. Rollins (1979) Fibronectin and proteoglycans as determinants of cell-substratum adhesion. J. Supramol. Struct. **11**: 401-427.

Delpech, B., and C. Halavent (1981) Characterization and purification from human brain of a hyaluronic acid-binding glycoprotein, hyaluronectin. J. Neurochem. **36**: 855-859.

Derby, M.A. (1978) Analysis of glycosaminoglycans within the extracellular environment encountered by migrating neural crest cells. Dev. Biol. **66**: 321-336.

Doillon, C.J., and F.H. Silver (1986) Collagen-based wound dressing: Effects of hyaluronic acid and fibronectin on wound healing. Biomaterials **7**: 3-8.

Ekblom, P. (1984) Basement membrane proteins and growth factors in kidney differentiation. In *The Role of Extracellular Matrix in Development*, R.L. Trelstad, ed., pp. 173-206. Alan Liss, New York.

Feinberg, R.N., and D.C. Beebe (1983) Hyaluronate in vasculogenesis. Science **220:** 1177-1179.

Fisher, M., and M. Solursh (1979) The influence of the substratum on mesenchyme spreading in vitro. Exp. Cell Res. **123:** 1-14.

Forrester, J.V., and E.A. Balazs (1980) Inhibition of phagocytosis by high molecular weight hyaluronate. Immunology **40:** 435-446.

Forrester, J.V., and J.M. Lackie (1981) Effect of hyaluronic acid on neutrophil adhesion. J. Cell Sci. **50:** 329-344.

Forrester, J.V., and P.C. Wilkinson (1981) Inhibition of leukocyte locomotion by hyaluronic acid. J. Cell Sci. **48:** 315-331.

Fraser, J.R., and B.J. Clarris (1970) On the reactions of human synovial cells exposed to homologous leucocytes in vitro. Clin. Exp. Immunol. **6:** 211-225.

Goldberg, R.L., and B.P. Toole (1984) Pericellular coat of chick embryo chondrocytes: Structural role of hyaluronate. J. Cell Biol. **99:** 2114-2122.

Goldberg, R.L., J.D. Seidman, G. Chi-Rosso, and B.P. Toole (1984) Endogenous hyaluronate-cell surface interactions in 3T3 and simian virus-transformed 3T3 cells. J. Biol. Chem. **259:** 9440-9446.

Grillo, H.C., C.M. Lapiere, M.H. Dresden, and J. Gross (1968) Collagenolytic activity in regenerating forelimbs of the adult newt. Dev. Biol. **17:** 571-583.

Hakansson, L., and P. Venge (1985) The combined action of hyaluronic acid and fibronectin stimulates neutrophil migration. J. Immunol. **135:** 2735-2739.

Hakansson, L., R. Hallgren, and P. Venge (1980) Regulation of granulocyte function by hyaluronic acid: In vitro and in vivo effects on phagocytosis, locomotion and metabolism. J. Clin. Invest. **66:** 298-305.

Hopwood, J.J., and A. Dorfman (1977) Glycosaminoglycan synthesis by cultured human skin fibroblasts after transformation with simian virus 40. J. Biol. Chem. **252:** 4777-4785.

Hopwood, J.J., and A. Dorfman (1978) Glycosaminoglycan synthesis by Wilm's tumor. Pediatr. Res. **12:** 52-56.

Horai, T., N. Nakamura, R. Tateishi, and S. Hattori (1981) Glycosaminoglycans in human lung cancer. Cancer **48:** 2016-2021.

Iozzo, R.V. (1985) Neoplastic modulation of extracellular matrix: Colon carcinoma cells release polypeptides that alter proteoglycan metabolism in colon fibroblasts. J. Biol. **260:** 7464-7473.

Johnson-Muller, B., and J. Gross (1978) Regulation of corneal collagenase production: Epithelial stromal cell interactions. Proc. Natl. Acad. Sci. (USA) **75:** 4417-4421.

Johnson-Wint, B., and E.A. Bauer (1985) Stimulation of collagenase synthesis by a 20,000-dalton epithelial cytokine: Evidence for pretranslational regulation. J. Biol. Chem. **260:** 2080-2085.

Kawai, T., M. Suzuki, M. Shinmei, Y. Maenaka, and K. Kageyama (1985) Glycosaminoglycans of malignant diffuse mesothelioma. Cancer **56:** 576-574.

Kimata, K., Y. Honma, M. Okayama, K. Oguri, M. Hozumi, and S. Suzuki (1983) Increased synthesis of hyaluronic acid by mouse mammary carcinoma cell variants with high metastatic potential. Cancer Res. **43:** 1347-1352.

Knudson, C.B., and B.P. Toole (1984) Hyaluronate-cell surface interactions during differentiation of limb bud mesoderm. J. Cell Biol. **99:** 162a.

Knudson, C.B., and B.P. Toole (1985a) Changes in the pericellular matrix during differentiation of limb bud mesoderm. Dev. Biol. **112:** 308-318.

Knudson, C.B., and B.P. Toole (1985b) Modulation of pericellular hyaluronate during limb chondrogenesis. J. Cell Biol. **101:** 99a.

Knudson, W., C. Biswas, and B.P. Toole (1984a) Stimulation of glycosaminolglycan production in murine tumors. J. Cell Biochem. **25:** 183-196.

Knudson, W., C. Biswas, and B.P. Toole (1984b) Interactions between human tumor cells and fibroblasts stimulate hyaluronate synthesis. Proc. Natl. Acad. Sci. (USA) **81:** 6767-6771.

Knudson, W., and B.P. Toole (1985) Properties of a membrane-bound tumor cell factor

which stimulates hyaluronate synthesis in fibroblasts. J. Cell Biol. **101**: 339a.
Kojima, J., N. Nakamura, M. Kanatani, and K. Ohmori (1975) The glycosaminoglycans in human hepatic cancer. Cancer Res. **35**: 542-547.
Kujawa, M.J., and A.I. Caplan (1986) Hyaluronic acid bonded to cell culture surfaces stimulates chondrogenesis in stage 24 limb mesenchyme cell cultures. Dev. Biol. **114**: 504-518.
Kujawa, M.J., and K. Tepperman (1983) Culturing chick muscle cells on glycosaminoglycan substrates: Attachment and differentiation. Dev. Biol. **99**: 277-286.
Kujawa, M.J., D.J. Pechak, M.Y. Fizman, and A.I. Caplan (1986a) Hyaluronic acid bonded to cell culture surfaces inhibits the program of myogenesis. Dev. Biol. **113**: 10-16.
Kujawa, M.J., D.A. Carrino, and A.I. Caplan (1986b) Substrate-bonded hyaluronic acid exhibits a size-dependent stimulation of chondrogenic differentiation of stage 24 limb mesenchymal cells in culture. Dev. Biol. **114**: 519-528.
Kvist, T.N., and C.V. Finnegan (1970) The distribuiton of glycosaminoglycans in the axial region of the developing chick embryo. I. Histochemical analysis. J. Exp. Zool. **175**: 221-240.
Love, S.H., B.T. Shannon, Q.N. Myrvik, and W.S. Lynn (1979) Characterization of macrophage agglutinating factor as a hyaluronic acid-protein complex. J. Reticuloendothelial Soc. **25**: 269-282.
Mailman, M.L., and M.H. Dresden (1979) Denervation effects on newt limb regeneration: collagen and collagenase. Dev. Biol. **71**: 60-70.
Margolis, R.U., R.K. Margolis, L.B. Chang, and C. Preti (1975) Glycosaminoglycans of brain during development. Biochemistry **41**: 85-88.
Marks, M.S., and B.P. Toole (1985) Hyaluronate-binding protein of chick and mouse brain. J. Cell Biol. **101**: 334a.
Markwald, R.R., T.P. Fitzharris, H. Bank, and D.H. Bernanke (1978) Structural analysis on the matrical organization of glycosaminoglycans in developing endocardial cushions. Dev. Biol. **62**: 292-316.
Maurer, P.H., and S.S. Hudack (1952) Isolation of hyaluronic acid from callus tissue during early healing. Arch. Biochem. Biophys. **38**: 49-53.
McBride, W.H., and J.B. Bard (1979) Hyaluronidase-sensitive halos around adherent cells. Their role in blocking lymphocyte-mediated cytolysis. J. Exp. Med. **149**: 507-515.
McCarthy, M.T., and B.P. Toole (1985) Association of hyaluronate and proteoglycan with the surface of rat chondrosarcoma chondrocytes. J. Cell Biol. **101**: 334a.
Merrilees, M.J., and G.J. Finlay (1985) Human tumor cells in culture stimulate glycosaminoglycan synthesis by human skin fibroblasts. Lab. Invest. **53**: 30-36.
Merrilees, M.J., and L. Scott (1980) Interaction of epithelial cells and fibroblasts in culture: Effect on glycosaminoglycan levels. Dev. Biol. **76**: 396-409.
Merrilees, M.J., and L. Scott (1981) Interaction of aortic endothelial cells and smooth muscle cells in culture: Effect on glycosaminoglycan levels. Atherosclerosis **39**: 147-161.
Merrilees, M.J., J. Sodek, and J.E. Aubin (1983) Effect of cells of epithelial rests of Mallassez and endothelial cells on synthesis of glycosaminoglycans by periodontal ligament fibroblasts in vitro. Dev. Biol. **97**: 146-153.
Mescher, A.L., and S.I. Munaim (1986) Changes in the extracellular matrix and glycosaminoglycan synthesis during the initiation of regeneration in adult newt forelimbs. Anat. Rec. **214**: 424-431.
Meyer, F.A. (1983) Macromolecular basis of globular protein exclusion and of swelling pressure in loose connective tissue (umbilical cord). Biochim. Biophys. Acta **755**: 388-399.
Meyer, F.A., Z. Laver-Rudich, and R. Tanenbaum (1983) Evidence for a mechanical coupling of glycoprotein microfibrils with collagen fibrils in Wharton's jelly. Biochim. Biophys. Acta **755**: 376-387.
Mikuni-Takagaki, Y., and B.P. Toole (1979) Shedding of hyaluronate from the cell surface of Rous sarcoma virus-transformed chondrocytes. J. Biol. Chem. **254**: 8409-8425.
Mikuni-Takagaki, Y., and B.P. Toole (1980) Cell-substratum attachment and cell surface

hyaluronate of Rous sarcoma virus-transformed chondrocytes. J. Cell Biol. **85**: 481-488.
Normand, G., J. Clos, G. Vincendon, and G. Gombos (1985) Postnatal development of rat cerebellum: Glycosaminoglycan changes related to variation in water content, cell formation and organ growth. Int. J. Devl. Neurosc. **3**: 245-256.
Orkin, R.W., C.B. Underhill, and B.P. Toole (1982) Hyaluronate degradation in 3T3 and Simian virus-transformed 3T3 cells. J. Biol Chem. **257**: 5821-5826.
Orkin, R.W., W. Knudson, and B.P. Toole (1985) Loss of hyaluronate-dependent coat during myoblast fusion. Dev. Biol. **107**: 527-530.
Pessac, B., and V. Defendi (1972) Cell aggregation: Role of acid mucopolysaccharides. Science **175**: 898-900.
Pintar, J.E. (1978) Distribution and synthesis of glycosaminoglycans during quail neural crest morphogenesis. Dev. Biol. **67**: 444-464.
Polansky, J.R., B.P. Toole, and J. Gross (1974) Brain hyaluronidase: Changes in activity during chick development. Science **183**: 862-864.
Pratt, R.M., M.A. Larsen, and M.C. Johnston (1975) Migration of cranial neural crest cells in a cell-free hyaluronate-rich matrix. Dev. Biol. **44**: 298-305.
Reid, T., and M.H. Flint (1974) Changes in glycosaminoglycan content of healing rabbit tendon. J. Embryol. Exp. Morphol. **31**: 489-495.
Singley, C.T., and M. Solursh (1981) The spatial distribution of hyaluronic acid and mesenchymal condensation in the embryonic chick wing. Dev. Biol. **84**: 102-120.
Smith, G.N., B.P. Toole, and J. Gross (1975) Hyaluronidase activity and glycosaminoglycan synthesis in the amputated newt limb. Comparison of denervated, nonregenerating limbs with regenerates. Dev. Biol. **42**: 221-232.
Solursh, M., M. Fisher, S. Meier, and C.T. Singley (1979) The role of extracellular matrix in the formation of the sclerotome. J. Embryol. Exp. Morphol. **54**: 75-98.
Solursh, M., T.E. Hardingham, V.C. Hascall, and J.H. Kimura (1980) Separate effects of exogenous hyaluronic acid on proteoglycan synthesis and deposition in pericellular matrix by cultured chick embryo limb chondrocytes. Dev..Biol. **75**: 121-129.
Solursh, M., C.T. Singley, and R.S. Reiter (1981) The influence of epithelia on cartilage and loose connective tissue formation by limb mesenchyme cultures. Dev. Biol. **86**: 471-482.
Takeuchi, J., M. Sobue, E. Sato, M. Shamato, K. Miura, and Nakagaki (1976) Variation in glycosaminoglycan components of breast tumors. Cancer Res. **36**: 2133-2139.
Thorogood, P.V., and J.R. Hinchliffe (1975) An analysis of the condensation process during chondrogenesis in the embryonic chick hind limb. J. Embryol. Exp. Morphol. **33**: 581-606.
Toole, B.P. (1973) Hyaluronate and hyaluronidase in morphogenesis and differentiation. Am. Zool. **13**: 1061-1065.
Toole, B.P. (1976) Morphogenetic role of glycosaminoglycans in brain and other tissues. In *Neuronal Recognition*, S.H. Barondes, ed., pp. 275-329. Plenum Press, New York.
Toole, B.P. (1981) Glycosaminoglycans in morphogenesis. In *Cell Biology of the Extracellular Matrix*, E.D. Hay, ed., pp. 259-294. Plenum Press, New York.
Toole, B.P., and J. Gross (1971) The extracellular matrix of the regenerating newt limb: Synthesis and removal of hyaluronate prior to differentiation. Dev. Biol. **25**: 57-77.
Toole, B.P., and R.L. Trelstad (1971) Hyaluronate production and removal during corneal development in the chick. Dev. Biol. **26**: 28-35.
Toole, B.P., C. Biswas, and J. Gross (1979) Hyaluronate and invasiveness of the rabbit V2 carcinoma. Proc. Natl. Acad. Sci. (USA) **76**: 6299.6303.
Toole, B.P., R.L. Goldberg, G. Chi-Rosso, C.B. Underhill, and R.W. Orkin (1984) Hyaluronate-cell interactions. In *The Role of Extracellular Matrix in Development*, R.L. Trelstad, ed., pp. 43-66. Alan R. Liss, Inc., New York.
Turley, E.A., and S. Roth (1979) Spontaneous glycosylation of glycosaminoglycan substrates by adherent fibroblasts. Cell **17**: 109-115.
Turley, E.A., and J. Torrance (1984) Localization of hyaluronate and hyaluronate-binding

protein on motile and non-motile fibroblasts. Exp. Cell Res. **161**: 17-28.
Turley, E.A., P. Bowman, and M.A. Kytryk (1985) Effects of hyaluronate and hyaluronate-binding proteins on cell motile and contact behavior. J. Cell Sci. **78**: 133-145.
Ullrich, S.J., and S.P. Hawkes (1983) The effect of the tumor promoter phorbol myristate acetate on hyaluronic acid synthesis by chicken embryo fibroblasts. Exp. Cell Res. **148**: 377-386.
Underhill, C.B. (1982) Interaction of hyaluronate with the surface of Simian virus 40-transformed 3T3 cells: Aggregation and binding studies. J. Cell Sci. **56**: 177-189.
Underhill, C.B., and A. Dorfman (1978) The role of hyaluronic acid in intercellular adhesion of cultured mouse cells. Exp. Cell Res. **117**: 155-164.
Underhill, C.B., and B.P. Toole (1979) Binding of hyaluronate to the surface of cultured cells. J. Cell Biol. **82**: 475-484.
Underhill, C.B., and B.P. Toole (1980) Physical characteristics of hyaluronate binding to the surface of Simian virus 40-transformed 3T3 cells. J. Biol. Chem. **255**: 4544-4549.
Underhill, C.B., and B.P. Toole (1981) Receptors for hyaluronate on the surface of parent and virus-transformed cell lines. Binding and aggregation studies. Exp. Cell Res. **131**: 419-423.
Underhill, C.B., and B.P. Toole (1982) Tranformation-dependent loss of the hyaluronate-containing coats of cultured cells. J. Cell Physiol. **110**: 123-128.
Underhill, C.B., G. Chi-Rosso, and B.P. Toole (1983) Effects of detergent solubilization on the hyaluronate-binding protein from membranes of simian virus 40-transformed 3T3 cells. J. Biol. Chem. **258**: 8086-8091.
Underhill, C.B., A.L. Thurn, and B.E. Lacey (1985) Characterization and identification of the hyaluronate-binding site from membranes of SV-3T3 cells. J. Biol. Chem. **260**: 8128-8133.
Wasteson, A., B. Estermark, U. Lindahl, and J. Ponten (1973) Aggregation of feline lymphoma cells by hyaluronic acid. Int. J. Cancer **12**: 169-178.
Werz, W., G. Fischer, and M. Schachner (1985) Glycosaminoglycans of rat cerebellum: A developmental study. J. Neurochem. **44**: 907-910.
Wiebkin, O.W., and H. Muir (1973) The inhibition of sulphate incorporation in isolated adult chondrocytes by hyaluronic acid. FEBS Lett. **37**: 42-46.
Wright, T.C., C.B. Underhill, B.P. Toole, and M.J. Karnovsky (1981) Divalent cation-independent aggregation of rat-1 fibroblasts infected with a temperature-sensitive mutant of Rous Sarcoma virus. Cancer Res. **41**: 5107-5113.

SECONDARY NEUROEPITHELIAL STEM CELLS OF THE CEREBELLUM AND THE DENTATE GYRUS ARE ATTACHED TO THE BASAL LAMINA DURING THEIR MIGRATION AND PROLIFERATION

B. Hausmann, D. Hartmann, and J. Sievers
Department of Anatomy, University of Kiel
Federal Republic of Germany

In brain regions with prolonged neurogenesis like the cerebellum and the hippocampus, the ventricular stem cell population divides early in ontogenesis into two spatially separated germinal zones, the primary ventricular zone and the secondary subpial germinal zone of translocated neuroepithelial stem cells (reviews in Hausmann and Sievers, 1985; Cowan et al., 1980, 1981).

The secondary stem cell populations are of special interest because they differ from the primary ventricular cells in their unique pattern of an intermitotic displacement and their prolonged proliferative activity. Since extracellular matrix is involved in cell movement and cell proliferation, we suspected that the basal lamina might participate in migration and proliferation of secondary neuroepithelial stem cells. We therefore investigated the morphology of developing cerebellar external granule cells and dentate precursor cells focussing on their relationship to the overlying basal lamina.

Intermitotic Migration and Proliferation of Cerebellar External Granule Cells

In the early rat cerebellum the ventricular cells display interkinetic nuclear migration characteristic for the pseudostratified neuroepithelium. Starting on embryonic day (E)14 the external granule cells separate from the caudo-lateral margin of the ventricular neuroepithelium and spread tangentially over the surface of the cerebellar anlage thereby constituting the secondary germinal zone, the external granular layer (EGL) (Fig. 1). It is still unresolved whether this tangential movement is a true active migration as proposed by Herrick (1891), Schaper (1894), Jakob (1928), Rüdeberg (1961), Korneliussen (1968), Rakic and Sidman (1970), and Gona (1976, 1978), or a passive displacement caused by the pressure of either intercalated mitoses or the continous cell proliferation at the caudolateral margin of the cerebellar anlage (Sidman and Rakic, 1973; Altman and Bayer, 1978; Kappel, 1981). The factors responsible for the prolonged period of neurogenesis in the EGL (which starts when neuronal cell proliferation in the ventricular neuro-

NATO ASI Series, Vol. H5
Mesenchymal-Epithelial Interactions in Neural Development
Edited by J. R. Wolff et al.

epithelium has ceased, and continues almost up to adulthood) and unexplained so far.

Two types of cell appear during development of the EGL in the rat between E14 and postnatal day (P)20 (Hausmann and Sievers, 1985). The first type is present in the period of intermitotic translocation, the second is predominant in the proliferative period of the EGL.

Migratory cell type of tangential orientation

The first EGL cells appear at the onset of the translocation period on E14 (Fig. 2). At this stage, the cerebellar anlage consists of a pseudostratified neuroepithelium of bipolar ventricular cells, which are arranged in prominent cell columns. In the caudolateral germinal trigone of the cerebellar anlage, which is the source of EGL cells (Altman and Bayer, 1978), the formerly radially orientated ventricular cells become reorientated. While their internal process remains in a radial position, the external process and cell body lies parallel to the pial surfacae below the basal lamina (BL) pointing in a rostro-medial direction. Further rostrally and medially, these tangential EGL cells appear as oval or round cells (Fig. 2b, 3) with a prominent external process, attached to the BL

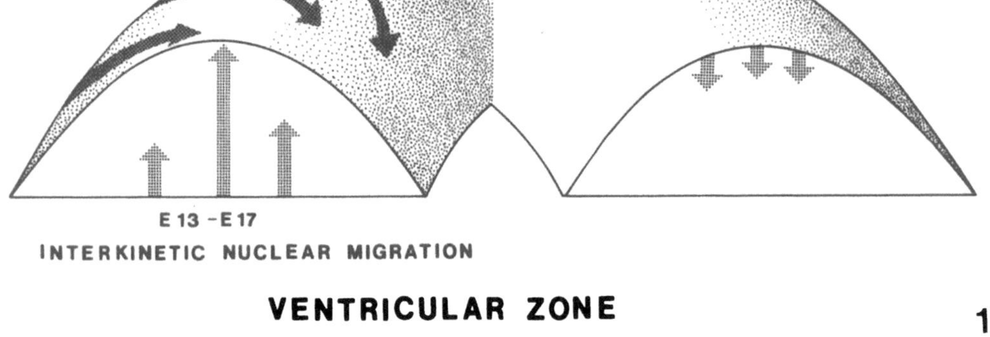

Figure 1: Schematical illustration of the developing cerebellar anlage, showing the tangential intermitotic migration of the EGL from the caudolateral germinal trigone across the cerebellar surface (dark curved arrows) and the radial interkinetic and postmitotic migration of the ventricular cells (light stippled arrows, left) and the EGL cells (arrows right).

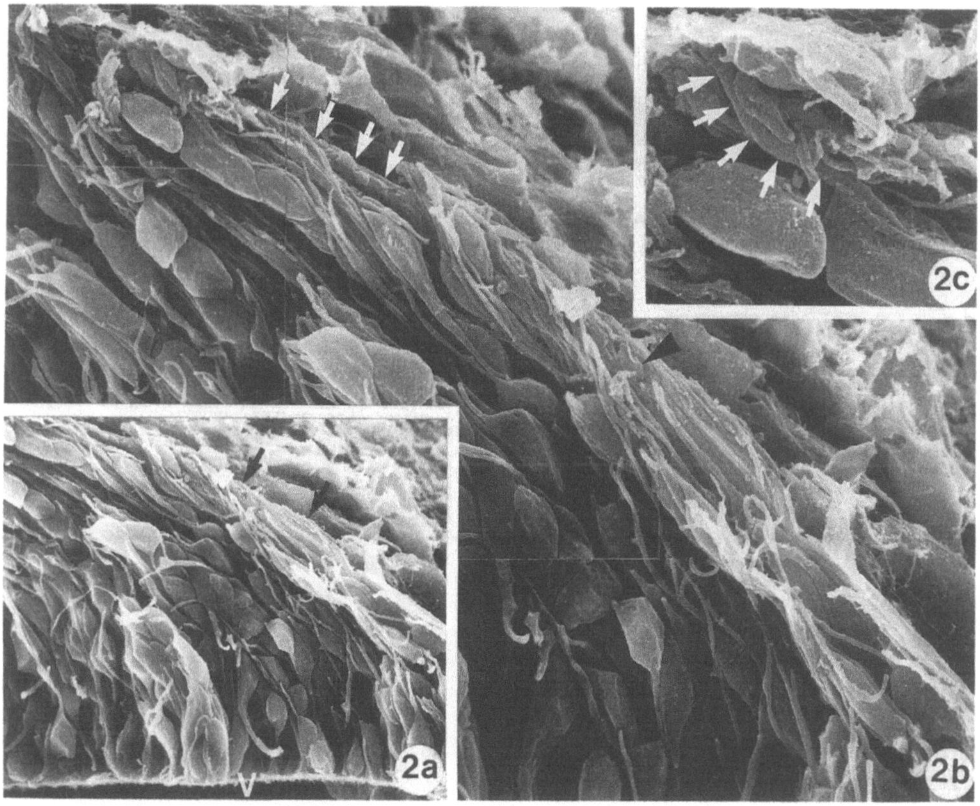

Figure 2: a) Sagittal paramedian section of the caudal part of the cerebella anlage on E 14. V = ventricle, arrows delineate the pial BL. The long axis of the rostral ventricular cells (left margin) is orientated radially, the caudalmost ventricular cells are aligned tangentially to the BL. x 920. b) Higher magnification of the cerebellar surface whown in (a). Tangential fascicles of the external processes run below the BL (arrows), where they attach with end-feet (arrowhead). x 1564. c) Tangential EGL cell (E14) whose external process (arrows) is attached to the BL which is covered by a thin feltwork of collagen fibrils. x 2760.

with end-feet (Fig. 2). The caudal cells are seen to retract their internal process, remnants of which are seen as classical retraction bulbs bearing a single cilium (Fig. 4). During the process of retraction the single cilium moves towards the soma (Fig. 5). According to Hinds and Ruffet (1971), Hinds and Hinds (1974), and Shoukimas and Hinds (1978), the retraction of the ventricular process and the translocation of the cilium are constant features of migratory cells detaching from the ventricular intercellular junction, for example postmitotic neurons. In contrast to postmitotic neurons, however, which also detach from the outer neuroepithelial surface, the tangential cells maintain

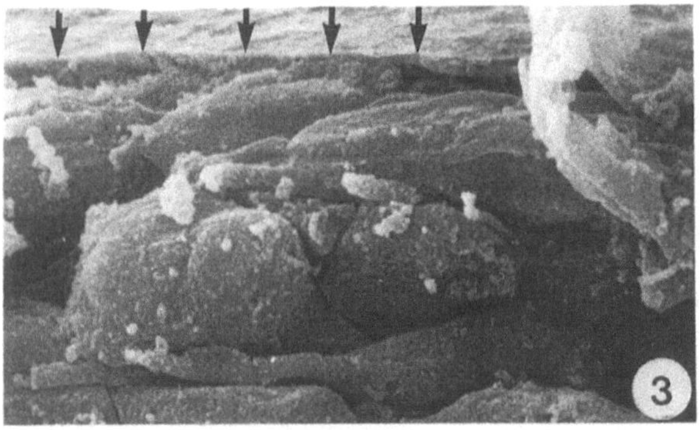

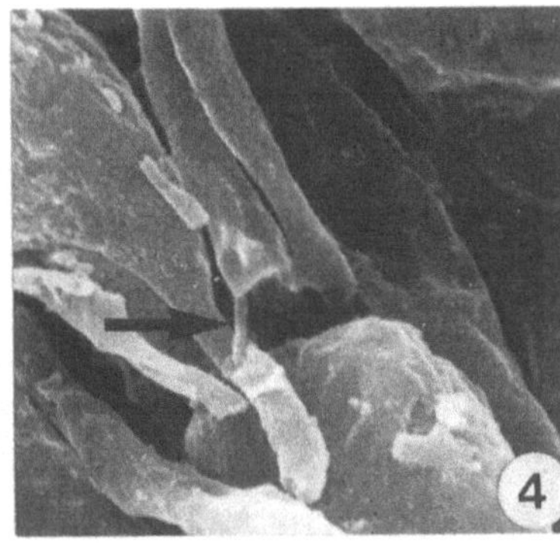

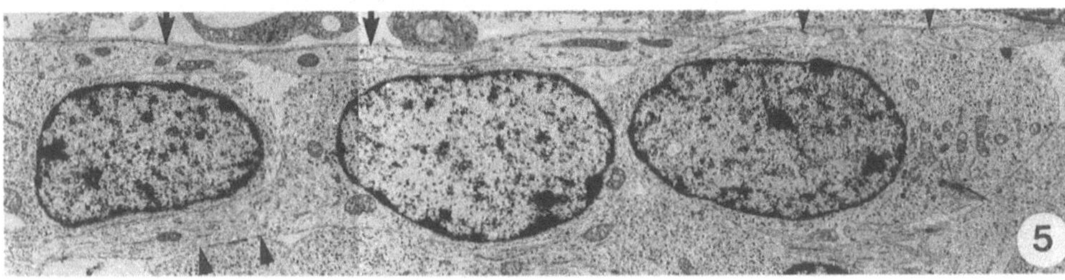

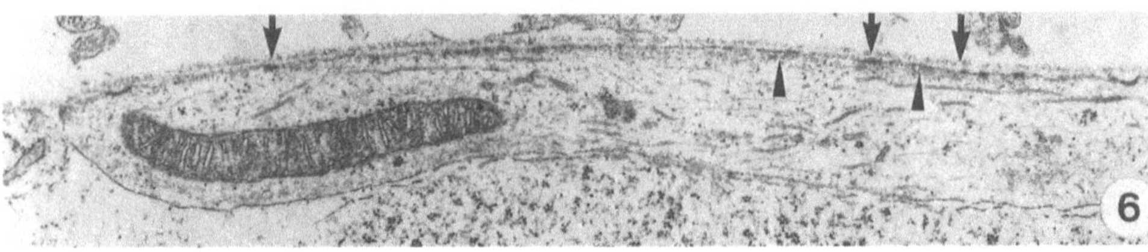

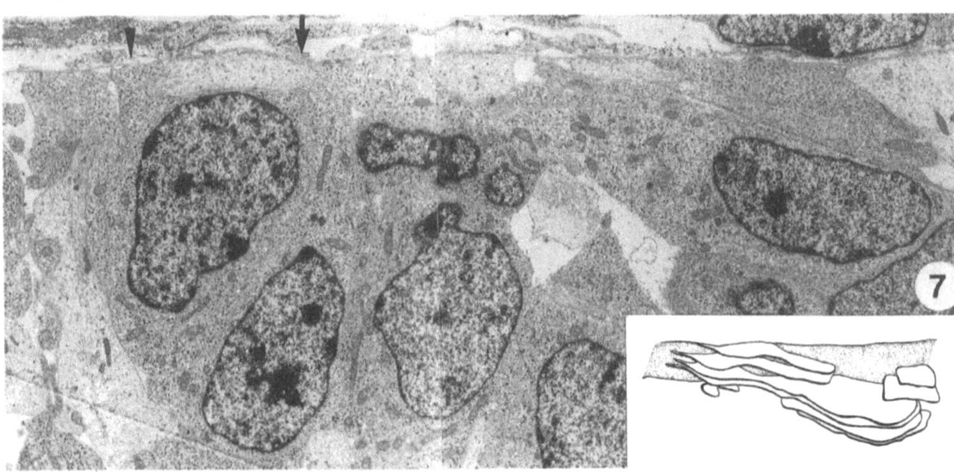

Figure 3: Densely packed cells in the rostral part of the EGL (E14) viewed from above. BL is marked by arrows. x 1750.

Figure 4: Retracting internal process of a caudal EGL cell (E14) (arrow), featuring a single cilium on the end bulb. x 7140.

Figure 5: Group of tangential EGL cells (sagittal paramedian section, E14), covered by long segments of BL-attached external processes (arrows). The right cell is bearing a somatic cilium. Note the prominent cytoskeleton in the proximal portion of the external process of the middle cell (arrowheads). x 4320.

Figure 6: Higher magnification of a tangential external leading process of a migrating EGL cell. Note the presence of a prominent cytoskeleton, the hypolemmal filament network (arrowheads) and electron dense attachment plaques (arrows) on the side facing the BL. x 16,000.

Figure 7: Tangential EGL cell among radial EGL cells on E17. The tip of the external leading process touches the BL and is drawn out into a filopodial process (arrow), which on serial reconstructions constitutes part of a lamellipodium (inset). On the left, a radial EGL cell is connected to the BL by a short extension of the cell body (arrowhead). x 4320.

their contact to the BL. Computer reconstructions of serial thin sections confirm the regular attachment of the tangential cells to the BL and show (Fig. 5) that the contacting leading process often is drawn out into a lamellipodial tip. The ultrastructure of these leading processes is also indicative of true migration (Fig. 6), in that it contains a prominent cytoskeleton, characterized by numerous parallel orientated microtubules and filaments. At the sites of contact with the BL a subsurface web of fine filaments and electron-dense attachment-plaques are visible. Such morphological specialisations are thought to mediate cell adhesion to the substrate as a prerequisite for migration (Brunk et al., 1971; Abercrombie et al., 1971; Overton and Mapp, 1974; Trinkaus, 1976; Goldman et al., 1976; Löfberg and Ahlfors, 1978).

These findings suggest that the translocation of the tangential EGL cells is achieved by active migration guided by the BL. This hypothesis is supported by the observation that other derivatives of the neural plate, the neural crest cells, as well as central and peripheral axons, also seem to use extracellular matrix as substrate and guidance structure for migration (reviews in Löfberg and Ahlfors, 1978; Scholes, 1979; Sanes, 1983; Carbonetto, 1984; Easter et al., 1984; Collins and Lee, 1984). It would seem an attractive hypothesis therefore, that the EGL cells use the overlying BL as a guidance structure, because such a relationship could greatly facilitate the mass migration of the EGL from the caudolateral margin to the rostral part of the anlage. However, we do not know whether the BL acts as a simple mechanical guidance structure or whether it mediates specific directional guidance cues or provides for increased adhesion of EGL cells. Also unexplained is the mechanism that terminates this intermitotic migration.

Alternatively, it seems not very likely that glial processes guide the tangential migration of EGL cells, as proposed for young postmitotic neurons (Rakic, 1971, 1972,

1981, 1983). Glial fibers of the cerebellar anlage are orientated radially and not tangentially to the surface and, thus, migrating EGL cells would contact these fibers orthogonally at very restricted sites. Such a relationship does not favor strong adhesion between the cells which normally is prerequisite for migration. Moreover, it is unlikely that afferent axons serve as a substrate for the EGL cells as proposed by Hynes et al. (1986), because transplantation experiments have shown that deafferentated cerebellar primordia develop a quite normal EGL (Wells and McAllister, 1982; Ezerman and Kromer, 1984). However, the contact of EGL cells to the BL per se does not prove definitively that the BL is the substrate for the translocating cells, because this relationship could also simply indicate that the EGL cells secrete BL, like the adjacent glial end-feet, although there is no evidence that EGL cells are capable of producing BL molecules (see Sievers et al., this volume).

Proliferative cell type of radial orientation

The later stages of development of the EGL are characterized by an intense proliferation of the EGL cells starting on E17. At this stage a new, radially orientated cell type emerges between the migrating tangential cells, exhibiting the typical ultrastructure of proliferating EGL cells (Fig. 7). Although these radially oriented cells are lacking a prominent tangential leading process, they also make contact with the BL either directly with their soma or via short processes without a pronounced cytoskeleton. In the next two days, towards the end of gestation, the number of these radial cells increases rapidly, in contrast to the tangential cells, which are mainly restricted to the caudal and rostral poles of the EGL.

The contacts of these radial cells with the BL are present throughout most of the lifespan of the EGL. At birth, they add up to almost one third of the internal laminal surface; later on they are somewhat reduced. Computer reconstructions of single radial cells define the extent of these contact zones. EGL cells from P15, for example, regularly show wide zones of contact with the BL, either by a short extension of their cytoplasm (Fig. 9) or via one or two processes of various length (Fig. 8). Towards the end of the proliferative period, on P20 (Chanda et al., 1973), the number of EGL cells contacting the BL decreases rapidly, and these rare contacts are of minute size (Fig. 10).

The persistent contact of EGL cells to the BL, which ceases with the end of proliferation, could indicate that the BL, in addition to its proposed guidance function, may sustain the mitotic activity of EGL cells. Increasing evidence indicates that the substratum is of major importance for cell proliferation (reviews in Weiss et al., 1975; Kleinman et al., 1981) and that the BL contains factors stimulating cell proliferation (Greenburg and Gospodarowicz, 1982). Moreover, in most other epithelia, including the olfactory neuroepithelium (Graziadei and Monti Graziadei, 1978), the proliferating cells are

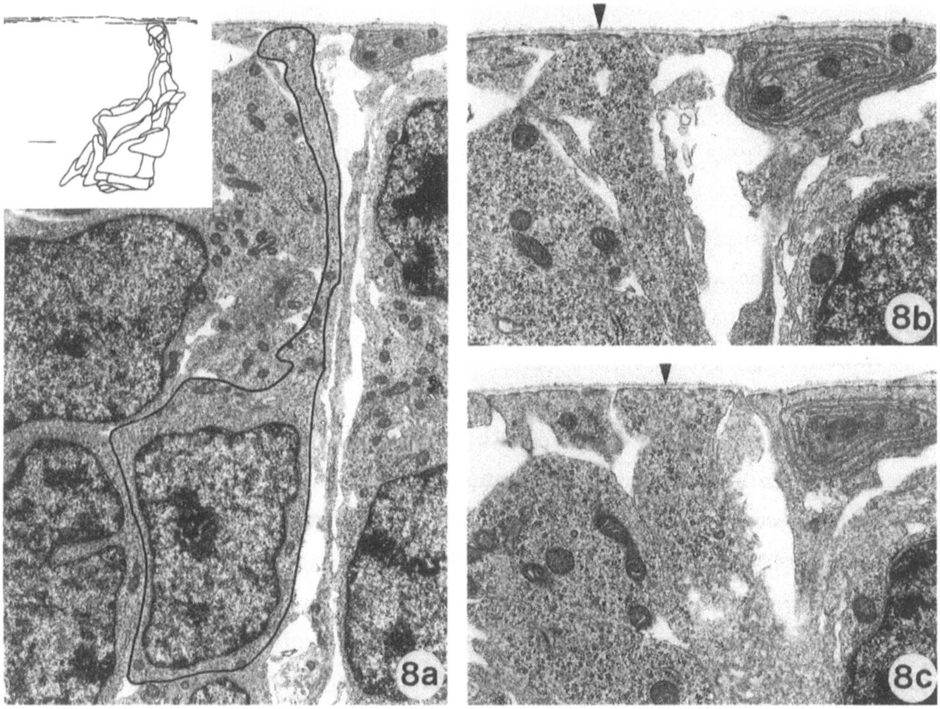

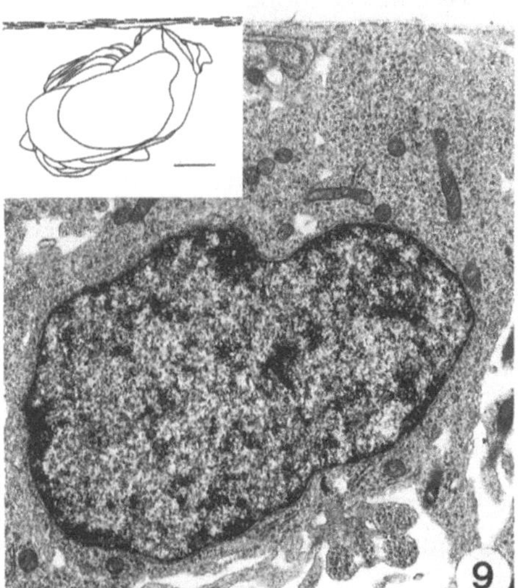

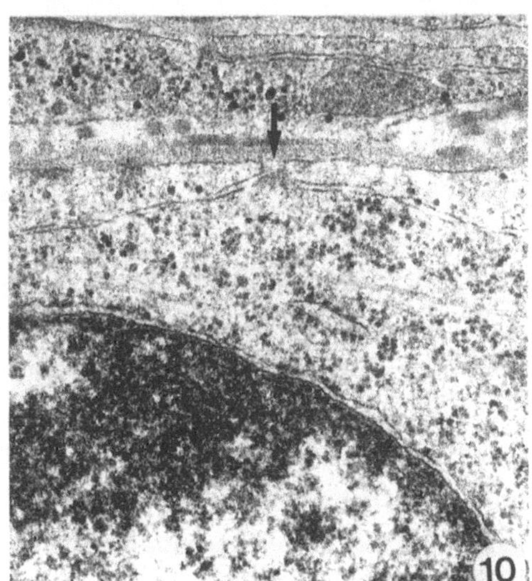

Figure 8: a) Radial EGL cell (outlined with black ink) on P15 contacting the overlying BL with a long process. x 8000. Inset: Reconstruction of the same cell, bar = 2 µm.
b) and c) Higher magnification of two sections of the tip (arrowhead) of the external process shown in a). x 16,000.

Figure 9: EGL cell on P15 contacting the BL with a short plump process. Inset: Reconstruction of the same cell. x 9600.

Figure 10: Minute BL contact (arrow) of an EGL cell on P20. x 40,000.

sited directly on the BL, and the experimental removal of the BL results in the cessation of cell division (Wessels, 1967). Similarly, the mitotic activity of the primary ventricular stem cells is also dependent on the BL: during each DNA synthetic phase the ventricular cells have to reestablish an intermittent contact with the BL (Sauer, 1935a, b, 1936; Fujita and Fujita, 1963; Meller et al., 1966; Hinds and Ruffet, 1971; Seymour and Berry, 1975).

To further support our hypothesis, we started to investigate the hippocampus, another brain region with a prolonged period of neurogenesis, which also is characterized by a secondary subpial proliferative zone of displaced neuroepithelial stem cells that forms the dentate gyrus.

Is the BL Also Involved in the Intermitotic Movement of Another Secondary Neuroepithelial Stem Cell Population, the Dentate Precursor Cells?

The dentate precursor cells which set up a secondary stem cell population below the hippocampal fissure that later gives rise to the two blades of the dentate gyrus, separate early in ontogeny from the ventricular neuroepithelium and move towards the pial surface of the medial cerebral hemisphere (Fig. 11). The mechanisms of this complicated displacement - which apparently takes place in two or three steps of different spatial orientation (radially along radial glial fibers in the fimbria fornicis, tangentially along the pial surface into the secondary proliferative zone, and radially from there into the blades of the dentate gyrus, Fig. 11) - are still not fully elucidated. Two modes of migration are discussed: Firstly, the intracellular translocation of the perikaryon of dentate precursor cells within their external process (Cowan et al., 1980, 1981), similar to the migration of cerebellar EGL cells, and, secondly, the migration of the dentate precursor cells in relation to radial glia fibers of the early hippocampal anlage, as proposed by Eckenhoff and Rakic (1984) and by Rickmann (this volume). In view of the contact of translocating EGL cells to the BL (see above), we studied, whether the migrating dentate stem cells also have contact to the overlying BL, which we suspected for the tangential subpial portion of their migratory path (Fig. 11).

At the time when the hippocampal fissure becomes visible (in the hamster on E14), a superficial stream of tangentially orientated cells is seen at the pial surface of the medial cerebral cortex, between the forming fimbria and the hippocampal fissure, where the secondary proliferative zone is forming (Fig. 12a). This purely tangential cell stream (which disappears in later developmental stages) is supplemented by another cell stream which runs radially or obliquely from the ventricular zone towards the pial surface (Fig. 12a, b). During these early ontogenetic stages between E14 and 15 (the day before birth in the hamster), many of the tangential cells exhibit the ultrastructure of undifferentiated neuroepithelial stem cells, like those of the EGL cells (Fig. 13). Also, they regu-

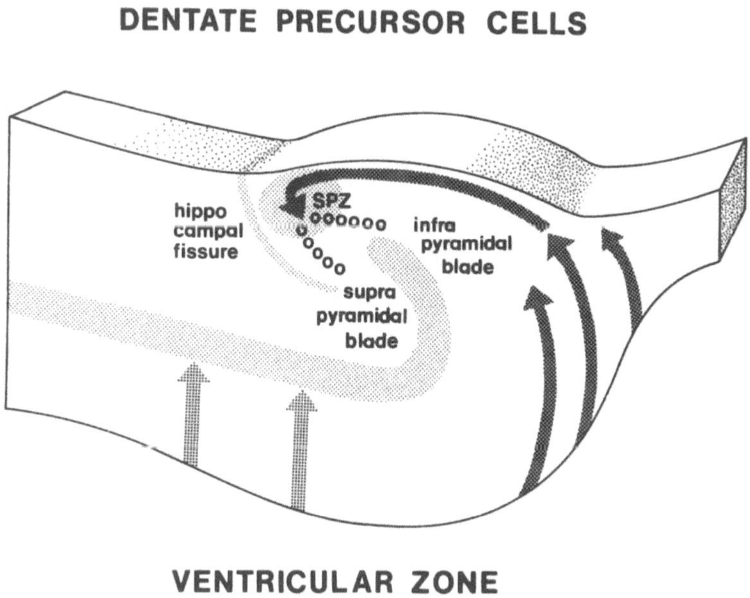

11

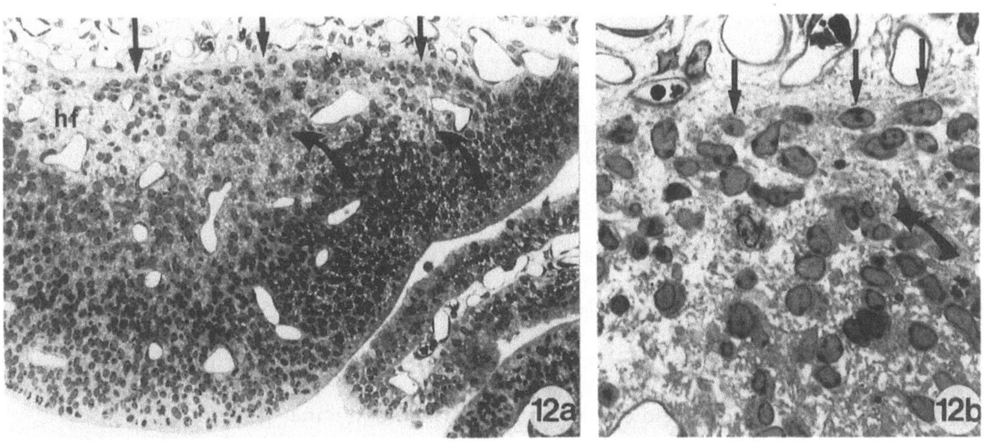

Figure 11: Schematical illustruction of the hippocampal anlage. The dark curved arrows mark the migratory path of dentate precursor cells from the ventricular neuroepithelium into the secondary proliferative zone (SPZ) below the hippocampal fissure. The first part of this path is radially or obliquely towards the pial surface, the second part runs tangentially to the pial surface, the third part is radially from the SPZ into the blades of the dentate gyrus.

Figure 12: a) Coronal section through the septal part of the hamster hippocampal anlage on E14. A superficial stream of tangentially orientated undifferentiated cells (arrows) is seen between the hippocampal fissure (hf) and the ventral tip of the neuroepithelium. The curved arrows outline the oblique path of additional undifferentiated cells feeding into the tangential cell stream. x 170. b) Higher magnification of (12a) demonstrating the tangential orientation of the superficial most undifferentiated cells (arrows), and the oblique orientation of the cells coming from more dorsal parts of the neuroepithelium (curved arrow). x 544.

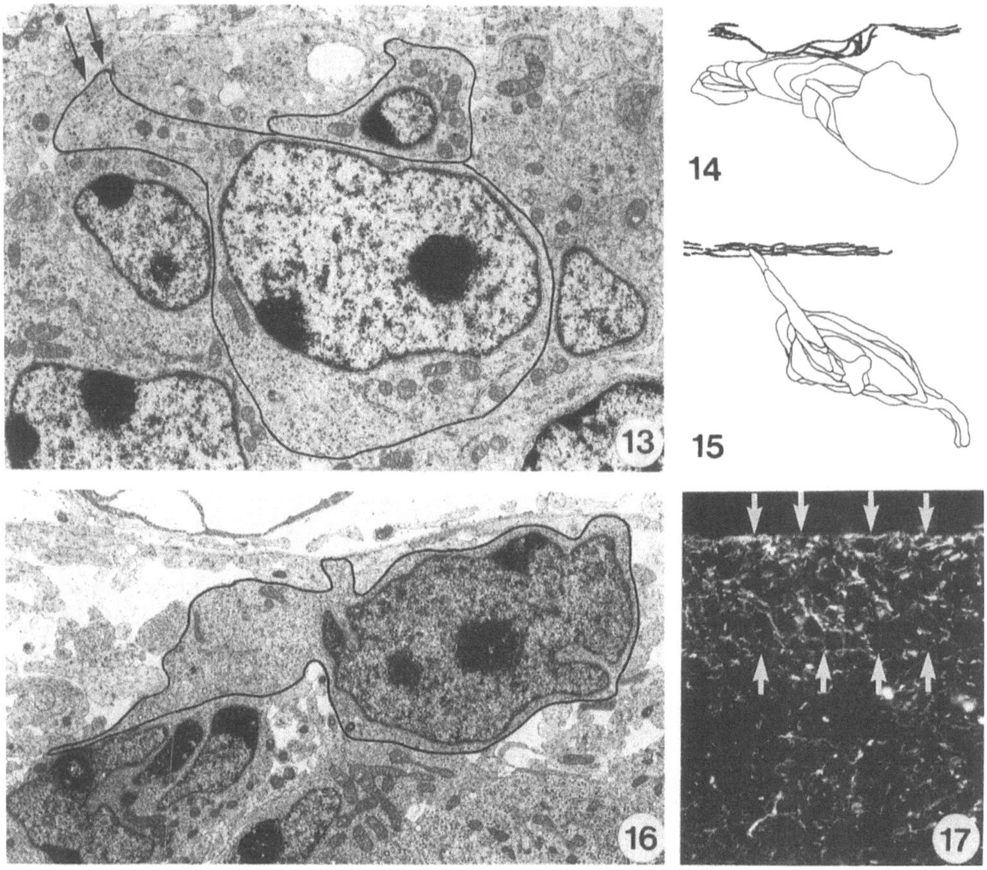

Figure 13: Tangential dentate precursor cell (outlined with black ink) from stage E14, contacting the overlying BL (arrows) via a short external process. The ultrastructure closely corresponds to that of early EGL cells (compare with Fig. 5). x 7200.

Figures 14 and 15: Reconstructions of two dentate precursor cells from E14 which contact the overlying BL either with their soma (Fig. 14) or an external leading process (Fig. 15). The cell in Fig. 15 also exhibits a retracting ventricular process.

Figure 16: Tangential dentate precursor cell (outlined with black ink) on P2 which is attached to the BL with a short process. The ultrastructure of this cell differs from that of earlier ontogenetic stages (Fig. 13) by possessing a dense cytoplasmic matrix and a preponderance of monoribosomes and lacking polyribosomes. x 6000.

Figure 17: Coronal section through the septal region of the hippocampus of a 5 day old hamster, reacted with antibodies against GFAP. Note the strong GFAP immunoreactivity in the subpial tangential migratory zone (between the arrows). x 350.

larly make contact with the BL. This relationship is confirmed by computer reconstructions of serial sections (Figs. 14, 15).

The tangential cells of later developmental stages (E16-P3) also contact the BL, but they show a different ultrastructure (Fig. 16), which is characterized by a higher electron density of their cytoplasmic matrix, a lack of polyribosomes and a preponderance of monoribosomes.

Nothing is known about the composition and determination of the migrating dentate stem cells; i.e., whether these cells are bipotential stem cells that are capable of producing both neurons and the glial cells of the dentate gyrus, or if they are two different stem cell populations which give rise to the neuronal and glial cell lines of the dentate gyrus. The differences in ultrstructure between the very early tangential cells and those in the later ontogenetic stages (see above) could indicate that they are different cell populations, the neuronal stem cells migrating first and the glial stem cells following later. This interpretation is supported by the finding that strong immunoreactivity for glial fibrillary acidic protein (GFAP) is found in the tangential cell stream of the later ontogenetic stages (Fig. 17). However, the evidence in support of this hypothesis is far from conclusive, and further investigations are needed to clarify this matter. Nevertheless, in the developing dentate gyrus too, neuroepithelial stem cells are in prolonged contact with the BL in the period of their translocation from the ventricular zone into the secondary proliferative zone, and it is possible that the BL influences both the migration and the proliferation of these cells.

In spite of the obvious differences between the EGL cells and the dentate precursor cells, both secondary stem cell populations share a constant crucial feature, their persistent attachment to the BL which, at least in the case of the EGL cells, is maintained until the end of their proliferative period. This relationship to the BL could indicate that extracellular matrix is not only involved in cell guidance but also in the regulation of mitotic activity of neuronal and perhaps glial stem cells.

Acknowledgement. We thank Rosemarie Sprang for excellent technical assistance.

References

Abercrombie, M., J.E.M. Heaysman, and S.M. Pegrum (1971) The locomotion of fibroblasts in culture. IV. Electron microscopy of the leading lamella. Exp. Cell Res. **67**: 359-367.

Altman, J., and S.A. Bayer (1978) Prenatal development of the cerebellar system in the rat. I. Cytogenesis and histogenesis of the deep nuclei and the cortex of the cerebellum. J. Comp. Neurol. **179**: 23-48.

Brunk, U., J.L.E. Ericsson, J. Pontén, and B. Westermark (1971) Specialization of cell surfaces in contact-inhibited human glia-like cells in vitro. Exp. Cell Res. **67**: 407-415.

Carbonetto, S. (1984) The extracellular matrix of the nervous system. TINS **7**: 382-387.

Chanda, R., D.J. Woodward, and S. Griffin (1973) Cerebellar development in the rat after

early postnatal damage by methylazoxymethanol: DNA, RNA and protein during recovery. J. Nerurochem. **21**: 547-555.
Collins, F., and M.R. Lee (1984) The spatial control of ganglionic neurite growth by the substrate-associated material from conditioned medium: An experimental model of haptotaxis. J. Neurosci. **4**: 2823-2829.
Cowan, W.M., B.B. Stanfield, and K. Kishi (1980) The development of the dentate gyrus. In *Current Topics in Developmental Biology*, vol. 15, A.A. Moscona and A. Monroy, eds., pp. 103-157. Academic Press, New York.
Cowan, W.M., B.B. Stanfield, and D.G. Amaral (1981) Further observations on the development of the dentate gyrus. In *Studies in Developmental Neurobiology*, W.M. Cowan, ed., pp. 395-435. Oxford University Press, Oxford, New York.
Easter, S.S., B. Bratton, and S.S. Scherer (1984) Growth-related order of the retinal fiber layer in goldfish. J. Neurosci. **4**: 2173-2190.
Eckenhoff, M.F., and P. Rakic (1984) Radial Organisation of the hippocampal dentate gyrus: A Golgi, untrastructural, and immunocytochemical analysis in the developing rhesus monkey. J. Comp. Neurol. **223**: 1-21.
Ezerman, E.B., and L.F. Kromer (1984) Cerebellar transplants: Models for development of foliar organisation and neuronal morphology. Abstract: Eric K. Fernström Symposium on Transplantation in the Mammalian CNS.
Fujita, H., and S. Fujita (1963) Electron mictroscopic studies on neuroblast differentiation in the central nervous system of domestic fowl. Z. Zellforsch. **60**: 463-478.
Goldman, R.D., J. Schloss, and J.M. Starger (1976) Organizational changes in actin-like microfilaments during animal cell movement. In *Cell Motility*, R.D. Goldman, T. Pollard, and J. Rosenbaum, eds., pp. 217-245. Cold Spring Harbor Laboratory, Book A, Cold Spring Harbor, New York.
Gona, A.G. (1976) Autoradiographic studies on cerebellar histogenesis in the bullfrog tadpole during metamprphosis: The external granular layer. J. Comp. Neurol. **165**: 77-88.
Gona, A.G. (1978) Ultrastructural studies on cerebellar histogenesis in the frog: The external granular layer and the molecular layer. Brain Res. **153**: 435-447.
Graziadei, P.P.C., and G.A. Monti Graziadei (1978) Continuous nerve cell renewal in the olfactory system. In *Handbook of Sensory Physiology*, vol. IX, M. Jacobson, ed., pp. 55-83. Springer, New York.
Greenburg, G., and D. Gospodarowicz (1982) Inactivation of a basement membrane component responsible for cell proliferation but not for cell attachment. Exp. Cell Res. **140**: 1-14.
Hausmann, B., and J. Sievers (1985) Cerebellar external granule cells are attached to the basal lamina from the onset of migration up to the end of their proliferative activity. J. Comp. Neurol. **241**: 50-62.
Herrick, C.J. (1891) Contribution of the comparative morphology of the central nervous system. J. Comp. Neurol. **1**: 5-37.
Hinds, J.W., and T.L. Ruffet (1971) Cell proliferation in the neural tube: An electron microscopic and Golgi analysis in the mouse cerebral vesicle. Z. Zellforsch. **115**: 226-264.
Hinds, J.W., and P.L. Hinds (1974) Early ganglion cell differentiation in the mouse retina. An electron microscopic analysis utilizing serial sections. Dev. Biol. **37**: 381-416
Hynes, R.O., R. Patel, and R.H. Miller (1986) Migration of neuroblasts along preexisting axonal tracts during prenatal cerebellar development. J. Neurosci. **6(3)**: 867-876.
Jacob, A. (1928) Das Kleinhirn. In *Möllendorffs Handbuch der mikroskopischen Anatomie des Menschen IV/I. Das Nervensystem*. Springer, Berlin.
Kappel, P.M. (1981) Development of the Cerebellum in Macaca mulatta. Thesis, University of Leiden, Leiden, Netherlands.
Kleinman, H.K., R.J. Klebe, and G.R. Martin (1981) Role of collagenous matrices in the adhesion and growth of cells. J. Cell Biol. **88**: 473-485.
Korneliussen, H.K. (1968) On the ontogenetic development of the cerebellum (nuclei, fissures, and cortex) of the rat, with special reference to regional variations in the

corticogenesis. J. Hirnforsch. **10**: 379-412.
Löfberg, J., and K. Ahlfors (1978) Extracellular matrix organization and early neural crest cell migration in the axolotl embryo. Zoon **6**: 87-101.
Meller, K., W. Breipohl, and P. Glees (1966) Early cytological differentiation in the cerebral hemisphere of mice. An electron microscopic study. Z. Zellforsch. **72**: 525-533.
Overton, J., and F.E. Mapp (1974) The fine structure of regeneration notochord in Anuran tadpoles. J. Exp. Zool. **187**; 103-120.
Rakic, P. (1971) Neuron-glia relationship during granule cell migration in developing cerebellar cortex: A Golgi and electron microscopic study in Macacus rhesus. J. Comp. Neurol. **141**: 283-312.
Rakic, P. (1972) Mode of cell migration to the superficial layers of fetal monkey neocortex. J. Comp. Neurol. **145**: 61-84.
Rakic, P. (1981) Neuronal-glial interaction during brain development. TINS **4**: 184-187.
Rakic, P. (1983) Early developmental events: Cell lineages, acquisition of neuronal positions, and areal and laminar development. Neurosci. Res. Prog. Bull. **20**: 439-451.
Rakic, P., and R.L. Sidman (1970) Histogenesis of cortical layers in human cerebellum, particularly in the lamina dissecans. J. Comp. Neurol. **139**: 473-500.
Rüdeberg, S.I. (1961) Morphogenetic studies on cerebellar nuclei and their homologization in different vertebrates including man. Thesis, Lund University, Lund, Sweden.
Sanes, J.R. (1983) Roles of extracellular matrix in neural development. Annu. Rev. Physiol. **45**: 581-600.
Sauer, F.C. (1935a) Mitosis in the neural tube. J. Comp. Neurol. **62**: 377-405.
Sauer, F.C. (1935b) The cellular structure of the neural tube. J. Comp. Neurol. **63**: 13-23.
Sauer, F.C. (1936) The interkinetic migration of embryonic epithelial nuclei. J. Morphol. **60**: 1-11.
Schaper, A. (1894) Die morphologische und histologische Entwicklung des Kleinhirns der Teleostier. Anat. Anz. **9**: 489-501.
Scholes, J. (1979) Nerve fibre topography in the retinal projection to the tectum. Nature **278**: 620-624.
Seymour, R.M., and M. Berry (1975) Scanning and transmission electron microscope studies of interkinetic nuclear migration in the cerebral vesicles of the rat. J. Comp. Neurol. **160**: 105-126.
Shoukimas, G.M., and J.W. Hinds (1978) The development of the cerebral cortex in embryonic mouse. An electron microscopic serial section analysis. J. Comp. Neurol. **179**: 795-830.
Sidman, R.L., and P. Rakic (1973) Neuronal migration with special reference to developing human brain. A review. Brain Res. **62**: 1-35.
Trinkaus, J.P. (1976) On the mechanism of Metazoan cell movements. In *The Cell Surface in Animal Embryogenesis and Development*, G. Poste and G.L. Nicholson, eds., pp. 222-329. North-Holland, Amsterdam.
Weiss, L., G. Poste, A. MacKearnin, and K. Willett (1975) Growth of mammalian cells on substrates coated with cellular microexudates. I. Effects on cell growth at low population densities. J. Cell Biol. **64**: 135-145.
Wells, J., and J.P. McAllister (1982) The development of cerebellar primordia transplanted to the neocortex of the rat. Dev. Brain Res. **4**: 167-179.
Wessells, N.K. (1967) Differentiation of epidermis and epidermal derivatives. N. Engl. J. Med. **227**: 21-33.

MIGRATION OF NEURONS IN VIVO

M. Rickmann
Zentrum Anatomie
Universität Göttingen
Göttingen
Federal Republic of Germany

Summary

Radial glial cells have been demonstrated by various methods in most developing mammalian brain regions, and their role in providing a suitable substrate for neuronal migration has become a generally accepted hypothesis. However, there remain several unresolved questions. The present morphological studies have dealt with two of these problems, the dynamics of the radial glial guidance system and the existence of other, non-radial glial cells which are connected to radial glial cells.

The rat dentate gyrus is characterized by two stages of cell production and migration, the first radial glial cells differentiate early enough to direct the stream of dentate precursor cells to set up the secondary proliferative zone. Here, new radial glial cells are produced together with the granule cells. This second set of radial glia is inserted into the lattice of its predecessor, but it is oriented radially with respect to the developing C-shape of the dentate gyrus. Migrating granule cells seem to migrate preferentially along the second, "dentate-intrinsic" set of radial glial cells.

During development of the neocortex of rat, attention was focused on non-radial glial cells. Since the time of their origin is difficult to determine using immunocytochemical markers or ^{3}H-thymidine autoradiography, the contact relationship was investigated of cells with the surfaces of the neuroepithelium. By using astroglial-like contact relationships (contact with basal lamina but not with the ventricle) as a criterion, it has been possible to detect immature glial cells which are among the very first cells leaving the neuroepithelial zone and which can be traced until they express GFAP. During development, there appears a variety of morphological types of immature glial cells, most of which appear to form specialized intercellular contacts with radial glial cells. The resulting complex framework is anchored at the ventricular surface and at the mesenchymal surfaces of the cortex. It appears possible that the non-radial component of this framework plays a role in rerouting, delaying or terminating neuronal migration and has a similarly important effect on the developing architecture of the neocortex as has been

NATO ASI Series, Vol. H5
Mesenchymal-Epithelial Interactions in Neural Development
Edited by J. R. Wolff et al.

demonstrated for the dynamics of the radial component in the case of the dentate gyrus.

Introduction

During the development of the central nervous system, neuronal migration is a rather ubiquitous phenomenon since the site of neuron production generally lies distant from the location where neurons differentiate. In most parts of the brain, the ventricular zone produces preneurons which have lost their mitotic capability and migrate to their sites of deployment.

The neocortex is a typical and well investigated example of such a system. When ^{3}H-thymidine is injected at various developmental stages, those neurons which have passed their last DNA-synthetic phase at the time of injection become labeled and can be traced during further development. Most cortical neurons migrate radially from the ventricular zone until they become positioned at the lower border of lamina I, which results in an inside-out layering of successive waves of neurons (Berry and Rogers, 1965; Rakic, 1974). However, there are two apparently systematic exceptions to this rule. First, the earliest neurons born migrate into the early marginal zone and are destined for both the deepest cortical layers and the most superficial lamina I (Rickmann et al., 1977). Second, in addition to the bulk of neurons which concentrates around the lamina just being formed there is a minor but constant insertion of neurons to all cortical layers present at a time (diffuse positioning; Chronwall and Wolff, 1981, see also J.R. Wolff, this volume).

In a few brain regions, however, the spatial relation between cell production and differentiation is more complex than in the neocortex. Quite distant from the primary neuroepithelium, a secondary matrix is built from proliferative precursor cells which have left the ventricular zone and are still capable of producing neurons for rather long periods of time (Jacobson, 1978). The dentate gyrus represents a typical example of such a brain region in which two migratory movements can be distinguished from each other, the migration of dentate precursor cells to the secondary proliferative zone and the subsequent migration of postmitotic neurons. The positioning of neurons in the dentate gyrus follows a complicated pattern. The neurons born before the secondary proliferative zone is set up predominantly lie in the hilus and apparently do not differentiate into granule cells (Lübbers et al., 1985), which constitute the principal type of neuron in the dentate gyrus and are apparently formed in the secondary proliferative zone (Bayer, 1980). The granule cells are deposited in the granule cell layer following three temporo-spatial gradients, from temporal to septal, suprapyramidal to infrapyramidal and outside to inside (Schlessinger et al., 1975; Rakic and Nowakowski, 1981).

The migration of neurons seems to be a most important link between cell production and the formation of brain architecture and has attracted much interest. At least morphologically, the mechanism of how migrating neurons are guided to their respective targets appears well understood. Although radial glial cells have been known since the end of the last century (Retzius, 1894), it was not before the 1970s that their role as guidance structures for migrating neurons was proposed (Rakic, 1971; Sidman and Rakic, 1973; Eckenhoff and Rakic, 1984). Since then, radial glial cells have been demonstrated in many developing mammalian brain regions and the neuronal guidance hypothesis has become generally accepted. Radial glial cells in the strict sense possess the original polarity of neuroepithelial cells in that they contact the ventricular surface, span the whole wall of the brain with their radial process and terminate at the pial basement membrane. However, there are also radial glial cells which have lost their ventricular contact and thus show the cellular polarity of astrocytes into which they may eventually transform. It has remained unclear whether the presence or absence of a contact with the ventricular surface exerts any influence on the capability of radial glial cells to guide neurons along their basal processes. Although radial glial cells are capable of proliferation (Levitt et al., 1983), many of them seem to undergo periods of proliferative arrest (Schmechel and Rakic, 1979). An outstanding feature of radial glial cells is that they apparently maintain their relative positions within the growing brain tissue. Thus, distortions in shape of radial glial processes appear to reflect the growth and shifts of the surrounding tissue (Rakic, 1971; Eckenhoff and Rakic, 1984).

Results and Discussion

While most studies on radial glial cells were restricted to their role in guiding migrating neurons in various brain regions, other important questions have remained largely unresolved. In a study on immature non-radial glial cells of the neocortex of rat (Rickmann and Wolff, 1985), it was tried to contribute to the question of how neuronal migration may be terminated. Another investigation on the developing dentate gyrus of rat focused on the influence which radial glial proliferation and the resulting dynamics of the guidance system exerts on neuronal migration and the construction of the granule cell layer (Rickmann, Amaral and Cowan, in preparation).

Non-radial glial cells of the neocortex

Radial glial cells have been demonstrated with antibodies directed against GFAP (glial fibrillary acidic protein; man, Antanitus et al., 1976; fetal monkey, Levitt and Rakic, 1980), in Golgi preparations (man, Choi and Lapham, 1978; monkey, Rakic, 1971, 1972) and by 3-dimensional reconstructions at the electron microscopic level (Rakic et al., 1974). In the neocortex of mice, radial glial cells become GFAP-positive by E18 (Woodhams et al., 1981) while in rat, they apparently do not express this antigen before postnatal stages (Bignami and Dahl, 1974), a time when neuronal migration has essentially finished. In contrast to radial glial cells, next to nothing is known about non-radial glial cells in the early development of neocortex. In a study on non-radial glial cells in the neocortex of rat (Rickmann and Wolff, 1985), it has been established that these cells originate prior to and exist during the time of neuronal migration. Here, we will summarize our evidence about their localization and discuss their possible involvement into the guidance system for migrating neurons.

It is difficult to study the birth dates of immature glial cells using ^{3}H-thymidine because these cells dilute their label due to continuous proliferation, although they have apparently been committed to the glial cell lineage (Korr et al., 1975). Also, the well established glial marker antigens and cytological markers seem to be expressed after glial cells have matured and thus do not identify their early stages of differentiation unequivocally (for references see Rickmann and Wolff, 1985). Therefore, the polarity of cells with respect to the surfaces of the neuroepithelium has been chosen as an identifying criterion. Neuroepithelial cells possess polarity in that they contact with their apical process the ventricular surface and with their basal process the mesenchymal surfaces. The same type of polarity is found in radial glial cells, the earliest type of glial cell identified so far. The principal cell types of the adult central nervous system, however, show a change in their contact relationship to the surfaces of the brain. Neurons and oligodendrocytes generally lose both their ventricular and mesenchymal contacts while astrocytes give up their contact to the ventricle but remain in contact with the surfaces of blood vessels or the pia mater. It is this typical polarity of astrocytes that we used to identify non-radial glial cells at early stages of differentiation.

In rat, histological differentiation of the neocortex is first observed at E13 when a few cells have migrated out of the ventricular layer and form a thin marginal layer beneath the pial surface. At E14, when a complete series of ultrathin, surface parallel sections was cut from the pial to the ventricular surface, we were able to detect the earliest changes in cellular polarity. As expected, our reconstructions showed that most neuroepithelial cells possessed both ventricular and pial contacts. Cells in the marginal zone were generally oriented horizontally, i.e., parallel to the pial surface, and had lost their ventricular contacts. Within the population of marginal cells, two cell types could be detected. One had lost its contacts to the pial surface and was, in that respect, similar to adult neurons, while the other cell type possessed clear endfeet at the pial surface, thus displaying the cellular polarity of marginal astrocytes.

Since there were virtually no other criteria to distinguish between these two cell types we tried to trace them until they had developed clear neuronal and glial characteristics. During further development, the cortical plate and the intermediate zone formed, and the marginal zone became split up into a superficial part, residing in lamina I, and a deep portion which contributes to subcortical layers. ^{3}H-thymidine injections and autoradiography after short survival times were used to label and follow up the marginal cells. After injections at E14, heavily labeled cells were found until E19 primarily in lamina I whereas, at subcortical levels, cells were progressively diluting their label, presumably due to proliferation. After E19, also the majority of cells in lamina I resumed proliferation and could no longer be regarded as fully labeled. Provided that differentiating neurons lose their proliferative capacity, this temporary mitotic arrest was a strong indication of a glial subpopulation in lamina I.

Ultrastructurally, the labeled subpial cells were acquiring more differentiated features. They accumulated large amounts of a rather typical rough endoplasmic reticulum, with wide cisternae enclosing a dense, granular matrix, and developed a thick main process containing the Golgi apparatus, numerous membrane bound structures in the form of either coated or uncoated vesicles and smooth endoplasmic reticulum. The main process was running tangentially towards the pial surface where it terminated in direct contact to the basement membrane. Endo- and/or exocytosis was observed on the plasma membrane of the process as it approached the pia. Close to their pial terminations, these processes were developing desmosomal connections with neighboring pial end-feet of neuroepithelial or radial glial cells. By E18, some of the presumptive astrocytic processes contained bundles of intermediate filaments which are also found in mature fibrous astrocytes. The progressive cytological specialization and the maintenance of direct contact with the basal lamina may indicate that the pial contacts of the presumptive glial subpopulation in lamina I are more than accidental.

In 3-dimensional reconstructions of serial electron micrographs of cells at E18, the presumptive astrocytes in lamina I could be distinguished from neurons. Cells with the typical ultrastructure, described above, were all characterized by basement membrane contacts of their primary process. Other processes of these cells were few in number and were restricted to the upper half of lamina I. In contrast, the neuronal population was showing different cytoplasmic characteristics, initial axon segments and various forms of synapses by E18. Although we reconstructed neuronal processes up to a total length of 130 µm, not a single pial contact could be detected. Other types of non-neuronal cells were residing predominantly in the lower portion of lamina I. They were devoid of any neuronal characteristics but instead showed cytoplasmic contours which were indented by other cells, and one class of pial endfoot belonged to cells of this type.

In order to complete the evidence on early glial differentiation, material of E18 was immunocytochemically reacted with antibodies against GFAP (glial fibrillary acidic protein; Eng and DeArmond, 1981) and against the C1-antigen (Sommer et al., 1981).

a very safe marker for cells of the ependymoastrocytic lineage, was found in a few subpial processes which, according to their orientation and location, were processes of the above described presumptive astrocytes. The low number of immunoreactive processes resembled those astrocytic processes which contained bundles of intermediate filaments. Far more presumptive astrocytes were expressing the C1-antigen. Although C1-immunoreactivity of a lower intensity was also seen in presumably undifferentiated and proliferative neuroepithelial cells, the strong immunofluorescence of presumptive astrocytes was characteristic, because these cells were mitotically inactive and differentiating. The lower portion of lamina I which does not contain presumptive astrocytes in the typical form described above also contained C1-positive processes indicating the existence of other types of glial precursors.

In conclusion, the first non-radial glial cells are apparently originated together with the first neurons. A special subpial type of presumptive astrocyte became temporarily arrested in its proliferative activity and could be shown to acquire astroglial characteristics. Among the various aspects used to identify these cells, the loss of the ventricular contact, maintenance of a mesenchymal contact relationship and the expression of GFAP and/or C1-antigen were the earliest and most reliable criteria.

After having established that non-radial glial cells originate that early and cover the whole period of neuronal migration, it was necessary to show that they not only appeared just beneath the pial surface but also in the deep regions through which neurons are migrating. The marginal zone containing the first non-radial glial cells becomes split up into a superficial part containing the presumptive astrocytes of lamina I and a subcortical portion in which proliferative cells reside. Therefore, the latter region was another place where to expect non-radial glial cells.

The predominant cell population found in the subcortical layers consisted of rather immature, dividing cells at various stages of their cell cycle. These we have named "polymorphous cells". Until E16, the cytoplasm of these cells contained little more than free ribosomes, polyribosomes and mitochondria.

However, they displayed already two rather "non-neuronal" features: during their interphase, they were adapting their shape to neighboring cellular elements, specifically, they could sheath other cells, presumably neuronal, with extended lamellar expansions. And they were forming intermediate junctions between themselves and with radial processes of neuroepithelial or radial glial cells. After E17, the polymorphous cells were acquiring additional features generally attributed to glial cells: a small proportion of them was undergoing postmitotic cell death, others were showing phagocytosis, and, most characteristically, some polymorphous cells were contacting blood vessels. These contacts appeared specific because they included rather large areas of contact or were shaped like endfeet. After E18, polymorphous cells could be demonstrated to contact blood vessels and/or radial processes of neuroepithelial or radial glial cells also in Golgi-preparations. At E18, GFAP-immunoreactivity appeared in the intermediate zone and strong C1-immunofluorescence in the intermediate and subventricular zones. In addition to the polymorphous cells which were developing glial features, cells resembling large glial precursors (Vaughn, 1969) were found by E18 and GFAP positive young astrocytes by E20.

The cortical plate, the precursor of laminae II-IV of the neocortex, begins to form after E16 in rat. By E18, the first signs of differentiation of non-radial glial cells was seen in the deeper, more differentiated part of the cortical plate which contained GFAP-positive and C1-positive non-radial processes. After E19, also the superficial cortical plate became populated first with immature glial cells and then also with GFAP-positive young astrocytes.

In summary, all cortical layers appeared to contain one or more glial cell types (Fig. 1). Lamina I and the subventricular and intermediate zones seem to inherit their glial population from the early marginal zone, whereas the cortical plate is progressively invaded by glial cells as it builds up. Apparently, a glial framework exists throughout the various layers and during the whole period of cortical differentiation (cf. His, 1889, 1890). This framework is anchored at the ventricular surface by radial glial cells and at the mesenchymal (meningeal and vascular) surfaces by radial and non-radial glial cells which are all interconnected with each other by specialized intercellular contacts.

When considering cell migration along radial glial cells, non-radial glial cells could be involved when neurons terminate their migration. Since the majority of cortical neurons are laid down in an inside-out sequence, the prefered location where neurons are positioned is the upper border of the cortical plate which is occupied by non-radial glial cells which first lie within the marginal zone and later in the lower portion of lamina I. Since non-radial glial cells occupy positions in the subcortical zone and the cortical plate as well, they could be involved also in the diffuse positioning mode by which part of the non-pyramidal neurons are added to already existent parts of the cortical plate (for more details see J.R. Wolff, this volume).

Development of radial glial cells in the dentate gyrus

Radial glial cells in the dentate gyrus of mice were demonstrated with GFAP-immunocytochemistry (Woodhams et al., 1981). These authors have noted their early differentiation in this brain region. In a study on the developing dentate gyrus of Rhesus monkeys, Eckenhoff and Rakic (1984) have emphasized the relation between radial glial cells and neuronal migration mainly at rather late developmental stages. However, the early changes in the radial glia population, the actual migratory pathways of neuronal precursor cells and the role of the secondary proliferative zone remained unclear. The author has tried to investigate whether radial glial cells could be responsible for the complex Migratory events of precurson cells and postmitotic granule cells and how they could

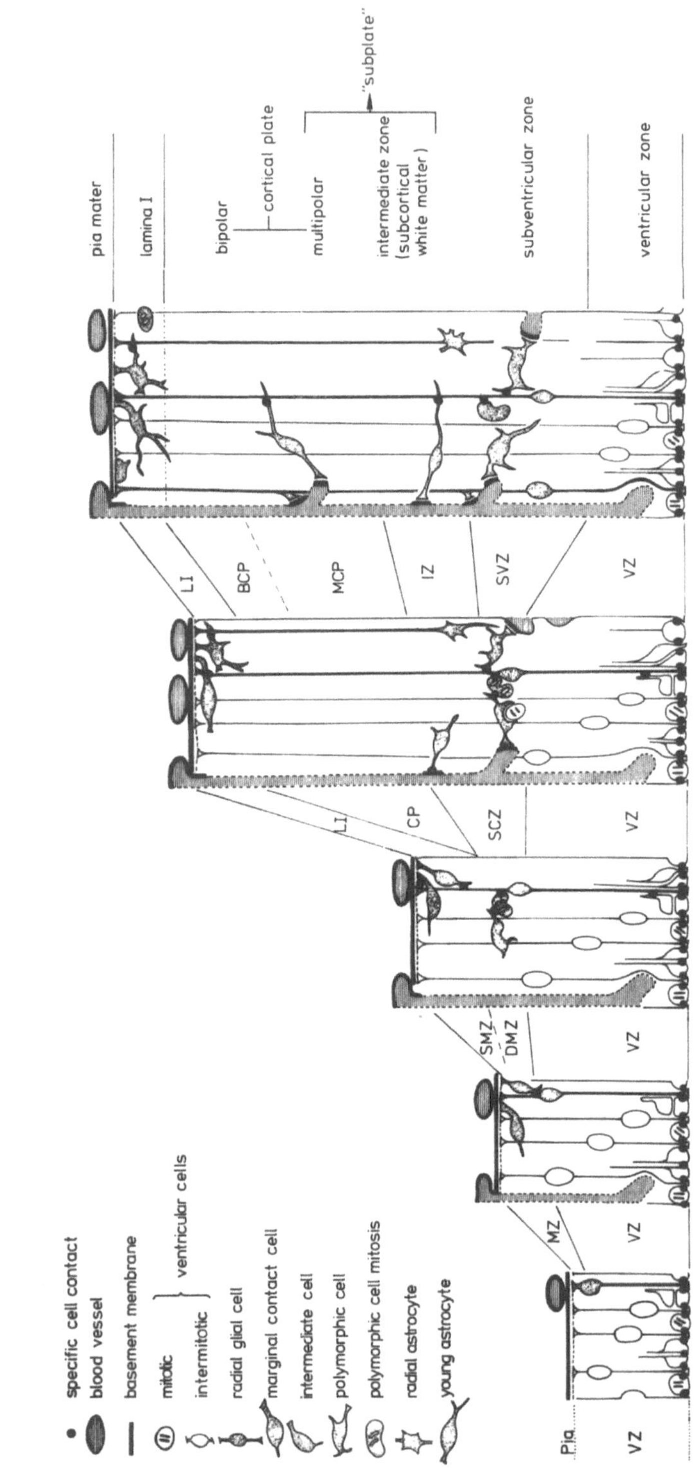
specific cell contact
blood vessel
basement membrane
mitotic
intermitotic
ventricular cells
radial glial cell
marginal contact cell
intermediate cell
polymorphic cell
polymorphic cell mitosis
radial astrocyte
young astrocyte
Pia
VZ
ED 11-12
MZ
VZ
ED 13-14
SMZ
DMZ
VZ
ED 15-16
LI
CP
SCZ
VZ
ED 17-18
LI
BCP
MCP
IZ
SVZ
VZ
ED 19-20
pia mater
lamina I
bipolar
cortical plate
multipolar
intermediate zone (subcortical white matter)
"subplate"
subventricular zone
ventricular zone

Figure 1: Summary of prenatal development of glial cell populations and their contact relations in the neopallium of rat. BCP: bipolar cortical plate. CP: cortical pate. DMZ: deep marginal zone. ED: embryonic day. IZ: intermediate zone. LI: lamina I. MCP: multipolar cortical plate. MZ: marginal zone. SCZ: subcortical zone. SMZ: superficial marginal zone. SVZ: subventricular zone. VZ: ventricular zone (from Rickmann and Wolff, 1985).

provide for the typical C-shape of the dentate gyrus (Rickmann, Amaral and Cowan, in preparation; see also Hausmann et al. and Sievers et al., this volume).

This study was mainly based on the immunocytochemical detection of two different antigens. Monoclonal R4-antibodies were kindly supplied by U.C. Dräger and rabbit anti human GFAP antibodies were a generous gift from L.F. Eng. In the adult mouse, the R4-antigen reacts with the ependymo-astroglial cell lineage (Dräger et al., 1984) and, in our developmental study, it has also proved to be sufficiently specific for cells of the ependymo-astroglial cell lineage including radial glial cells. The R4-antigen was especially useful because it appears very early during glial differentiation. GFAP, the constituent of the astrocytic type of intermediate filaments, however, is expressed only after the cells have reached a considerable degree of maturity.

The first sign of differentiation of the dentate gyrus was detected on E15 (when the postconceptional day is counted as E1) as R4-immunoreactivity appeared concentrated in the medial wall of the telencephalic vesicles (Fig. 2, E15). Staining was confined in radial processes terminating at the pial basal lamina. This early onset of glial differentiation was somewhat surprising because it occurred at a stage when histological differentiation of the dentate gyrus itself was hardly detectable. One day later at E16, transmission electron microscopy showed numerous glycogen granules in radially oriented cells, the distribution of which matched the region of strong R4-immunoreactivity. Since in mammalian brains, larger numbers of glycogen granules can be regarded as a safe glial marker, these findings confirmed the early differentiation of radial glial cells in that part of the neuroepithelium which gives rise to the dentate gyrus. During the subsequent stages, the anlage of the dentate gyrus became visible as the pial surface was bulged out by a densely packed cluster of cells which had apparently migrated out from the ventricular zone and accumulated beneath the pial basement membrane. From E17 onwards, also GFAP was expressed in radial processes and now both antigens demonstrated the precociousness of radial glial differentiation in the anlage of the dentate gyrus as compared with other parts of the pallium.

By E19, differentiating neurons had started to form the suprapyramidal tip of the granule cell layer. During this time the array of glial processes had become increasingly complex. Although we were using 50 µm thick sections, it was impossible to trace individual pro-

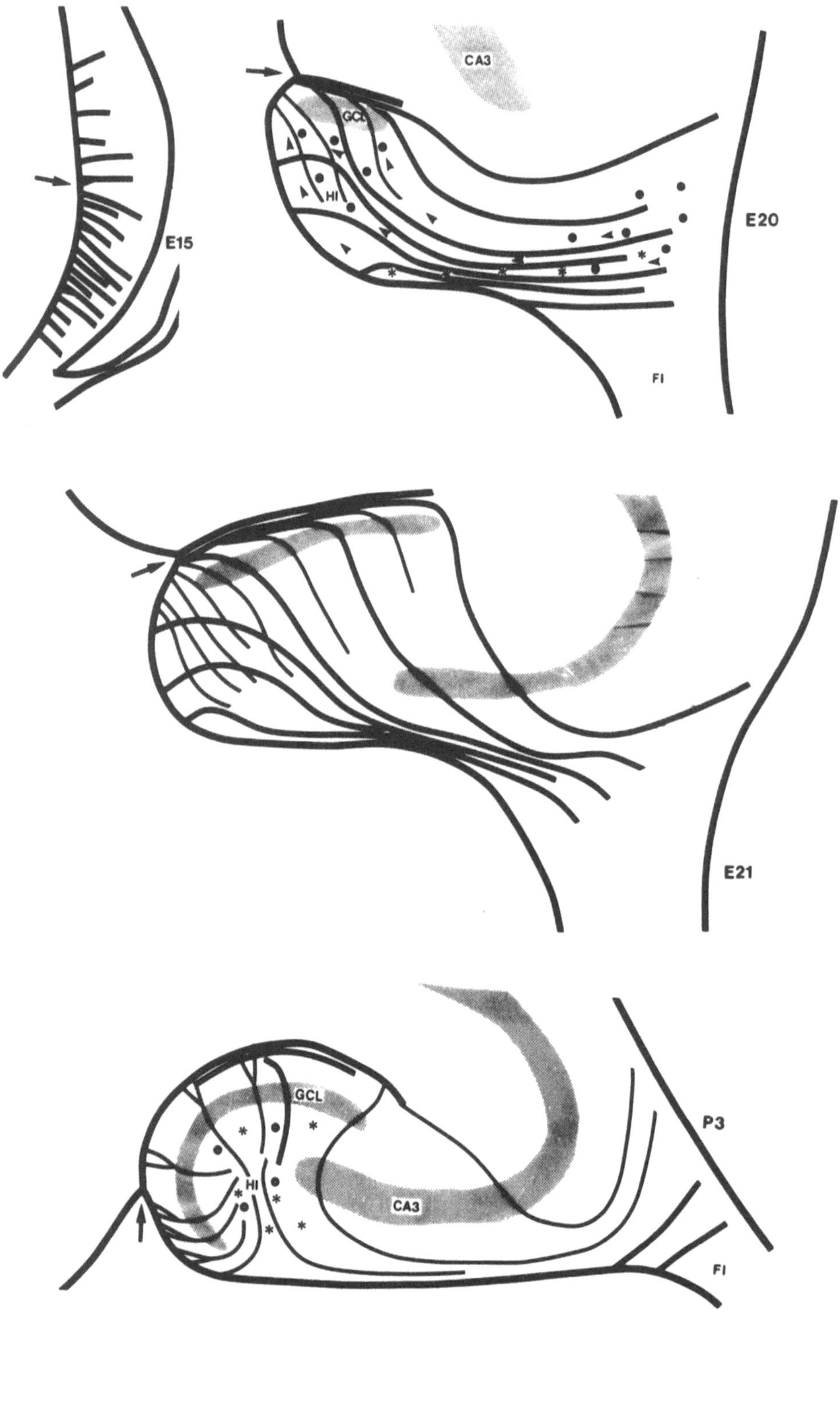
E15
CA3
GCL
HI
E20
FI
E21
GCL
HI
CA3
P3
FI

Figure 2: Diagram showing the development of radial glial cells in the dentate gyrus of rat. "Neuroepithelial" and "dentate-intrinsic" radial glial fibers are represented by thick and thin lines. At E20 (embryonic day 20) and P3 (postnatal day 3) the distribution of GFAP-positive (asterisks) and GFAP-negative (circles) mitotic figures (circles) is included. At E20, the possible directions of neuronal migration (arrow heads) has been judged from the orientation of all cell nuclei. Arrows indicate the point of invagination of the hippocampal fissure. CA3: field CA3 of the hippocampus proper. GCL: granule cell layer. HI: hilus. FI: fimbria.

cesses. Thus, the following description will be based mainly on the trajectories formed by whole arrays of radial glial processes. Given that radial glial cells maintained their pial contacts and that their processes became passively transformed in shape, the orientation and density of the radial glial processes were depicting the growth of the dentate gyrus (Fig. 2, E15-20): close to the ventricular surface of the fimbria, the glial processes were running radially with respect to the neuroepithelium. When approaching the fimbrio-dentate juncture, they were condensing into a bundle which ran increasingly parallel to the pial surface. In the dentate gyrus itself, the radial glial processes were fanning out and traversing both the entire presumptive hilus and the suprapyramidal granule cell layer, Mesenchmal endfeet of these processes were distributed all over the pial surface from the fimbrio-dentate juncture up to the point of invagination of the hippocampal fissure where they were especially concentrated. In the following text, we will refer to these radial glial cells as "neuroepithelial radial glia".

From E20 onwards, the analysis of radial glial cells was complicated in two respects: 1) GFAP-positive cells were proliferating. This was observed in cells which showed an immunoreactive cytoplasm but contained unstained areas resembling mitotic chromosomal condensations. We think that these mitotic cell bodies belonged to radial glial cells since the only GFAP-positive cells observed in prenatal stages were radial glial cells, and since some of the mitotic cells were bearing a process pointing towards the pia parallel with other radial glial processes. When recording the spatial distribution of mitotic figures (Fig. 2, E20, P3), at E20, GFAP-negative proliferative cells were already found in the secondary proliferative zone of the hilus, whereas GFAP-positive mitotic cells were lagging behind and concentrated around the fimbrio-dentate junction. During the following stages, the mitotic activity was shifted totally into the hilus of the dentate gyrus, and the secondary proliferative zone was composed by both GFAP-positive and GFAP-negative dividing cells. 2) Also from E20 onwards, increasing numbers of a second set of "dentate intrinsic" radial glial processes appeared in addition to the "neuroepithelial" set described above (Fig. 2, E20,P3). The "dentate-intrinsic" radial glial fibers could be distinguished from their precursors because they had a much smaller diameter and because their course suggested an origin from cell bodies lying in the hilus and not in the fimbria as was the case for the neuroepithelial radial glial processes. Our strongest argument for an additional set of fibers was that, in the hilus, two arrays of radial glial processes crossed each other. The intrinsic set was preferentially directed towards the growing granule cell layer. These fiber crossings can not be explained by a passive distortion of radial processes due to the growth of the tissue. Instead, we think that the proliferating glial cell were generating a second set of radial glial cells which no longer showed the neuroepithelial type of orientation but ran radially with respect to the developing C-shape of the dentate gyrus.

Once this concept of the existence of two sets of radial glial processes was adopted, their course and distribution became intelligible also later in development. The rather thick, "neuroepithelial" radial processes became increasingly diluted; up to P3 (postnatal day 3) they were still found in positions comparable to those at E20. Glial processes of the "dentate-intrinsic" set dramatically increased in number. They emerged from the hilus and ran radially towards and across the layer of granule cells. The highest density of "dentate intrinsic" radial glial processes was always found around the tip of the granule cell layer which grew to form the infrapyramidal blade (Fig. 2, P3). After P3, staining for GFAP also revealed increasing numbers of typical astrocytes. These had become the prevailing glial element by P7, when the growth of the dentate gyrus along its gradient from supra- to infrapyramidal blade is essentially complete. At this and later stages, the remaining radial glial cells seemed to have transformed into a special type of astrocyte which had perikarya lying just below or within the granule cell layer and radial processes extending into the overlying molecular layer.

Throughout the time covered by this study, the appearance and orientation of radial glial processes showed a good general correlation with the migration and depositioning of dentate precursor cells as well as postmitotic granule cells. Since neuronal migration was the main object of our research, the relation of migrating cells to radial glial processes was studied in more detail at E20 (Fig. 2), when migration of precursor cells toward the hilus and migration of granule cells toward the granule cell layer coexisted. The average orientation of all cell nuclei was used as a measure for possible migratory trajectories and was recorded together with the orientation of GFAP-positive radial glial processes in the same material. On the segment between fimbria and the hilus, which is mainly traversed by precursor cells, both orientations were found to be identical. However, it is unknown whether radial glial processes guide migrating precursor cells directly or whether they organize the general orientation of the tissue through which precursor cells translocate.

Migrating granule cells, however, appeared to be guided directly by radial glial cells. This was seen in the region of the hilus which possessed the two crossing arrays of radial processes and appeared to be mainly traversed by postmitotic neurons on their way from the secondary proliferative zone to the granule cell layer. Here, granule cells were confronted with two conflicting guidance cues, but their orientation showed that they clearly follow the "dentate intrinsic" radial glial fibers and neglect the "neuroepithelial" ones. This notion is also supported by the extremely dense array of radial glial fibers around the growing tip of the granule cell layer where newly formed neurons completed the C-shape of the dentate gyrus. At P3, we investigated this region electron microscopically in material stained for GFAP, and a close apposition between migrating granule cells and GFAP-positive radial precesses was found.

We hope that the data which we have compiled on the morphology and topography of developing glial cells will prove useful in further studies on neuronal migration. The non radial glial cells of the neocortex may provide a suitable in vivo system to test special molecules immunocytochemically for their possible involvement in the termination of neuronal migration. The two different sets of radial glial cells in the dentate gyrus may help determine which glial cell surface molecules actually attract migrating neurons.

References

Antanitus, D.S., B.H. Choi, and L.W. Lapham (1976) The demonstration of glial fibrillary acidic protein in the cerebrum of the human fetus by indirect immunofluorescence. Brain Res. **103**: 613-616.

Bayer, S. (1980) Development of the hippocampal region in the rat. II. Morphogenesis during embryonic and early postnatal life. J. Comp. Neurol. **190**: 115-134.

Berry, M., and A.W. Rogers, (1965) The migration of neuroblasts in the developing cerebral cortex. J. Anat. **99**: 691-709.

Bignami, A., and D. Dahl (1974) Astrocyte-specific protein and radial glia in the cerebral cortex of newborn rat. Nature **252**: 55-56.

Choi, B.H., and L.W. Lapham (1978) Radial glia in the human fetal cerebrum: A combined Golgi, immunofluorescent and electron microscopic study. Brain Res. **148**: 295-311.

Chronwall, B.M., and J.R. Wolff (1981) Non-pyramidal neurons in early developmental stages of the rat neocortex. Biblthca. anat. **19**: 147-151.

Dräger, U.C., D.L. Edwards, and C.J. Barnstable (1984) Antibodies against filamentous components in discrete cell types of the mouse retina. J. Neurosci. **4**: 2025-2042.

Eckenhoff, M.F., and P. Rakic (1984) Radial organization of the hippocampal dentate gyrus: A Golgi, ultrastructural and immunocytochemical analysis in the developing Rhesus monkey. J. Comp. Neurol. **233**: 1-21.

Eng, L.F., and S.J. DeArmond (1981) Glial fibrillary acidic (GFA) protein immunocytochemistry in development and neuropathology. Prog. Clin. Biol. Res. **59A**: 65-79.

His, W. (1889) Die Neuroblasten and deren Entstehung im embryonalen Mark. Abhandl. mat.-phys. Kl. königl. Sächs. Ges. Wiss. **26**: 313-372.

His. W. (1890) Histogenese und Zusammenhang der Nervenelemente. Arch. Anat. Physiol. Suppl. **95**: 95-117.

Jacobson, M. (1978) *Developmental Neurobiology*. Second Edition. Plenum Press, New York, London.

Korr, H., B. Schultze, and W. Maurer (1975) Autoradiographic investigations of glial proliferation in the brain of adult mice. II. Cycle time and mode of proliferation of neuroglia and endothelial cells. J. Comp. Neurol. **160**: 477-490.

Levitt, P., M.L. Cooper, and P. Rakic (1983) Early divergence and changing properties of neuronal and glial precursor cells in the primate cerebral ventricular zone. Dev. Biol. **96:** 472-484.

Levitt, P., and P. Rakic (1980) Immunoperoxidase localization of glial fibrillary acidic protein in radial glial cells and astrocytes of the developing rhesus monkey brain. J. Comp. Neurol. **193:** 815-840.

Lübbers, K., J.R. Wolff, and M. Frotscher (1985) Neurogenesis of GABAergic neurons in the rat dentate gyrus: A combined autoradiographic and immunocytochemical study. Neurosci. Lett. **62:** 317-322.

Rakic, P. (1971) Guidance of neurons migrating to the fetal monkey neocortex. Brain Res. **34:** 471-476.

Rakic, P. (1972) Mode of cell migration to the superficial layers of the fetal monkey cortex. J. Comp. Neurol. **145:** 61-84.

Rakic, P. (1974) Neurons in Rhesus monkey visual cortex: Systematic relation between time of origin and eventual disposition. Science **183:** 425-427.

Rakic, P., and R.S. Nowakowski (1981) The time of origin of neurons in the hippocampal region of the Rhesus monkey. J. Comp. Neurol. **196:** 99-128.

Rakic, P., L.J. Stensaas, E.P. Sayre, and R.L. Sidman (1974) Computer-aided three dimensional reconstruction and quantitative analysis of cells from serial electron microscopic montages of foetal monkey brain. Nature **250:** 31-34.

Retzius, G. (1894) *Die Neuroglia des Gehirns beim Menschen und bei Säugetieren*. Biologische Untersuchungen, neue Folge VI: 1-28.

Rickmann, M., B.M. Chronwall, and J.R. Wolff (1977) On the development of nonpyramidal neurons and axons outside the cortical plate: The early marginal zone as a pallial anlage. Anat. Embryol. **151:** 285-307.

Rickmann, M., and J.R. Wolff (1985) Prenatal gliogenesis in the neopallium of rat. Adv. Anat. Embryol. Cell Biol. **93.**

Schlessinger, A.R., W.M. Cowan, and D.I. Gottlieb (1975) An autoradiographic study of the time of origin and the pattern of granule cell migration in the dentate gyrus of the rat. J. Comp. Neurol. **159:** 149-176.

Schmechel, D.E., and P. Rakic (1979) Arrested proliferation of radial glial cells during midgestation in Rhesus monkey. Nature **227:** 303-305.

Sidman. R.L., and P. Rakic (1973) Neuronal migration with special reference to developing human brain: A review. Brain Res. **62:** 1-35.

Sommer, I., C. Lagenaur, and M. Schachner (1981) Recognition of Bergmann glial and ependymal cells in the mouse nervous system by monoclonal antibody. J. Cell Biol. **90:** 448-458.

Vaughn, J.E. (1969) An electron microscopic analysis of gliogenesis in rat optic nerves. Z. Zellforsch. **94:** 293-324.

Woodhams, P.L., E. Basco, F. Hajos, A. Csillag, and R. Balazs (1981) Radial glia in the developing mouse cerebral cortex and hippocampus. Anat. Embryol. **163:** 331-343.

Section 5

GROWTH OF NEURONAL PROCESSES

MESENCHYMAL-EPITHELIAL INTERACTIONS DURING NEURAL SEGMENTATION IN THE CHICK EMBRYO

Roger J. Keynes[1] and Claudio D. Stern[2]

[1] Department of Anatomy
Cambridge
U.K.

[2] Department of Human Anatomy
Oxford
U.K.

Abstract. When motor and sensory axons first grow out from the neural tube region in higher vertebrate embryos, they are confined to the rostral half of each successive mesodermal segment. The basis for this phenomenon, which also applies to migrating neural crest cells, lies in the sclerotomal cells derived from each somite. Differences must exist between rostral and caudal sclerotome halves, to which both growth cones and crest cells are responsive. The process of segmentation may also influence axon guidance during later stages of development, and we speculate on the possible mechanisms involved.

One of the most simple forms of neural organisation in vertebrates is the repeating pattern of the peripheral nerves as they emerge from the spinal cord. During the development of this arrangement, the cells concerned undergo a number of transitions between epithelium and mesenchyme (Table 1). For example, while the somite is epithelial, one of its derivatives, the sclerotome, is mesenchymal. The phenomenon of neural segmentation therefore provides an attractive experimental system for the study of early neural development and mesenchymal-epithelial interactions.

The Cellular Basis of Peripheral Neural Segmentaion

The earliest experimental study of neural segmentation was carried out by Lehmann (1927), who found that removal of several consecutive somites in urodeles leads to a loss of segmentation of sensory ganglia in the operated region. Detwiler (1934) then extended this approach by showing that grafting an additional somite produces an additional spinal nerve and ganglion. They both concluded that segmentation of the peripheral nerves is secondary to segmentation of the mesoderm; they also extrapolated to the conclusion

NATO ASI Series, Vol. H5
Mesenchymal-Epithelial Interactions in Neural Development
Edited by J. R. Wolff et al.

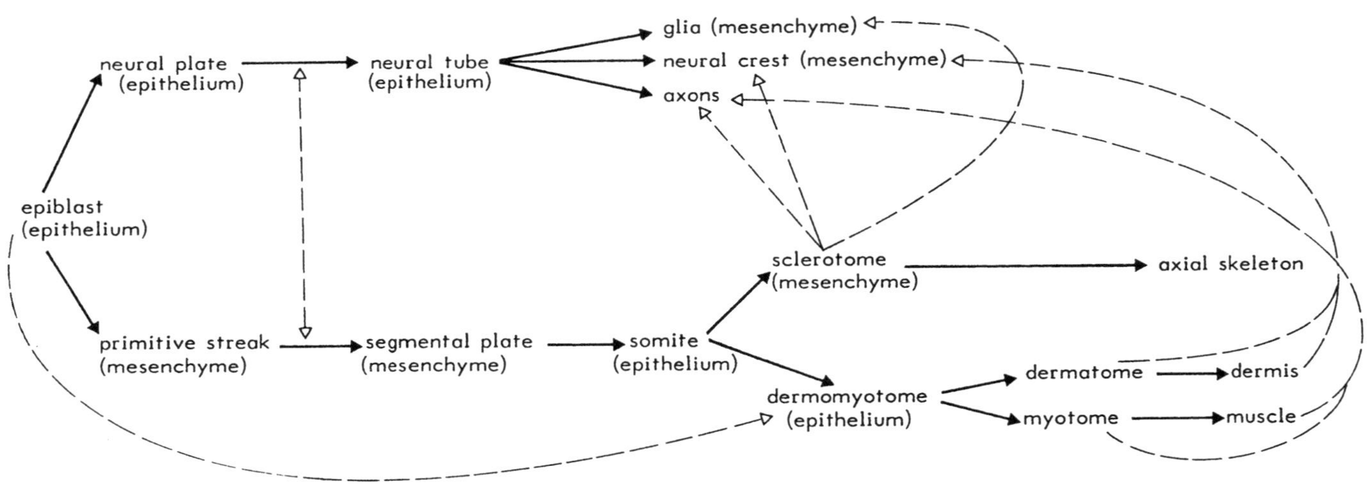

Table 1. Mesenchymal–epithelial interactions in neural segmentation. Interactions shown by interrupted hollow arrows.

that the spinal cord has no intrinsic segmentation, a theme to which we shall return later.

These pioneering studies were not particularly concerned with exactly where the spinal nerves develop in relation to the somites. Contemporary textbooks describe the nerves as growing either opposite the middle of each somite or between somites. We were therefore surprised to find, in zinc iodide-osmium tetroxide stained, whole-mounted chick embryos, that axons actually traverse the rostral half of each sclerotome (Keynes and Stern, 1984) (Fig. 1). To test whether this segmented outgrowth occurs because of intrinsic neural tube segmentation or because of some difference between rostral and caudal sclerotome halves, rotation experiments were carried out. First, a portion of neural tube opposite 2 or 3 somites was rotated 180° rostro-caudally prior to axon outgrowth, so that neural tube previously opposite rostral half-somite came to lie opposite the caudal half. After 2 days of further development axons had still grown out through the rostral halves of those somites opposite the rotated neural tube. Second, a portion of segmental plate mesoderm, 2-4 presumptive somites long, was rotated 180° rostro-caudally; this time, after further development, axons had traversed the caudal (original

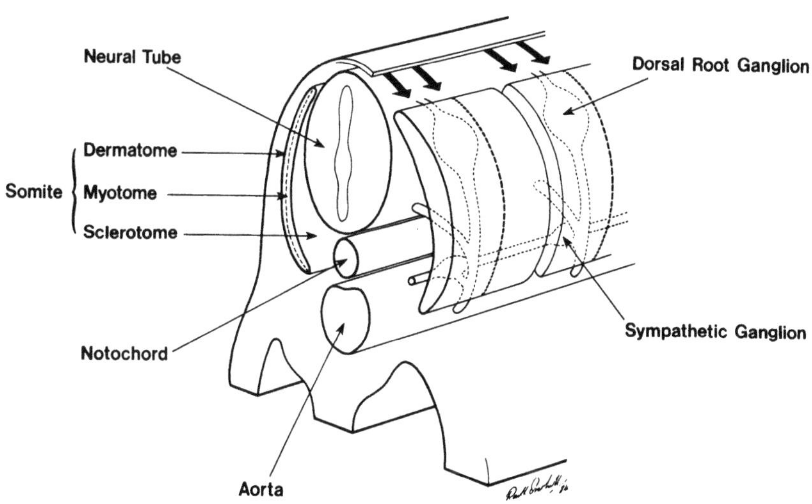

Figure 1: Diagram showing the major pathways of axon growth and trunk neural crest cell migration in the chick embryo. The somite has dispersed into its three components. All the components of the peripheral nervous system are confined to the rostral half of each sclerotome. Taken from Keynes and Stern, 1985.

rostral) halves of the grafted somites (Keynes and Stern, 1984).

Axons therefore grow through rostral half-sclerotome, regardless of its position relative to the neural tube or to the rostro-caudal axis of the whole embryo. These experiments confirm Lehmann's and Detwiler's conclusion that segmented axonal outgrowth is due to the somites. In addition, they show that in the chick segmentation is due to a difference between rostral and caudal sclerotome cells.

Does the rostro-caudal subdivision of each segment determine spinal nerve segmentation in all vertebrates? In fishes and amphibia, axons normally grow out at a stage when few or no sclerotome cells are present: motor axons have been described as being either between myotomes (Xenopus) or opposite the middle of the myotome. At later stages, though, both motor and sensory axons and sensory ganglia are found in the rostral half-sclerotome. In apodan amphibia, reptiles, birds and mammals, axons grow out at a stage by which the sclerotome is well-developed, and it is likely that in these vertebrates the rostro-caudal subdivision simultaneously determines axonal segmentation (Keynes and Stern, 1984).

The restriction of peripheral nerves to the rostral half-sclerotome was well known to embryologists of the late 19th century, after its initial description in 1855 by Remak, but in the first half of this century, it no longer received mention in embryological textbooks. Thus Tello (1923) described the phenomenon clearly, but suggested that motor axons first grow towards the myotome evenly along the length of each segment *"d'une manière homogène et diffuse"*, and then become confined to the rostral half *"peu-à-peu . . . à mésure que le mesenchyme vertébral se développe dans la moitié caudale de chaque segment"*. There are two problems with this description: 1) that the motor nerves, in fact, exit the neural tube in a punctuated manner along the longitudinal axis. First, axons exit opposite rostral half-sclerotomes. Later axons also exit opposite caudal half-sclerotomes but enter the rostral half by fasciculating on the earlier outgrowth. From the start, therefore, axon outgrowth is confined to the rostral half-segment (Keynes and Stern, 1984). 2) While the exact fate of the rostral half-sclerotome is still undecided, the extensive literature on the development of the vertebral column is nevertheless unanimous in ascribing a role for both sclerotome halves in vertebra formation.

Since sensory, as well as motor, axons develop in the rostral half-sclerotome, presumptive dorsal root ganglion cells, derived from the neural crest, should develop in this position. This is the case. Using the monoclonal antibodies NC-1 and HNK-1 to stain truncal crest cells, it is found that the ventral pathway of these cells is, like the motor axons, into the rostral half-sclerotome (Rickmann et al., 1985) (Figs. 1, 2). Crest cells precede motor axons in the rostral half-sclerotome, so one possibility would be that axons grow on crest cells. However, surgical removal of the neural crest does not alter the segmented outgrowth of motor axons (Rickmann et al., 1985). It would appear, then, that whatever differences exist between rostral and caudal half-sclerotome cells and/or their matrices, they can be detected independently by axons and neural crest cells.

The Molecular Basis of Peripheral Neural Segmentation

What might these differences be? One approach is to look for differences between rostral and caudal cells using antibodies to molecules, such as laminin and fibronectin, which are known to influence axon and crest cell growth in vitro. However, immunohistochemical studies with polyclonal antibodies have not revealed any differential distribution of laminin or fibronectin within the sclerotome: laminin is restricted to basal laminae, while fibronectin is additionally concentrated between adjacent segments (Rickmann et al., 1985). The epitope recognised by HNK-1 antibodies is found on a family of related molecules, among which are myelin-associated glycoprotein, L- and N-CAM (Kruse et al., 1984) and J1 (Kruse et al., 1985). Since there is no differential distribution of HNK-1 immunoreactivity in the sclerotome during the migration of neural crest cells, it also appears unlikely that any molecule on the sclerotomal cells bearing the HNK-1 epitope determines the path of either crest cells or motor axons through the rostral half-sclerotome (Rickmann et al., 1985).

In our present state of ignorance we can speculate that the differences are likely to residue in protein and/or glycoconjugate molecules on either the cell surface or in the extracellular matrix. Support for this comes from the observation that the caudal half-sclerotome cells are decorated with peanut lectin (PNA) receptor, while the rostral half-sclerotome cells are not (Stern et al., 1986). PNA has specificity for the disaccharide: galactose - N-acetyl D-galactosamine.

In the absence of somite-derived tissue, motor nerves are able to grow out of the neural tube but lose their segmental pattern (Lewis et al., 1981). This suggests that caudal half-sclerotome cells inhibit motor nerve outgrowth and neural crest cell migration (Stern et al., 1986). It is possible that the PNA receptor is itself an inhibitor of neural attachment to the surfaces of the caudal half-sclerotome cells. In support of this, we have found that: a) while cultured caudal cells in fact support neurite outgrowth, the distribution of the PNA receptor over the surfaces of these cells is patchy, unlike the pattern in vivo, where it covers the entire cell surface and b) in vitro, growth cones only make contact with those regions of the caudal sclerotome cell surfaces that are devoid of PNA receptor (Stern et al., 1986).

There is also evidence that sclerotome cells, whether from rostral or caudal halves, enhance neurite outgrowth in vitro (Stern et al., 1986). Although we have been unable to find evidence for neurite growth-stimulating factors confined to the rostral half-sclerotome, such factors may exist.

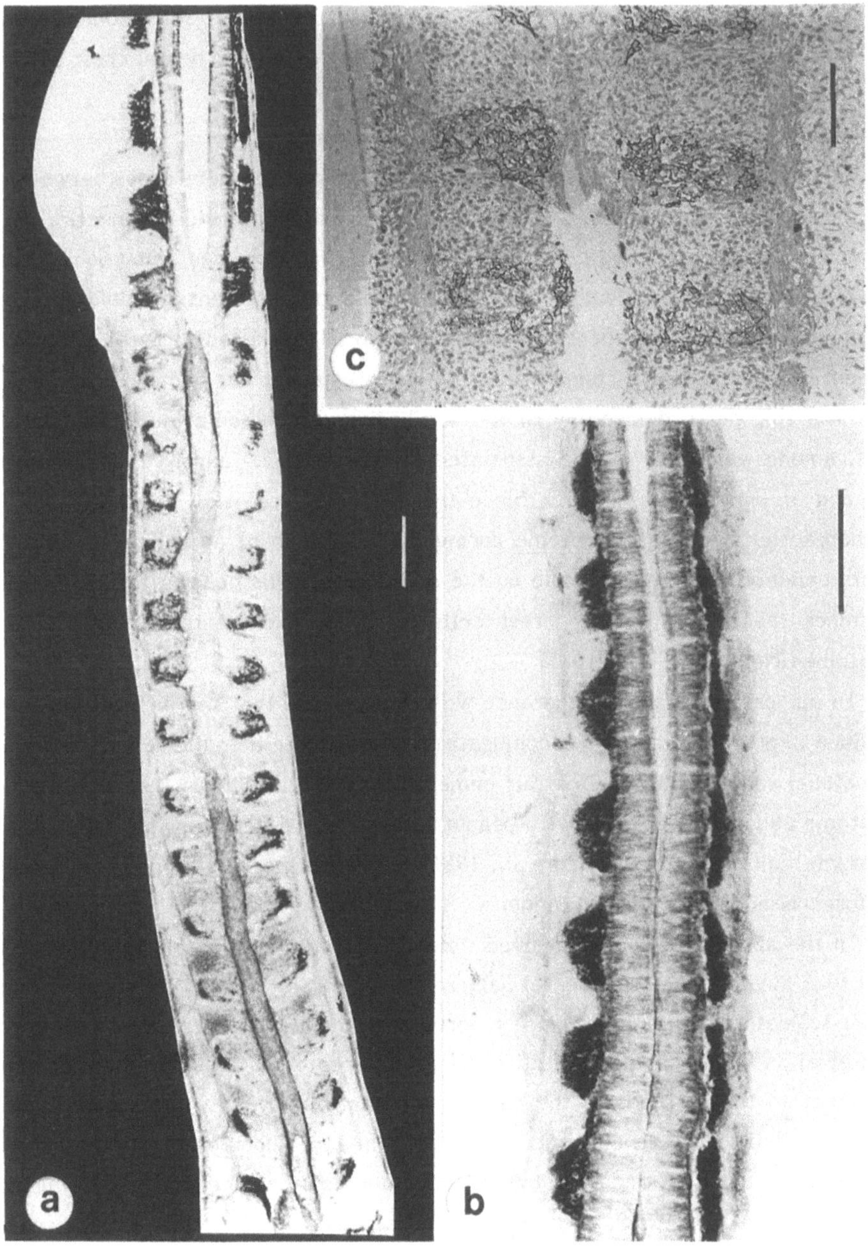

Figure 2: Horizontal sections through chick embryos stained by indirect immunoperoxidase with HNK-1. Rostral is to the left of each photograph. a: stage-17. The section is taken at the level of the notochord, which can be seen running along the midline. Bar = 200 µm. b: stage-17. This section is more dorsal. Bar = 200 µm. c: Horizontal semithin plastic section from a stage-18 embryo taken just ventral to the level of the notochord. The neural crest cells are clearly restricted to the rostral half of each sclerotome, intermingled with unstained cells. Bar = 100 µm. Taken from Rickmann et al., 1985.

Establishment of the Segmental Pattern

When do the two populations of cells diverge? We have approached this question by transplanting one or more halves of newly-formed epithelial somites in place of one whole somite, in both homotypic chick-chick and heterotypic chick-quail combinations (Stern and Keynes, in preparation). We find that each half-somite develops into a half-sclerotome, which retains its rostral or caudal characteristics. Thus, the sclerotome that develops from a transplanted rostral half allows neural crest cells and motor axons through and lacks the PNA receptor. Similarly, the caudal half inhibits neural growth and expresses the PNA receptor. These results show that by the time of segmentation, the somite is divided into two populations of cells that are committed to one of two fates, rostral or caudal, which will be expressed when the sclerotome forms, some 6-8 hours (4-6 somites) later.

Other considerations (Stern and Keynes, 1986) make it likely that pluripotent segmental plate cells are determined as rostral or caudal at the time of segmentation. Somite formation comprises a rapid epithelialisation of pre-somitic mesenchyme, accompanied by a radial contraction (Chernoff and Hilfer, 1982). During the formation of a somite, presumptive rostral cells lie adjacent to an inter-segmental border for longer than do the presumptive caudal cells of the same somite. In other words, the rostral half of a newly-forming somite can be identified at the same time as the caudal half of the somite that had just previously segmented. Thus, rostral and caudal cells in a newly-formed somite differ with respect to the length of time they spend adjacent to a border prior to their epithelialisation. Determination of rostral or caudal fates could be linked to this time difference (Stern and Keynes, 1986).

We have also analysed the interactions between a grafted half-somite and neighbouring host halves in multiple-rostral and -caudal chimaeric somites (Stern and Keynes, in preparation). We find that the interactions obey a precise rule: "cells from like halves (e.g., rostral/rostral) mix with each other, while those from unlike halves (e.g., rostral/caudal) do not, irrespective of their segmental levels of origin; when cells from unlike halves are adjacent to one another, a border is produced". This rule predicts the presence of a border, additional to the inter-segmental border, in the middle of each sclerotome, where unlike halves also meet. Such a border does exist, and was first described by von Ebner, in 1888 (Fig. 3) (see Keynes and Stern, 1984).

What is the function of such an interaction? At the time of segmentation, each somite is an epithelial sphere, surrounded by a basal lamina. The somite then develops into a mesenchymal sclerotome, and a dermomyotome which remains epithelial. In the absence of some interaction between cells at the inter-segmental border, the loosely packed mesenchymal cells of the sclerotome would tend to mix with each other, causing a loss of the segmental pattern: the immiscibility of rostral and caudal half-sclerotome

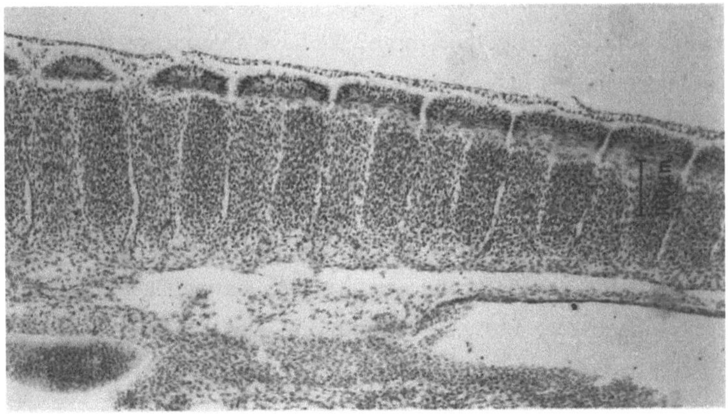

Figure 3: Sagittal section through a stage-19 chick embryo, stained with haematoxylin. The dermomyotomes span two half-sclerotomes. Von Ebner's fissure can be seen in the sclerotomes, opposite the middle of each dermomyotome. Rostral is to the left.

cells probably serves to maintain the segmental arrangement.

Are Segments Positionally Specified?

1. *Sclerotome.* So far we have discussed a property of the somite, namely the rostro-caudal subdivision of the sclerotome, which is repeated in every segment, and which is apparently identical from segment to segment. Presumptive sclerotome is, however, regionally specified with respect to its skeletal derivatives. For example, thoracic segmental plate mesoderm, implanted in place of cervical segmental plate, gives rise to ribs in the cervical region (Kieny et al., 1972). Is the sclerotome also regionally specified as far as nerve cells are concerned?

Since the late 19th century it has been known that, in higher vertebrate embryos, dorsal root ganglia are never fully formed in the region of the most rostral somites, immediately caudal to the otic vesicle. Those that do form are associated with ventral root fibres of the hypoglossal nerve, and are known as "Froriep's ganglia" (Froriep, 1882). In most species, such a ganglion develops only in relation to the segment immediately rostral to the first cervical nerve; the 3 to 5 segments still further rostral are usually devoid of dorsal root ganglion cells (Chiarugi, 1889; R.M. Hunter, 1935).

In the chick embryo, the ventral migration pathway is not available for crest cells at the level of the rostral five somites (LeDouarin, 1982; Thiery et al., 1982). Instead, cells migrate over the somite to colonise the developing gut. These authors have suggested that crest cells are unable to migrate ventrally at this level, because here the somitic mesenchyme dissociates earlier in relation to the emigration of the neural crest.

One could suggest an alternative possibility: that suitable positive cues for crest migration do not exist in these sclerotomes. The crucial observation is that, despite the reluctance of crest cells to enter these rostral somites, motor axons still do, traversing each rostral half-sclerotome and then fasciculating to form the hypoglossal nerve. This could imply the existence of separate cues for motor axons and neural crest cells within the sclerotome, which can be differentially expressed along the embryonic axis.

2. *Dermomyotome.* Like the sclerotome, the dermatome is regionally determined along the rostro-caudal axis prior to somite formation. Transposition of thoracolumbar segmental plate to the posterior cervical region results in a thoracolumbar dorsal plumage pattern in this new position (Mauger, 1972). The myotome, however, is not determined to form muscles of a particular pattern in this way (Chevallier et al., 1977), at least in limb regions.

We should also consider the possibility that cells of the dermomyotome are regionally specified with respect to neural labels. Motor axons innervating the limb can be guided by specific cues to their correct muscles; for example, after 180° rostro-caudal rotation of a length of neural tube opposite 3 to 5 leg somites, motor axons are still able to project to their correct muscles (Lance-Jones and Landmesser, 1981b). Using orthograde and retrograde horseradish peroxidase tracing, Lance-Jones and Landmesser (1981a) have shown that while motor axons destined for different limb muscles are mixed with each other within each ventral root, they normally sort out at the root of the limb, in the region of the developing nerve plexus. This is also the region where axons sort out following a variety of experimental manipulations (Lance-Jones and Landmesser, 1981a, b; Ferguson, 1983; Stirling and Summerbell, 1985). As a result of this process, axons from the motor pool of a given muscle, which project out in more than one ventral root, are collected together, and they remain together as they grow towards the developing muscle.

In considering possible mechanisms of specific guidance for motor axons it would be interesting to know what cell types axons encounter as they undergo this sorting process. Muscle cells themselves could form cellular trails for axons to follow into the limb muscle masses (Keynes and Stern, 1985). Such a muscle cell trail, also somite derived, precedes the outgrowth of axons of the hypoglossal nerve, in what could be an analogous developmental system R.P. Hunter, 1935). For muscle cells to provide specific cues requires that they be appropriately labelled. They might, for example, be specified on a rostro-caudal basis according to their somite of origin, and carry this label into the limb: motor axons of the same segmental level may then associate with these cells in preference to those of a different segmental origin. This idea, however, is ruled out by the observation that motor axons still project to their correct wing muscles after 180°

rostro-caudal rotation of wing somitic mesoderm 3-4 presumptive somites in length: wing muscles are populated by somite cells from an abnormal segmental level, yet the pattern of motor innervation is unperturbed (Keynes, Stern, Stirling and Summerbell, in preparation). The experiments of Wigston and Sanes (1982) nevertheless suggest that the segmentally derived intercostal muscles retain some label, perhaps segmentally determined, which biases the innervation they receive. It may be that such axial muscle labels are determined in the somite. Limb muscle cell labels, if they exist, could be acquired after muscle cells have left the somite, as they enter the limb. Alternatively, non-muscle cells in the limb could provide guidance cues for both muscle cells and axons (Lewis et al., 1981).

The Genetic Basis of Segmentation

If there are segmental labels in the vertebrate embryo, how are they specified? in Drosophila, three classes of gene affecting different aspects of segmentation have been identified: homoeotic (e.g., bithorax, bx), segment polarity (e.g., engrailed, en) and pair-rule genes (e.g., fushiterazu, ftz) (Nüsslein-Volhardt and Wieschaus, 1980). If genes homologous to those of Drosophila exist in the vertebrate embryo (McGinnis et al., 1984; Joyner et al., 1985), it is possible to speculate on their functions. Homeotic-like genes might be concerned with the differences between, say, thoracic and cervical vertebrae. Segment-polarity genes like engrailed could code for the molecular differences between the sclerotome halves, and be responsible for the initial pattern of neural segmentation. Genes expressed repeatedly every two or more segments might help to specify positional addresses using a minimum number of different loci.

Is the Neural Tube Intrinsically Segmented?

The results of embryonic manipulations suggest that motor axons are also labelled to allow them to recognise their guidance cues, but the basis for this is unknown. As far as the rostro-caudal axis is concerned, there are perhaps two major possibilities. First, that there is a graded, continuous labelling system extending down the neural tube, and second, that there is a discontinuous system, arising on the basis of intrinsic segmentation within the neural tube. Detwiler (1934) concluded that intrinsic segmentation of the spinal cord is non-existent, but neither his nor our experiments on peripheral nerve outgrowth in fact exclude intrinsic segmentation particularly with respect to the development of neurons within the neural tube.

Morphological segments in the neural tube were first noticed by von Baer in 1828, and came to be called neuromeres. They were taken by some authors as evidence that vertebrates evolved from a primitive segmented ancestor, and that the vertebrate head has a segmental origin. Certainly, such divisions of the rostral neural tube as forebrain,

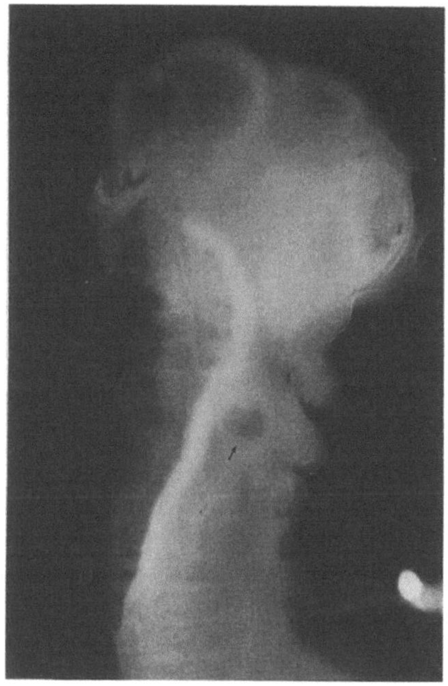

Figure 4: Neuromeres in an unfixed, stage-21 chick embryo. The hindbrain region of the neural tube was opened by a dorsal cut along the midline, and several segments (rhombomeres) are visible on each side, rostral to the otic vesicle (arrow). Photographed with reflected light, x100.

midbrain, hindbrain, and their further subdivision (see Fig. 4), may reflect genuine intrinsic segmentation at these rostral levels. What is less clear, and what has always been more contentious, is whether the presumptive spinal cord is also segmented.

Periodic swellings of the neural tube are visible in truncal regions of higher vertebrate embryos, each swelling being thickest between somites. Such swellings, out of register with the somites by a distance of half a somite, have been called myelomeres. Neal (1918), however, pointed out that they could represent merely "the passive result of the mechanical pressure of the adjacent somites", and there is still no firm evidence either way.

More impressive are descriptions of segmentally arranged neurons within the spinal cords of Amphioxus and several vetebrates (Whiting, 1948; Bone, 1960; Myers, 1985), which inevitably seem to imply the existence of some form of periodically expressed cellular specification mechanism for at least some neurons in the spinal cord. Such a system could, for example, provide a basis for the labelling of motor neurons to allow them to recognise specific guidance cues. It should also be pointed out that an apparently homogeneous column of neurons, extending down the neural tube, could still be determined on a segmental basis: lack of visible boundaries between neurons does not necessarily mean lack of segments.

Mesenchymal-epithelial Interactions That Generate the Segmental Pattern. How is the segmentation pattern generated in the first place? If we accept the evidence for intrinsic neural tube segmentation, it is interesting to speculate on the relationship between neural tube and somite segmentation. We have seen how the segmented mesoderm imposes segmentation on the peripheral nervous system. It has also been suggested that, at an earlier stage in development, the developing nervous system might itself impose segmentation on the mesoderm (Child, 1921; Fraser, 1960), although the evidence is not compelling (see Bellairs, 1963). Interestingly, brief exposure of chick embryos to cytotoxic drugs affecting the cell cycle causes anomalies of both the somites and the neural tube at discrete, identical segmental levels (Primmett, Stern and Keynes, in preparation). Rather than one element providing the dominating pattern, it is possible that the interactions between neural tube epithelium and somitic mesenchyme that lead to the segmental pattern are reciprocal. Finally, the differentiation of the somite into dermomyotome and sclerotome arises as a result of inductive interactions with the adjacent epiblast and endoderm (Gallera, 1966). Further study of such interactions, using the techniques of cell and molecular biology, should help elucidate the mechanisms of pattern formation in vertebrate embryos.

References

Bellairs, R.. (1963) The development of somites in the chick embryo. J. Embryol. exp. Morph. **11**: 697-714.

Bone, Q. (1960) The central nervous system in Amphioxus. J. Comp. Neurol. **115**: 27-64.

Chernoff, E.A.G., and S.R. Hilfer (1982) Calcium dependence and contraction in somite formation. Tiss. Cell. **14**: 435-450.

Chevallier, A., M. Kieny, and A. Mauger (1977) Limb-somite relationships: Origin of the limb musculature. J. Embryol. exp. Morph. **41**: 245-258.

Chiarugi, G. (1889) Lo sviluppo dei nervi vago accessorio, ipoglosso è primi cervicali nei sauropsidi è nei mammiferi. Atti. Soc. Toscana Sc. Nat. in Pisa, Memorie. **10**: 149-250.

Child, C.M. (1921) *The Origin and Development of the Nervous System*. Univ. Chicago Press, Illinois.

Detwiler, S.R. (1934) An experimental study of spinal nerve segmentation in Amblystoma with reference to the plurisegmental contribution to the brachial plexus. J. exp. Zool. **67**: 395-441.

Ferguson, B.A. (1983) Development of motor innervation of the chick following dorsoventral limb bud rotation. J. Neurosci. **3**: 1760-1770.

Fraser, R.C. (1960) Somite genesis in the chick. III. The role of induction. J. exp. Zool. **145**: 151-167.

Froriep, A. (1882) Über ein Ganglion des Hypoglossus und Wirbelanlagen in der Occipitalregion. Beitrag zur Entwicklungsgeschichte des Säugethierkopfes. Arch. f. Anat. u. Physiol. 279-302.

Gallera, J. (1966) Mise en évidence du role de l'ectoblaste dans la differenciation des somites chez les oiseaux. Rev. Suisse Zool. **73**: 492-503.

Hunter, R.M. (1935) The development of the anterior post-otic somites in the rabbit. J.

Morph. **57**: 501-531.
Hunter, R.P. (1935) The early development of the hypoglossal musculature in the chick. J. Morphol. **57**: 473-499.
Joyner, A.L., T. Kornberg, K.G. Coleman, D.R. Cox, and G.R. Martin (1985) Expression during embryogenesis of a mouse gene with sequence homology to the Drosophila engrailed gene. Cell **43**: 29-37.
Keynes, R.J., and C.D. Stern (1984) Segmentation in the vertebrate nervous system. Nature (Lond.) **310**: 786-789.
Keynes, R.J., and C.D. Stern (1985) Segmentation and neural development in vertebrates. Trends Neurosci. **8**: 220-223.
Kieny, M., A. Mauger, and P. Sengel (1972) Early regionalisation of the somitic mesoderm as studied by the development of the axial skeleton of the chick embryo. Develop. Biol. **28**: 142-161.
Kruse, J., R. Maihammer, H. Wernicke, A. Faissner, I. Sommer, C. Goridis, and M. Schachner (1984) Neural cell adhesion molecules and myelin-associated glycoprotein share a common carbohydrate moiety recognised by monoclonal antibodies L2 and HNK-1. Nature (Lond.) **311**: 153-155.
Kruse, J., G. Keilhauer, A. Faissner, R. Timpl, and M. Schachner (1985) The J1 glycoprotein - a novel nervous system cell adhesion molecule of the L2/HNK-1 family. Nature (Lond.) **316**: 146-148.
Lance-Jones, C., and L. Landmesser (1981a) Pathway selection by chick lumbosacral motoneurons during normal development. Proc. Roy. Soc. Lond. B. **214**: 1-18.
Lance-Jones, C., and L. Landmesser (1981b) Pathway selection by embryonic chick motoneurons in an experimentally altered environment. Proc. Roy. Soc. Lond. B. **214**: 19-52.
LeDouarin, N.M. (1982) *The Neural Crest*. Cambridge University Press, Cambridge.
Lehmann, F. (1927) Further studies on the morphogenetic role of the somites in the development of the nervous system of amphibians. J. exp. Zool. **49**: 93-131.
Lewis, J., A. Chevallier, M. Kieny, and L. Wolpert (1981) Muscle nerve branches do not develop in chick wings devoid of muscle. J. Embryol. exp. Morph. **64**: 211-232.
McGinnis, W., R.L. Garber, J. Wirz, A. Kuroiwa, and W.J. Gehring (1984) A homologous protein-coding sequence in Drosophila homeotic genes and its conservation in other metazoans. Cell **37**: 403-408.
Mauger, A. (1972) Rôle du mésoderm somitique dans le développement du plumage dorsal chez l'embryon de poulet. II. Régionalisation du mésoderme plumigène. J. Embryol. exp. Morph. **28**: 343-366.
Myers, P.Z. (1985) Spinal motoneurons of the larval zebrafish. J. Comp. Neurol. **236**: 555-561.
Neal, H.V. (1918) Neuromeres and metameres. J. Morph. **31**: 293-315.
Nüsslein-Volhard, C., and E. Wieschaus (1980) Mutants affecting segment number and polarity in Drosophila. Nature (Lond.) **287**: 795-801.
Remak, R. (1855) *Untersuchungen über die Entwicklung der Wirbelthiere*. Reimer, Berlin.
Rickmann, M., J.W. Fawcett, and R.J. Keynes (1985) The migration of neural crest cells and the growth of motor axons through the rostral half of the chick somite. J. Embryol. exp. Morph. **90**: 437-455.
Stern, C.D., and R.J. Keynes (1986) Cell lineage and the formation and maintenance of half-somites. In *Somite Development*, R. Bellairs, D. Ede, and J. Lash, eds. Plenum Press, New York.
Stern, C.D., S.M. Sisodiya, and R.J. Keynes (1986) Interactions between neurites and somite cells: Inhibition and stimulation of nerve growth in the chick embryo. J. Embryol. exp. Morph. **91**: 209-226.
Stirling, R.V., and D. Summerbell (1985) The behaviour of growing axons invading developing chick wing buds with dorsoventral or anteroposterior axis reversal. J. Embryol. exp. Morph. **85**: 251-269.
Tello, J.F. (1923) Les differenciations neuronales dans l'embryon du poulet pendant les premiers jours de l'incubation. Trav. Lab. de Rech. Biol. **21**: 1-93.

Thiery, J.P., J.L. Duband, and A. Delouvée (1982) Pathways and mechanisms of avian trunk neural crest cell migration and localization. Develop. Biol. **93:** 324-343.
von Baer, K.E. (1828) *Über die Entwicklungsgeschichte der Thiere*. Königsberg.
von Ebner, V. (1888) Urwirbel und Neugliederung der Wirbelsäule. Sitzungsber. Akad. Wiss. Wien (Physiol. Anat. Med.) **97:** 194-206.
Whiting, H.P. (1848) Nervous structure of the spinal cord of the young larval Brook-lamprey. Q. J. Microsc. Sci. **89:** 359-384.
Wigston, D.J., and J.R. Sanes (1982) Selective reinnervation of adult mammalian muscle by axons from different segmental levels. Nature Lond.) **299:** 464-467.

CHEMOTROPIC INFLUENCE OF SPECIFIC TARGET EPITHELIUM ON THE GROWTH OF EMBRYONIC SENSORY NEURITES

A.G.S. Lumsden[1] and A.M. Davies[2]

[1] Department of Anatomy
Guy's Hospital Medical School, and

[2] Department of Anatomy
St. George's Hospital Medical School

London, England

Introduction

Growing nerve fibres navigate over considerable distances along particular routes to reach their peripheral targets during development. Their outgrowth is directed by extrinsic cues which lead them to an appropriate region of the target with a high degree of accuracy (Scott, 1982; Landmesser, 1984; Jacobson and Huang, 1985; Eisen et al., 1986). It has been reasoned, because of the highly stereotyped patterns of nerve fibre outgrowth, that pioneering axons may be guided by cues confined to specific pathways through intervening tissue (Rogers et al., 1986; Letourneau, this volume). However, axons have the ability to reach their correct destinations by aberrant paths when displaced or misdirected at the outset (Lance-Jones and Landmesser, 1980a,b, 1981a,b; Ferguson, 1983; Landmesser, 1984); pioneering axons from ectopic transplanted eyes in Xenopus, for example, are able to home in on the tectum along quite novel routes (Harris, 1986). Such experimental evidence argues against preformed guidance pathways and indicates that axons are specified with respect to their innervation territory and are able to seek out their appropriate targets by responding to specific cues which are detectable over a wider area. The molecular nature of these cues, however, and the mechanisms by which growth cones respond to them, remain conjectural. One possibility is that growth cones can detect cell surface molecules which signal positional information (Wolpert, 1969; Bonhoeffer and Huf, 1982; Harris, 1986), another possibility is that growth cones can detect and respond directionally to gradients of specific diffusable agents which may emanate from their targets (Cajal, 1910, 1919).

NATO ASI Series, Vol. H5
Mesenchymal-Epithelial Interactions in Neural Development
Edited by J. R. Wolff et al.

Although there is no direct evidence for the existence of chemotropic molecules in embryos, it has been supposed that nerve growth factor (NGF), produced by target tissues, attracts developing sensory and sympathetic fibres (Levi-Montalcini, 1982). The growth of sympathetic axons towards an artificial source of NGF in vivo (Menesini-Chen et al., 1978) and of sensory neurites either up a concentration gradient of NGF (Letourneau, 1978) or towards a source of NGF (Gundersen and Barrett, 1979, 1980) in vitro, are phenomena which have lent support to this hypothesis.

The targets of NGF-responsive neurones in the embryo, however, are virtually ubiquitous; it is hard to conceive how multiple gradients could result in the highly organised and stereotyped patterns of sensory and sympathetic nerves. Furthermore, NGF has a well-established developmental role as a trophic agent mediating selective survival during the period of neuronal death (Thoenen and Barde, 1980; Yip and Johnson, 1984; Johnson and Yip, 1985). Selective maintenance implies restricted availability whereas chemotropism implies free diffusion and the establishment of gradients; it is dufficult to see how such apparently conflicting requirements could both be met by a single molecule. These considerations, together with the high concentrations of NGF required to demonstrate the effect (compared with the very low levels in embryonic targets, Harper et al., 1980; Korsching and Thoenen, 1983; A.M. Davies and S. Korsching, unpublished results), suggest that NGF is unlikely to have a chemotropic role in vivo.

A second line of evidence which has been held to support Cajal's theory comes from observations of neurite outgrowth from explanted neural tissue in co-culture with peripheral organs. For example, an increased growth of neurites was observed from explants of rat sympathetic ganglia towards vas deferens (Chamley et al., 1973; Chamley and Dowel, 1975), from frog spinal cord to limb bud mesenchyme (Pollack et al., 1981), from mouse submandibular ganglion to submandibular salivary gland (Coughlin, 1975), and from various sensory, sympathetic and parasympathetic chick embryo ganglia to a number of tissues (Ebendal and Jacobson, 1977; Ebendal et al., 1978; Ebendal, 1981). In Coughlin's study, parasympathetic nerve fibres grew through capsular mesenchyme towards the target glandular epithelium. In the other co-culture studies, however, nerve outgrowth was not confined to the target-facing side of the neural explant nor was the outgrowth on the target facing-side consistently directed toward the target tissue; rather, the presence of the target caused a local increase in the vigour or profusion of radially extending neurites. Such outgrowth patterns are suggestive of a concentration-dependent response to a scalar field of neurite-promoting activity, enhancing either the survival or growth of neurones which happen to be nearest its source. In chemotropism (or chemotaxis in translocating cells) the cellular response is to the vector field, which is described by both the magnitude and the direction of the concentration gradient (Dunn, 1981); proof of chemotropism requires the demonstration that the growth of neurites has been preferentially oriented towards the target by a chemical gradient.

One reason for the conflation of trophism with tropism is that both the above co-culture studies, and the in vitro studies of the chemotropic response of neurites to NGF (Letourneau, 1978; Gundersen and Barrett, 1979, 1980) were conducted on regenerating neurones which had already encountered their targets and had become dependent on HGF or target-derived trophic factor for survival. Similarly, the peripheral organs used in the co-culture studies had already become innervated in situ and therefore cannot be regarded as true developmental targets. At least some primary sensory neurones (e.g., trigeminal ganglion neurones) are not dependent on NGF for their survival and growth during the earlier period when their fibres are growing to the periphery (Davies and Lumsden, 1984). Experiments designed to reveal chemotropism should examine the effects of virgin target tissue on such neurones as are commencing or undergoing initial fibre outgrowth.

The Trigeminal System in the Mouse Embryo

The trigeminal system of the mouse presents several advantages for the study of target influences on nerve growth ab initio. First, both neuronal and target-cell populations are well-defined; the trigeminal ganglion consists of primary sensory neurones whose target periphery includes the rostral part of the maxillary process, the whisker field. Approximately 60% of neurones in the trigeminal ganglion innervate the whiskers field (Davies and Lumsden, 1984). Secondly, the maxillary nerve which projects to this target is large and contains only sensory fibres; all trigeminal motor fibres, whose cell bodies lie in the floor of the hindbrain, travel in the mandibular nerve. The analysis of growth is therefore not complicated by the presence of a contingent of motor fibres which may differ in their growth characteristics from sensory fibres (see Tosney and Landmesser, 1985). Thirdly, the specific target of maxillary nerve fibres within the whisker field is the prospective sensory epithelium; in the adult, fibres terminate as intra-epithelial free nerve endings and disc endings associated with Merkel cells in the epithelial external root sheathes of whiskers (Vincent, 1913; Patrizi and Munger, 1966). Lastly, the time course for development of the trigeminal-whisker field projection is known (Davies and Lumsden, 1984): the trigeminal ganglion appears at embryonic day 9 (E9, day of vaginal plug is E0) and by E10 both the ganglion and the whisker field become discernable. Nerve fibres first emerge from the ganglion early in E9, and by E10 about 1,200 fibres extend distally towards the maxillary process. Fibres first reach the whisker field epithelium during E11 and invade the differentiating whisker follicles during E12-14. At this time, fibre number in the maxillary nerve reaches its peak of 26,000 and thereafter declines to a stable complement of 13,000 by birth. The feasibility of explanting both the trigeminal ganglion and its major target, the whisker field, at the early embryonic stages before and during nerve fibre outgrowth to the periphery enabled the influence of the latter on the growth of these early neurites to be studied at the appropriate stages in vitro.

Chemotropic Effect of the Maxillary Process Target

In a series of experiments (Lumsden and Davies, 1983; Fig. 1) early trigeminal ganglia were co-cultured in three dimensional collagen gels, in both control medium and medium supplemented with isogeneric anti-NGF, with two cutaneous tissues of the same developmental age - the whisker field (comprising the superficial 100 µm of the rostral half of the maxillary process) and the forelimb bud (distal 200 µm). The maxillary process and limb bud are broadly homotypic in that both are composed of ectodermal epithelia covering undifferentiated mesenchyme and both develop structures which receive cutaneous innervation (the limb bud receives its innervation from DRG). The limb bud was thus considered to provide an alternative, inappropriate target for growing trigeminal neurites and also acted as a control for possible non-specific contact guidance effects mediated by alignment of fibrils in the collagen gel matrix which can result from mechanical stresses (Ebendal, 1976) and cell traction (Stopak and Harris, 1982).

Explant co-cultures were established at E10, E11 and E12, representing, respectively, the periods of initial outgrowth, mass outgrowth, and target encounter. In both E10 and E11 co-cultures in control medium, there was a clear prevalence of exclusively directed outgrowth from the trigeminal ganglia towards their natural target tissue, the maxillary process. In none of the cultures at these ages were neurites extended towards the control tissue, the forelimb bud. This pattern of exclusively target-directed neurite outgrowth persisted throughout the period of observation (up to 72 h). In E12 co-cultures, there was a lower incidence of exclusively target-directed outgrowth, a higher incidence of outgrowth in lateral quadrants and, for the first time, there was outgrowth towards the control tissue. There was no neurite outgrowth from ganglia cultured in isolation at any stage ($n = 18$). Anti-NGF did not significantly influence the pattern or magnitude of neurite outgrowth in E10 co-cultures. Fig. 2 shows representative E10 co-cultures after 48 h incubation in the presence of anti-NGF. In E11 co-cultures, anti-NGF resulted in both a small reduction in the overall number of ganglia exhibiting outgrowth and a small but significant reduction in the length of target-directed neurites (control medium $\bar{x}$ = 380 +/- 20 µm, anti-NGF $\bar{x}$ = 280 +/- 30 µm, $P < 0.01$ Student's t-test). In E12 co-cultures, however, anti-NGF effected a large reduction in the number of explants exhibiting outgrowth and greatly reduced the lengths of neurites in the remainder (control meduim $\bar{x}$ = 350 +/- 30 µm, anti-NGF $\bar{x}$ = 50 +/- 10 µm, $P < 0.001$). Fig. 3 shows a representative E12 co-culture after 48 h incubation in control medium.

These results demonstrate the existence of an agent, immunochemically unrelated to NGF, which is produced by the peripheral region of the maxillary process and which

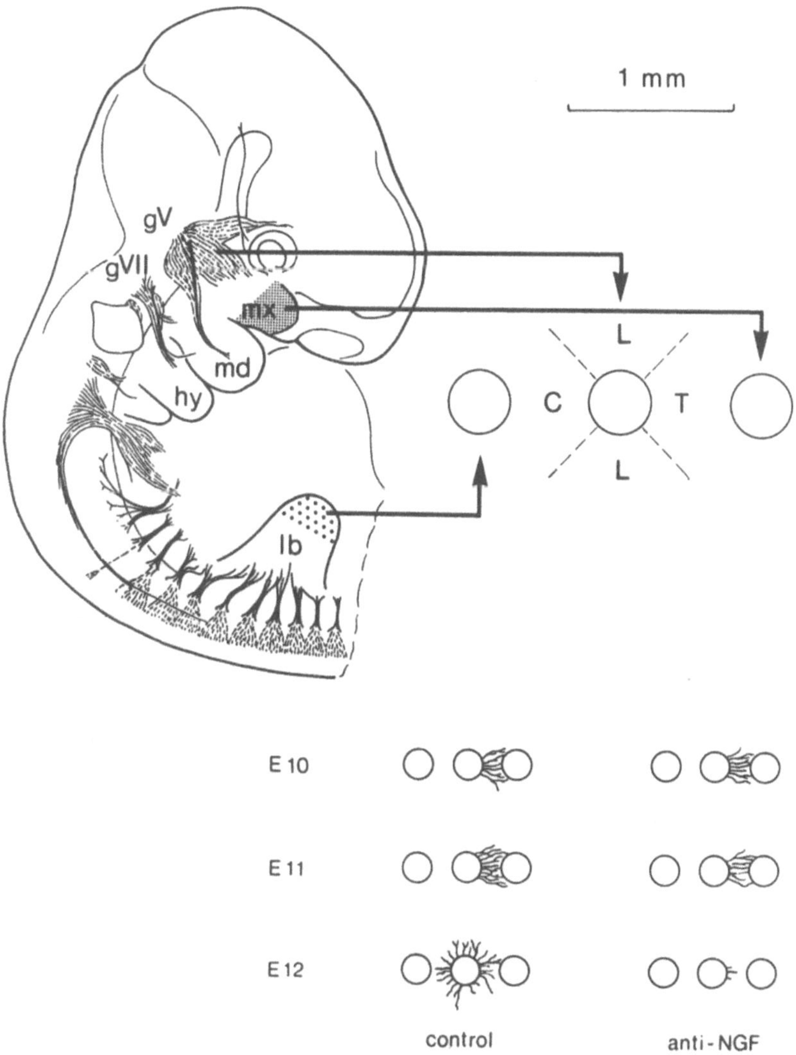

Figure 1: Scheme of procedure (to scale) in which the trigeminal ganglion (gV), maxillary process whisker field target tissue (mx) and limb bud control tissue (lb) were explanted from E10 (shown), E11 and E12 mouse embryos, and co-cultured in 3-D hydrated collagen gels. The three explants, each one oriented randomly, were aligned with the ganglion flanked by maxillary process and limb bud explants with an average spacing of 0.5 mm (+/- 0.15 mm). These were submerged in collagen gel medium (for method see Lumsden and Davies, 1983) with or without the inclusion of anti-mouse submandibular gland 2.5S NGF antiserum (an amount sufficient to block the activity of 2 BU/ml 2.5S NGF in the standard biological assay). Cultures were examined after 24, 48, and 72 h incubation at 37°C and the orientation, length and magnitude of neurite outgrowth in each quadrant of the ganglion measured (T, target quadrant; L, lateral quadrants; C, control quadrant). For quantitation methods see Lumsden and Davies, (1983), and Davies and Lumsden (1984). A cartoon summary of the results, showing patterns of neurite outgrowth characteristic of each of the six age/medium groups, is also shown. md, mandibular process; gVII, geniculate ganglion; hy, hyoid process.

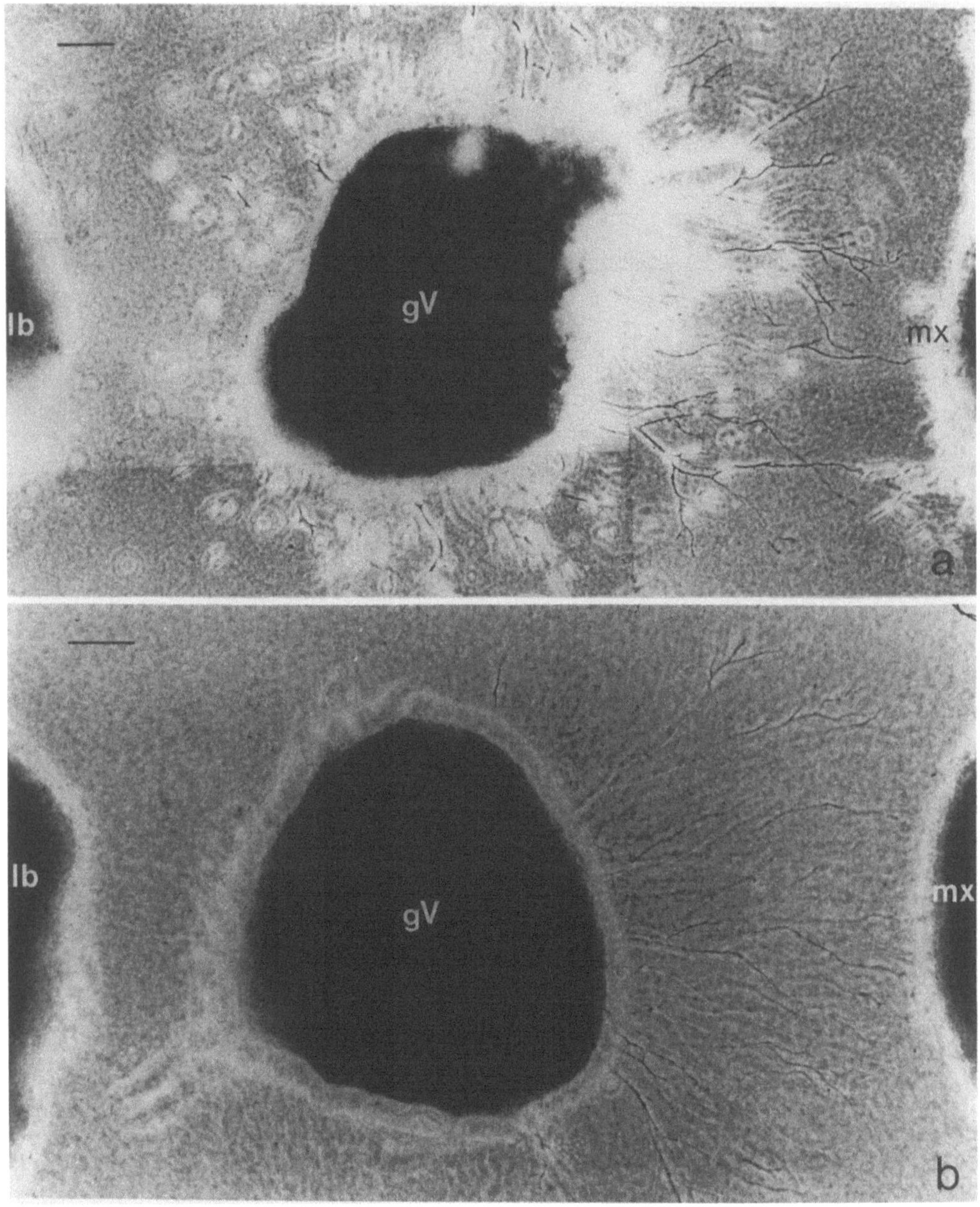

Figure 2: Phase contrast photomontage (a) and photomicrograph (b) of representative collagen gel co-cultures of E10 trigeminal ganglion (gV), maxillary process (mx) and limb bud (lb) after 48 h incubation in the presence of anti-NGF. Cytosine arabinoside (ARA-C, 10^{-5}M) was added to some cultures (e.g., b) to suppress non-neuronal cell proliferation. In both examples neurites have extended directly and exclusively to the maxillary process target but none has emerged int he quadrants of the ganglion surface which face laterally or towards the limb bud control tissue. Scale bars = 100 µm.

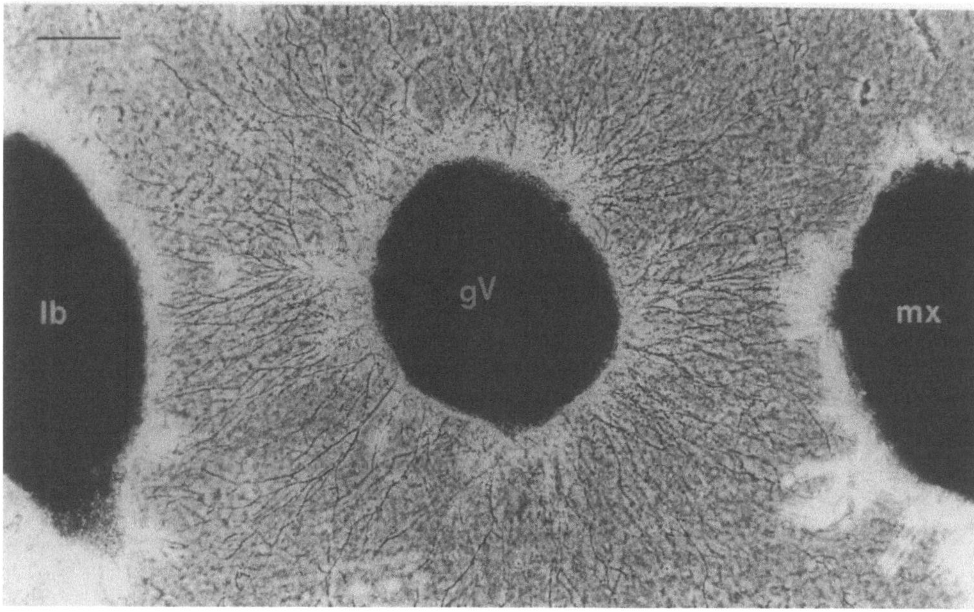

Figure 3: Phase contrast photomicrograph of representative collagen gel co-culture of E12 trigeminal ganglion (gV), maxillary process (mx) and limb bud (lb) after 48 h incubation in control medium (including ARA-C, 10^{-5}M). Neurites have extended radially from all quadrants of the ganglion. In co-cultures of the same age tissues incubated with anti-NGF, there was little or no outgrowth. Scale bar = 200 µm.

both elicits and directs the growth of very early trigeminal neurites. The production, release or response to this target-derived agent (the signal) declines after E11, when nerve fibres reach the periphery during normal development.

The pattern of exclusively target-directed outgrowth in E10 and E11 co-cultures, together with the lack of outgrowth in the absence of this target, suggested that the signal has both neurite-promoting and orienting (tropic) effects. Thus, neurite tips displayed a consistently preferred orientation towards the target (Fig. 4) and the trajectories of neurites, being a record of the progressive changes in orientation of neurite tips as they course through the fibrillar matrix, also displayed an overall vector towards the target. The observation that the pattern of exclusively target-directed outgrowth in E10 and E11 co-cultures persisted throughout the period of study (24-72 h) is also indicative of a tropic influence since it would be expected that an agent with solely trophic influence would elicit (radial) neurite outgrowth from the lateral and control quadrants as it diffused away from the source. Further support for a chemotropic interpretation of this phenomenon was given by a series of experiments in which E10 and E11 whisker field

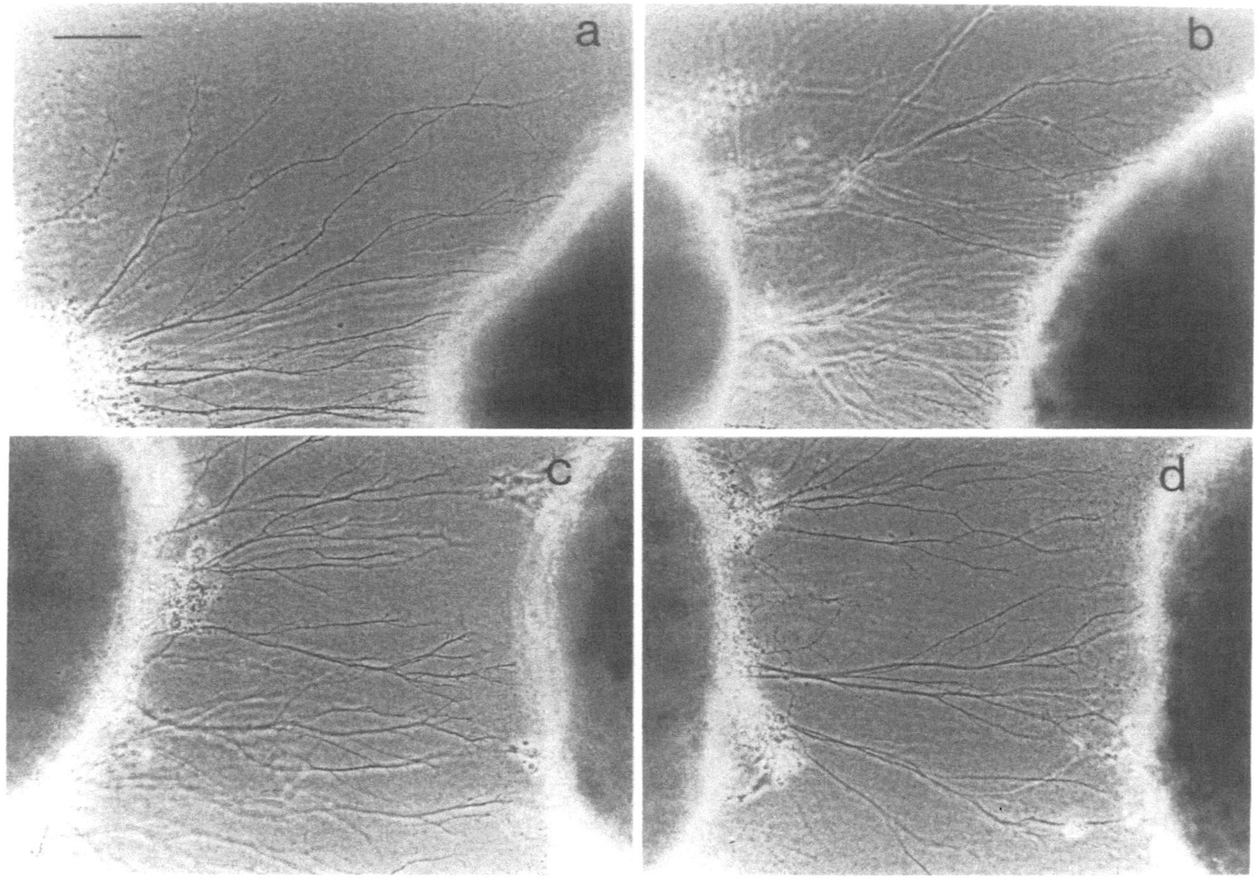

Figure 4: Phase contrast photomicrograph of E10 (a, c) and E11 (b, d) co-cultures after 48 h incubation in the presence of anti-NGF, showing the pattern of maxillary process target-directed neurite outgrowth in the target-facing quadrant. Scale bar = 100 μm.

targets were co-cultured with two ganglia in tandem (Lumsden and Davies, 1983). In these co-cultures, the magnitude of neurite outgrowth from the target quadrant of the distal ganglion was significantly greater than that from the away facing quadrant of the proximal ganglion. This result demonstrated that the signal had effectively diffused beyond the proximal ganglion and that neurite outgrowth was directed up a concentration gradient towards the target.

The effect of anti-NGF on E11 and E12 co-cultures revealed a decline in production of the signal and the development of NGF promoted outgrowth. This was taken to indicate the inception of NGF production by normal target and control tissues rather than the maturation of neuronal responsiveness to NGF since both E10 and E11 trigeminal ganglion neurones are known to respond to NGF in vitro with neurite outgrowth (Davies et al., 1981). Recently, this indication has been confirmed by 2-site enzyme immunoassay of NGF in the developing mouse maxillary process; NGF is present at E12 and first becomes detectable during E11 (A.M. Davies and S. Korsching, unpublished results). That the commencement of NGF production by peripheral tissues coincides with their innervation and just precedes the onset of neuronal death (Davies and Lumsden, 1984) adds to the considerable body of evidence that the biological role of NGF is to mediate selective survival of sensory (and sympathetic) neurones during the period of natural neuronal death (Hamburger et al., 1981; Johnson et al., 1980; Hendry and Campbell, 1976).

The above evidence for the production of a target signal with both neurite-promoting and tropic effects which directs neurites at the stage during which they normally grow to their targets, together with the demonstration that NGF is not produced until fibres reach their targets, provides strong support for the contention that NGF is not involved in nerve guidance.

Neurite Promoting Activities of Medium Conditioned by Maxillary Process Explants

Media conditioned by a variety of different cell types contain factors permissive for neurite extension, which act via association with the substratum (Adler et al., 1981; Collins and Dawson, 1982; Lander et al., 1982) and which have been proposed to play a role in neurite guidance (Collins and Garrett, 1980; Adler and Varon, 1981). Recent studies by Lander et al. (1985a, b) and Davis et al. (1985) have shown that the activity of these conditioned medium (CM) factors is due to laminin, a major glycoprotein of basal laminae which, in purified form, permits outgrowth from cultured neurones to occur on a variety of substrates (Baron van Evercooren et al., 1982;Manthorpe et al., 1983). For reasons which are not yet clear, antisera directed against purified laminin that completely block neuronal responses to laminin-treated substrata, fail to affect the activities of most CM-treated substrata (Sanes, 1985).

The possibility exists that the neurite-promoting effect of maxillary process on co-cultured trigeminal ganglia could also be due to laminin. Since the simple addition of anti-tumour laminin to co-cultures cannot be expected to block endogenous laminin, this possibility was approached via media conditioned by either maxillary process or limb bud explants (Fig. 5); the laminin contents of these CM were measured by enzyme-linked immunosorbent assay (ELISA) and their neurite-promoting activity assayed, using trigeminal ganglion explants, both before and after affinity removal of laminin by solid phase-bound anti-laminin. The resluts indicated that the neurite-promoting effect of

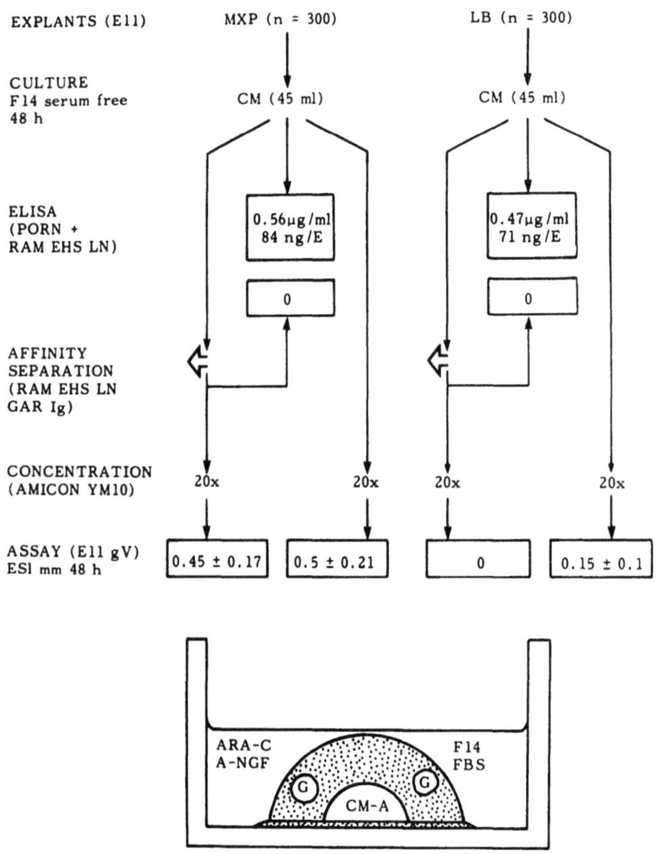

Figure 5: Preparation and assay of serum free medium conditioned (CM) by E11 maxillary process (mxp) and limb bud (lb) explants. Laminin contents of the CM were measured by ELISA using rabbit anti-EHS laminin on samples applied to polyornithine coated plastic wells. Laminin was removed from CM by incubation with anti-laminin coated beads. Concentrated CM were assayed for neurite-promoting activity using E11 trigeminal ganglion explants in collagen gels. The mean length of neurites (Esl) after 48 h culture are given.

maxillary process explants is not due to laminin: both maxillary process and limb bud explants produced laminin during the 48 h culture period, and in equivalent amounts; neurite outgrowth occurred from ganglia cultured with medium conditioned by maxillary process explants but not by limb bud explants; treatment of the CM over solid-phase anti-laminin completely removed the laminin from CM (as measured by ELISA) but did not significantly deplete the neurite-promoting activity of maxillary process CM.

Specificity

To understand the role of the maxillary process chemotropic signal and the mechanism by which it acts, it is necessary to establish, first, whether there is specificity in the regional extent of its production and in the neuronal populations which respond, and secondly, the identity of the cells which synthesise it.

The foregoing experiments have indicated that the signal has some degree of regional specificity as it is not produced by the limb bud which, though also a presumptive cutaneous field, derives sensory innervation from a different source. Regional specificity of the signal was further investigated by co-culturing the trigeminal ganglion with part of its cutaneous target field (either the maxillary or mandibular process of the first branchial arch) and the adjoining cutaneous target field of the geniculate ganglion (the hyoid process of the second branchial arch), and by co-culturing the geniculate ganglion with both its own and trigeminal targets (see Fig. 1 for location of structures). In all trigeminal ganglion co-cultures in which neurite outgrowth was observed (54% at E10, n = 85; 63% at E11, n = 65), the neurites were confined to the quadrant facing the first arch explant; in no co-cultures were neurites seen in the quadrant facing the second arch explant (Fig. 6a). Trigeminal ganglia cultured alone in a collagen gel (as before) did not extend neurites. Geniculate ganglia, however, extended neurites when cultured in isolation at both E10 and E11, and this outgrowth was neither augmented by nor oriented towards co-cultured target tissue of either arch (Fig. 6b). These findings (Lumsden and Davies, 1986) show that production of the chemotropic signal is precisely limited to the trigeminal field; it is not produced by the adjoining peripheral field, and it effects only the population of neurones appropriate to its field.

The identity of the cells which synthesise the signal was preliminarily investigated by co-culturing trigeminal ganglia with the epithelial and mesenchymal components of the maxillary process after their separation by treatment with trypsin (Lumsden and Davies, 1986). After 48 h incubation, nurite outgrowth was observed in 15% of E10 co-cultures (n = 160) and in 27.8% of E11 co-cultures (n = 90); in all cases, neurites were present only in the epithelium quadrant, and no neurites were present in the mesenchyme quadrant or in lateral quadrants (Fig. 7). Neurites first appeared at 24 h and, by 48 h, the majority were either close to or in contact with epitheilium.

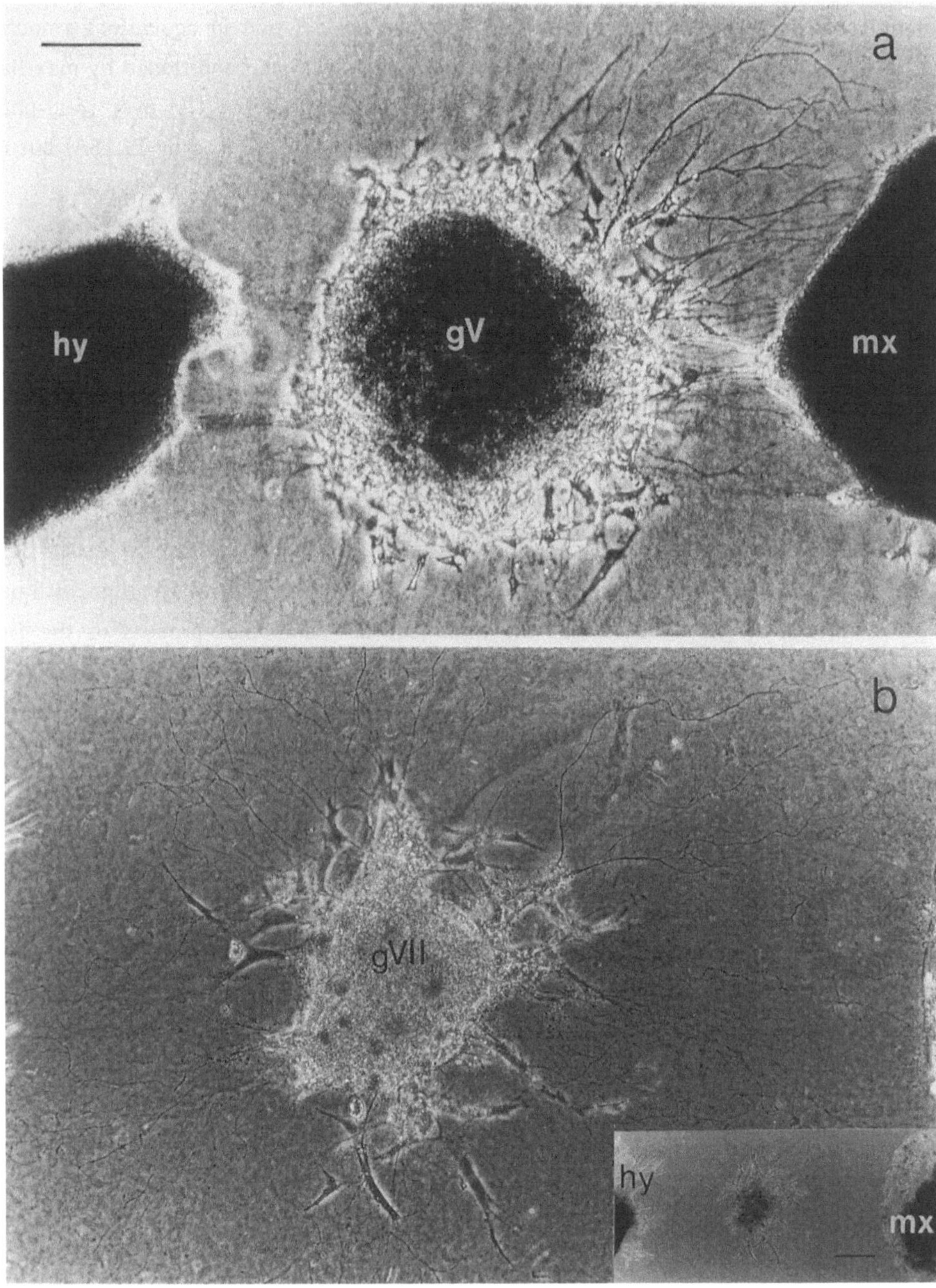
a
hy
gV
mx
b
gVII
hy
mx

Figure 6 (facing page): a: Phase contrast photomicrograph of a representative collagen gel co-culture of an E11 trigeminal ganglion (gV), maxillary process (mx) and hyoid process (hy) after 48 h incubation in the presence of anti-NGF. Neurites have extended towards the first arch explant but none has appeared in the quadrant facing the second arch explant. Collagen fibrils have become alligned in the hyoid-facing quadrant. Scale bar = 200 µm. b: Phase contrast photomicrographs of a representative collagen gel co-culture of an E10 geniculate ganglion (gVII), hyoid process (hy) and maxillary process (mx) after 48 h incubation in the presence of anti-NGF. Neurites have appeared around the entire periphery of the ganglion and extend radially with no apparent orientation towards either trget tissue. Inset figure shows the positions and spacing of the explants. Scale bar = 200 µm.

Epithelium-directed neurite outgrowth was enhanced in cases where epithelium had been able to contact and spread on the plastic surface beneath the collagen matrix. By reason of this observation, co-cultures were set up in which the mesenchymal masses and epithelial sheets were first allowed to adhere to the bottom of the dish for a period of 4 h before being overlain with collagen solution (in which ganglia were positioned as before), medium and sera. In these experiments, epithelium-directed outgrowth occurred in 60% of cultures (n = 20) and, as before, there was no neurite outgrowth towards mesenchyme. Outgrowth in these cultures was generally more profuse (Fig. 8, inset) than that which grew towards epithelium explants suspended in the collagen matrix.

Developing nerve fibres may be guided to their targets by two general mechanisms: either they respond to cues laid down in the intervening tissues, as has been suggested in some systems (Tosney and Landmesser, 1984; Bonhoeffer and Huf, 1980; Easter et al., 1984; Harris, 1986), or they respond by chemotropism to a long-range target-derived signal (Cajal, 1919). The finding that early trigeminal but not geniculate neurites respond by directional growth to a signal which is produced during the appropriate period of development, is precisely restricted to the trigeminal territory, and which emanates from the very peripheral region of this territory (epithelium) rather than the tissue through which the nerve fibres grow (mesenchyme), provides strong evidence for the operation of the second of these mechanisms. These phenomena suggest that trigeminal epithelium is specified to attract its correct sensory innervation at a distance and that mesenchyme, through which nerve fibres grow to reach epithelial targets and in which preformed guidance pathways have been postulated, does not provide directional information.

Is the Trigeminal a Special Case?

In addition to our evidence that there is no serial homologue of first arch chemotropism in the second arch, and the lack of evidence for regionally specific chemotropism elsewhere in the embryo, the trigeminal system displays a number of distinctive features which indicate that it may be a special case. The trigeminal territory plays a very important role in tactile exploration, especially in the many orders of mammal in which

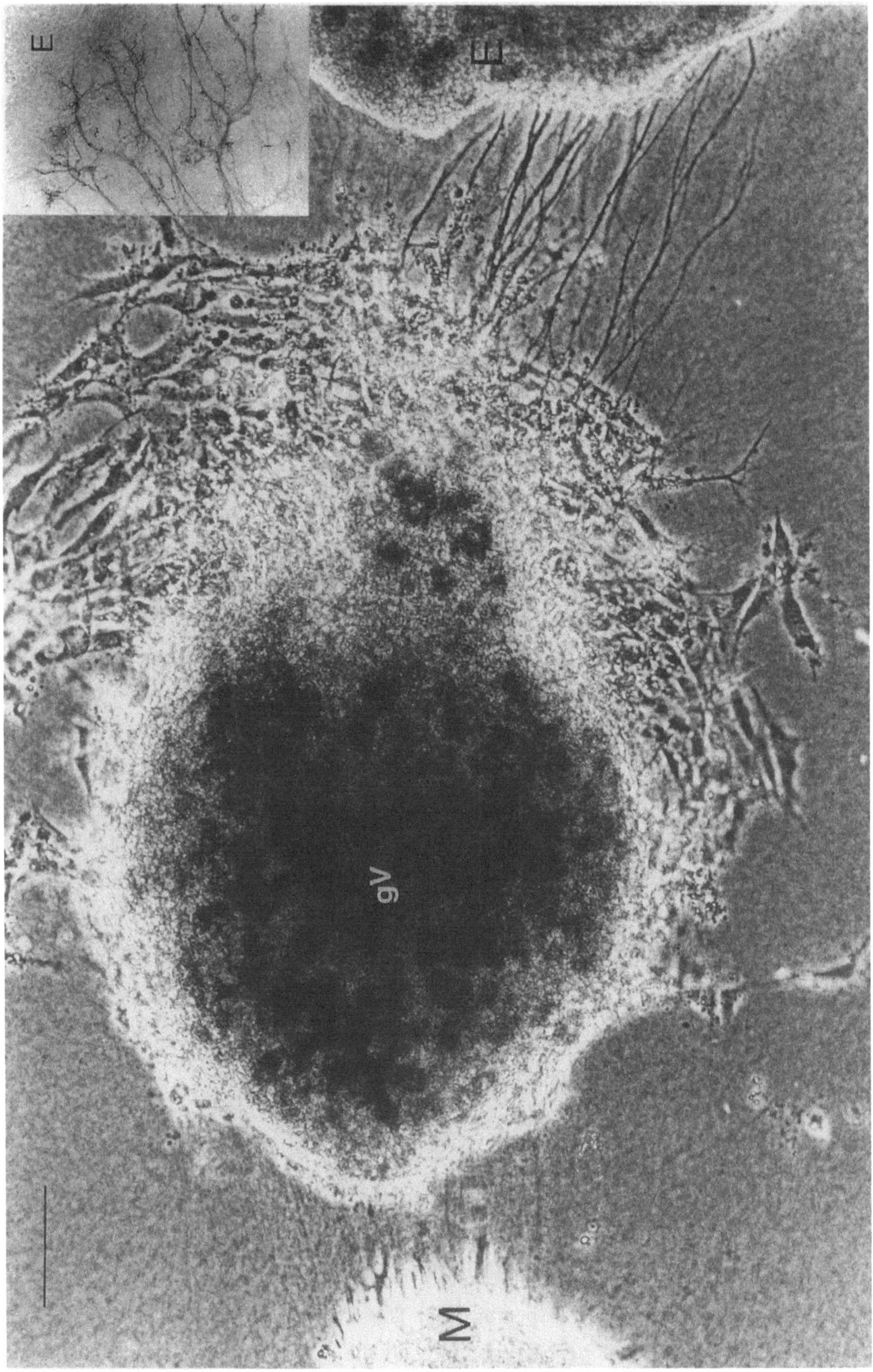
E
E
gv
M

Figure 7 (facing page): Phase contrast photomicrograph of a representative collagen gel co-culture of an E10 trigeminal ganglion (gV), maxillary process mesenchyme (M) and epithelium (E). Epithelium and mesenchyme were separated by treatment with 1% crude trypsin in Ca free PBS for 45 min. at 4°C, before being submerged in collagen gels. For culture method see Lumsden and Davies (1986). Neurites have extended exclusively and directly to the epithelium but none has emerged from the surface of the ganglion which faces laterally and towards the mesenchyme. In this particular culture, collagen fibrils have become aligned on the mesenchyme side of the ganglion, presumably through cell traction, but fibrils have not been aligned on the epithelium side. Inset figure shows epithelium directed E10 gV neurites growing towards and onto a monolayer of E11 maxillary process epithelium (E); the neurites have been stained by indirect immunoperoxidase using a monoclonal anti-neurofilament antibody (RT 97, courtesy of Dr. John Wood). Scale bar = 100 μm.

whiskers are prominent; it is by far the most densely innervated region of the integument. In spite of the very large number of trigeminal nerve fibres which grow to the periphery during development, the receptive field of the trigeminal ganglion is precisely circumscribed, in contrast to the considerable overlap between the receptive fields of DRG. Furthermore, in spinal nerves, motor axon outgrowth from the cord precedes outgrowth from DRG and has been proposed to provide a guidance pathway for sensory fibre outgrowth (Tosney and Landmesser, 1985). In the trigeminal nerve, however, the maxillary and ophthalmic divisions (60% and 15% of trigeminal ganglion fibres repsectively, Davies and Lumsden, 1984) are completely devoid of motor axons and in the mandibular division (25%) the sensory fibres are only associated with motor axons for a short part of their course before entering the mandibular process. In the trigeminal system a chemotropic signal emanating from the periphery may thus substitute for the stereotropic guidance cues provided elsewhere by motor fibres.

A number of recent findings in other systems are, however, quite consistent with chemotropism, and give reason to believe that the trigeminal nerve may not be unique. Homing behaviour by axons of displaced neurones (Lance-Jones and Landmesser, 1980a, b; 1981a, b; Harris, 1986) indicates that specific directional cues are not confined to particular pathways but are more widely distributed in the embryo. Although discussion of the nature of distributed guidance mechanisms has considered the notion of positional information to be a more favourable option than chemotropism, this is a bias which reflects the previous lack of direct evidence for the latter. Several studies of the effects of target organs on co-cultured neural explants have claimed to show chemotropism, but have failed to distinguish trophic effects, and in systems for which target-derived trophic effects are recognised (such as the retino-tectal system) there has been no in vitro demonstration of chemotropism (Carri and Ebendal, 1983; Harris et al., 1985). Thus, our evidence for region-specific chemotropism in guiding trigeminal fibres gives substance to the possibility that other nerves may be guided by a similar mechanism. As Trinkaus (1985) has argued, directional cell (and growth cone) movements are so important for normal development, and chemotactic mechanisms are so efficient for giving directionality to moving cells, that it is "not improbable that selection would have favoured such a mechanism, wherever it appeared."

References

Adler, R., and S. Varon (1981) Neuritic guidance by polyornithine-attached materials of ganglionic origin. Dev. Biol. **81:** 1-11.

Adler, R., M. Manthorpe, S.D. Skaper, and S. Varon (1981) Polyornithine-attached neurite-promoting factors; Culture sources and responsive neurons. Brain Res. **206:** 129-144.

Baron-van Evercooren, A., H.K. Kleinman, S. Ohno, P. Marangos, J.P. Schwartz, and M.E. Dubois-Dalcq (1982) Nerve growth factor, laminin, and fibronectin promote neurite growth in human fetal sensory ganglion cultures. J. Neurosci. Res. **8:** 179-193.

Bonhoeffer, F., and J. Huf (1980) Recognition of cell types by axonal growth cones in vitro. Nature **288:** 162-164.

Bonhoeffer, F., and J. Huf (1982) In vitro experiments on axon guidance demonstrating an anterior-posterior gradient on the tectum. Embo J. **1:** 427-431.

Cajal, S. Ramon y (1910) Algunas observaciones favorables a la hipothesis neurotropica. Trab. Lab. Invest. biol. Univ Madr. **8:** 63-134.

Cajal, S. Ramon y (1919) Accion neurotropica de los epithelios. Trab. Lab. Invest. biol. Univ. Madr. **17:** 191-228.

Carri, N., and T. Ebendal (1983) Organotypic cultures of neural retina: Neurite outgrowth stimulated by brain explants. Dev. Brain Res. **6:** 219-229.

Chamley, J.H., and J.J. Dowel (1975) Specificity of nerve fibre 'attraction' to autonomic effector organs in tissue culture. Exper. Cell Res. **90:** 1-7.

Chamley, J.H., I. Goller, and G. Burnstock (1973) Selective growth of sympathetic nerve fibers to explants of normally densely innervated autonomic effector organs in tissue culture. Dev. Biol. **31:** 362-379.

Collins, F., and A. Dawson (1982) Conditioned medium increases the rate of neurite elongation: Separation of this activity from the substratum-bounde inducer of neurite outgrowth. J. Neurosci. **2:** 1005-1010.

Collins, F., and J.E. Garrett (1980) Elongating nerve fibers are guided by a pathway of material released from embryonic non-neuronal cells. Proc. Natl. Acad. Sci. (USA) **77:** 6226-6228.

Coughlin, M.D. (1975) Target organ stimulation of parasympathetic nerve growth in the developing mouse submandibular gland. Dev. Biol. **43:** 140-158.

Davies, A.M., and A.G.S. Lumsden (1984) Relation of target encounter and neuronal death to nerve growth factor responsiveness in the developing mouse trigeminal ganglion. J. Comp. Neurol. **223:** 124-137.

Davies, A.M., A.G.S. Lumsden, H.C. Slavkin, and G. Burnstock (1981) Influence of nerve growth factor on the embryonic mouse trigeminal ganglion in culture. Dev. Neurosci. **4:** 150-156.

Davies, G.E., M. Manthorpe, E. Engvall, and S. Varon (1985) Isolation and characterisation of rat schwannoma neurite-promoting factor: Evidence that the factor contains laminin. J. Neurosci. **5:** 2662-2671.

Dunn, G.A. (1981) Chemotaxis as a form of directed cell behaviour: Some theoretical considerations. In *Biology of the Chemotactic Response*, J.M. Lackie and P.C. Wilkinson, eds., pp. 1-26. Cambridge Univ. Press.

Easter, S.S., B. Bratton, and S.S. Scherer (1984) Growth-related order of the retinal fiber layer in goldfish. J. Neurosci. **4:** 2173-2190.

Ebendal, T. (1976) Experiments in vitro on neuron development and axon orientation in the chick embryo. Exp. Cell Res. **98:** 159-169.

Ebendal, T. (1981) Control of neurite extension by embryonic heart explants. J. Embryol. exp. Morph. **61:** 289-301.

Ebendal, T., and C.O. Jacobson (1977) Tissue explants affecting extension and orientation of axons in cultured chick embryo ganglia. Exper. Cell Res. **105:** 379-387.

Ebendal, T., A. Jordell-Kylberg, and S. Soderstrom (1978) Stimulation by tissue explants on nerve fibre outgrowth in culture. Zoon **6:** 235-243.

Eisen, J.S., P.Z. Myers, and M. Westerfield (1986) Pathway selection by growth cones of identified motoneurones in live zebra fish embryos. Nature **320:** 269-271.

Ferguson, B.A. (1983) Development of motor innervation of the chick following dorso-ventral limb bud rotations. J. Neurosci. **3:** 1760-1772.

Gundersen, R.W., and J.N. Barrett (1979) Neuronal chemotaxis: Chick dorsal root axons turn towards high concentrations of nerve growth factor. Science **206:** 1079-1080.

Gundersen, R.W., and J.N. Barrett (1980) Characterisation of the turning response of dorsal root neurites towards nerve growth factor. J. Cell Biol. **87:** 546-554.

Hamburger, V., J.K. Brunso-Bechtold, and J.W. Yip (1981) Neuronal death in the spinal ganglia of the chick embryo and its reduction by nerve growth factor. J. Neurosci. **1:** 60-71.

Harper, G.P.; F.L. Pearce, and C. Vernon (1980) The production and storage of nerve growth factor in vivo by tissues of the mouse, rat, guinea pig, hamster, and gerbil. Dev. Biol.: 391-402.

Harris, W.A. (1986) Homing behaviour of axon in the embryonic vertebrate brain. Nature **320:** 266-269.

Harris, W.A., C.E. Holt, T.A. Smith, and N.J. Gallenson (1985) Growth cones of developing retinal cells in vivo, on culture surfaces, and in collagen matrices. J. Neurosci. Res. **13:** 101-122.

Hendry, I.A., and J. Campbell (1976) Morphometric analysis of rat superior cervical ganglion after axotomy and nerve growth factor treatment. J. Neurocytol. **5:** 351-360.

Jacobson, M., and S. Huang (1985) Neurite outgrowth traced by means of horseradish perioxidase inherited from neuronal ancestral cells in frog embryos. Dev. Biol. **110:** 102-113.

Johnson, E.M., P.D. Gorin, L.D. Brandeis, and J. Pearson (1980) Dorsal root ganglion neurons are destroyed by exposure in utero to maternal antibody to nerve growth factor. Science **210:** 916-918.

Johnson, E.M., and H.K. Yip (1985) Central nervous system and peripheral nerve growth factor provide trophic support critical to mature sensory neuronal survival. Nature **314:** 751-752.

Korsching, S., and H. Thoenen (1983) Nerve growth factor in sympathetic ganglia and corresponding target organs of the rat: Correlation with density of sympathetic innervation. Proc. Natl. Acad. Sci. (USA) **80:** 3513-3516.

Lance-Jones, C., and L. Landmesser (1980a) Motoneuron projection patterns in the chick hind limb following early spinal cord deletions. J. Physiol., Lond. **302:** 559-580.

Lance-Jones, C., and L. Landmesser (1980b) Motoneuron projection patterns in the chick hind limb following early partial spinal cord reversals. J. Physiol., Lond. **302:** 581-602.

Lance-Jones, C., and L. Landmesser (1981a) Pathway selection by chick lumbrosacral motoneurons during normal development. Proc. R. Soc. Lond. **B214:** 1-18.

Lance-Jones, C., and L. Landmesser (1981b) Pathway selection by embryonic chick motoneurons in an experimentally altered environment. Proc. R. Soc. Lond. **B214:** 19-52.

Lander, A.D., D.K. Fujii, D. Gospodarowicz, and L.F. Reichardt (1982) Characterisation of a factor that promotes neurite outgrowth: Evidence linking activity to a heparan sulfate proteoglycan. J. Cell Biol. **94:** 574-585.

Lander, A.D., D.K. Fujii, and L.F. Reichardt (1985a) Purification of a factor that promotes neurite outgrowth: Isolation of laminin and associated molecules. J. Cell

Biol. **101**: 898-913.
Lander, A.D., D.K. Fujii, and L.F. Reichardt (1985b) Laminin is associated with the "neurite outgrowth-promoting factors" found in conditioned media. Proc. Natl. Acad. Sci (USA) **82**: 2183-2187.
Landmesser, L. (1984) The development of specific motor pathways in the chick embryo. Trends Neurosci. **7**: 336-339.
Letourneau, P.C. (1978) Chemotactic response of nerve fiber elongation to nerve growth factor. Dev. Biol. **66**: 183-196.
Levi-Montalcini, R. (1982) Developmental neurobiology and the natural history of nerve growth factor. A. Rev. Neurosci. **5**: 341-362.
Lumsden, A.G.S., and A.M. Davies (1983) Earliest sensory nerve fibres are guided to peripheral targets by attractants other than nerve growth factor. Nature **306**: 786-788.
Lumsden, A.G.S., and A.M. Davies (1986) Chemotropic effect of specific target epithelium in development of the mammalian nervous system. Submitted.
Manthorpe, M., E. Engvall, E. Ruoslahti, F.M. Longo, G.E. Davis, and S. Varon (1983) Laminin promotes neuritic regeneration from cultured peripheral and central neurons. J. Cell Biol. **97**: 1882-1890.
Mensini-Chen, M.G., J.S. Chen, and R. Levi-Montalcini (1978) Sympathetic nerve fiber ingrowth in the central nervous system of neonatal rodents upon intracerebral NGF injections. Arch. Ital. Biol. **116**: 53-84.
Patrizi, G., and B.L. Munger (1966) The ultrastructure and innervation of rat vibrissae. J. Comp. Neurol. **126**: 423-436.
Pollack, E.D., W.L. Muhlach, and V. Liebig (1981) Neurotropic influence of mesenchymal limb target tissue on spinal cord neurite growth in vitro. J. Comp. Neurol **200**: 393-405.
Rogers, S.L., K.J. Edson, P.C. Letourneau, and S.C. McLoon (1986) Distribution of laminin in the developing peripheral nervous system of the chick. Dev. Biol. **113**: 429-435.
Sanes, J.R. (1985) Laminin for axonal guidance? Nature **315**: 714-715.
Scott, S.A. (1982) The development of the segmental pattern of skin sensory innervation in embryonic chick hind limb. J. Physiol., Lond. **330**: 203-230.
Stopak, D., and A.K. Harris (1982) Connective tissue morphogenesis by fibroblast traction. I. Tissue culture observations. Dev. Biol. **90**: 383-398.
Thoenen, H., and Y.A. Barde (1980) Physiology of nerve growth factor. Physiol. Rev. **60**: 1284-1335.
Tosney, K.W., and L. Landmesser (1984) Pattern and specificity of axonal outgrowth following varying degrees of chick limb bud ablation. J. Neurosci. **4**: 2518-2527.
Tosney, K.W., and L. Landmesser (1985) Growth cone morphology and trajectory in the lumbrosacral region of the chick embryo. J. Neurosci. **5**: 2345-2358.
Trinkaus, J.P. (1985) Further thoughts on directional cell movement during morphogenesis. J. Neurosci. Res. **13**: 1-19.
Vincent, S.B. (1913) The tactile hair of the white rat. J. Comp. Neurol. **23**: 1-34.
Wolpert, L. (1969) Positional information and the spatial pattern of cellular differentiation. J. Theor. Biol. **25**: 1-47.
Yip, H.K., and E.M. Johnson (1984) Developing dorsal root ganglion neurons require trophic support from their central processes: Evidence for a role of retrogradely transported nerve growth factor from the central nervous system to the periphery. Proc. Natl. Acad. Sci. (USA) **81**: 6245-6249.

SUBSTRATES INFLUENCING THE SURVIVAL AND DEVELOPMENT OF NEURAL CELLS IN VITRO

D. Edgar, V. Nurcombe, and U. Dohrmann
Abteilung Neurochemie
Max-Planck-Institut für Psychiatrie
Martinsried
Federal Republic of Germany

The survival and development of neurons is regulated by epigenetic factors (for review see Thoenen and Edgar, 1985). For example, the protein, Nerve Growth Factor (NGF), has been demonstrated to play a role in the nervous system: both peripheral sympathetic and sensory neurons derived from the neural crest need NGF for survival during development, and for maintenance during adulthood (see Thoenen and Edgar, 1985). Furthermore, it now seems that NGF plays a trophic role in the maintenance of magnocellular, cholinergic neurons of the central nervous system (Otten et al., 1985). The role of NGF in vivo is also reflected by the necessity for its presence for the survival of NGF-responsive neurons in culture (Thoenen and Edgar, 1985). Recently, it has become apparent that a variety of neurons that do not respond to NGF require other soluble trophic factors if they are to survive in vitro (see Barde et al., 1983). Two putative trophic molecules have been isolated by virtue of their ability to support neuronal survival in culture (Barde et al., 1982; Barbin et al., 1984). The role of these factors in vivo remains to be determined.

The use of tissue culture experiments has thus been essential in the detection, purification and analysis of the function of neurotrophic factors (see Barde et al, 1983). Such culture experiments also reveal that neuronal survival can be profoundly influenced not only by soluble molecules, but also by components of the extracellular matrix.

Effects of Molecules of the Extracellular Matrix on Neurons

A variety of different types of cell produce molecules that can attach to polycationic tissue culture substrates and thereby stimulate the rapid growth of neurites from cultured neurons (Collins, 1978; Adler et al., 1981; Barde et al., 1983). Subsequently, it was shown that, although such substrate-attached molecules can not themselves support neuronal survival, they do potentiate survival in response to soluble trophic factors like NGF: e.g., NGF can only maintain about 40% of the total population of neurons dissoci-

NATO ASI Series, Vol. H5
Mesenchymal-Epithelial Interactions in Neural Development
Edited by J. R. Wolff et al.

ated from embryonic chick sympathetic ganglia and cultured on a synthetic (polycationic) substrate. If, however, the substrate is coated with molecules produced by either cultured embryonic heart cells (Edgar and Thoenen, 1982) or cultured Schwann cells (Edgar, 1985), then NGF can support the survival of essentially all the neurons brought into culture. Attempts have been made to isolate the molecules present in such complex cell exudates and responsible for their effects on neurons, and partial purifications have indicated that a heparan sulfate proteoglycan (Lander et al., 1982) complexed with a laminin-like molecule is the active agent (Davis et al., 1985; Edgar, 1985; Lander et al., 1985). Indeed, the purified basement membrane protein laminin (Timpl et al., 1979) had already been shown to stimulate rapid neurite outgrowth in an apparently similar fashion to the cell exudates (Baron-van Evercooren et al., 1982).

Interaction of Laminin with Neural Cells

In addition to its ability to stimulate rapid neurite outgrowth (Baron-van Evercooren et al., 1982), laminin also potentiates the survival-promoting effects of soluble neurotrophic factors (Edgar et al., 1984), indicating that one molecule can be responsible for the several effects of cell-derived culture substrates. Furthermore, laminin substrates can evoke an increase in the activity of the catecholamine-synthetic enzyme, tyrosine hydroxylase (TH), in cultures of calf adrenal medullary chromaffin cells (Acheson et al., 1986). As laminin is not required for the survival of these cells, then it is possible to perform adequate control experiments to analyze the biochemical mechanism of action of laminin. It could thus be shown that laminin stimulates the TH activity of these cells by a two-step process: in the short term (up to 6 hr after start of culture) the pre-existing molecules of TH in the cells are activated, so that enzymic activity increases with no change in number of molecules, whereas in the long term (after 2 days in culture) laminin evokes an increase in the number of TH molecules per chromaffin cell (Acheson et al., 1986). Thus, the consequences of the interaction of laminin with these cells are initially the stimulation of a post-translational modification of TH, by an as yet unknown mechanism. Subsequently, the increased levels of TH result from an increased rate of transcription of the TH gene (see Acheson et al., 1986, for discussion).

The details of the mechanism(s) whereby laminin evokes these biochemical responses remain to be elucidated. It is clear, however, that several types of cell possess high affinity receptors for laminin (for review see Von der Mark and Kühl, 1985). By use of ^{125}I-labelled laminin we have been able to show that calf chromaffin cells are able to bind laminin with high affinity and low capacity. Thus, the K_D for laminin binding was estimated to be 2.4×10^{-9} M, with 16,000 putative receptors per cell (Fig. 1).

Such values for high affinity laminin binding to chromaffin cells should, however, be interpreted with caution; it is unlikely that the interaction of soluble laminin with its

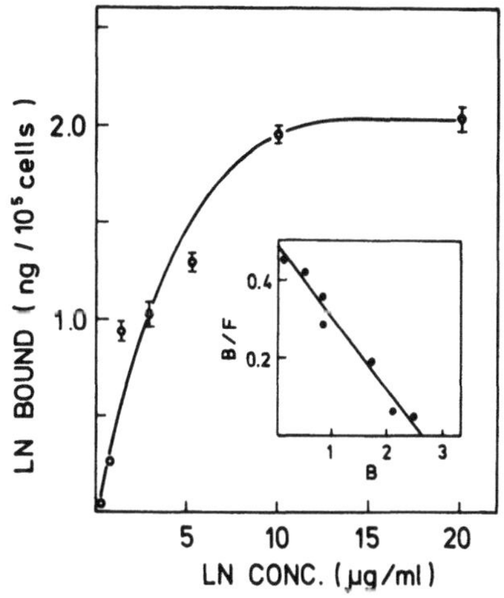

Figure 1: Equilibrium binding of ^{125}I-labelled laminin (LN) to calf adrenal medullary cells. Specific high-affinity binding was determined by subtracting non-specific binding (in the presence of excess unlabelled laminin) from total binding. The insert shows a Scatchard plot of the data with ordinate: percentage of bound laminin, and abscissa: amount of bound laminin (ng per 10^5 cells).

receptor accurately reflects those interactions taking place when the laminin molecules are anchored in an insoluble matrix with, presumably, a fixed orientation. The immediate consequences of laminin-receptor interaction (e.g., ion flux, protein kinase activation or stimulation of second messengers such as cyclic AMP or inositol phosphates) remain to be established.

In contrast to this lack of knowledge of the intracellular consequences of laminin-receptor interaction, considerable progress has been made in the analysis of the structure of the laminin molecule, in order to define which parts of it interact with neural cells. A prerequisite for such studies was the availability of well defined proteolytic fragments of laminin (see Ott et al., 1982), together with the corresponding affinity purified fragment-specific antibodies. Inhibition studies using such antibodies (Edgar et al., 1984) showed that while whole anti-laminin antiserum blocked the effects of laminin on neural cells, only those fragment-specific antibodies directed against a fragment derived from the globular domain at the end of the long arm of the molecule were inhibitory (see Fig. 2). Furthermore, these antibodies were able to block not only neurite outgrowth and potentiation of neuronal survival (Edgar et al., 1984) but also the stimulatory effects of laminin on chromaffin cell TH activity (Acheson et al., 1986). Similarly, the same antibodies were able to block the spreading of Schwann cells on laminin substrates (unpublished result), indicating that the same domain of laminin may be involved in its interaction with a wide variety of neural cells.

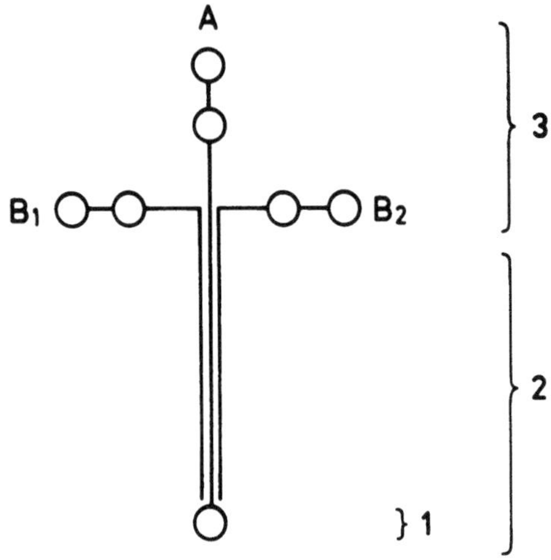

Figure 2: Schematic diagram of laminin (not to scale). The molecule consists of 3 polypeptide chains: B_1 and B_2 of 220 KD, and A of 440 KD, linked by disulfide bonds. 1, the globular domain, recognized by antibodies that block the effects of laminin on neural cells; 2, the long arm of laminin that comprises the globular domain (1) together with a long rod-containing α-helical structure. This part of the molecule roughly corresponds to the fragment shown to interact with neural cells (Edgar et al., 1984); 3, the short arms of laminin, reported to interact with a variety of tumor cells (see Terranova et al., 1983).

The results of the antibody inhibition studies do not, however, indicate if the globular domain at the end of the long arm of laminin is itself responsible for neural cell interactions, or if the antibody blockade is simply a consequence of steric hindrance, the active site being elsewhere (but presumably spatially close to the blocking epitope). This consideration is of particular importance because previous studies had located the binding domains for a variety of non-neuronal cells at the cross formed by the short arms of laminin (Terranova et al., 1983; see Fig. 2). By use of mild proteolytic fragmentation, however, it was possible to isolate a large fragment of laminin of 300 kD, that consisted of the long arm together with its globular domain, but which lacked any short arm structures (Edgar et al., 1984). The fact that this fragement could also evoke rapid neurite outgrowth and potentiate neuronal survival indicated directly that the site on laminin that interacts with neural cells is indeed associated with the long arm (Edgar et al., 1984). However, the fact that several laminin-like molecules appear to lack the epitope recognized by blocking antibodies (see below) would indicate that antibody inhibition was via an indirect steric hindrance, and that the site for neural cell interactions lies somewhere on the long arm of laminin, in the vicinity of its globular domain. A further implication of these results is that because neural and non-neural cells appear to interact with differing sites on the laminin molecule, then neural cells would have laminin binding proteins in their membranes that are distinct from those previously characterized (see von der Mark and Kühl, 1985). This hypothesis is currently being tested.

Laminin-like Molecules Affecting Neural Cells

As previously mentioned, a variety of cultured cells produce molecules that have effects similar to those of laminin on neural cells. Such molecules can be shown to be immunologically related to laminin in that they are immunoprecipitable with antibodies against laminin (Lander et al., 1985; Davis et al., 1985; Edgar, 1985). They differ from laminin, however, because their effects on neurons are not blockable with anti-laminin antibodies (see e.g., Edgar et al., 1984). The reason for this is that such molecules are not reconized by specifically those antibodies present in anti-laminin antiserum that are capable of blocking effects of laminin (Dziadek et al., 1986). The fact that such laminin-like molecules do, however, affect neural cells indicates that the epitope recognized by blocking antibodies is not directly involved in neural cell interactions (see above). One hypothesis to account for lack of recognition of the blocking epitope is that the laminin-like molecules produced by cultured cells are complexed with a heparan sulfate proteogylcan (Lander et al., 1982) and this could mask the epitope (laminin has been shown to contain a haparin binding site; Sakashita et al., 1980; Ott et al., 1982). An alternative hypothesis to account for lack of recognition of blocking antibodies is simply that the blocking epitope is absent from the laminin-like molecules. In support for this hypothesis, it has been shown that laminin produced by Schwann cells has a very low A chain content (see Cornbrooks et al., 1983; Edgar, 1985; Dziadek et al., 1986), and current models of the laminin molecule indicate that it is the A chain of laminin that forms the globular domain at the end of the long arm (see Yamada et al., this volume). Whether a lack of A chain from the Schwann cell laminin is a consequence of lack of synthesis (see Yamada et al., this volume) or specific degradation, remains to be elucidated.

The significance of such soluble forms of laminin produced by cultured cells for neuronal development is open to question; in vivo, laminin appears to be exclusively located in insoluble form in basal laminae (but see Rogers et al., 1986, and Cohen, this volume), and a significant interaction of developing axons with basal laminae during development has yet to be demonstrated (but see Easter et al., this volume). In regeneration (see chapter by Berry et al., this volume), however, it is now clear that the initial re-growth of peripheral axons occurs along the inside of pre-existing Schwann cell basal laminae, distal to the lesion (Ide et al., 1984). Recently, however, we have detected the presence of large amounts of apparently freely soluble laminin in extracts of embryonic chick muscle (Fig. 3), and this laminin is at least in part responsible for the ability of these extracts to stimulate rapid neurite outgrowth (Dohrmann et al., 1986). It is therefore possible that during development, the location and functions of laminin differ from those in the mature animal.

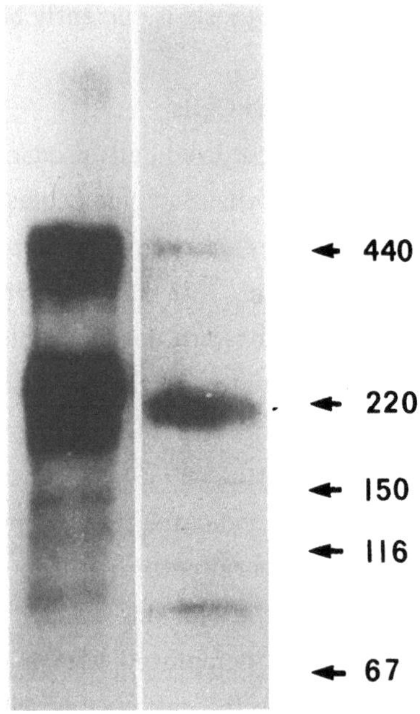

Figure 3: Immunoblot of a soluble extract of embryonic chick muscle. Skeletal muscle from the legs of 17-day-old chick embryos was extracted and subjected to SDS-gel electrophoresis under reducing conditions. The gel was electroblotted and stained with anti-laminin antiserum followed by ^{35}S-labelled protein A. The right hand lane is an authentic laminin standard, the left hand lane is the muscle extract. Positions of molecular weight markers are shown.

Acknowledgements. V.N. is the recipient of a C.J. Martin Fellowship from the NH and MRC of Australia. U.D. is supported by a grant from the International Spinal Research Trust, London. We thank Hans Thoenen and Rupert Timpl for their help and advice, and we are grateful to Andrea Kussmaul for her technical assistance.

References

Acheson, A., D. Edgar, R. Timpl, and H. Thoenen (1986) Regulation of enzyme synthesis in chromaffin cells by contact with the basement membrane protein, laminin. J. Cell Biol. **102:** 151-159.

Adler, R., M. Manthorpe, S.D. Skaper, and S. Varon (1981) Polyornithine neurite promoting factors. Brain Res. **206:** 129-144.

Barbin, G., M. Manthorpe, and S. Varon (1984) Purification of the chick eye ciliary neurotrophic factor. J. Neurochem. **43:** 1468-1478.

Barde, Y.-A., D. Edgar, and H. Thoenen (1982) Purification of a new neurotrophic factor from mammalian brain. EMBO J. **1:** 549-553.

Barde, Y.-A., D. Edgar, and H. Thoenen (1983) New neurotrophic factors. A. Rev. Physiol. **45:** 601-612.

Baron-van Evercooren, A., H.K. Kleinman, S. Ohno, P. Marangos, J.P. Schwartz, and M.E. Dubois-Dalcq (1982) Nerve growth factor, laminin, and fibronectin promote neurite growth in human fetal sensory ganglia cultures. J. Neurosci. Res. **8:** 179-193.

Collins, F. (1978) Induction of neurite outgrowth by a conditioned medium factor bound to the culture substratum. Proc. Natl. Acad. Sci. (USA) **75**: 5210-5213.

Cornbrooks, C.J., D.J. Carey, J.A. McDonald, R. Timpl, and R.P. Bunge (1983) In vivo and in vitro observations on laminin production by Schwann cells. Proc. Natl. Acad. Sci. (USA) **80**: 3850-3854.

Davis, G.E., M. Manthorpe, E. Engvall, and S. Varon (1985) Isolation and characterization of rat schwannoma neurite promoting factor: Evidence that the factor contains laminin. J. Neurosci. **5**: 2662-2671.

Dohrmann, U., D. Edgar, M. Sendtner, and H. Thoenen (1986) Muscle-derived factors that support survival and promote fiber outgrowth from embryonic chick motor neurons in culture. Dev. Biol., in press.

Dziadek, M., D. Edgar, M. Paulsson, R. Timpl, and R. Fleischmajer (1986) Basement membrane proteins produced by Schwann cells and in Neurofibromatosis. Annals N.Y. Acad. Sci., in press.

Edgar, D. (1985) Nerve growth factor and molecules of the extracellular matrix in neuronal development. J. Cell Sci. Suppl. **3**: 107-113.

Edgar, D., and H. Thoenen (1982) Modulation of NGF-induced survival of chick sympathetic neurons. Dev. Brain Res. **5**: 89-92.

Edgar, D., R. Timpl, and H. Thoenen (1984) The heparin-binding domain of laminin is responsible for its effects on neurite outgrowth and neuronal survival. EMBO J. **3**: 1463-1468.

Ide, C., K. Tohyama, R. Yokota, T. Nitatori, and S. Onodera (1983) Schwann cell basal lamina and nerve regeneration. Brain Res. **288**: 61-75.

Lander, A.D., D.K. Fujii, D. Gospodarowicz, and L.F. Reichardt (1982) A heparan sulfate proteoglycan induces rapid neurite outgrowth in vitro. J. Cell Biol. **94**: 574-582.

Lander, A.J., D.K. Fujii, D. Gospodarawicz, and L.F. Reichardt (1985) Neurite outgrowth promoting factors in conditioned medium are complexes containing laminin. Proc. Natl. Acad. Sci. (USA) **82**: 2183-2187.

Ott, U., E. Odermatt, J. Engel, H. Furthmayr, and R. Timpl (1982) Protease resistance and conformation of laminin. Eur. J. Biochem. **123**: 63-72.

Otten, U., G. Weskamp, M. Schlumpf, W. Lichtensteiger, and W.C. Mobley (1985) Effects of antibodies against NGF on developing cholinergic forebrain neurons in rat. Soc. Neurosci. Abst. Vol. 11, Part 1, p. 661.

Rogers, S.L., K.J. Edson, P.C. Letourneau, and S.C. McLoon (1986) Distribution of laminin in the developing peripheral nervous system of the chick. Dev. Biol. **113**: 429-435.

Sakashita, S., E. Engvall, and E. Ruoslahti (1980) Basement membrane protein laminin binds to heparin. FEBS Lett. **116**: 243-246.

Terranova, V.P., C.N. Rao, T. Kalebic, I.M. Margulies, and L.A. Liotta (1983) Laminin receptor on human breast carcinoma cells. Proc. Natl. Acad. Sci. (USA) **80**: 444-448.

Thoenen, H., and D. Edgar (1985) Neurotrophic factors. Science **229**: 238-242.

Timpl, R., H. Rohde, P.G. Robey, S.I. Rennard, S.M. Foidart, and G.R. Martin (1979) Laminin - a glycoprotein from basement membranes. J. Biol. Chem. **254**: 9933-9937.

Von der Mark, K., and U. Kühl (1985) Laminin and its receptor. Biochim. Biophys. Acta **823**: 147-160.

THE ROLE OF GROWTH CONE ADHESION IN NEURONAL MORPHOGENESIS, AS DEMONSTRATED BY INTERACTIONS WITH FIBRONECTIN AND LAMININ

Paul Letourneau[1], Sherry Rogers[3], James Hammarback[4], Anne Madsen[1], Sally Palm[2], James McCarthy[2], Leo Furcht[2], Donna Bozyczko[5], and Alan Horwitz[5]

[1] Department of Cell Biology and Neuroanatomy
[2] Department of Laboratory Medicine and Pathology
University of Minnesota
Minneapolis, MN

[3] Department of Anatomy
University of New Mexico
Albuquerque, NM

[4] Worcester Foundation for Experimental Biology
Shrewsbury, MA

[5] Department of Biochemistry and Biophysics
University of Pennsylvania
Philadelphia, PA

U.S.A.

Introduction

The embryonic formation of nerve fibers (axons and dendrites) by neurons is both complex and regular. For example, the arrangement of nerves at the brachial plexus and beyond in the vertebrate forelimb is predictable within members of a species, yet is distinct from the pattern of other species. Such precise axonal pathways are forged by the activities of extending nerve fiber tips, first named the growth cone by Santiago Ramon y Cajal (1890). As holds for all cell movements, growth cones can be studied in terms of distinct questions: what starts, what maintains, what regulates the directions of and what stops growth cone movements (Trinkaus, 1984)? Each question probes a different facet of growth cone behavior, and answers to each may involve different intrinsic and extrinsic factors. In the case of the growth cone, morphogenetic behavior can be divided into five distinct activities; neurite elongation, turning, branching, retraction and synaptogenesis. Consistent patterns of nerve fiber pathways, such as in the brachial plexus, arise from the readout of developmental programs that determine these five growth cone activities.

This chapter discusses the interactions of nerve growth cones with two glyco-

NATO ASI Series, Vol. H5
Mesenchymal-Epithelial Interactions in Neural Development
Edited by J. R. Wolff et al.

proteins, fibronectin (FN) and laminin (LAM), that may be adhesive ligands for extending neurites. A growth cone must adhere to other surfaces in order to be pulled forward via tensile forces exerted within filopodial and lamellipodial protrusions (Bray, 1982; Letourneau, 1982). The spatial and temporal regulation of growth cone anchorage may be critical to each growth cone activity that determines neuronal shape. It will advance understanding of this regulation to elucidate both the cell surface molecules on growth cones that participate in adhesive bonds and the specific molecular sequences or domains that are recognized by receptors on growth cones to get anchorage for elongation.

In Vivo Interactions of Growth Cones With FN and LAM

If, in fact, FN and LAM are ligands for growth cone traction in embryos, then we should find these glycoproteins near elongating axons during neurodevelopment. FN is present in embryonic connective tissues, where the glycoprotein may be available to the growth cones of peripheral nerve fibers (Mayer et al., 1981; Kurkinen et al., 1979; Wartiovaara et al., 1980). FN does not seem to be concentrated along particular nerve fiber pathways, although its ubiquitous presence in connective tissues may support the extension of sensory or autonomic nerve fibers into the vicinity of epithelial target tissues. More controversial and puzzling is the possibility that FN is transiently expressed in developing brain tissue (Hatten et al., 1982; Hynes et al., 1986; Pearlman et al., 1984; Schachner et al., 1978).

LAM is concentrated in the basal laminae of adult tissues (Timpl et al., 1983), although a significant possibility during embryogenesis is that LAM is deposited on cell surfaces in a different form than as a recognizable lamina densa (Leivo et al., 1980; Liesi et al., 1983). Anti-LAM antibodies are concentrated at the surface of the developing neural tube, as early as day 8 in the mouse embryo (Madsen and Letourneau, unpublished observation). The earliest nerve fibers in the brain and spinal cord run at the surface of the developing CNS, where their growth cones may contact the LAM-rich external limiting membrane (Fig. 1; Rogers et al., 1986). However, immunoreactivity for LAM also extends inward from the surface of the developing brain and spinal cord, and is seen as a punctate localization of flourescence on the basolateral surfaces of the cells that comprise the primitive CNS. Thus, growth cones of these early axons may not have to contact the external basal lamina to encounter LAM-containing surfaces. Outside the CNS, LAM immunoreactivity is present in the basement membranes of epithelial tissues, as expected (Leivo et al., 1980). Punctate flourescence with anti-LAM is also seen in the sclerotome, through which motor and sensory axons extend, and perhaps a significant finding is that developing ventral roots show enhanced staining with anti-LAM (Fig. 2; Rickmann et al., 1985; Rogers et al., 1986). Sheath cells, derived from the neural crest, may synthesize the LAM that is concentrated in the ventral root, and electron micro-

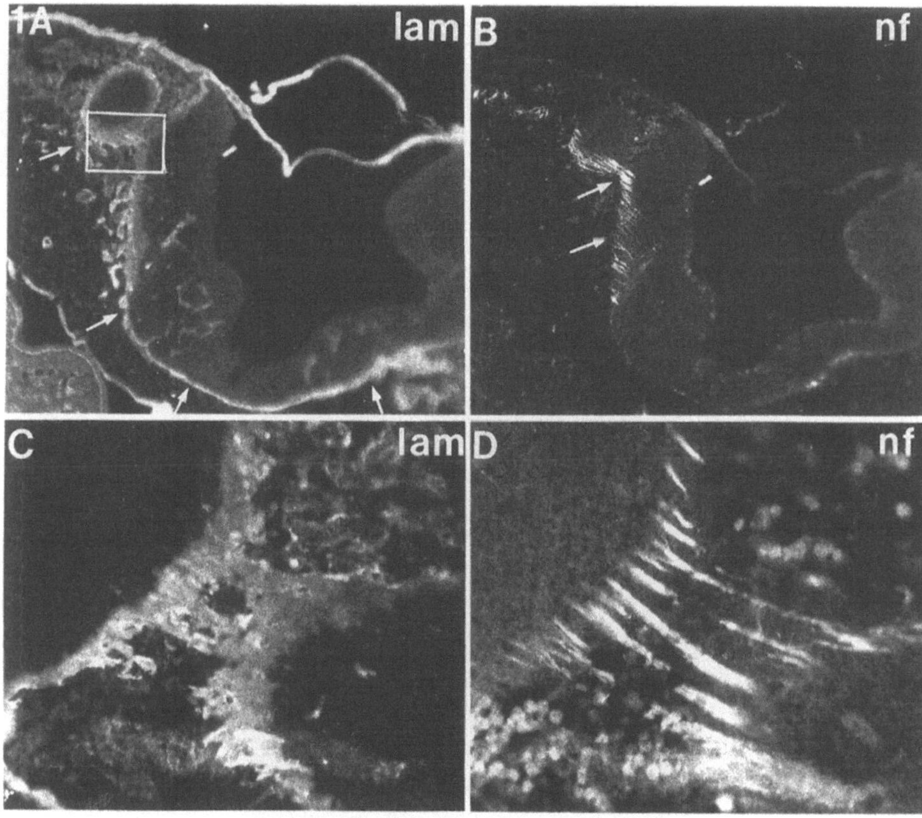

Figure 1. A sagittal section from a 9 day mouse embryo double-stained with a polyclonal anti-laminin (A, C) and a monoclonal anti-200 kD neurofilament protein (B, D). In (A) structures that stain for laminin include the epidermis, endoderm, endothelial lining of blood vessels and the surface of the developing brain (arrows). Anti-laminin staining in the area of the rectangle in (A) includes an oblique face view of a portion of the surface of the brain. This region at the surface of the brain is shown at higher magnification in (C). In (B) early axons are stained extending in the brain of a 9 day embryo. Most of these initial fibers lie at the surface of the brain, and strong anti-neurofilament staining is present at the same focal plane as the anti-laminin staining at the surface of the brain (arrows). (D) shows neurofilament staining of the same section shown in (C). Many individual nerve fibers and small fascicles run in a rostro-caudal direction and lie in the same focal plane as the laminin-containing surface of the brain.

scopic reconstructions of axons and growth cones of the ventral root fibers show interactions of growth cones with mesenchymal cells that may contain LAM on their surfaces (Tosney and Landmesser, 1985).

These immunohistochemical studies are consistent with the possibilities that growth cones do use FN- and LAM-containing surfaces to get traction for the elongation of neurites. Direct evidence has not been presented to demonstrate that these glyco-

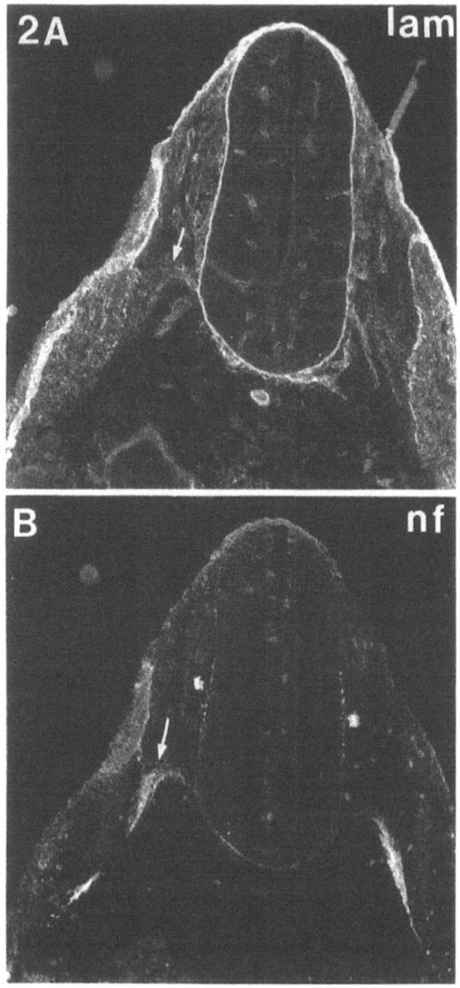

Figure 2. A cross section from the level of the spinal cord of a 10 day mouse embryo, double-stained with anti-laminin (A) and anti-200 kD neurofilament protein (B). Anti-laminin staining is strongest in the epidermis, along the medial edge of a somite, in the walls of blood vessels, and at the surface of the spinal cord. Binding of anti-laminin is also present along the course of the ventral root, extending from the spinal cord (arrow). In (B) staining of nerve fibers is seen in the ventral root (arrow) and individual fibers or small fascicles, shown in cross section, run rostro-caudally near the surface of the spinal cord.

proteins serve as adhesive ligands in vivo. However, a possible approach to this question is indicated by findings that a related morphogenetic movement, the migration of avian neural crest cells, is perturbed by injection into neural crest pathways of molecules that interfere with cell binding to FN (Boucaut et al., 1984; Bronner-Fraser, 1985).

In Vitro Interactions of Growth Cones with FN and LAM

Central and peripheral neurons from a wide variety of sources attach to and extend neurites on LAM-containing surfaces (Baron-van Evercooren et al., 1982; Edgar et al., 1984; Liesi et al., 1984; Rogers et al., 1983). Peripheral neurons also extend neurites on FN-coated surfaces, but central neurons do not attach well to FN-coated surfaces. Instead, dissociated cells from embryonal CNS tissues form cell aggregates that are loosely attached to a FN-treated surface and occasionally extend a few neurites onto the substratum (Rogers et al., 1983). This difference in the behavior of populations of central vs. peripheral neurites on FN may be related to the apparent absence of FN from most axonal pathways in the CNS. Perhaps, central and peripheral neurons are distinguised by differences in the expression of growth cone binding sites for FN. Such differences may comprise part of the developmental regulation of neuronal shape.

Guidance of Growth Cones by LAM-Containing Pathways

It has not been established what pathways for growth cones in vivo are marked by the presence of LAM. However, pathways of LAM-containing surfaces can be constructed in tissue culture dishes in order to investigate the ability of growth cones to be guided by LAM. The results show that both continuous and discontinuous pathways of LAM can guide growth cones extended by peripheral neurons.

A patterned surface containing a mosaic of LAM and ultraviolet light-irradiated LAM was constructed by covering regions of an LAM-coated coverslip with electron microscope grids (Hammarback et al., 1985). The coverslips were irradiated with ultraviolet light, and then the grids were removed prior to the culture of chick neurons on the patterned surfaces. The neurons preferentially bound to the protected LAM, and a dense outgrowth of neurites revealed the gridded network of the LAM that was protected by irradiation (Fig. 3). Antibodies to LAM also bind more to the protected LAM, allowing immunoflourescence to demonstrate the coincidence of protected LAM with the patterned neurite outgrowth (Fig. 3).

The most significant feature of this phenomenon is that neuronal attachment to the protected LAM is more than attachment to the irradiated LAM. This difference was greater when more LAM was initially deposited on the coverslip (Table 1), and a greater differential adhesion of neurons to protected vs. UV-irradiated LAM correlated with a stronger preference of growth cones for pathways of protected LAM. Thus, growth cones are guided for hundreds of microns along LAM pathways, as narrow as 7-10 micra. This navigational ability may depend on the ability of growth cones and their protrusions to sense the more adhesive nature of the protected LAM.

Growth cones in embryos traverse non-planar connective tissues and other spaces that are unlike tissue culture surfaces. In such cases, cues for guidance of growth cones

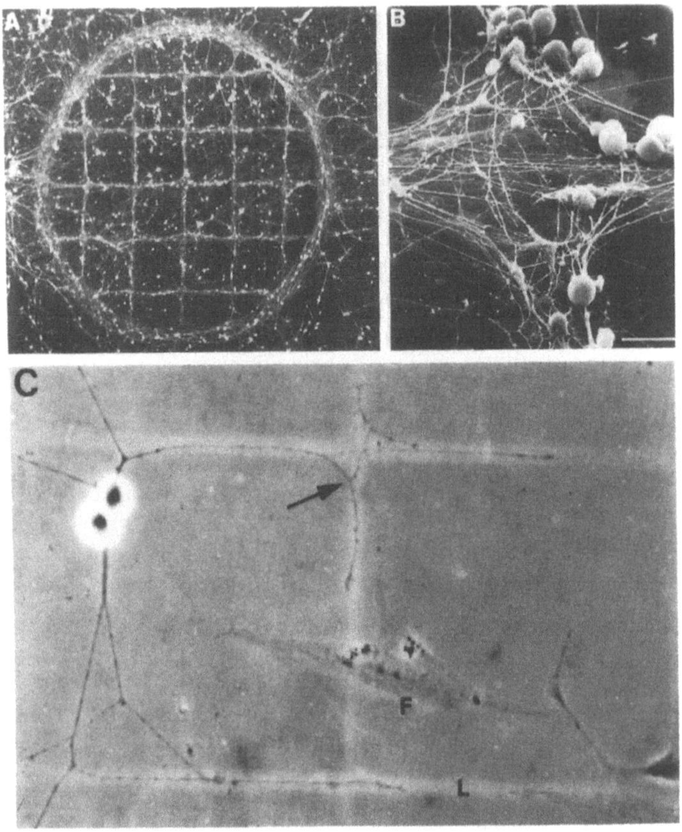

Figure 3. (A): Scanning electron microscopy of dissociated dorsal root ganglion cells on an ultraviolet light (UV)-patterned laminin substrate after 2 days culture in serum-free medium. Neurite outgrowth is guided by laminin protected from UV light, resulting in a pattern of dense neurite outgrowth resembling an electron microscope grid. Diameter of grid 3 mm. (B): High magnification of center of grid pattern in (A). There is no evidence of a mechanical barrier at the border between protected laminin and UV-exposed laminin. Neurites do not accumulate at the border and they do not tend to fasciculate. (C): Simultaneous phase contrast and epi-fluorescence microscopy identified the spatial relationship between neurites and the unirradiated laminin paths. These 7 µm wide paths (L) are precisely followed by extending neurites that often branch near the intersections of paths (arrow). Many fibroblastic cells (F) do not seem to be aligned by the laminin paths. Reprinted with permission from Hammarback et al. (1985).

may be discontinuously arranged. In grasshopper embryos, growth cones of sensory neurons make selective contacts with strategically placed guidepost cells that mark critical points for turning or advancement along the pathway from the limb into the CNS (Goodman et al., 1984). In vertebrates growth cones take a course dorsally or ventrally at the base of a developing limb that is determined by the origin of the neuronal somata (Lance-Jones and Landmesser, 1981a, b). This behavior may involve a choice determined by inter-

Table 1: Effects of laminin concentration on DRG cell adhesion to protected vs. UV-irradiated laminin

Laminin applied to Coverslip	Ratio Cells on Protected Laminin Cells on UV'd Laminin
23 μm/ml	0.88 ± 0.09
50 μg/ml	1.88 ± 0.09
100 μg/ml	3.01 ± 0.72

Laminin was applied to coverslips, and then half of each coverslip was UV irradiated, as described in Hammarback et al. (1985). Dissociated DRG cells were plated on the coverslips for one hour before removal of unattached cells, as described in Hammarback et al. (1985). Fixed cells remaining on the coverslips were counted in sample areas of the same size on both substrates.

actions of growth cones with specific cells at the decision site (Tosney and Landmesser, 1984).

Chick embryo neurons were cultured on mosaic substrata of LAM-containing islands surrounded by nonadhesive agarose-albumin gaps of varying widths (Hammarback and Letourneau, 1986). When gaps were 10 microns wide, many neurites crossed between the LAM islands, but when the gaps were 40 microns wide or more, growth cones rarely crossed the nonadhesive gaps (Fig. 5). Time lapse cinematography revealed a sequence of events leading to growth cones crossing the nonadhesive gaps. A growth cone advanced to the edge of a LAM island, halted and continued to extend filopodia across the agarose-albumin gap. Eventually, a single filopodium spanned the gap to a neighboring LAM island and made adhesive contact. Very rapidly, the growth cone crossed the gap, anchored by the attached filopodium, and then flattened when it gained contact with the LAM-containing surface. An interesting aspect of this behavior is that average filopodial length of a chick sensory growth cone is about 10 microns, while growth cones can cross gaps of more than 40 microns. Apparently, this behavior depends upon the uncommon protrusion of long filopodia.

These results may apply not only to LAM-containing surfaces, but to adhesive interactions, in general, and they have several implications for the guidance of growth cones by discontinuous substrata. Firstly, protrusive characteristics may limit the ability of growth cones to encounter appropriate substrata. If filopodia are too short to reach an alternative surface across a nonadhesive space, then a growth cone may remain on one

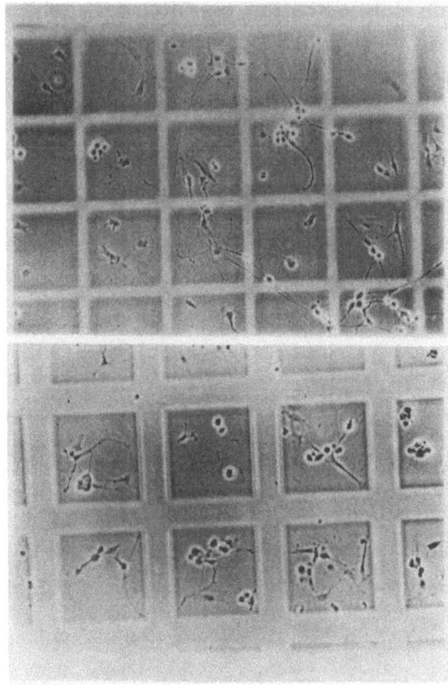

Figure 4. Dissociated DRG cells cultured on areas of adhesive laminin substrata separated by non-adhesive agarose substrata. An agarose-albumin mixture was dried onto glass coverslips that were masked with electron microscope grids and UV-irradiated. After irradiation, grids were removed, laminin was applied to the substrata, and laminin bound only to the irradiated agarose-albumin. The relationship of cells to the irradiated areas can be seen, because these areas are phase-dark. If the unirradiated gaps between UV-irradiated substrata are not too wide (top; 10 μm gaps), neurites can extend across them. If these gaps are wider, neurites do not extend across them (bottom; 40 μm gaps). Cells do not attach or extend neurites on unirradiated agarose albumin alone. Reprinted with permission from Hammarback and Letourneau (1986).

pathway vs. another. Similarly, a factor that could induce growth cones to extend longer filopodia would allow growth cones to sense new adhesive surfaces and perhaps follow new pathways. The ability of growth cones to cross wider gaps when adhesivity to the islands is greater indicates another component of navigational behavior. Adhesive guidepost cells could be positioned away from an axonal pathway so that only growth cells able to make sufficiently strong adhesions to the guidepost would be able to turn towards the cell. These tissue culture experiments provide basic information that is useful in formulating models to explain the choices and morphological changes that growth cones exhibit. In particular, these data indicate how a program of spatial and temporal variations in a finite set of adhesive and other properties of growth cones can generate the diversity and complexity of axonal pathways.

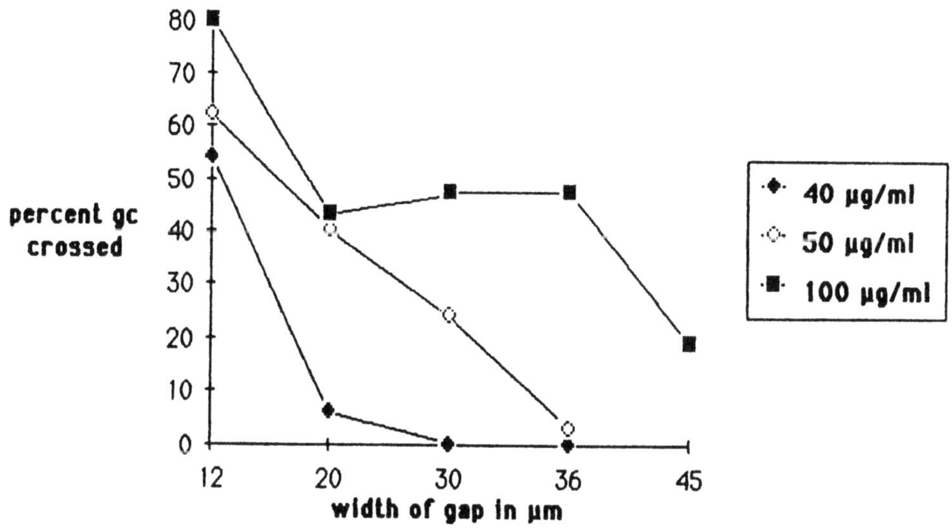

Figure 5. The relationship between adhesivity of laminin islands and widths of non-adhesive gaps that growth cones can cross. Laminin at the 3 concentrations indicated was applied to coverslips coated with agarose-albumin and UV-irradiated as described in legend for Fig. 4. The percent of growth cones that had reached the boundary of an island and crossed the nonadhesive agarose-albumin gap is plotted against gap width, a function of the EM grid that was used as a mask. The data show that growth cones cross wider gaps when the adhesivity of the laminin-coated islands is higher. Data from Hammarback and Letourneau (1986).

Biochemical Bases of Growth Cone Adhesion to FN and LAM

What is the nature of the adhesive interactions of nerve growth cones with the extracellular matrix? Significant issues include definition of the portions of FN and LAM that mediate adhesion of growth cones and characterization of cell surface components of the growth cone that make adhesive bonds to FN and LAM.

An important feature of FN and LAM is that these large glycoproteins interact with a multiplicity of cellular and extracellular components at specific conformational sites in the glycoproteins (Hynes and Yamada, 1982; Timpl et al., 1983). A combination of treatment with proteolytic enzymes and affinity chromatography with various ligands has allowed the purification from plasma FN of fragments that separate some of these binding affinities. Peripheral and central neurons from chick embryos were cultured on surfaces coated with each of five fragments from different regions of FN (Rogers et al., 1985). DRG ganglion neurons attach to and extend neurites on surfaces coated with a 75 kd fragment that contains a previously described cell binding domain of FN (Pierschbacher and Rouslahti, 1984). Spinal cord cells attach to but do not extend stable neurites on this fragment, so the response is like their unfavorable growth on whole FN. However, both DRG and spinal cord neurons extend neurites on a fragment from the carboxyl end of FN, that also binds strongly to heparin, a glycosaminoglycan (GAG) with structure similar to the extracellular GAG, heparan sulfate. These results indicate that FN may contain at least two regions that mediate cell attachment, and the different responses of DRG and spinal cord neurons indicate that neurons may differ in the presence of cell surface receptors for components of the extracellular matrix or other cell surfaces.

Other work has probed the involvement of a cell surface glycoprotein that has been defined by the production of monoclonal antibodies that clock cell attachment to extracellular matrices (Buck et al., 1985). The antibody CSAT precipitates a complex of three glycoproteins that may mediate cell attachment to FN and perhaps also to LAM (a different monoclonal, JG22, recognizes the same glycoprotein complex; Hasegawa et al., 1985). In biochemical studies the glycoprotein complex binds to both FN and LAM (Horwitz et al., 1985), and the binding to FN can be disrupted by presence of the tetrapeptide, Arg-Gly-Asp-Ser, a FN sequence that is recongized by cell surface binding sites (Akiyama et al., 1986; Pierschbacher and Rouslahti, 1984). Bozyczko and Horwitz (1986) have shown that neurite elongation on FN and LAM is disrupted by the addition of CSAT to culture medium. Their immunohistochemical work also indicates that the CSAT glycoprotein is on neurites and growth cones. We have found that when CSAT is added, DRG growth cones stop moving on FN and LAM surfaces, although filopodial activity continues. Growth cone extension on FN is also stopped by the FN tetrapeptide, but the tetrapeptide does not stop growth cones on an LAM-coated surface. These preliminary studies indicate that growth cones interact with the same sequence of fibronectin that has been described by others as a cell binding region, and the CSAT glycoprotein complex on growth cones may mediate adhesive interactions with FN and LAM. The relationship of these initial findings to the interaction of neurons with the distinct heparin-binding fragment of FN needs further study, as do the adhesive interactions of the growth cones of spinal cord neurons with FN and LAM.

Summary and Conclusions

FN and LAM are present in the extracellular matrices of embryos, where they may mediate adhesive interactions that are required for the traction that permits growth cone motility. Evidence also exists for the presence of these glycoproteins in extracellular spaces of the developing CNS, a region that has not been considered to contain extracellular matrix. A specific complex of glycoproteins on the growth cones of DRG neurons may mediate adhesive interactions with FN and LAM.

It is becoming clear that growth cones trace stereotypic pathways to generate the mature nervous system. Growth cones are sensory-effector systems that interact with the environment and respond with stereotypic elongation, turning, directionality, etc.

The presence of cell surface ligands for adhesion of growth cones to the extracellular matrix and to the surfaces of other nerve fibers and cells is a crucial component of this sensory-effector system. The specific distribution of these ligands on growth cones may help to differentiate the navigation of growth cones from different neurons. We will learn much by describing the ligands and receptors on growth cones, but it will be a further challenge to understand how these sensory capabilities are actually integrated to produce characteristic and distinct growth cone behaviors in a complex and dynamic environment.

Acknowledgements. The research on neuronal interactions with fibronection and laminin in the senior author's laboratory has been supported by NIH grants HD17192 and 19950, NSF grant PCM8203855 and a grant from the Spinal Cord Society. The technical assistance of Terri Shattuck and Jerry Sedgewick is greatly appreciated.

References

Akiyama, S.K., S.S. Yamada, and K.M. Yamada (1986) Characterization of a 140-kD avaian cell surface antigen as a fibronection-binding molecule. J. Cell Biol. **102:** 442-448.

Baron-van Evercooren A., H.K. Kleinman, S. Ohno. P. Marangos, J.P. Schwartz, and M. Dubois-Dalcq (1982) Nerve growth factor, laminin, and fibronection promote neurite growth in human sensory ganglia cultures. J. Neurosci. Res. **8:** 179-193.

Boucaut, J.C., T. Darribere, T.J. Poole, H. Aoyama, K.M. Yamada, and J.P. Thiery (1984) Biologically active synthetic peptides as probes of embryonic development: A competitive peptide inhibitor of fibronectin function inhibits gastrulation in amphibian embryos and neural crest migration in avian embryos. J. Cell Biol. **99:** 1822-1830.

Bozyczko, D., and A.F. Horwitz (1986) The participation of a putative cell surface receptor for laminin and fibronection in peripheral neurite extension. J. Neurosci., in press.

Bray, D. (1982) Filopodial contraction and growth cone guidance. In *Cell Behavior,* R. Bellair, A. Curtis, and G. Dunn, eds. pp. 299-318. Cambridge Univ. Press, Cambridge, UK.

Bronner-Fraser, M. (1985) Alterations in neural crest migration by a monoclonal antibody that affects cell adhesion. J. Cell Biol. **101:** 610-617.

Buck, C.A., K.A. Knudsen, C.H. Damsky, C.L. Decker, R.R. Greggs, K.E. Duggan, D. Bozyczko, and A.F. Horwitz (1985) Integral membrane protein complexes in cell-matrix adhesion. In *The Cell in Contact,* G.M. Edelman and J.P. Thiery, eds., pp. 345-364. John Wiley and Sons. New York.

Edgar, D., R. Timpl, and H. Thoenen (1984) The heparin binding domain of laminin is responsible for its effects on neurite outgrowth and neuronal survival. EMBO J. **3:** 1463-1468.

Goodman, C.S., M.J. Bastiani, C.Q. Doe, S. du Lac, S.L. Helfand, J.Y. Kuwada, and J.B. Thomas (1984) Cell recognition during neuronal development. Science **225:** 1271-1279.

Hammarback, J A., S.L. Palm, L.T. Furcht, and P.C. Letourneau (1985) Guidance of neurite outgrowth by pathways of substratum-absorbed laminin. J. Neurosci. Res. **13:** 213-220.

Hammarback, J.A., and P.C. Letourneau (1986) Neurite extension across regions of low adhesivity: Implications for the guidepost hypothesis of axonal pathfinding. Dev. Biol., in press.

Hasegawa, T., E. Hasegawa, W.T. Chen, and K.M. Yamada (1985) Characterization of a membrane-associated glycoprotein complex implicated in cell adhesion to fibronectin. J. Cell. Biochem. **28:** 307-318.

Hatten, M.E., M.B. Furie, and D.B. Rifkin (1982) Binding of developing mouse cerebellar cells to fibronectin: A possible mechanism for the formation of the external granule layer. J. Neurosci. **2:** 1195-1206.

Horwitz, A., K. Duggan, R. Greggs, C. Decker, and C. Buck (1985) The CSAT antigen has properties of a receptor for laminin and fibronectin. J. Cell Biol. **101:** 2134-2144.

Hynes, R.O. and K.M. Yamada (1982) Fibronectins: Multifunctional modular glycoproteins. J. Cell Biol. **95:** 369-378.

Hynes, R.O., R. Patel, and R.H. Miller (1986) Migration of neuroblasts along preexisting axonal tracts during prenatal cerebellar development. J. Neurosci. **6:** 867-876.

Kurkinen, M., K. Alitalo, A. Vaheri, S. Stenman, and L. Saxen (1979) Fibronectin in the development of the embryonic chick eye. Dev. Biol. **69:** 589-600.

Lance-Jones, C., and L. Landmesser (1981a) Pathway selection by chick lumbrosacral motoneurones during normal development. Proc. R. Soc. Lond. (Biol.) **214:** 1-18.

Lance-Jones, C. and L. Landmesser (1981b) Pathway selection by chick lumbrosacral motoneurones in an experimentally altered environment. Proc. R. Soc. Lond. (Biol.) **214:** 19-52.

Letourneau, P.C. (1982) Nerve fiber growth and its regulation by extrinsic factors. In Neuronal Development, N.C. Spitzer, ed. pp. 213-254. Plenum Press, New York.

Liesi, P., D. Dahl, and A. Vaheri (1983) Laminin is produced by early rat astrocytes in primary culture. J. Cell Biol. **96:** 920-924.

Liesi, P., D. Dahl, and A. Vaheri (1984) Neurons cultured from developing rat brain attach and spread preferentially on laminin. J. Neurosci. Res. **11:** 241-251.

Lievo, I., A. Vaheri, R. Timpl, and J. Wartivovaara (1980) Appearance and distribution of collagens and laminin in the early mouse embryo. Dev. Biol. **76:** 100-114.

Mayer, B.W., E.D. Hay, and R.O. Hynes (1981) Immunocytochemical localization of fibronectin in embryonic chick trunk and area vasculosea. Dev. Biol. **82:** 267-286.

Pierschbacher M., and E. Rouslahti (1984) Cell attachment activity of fibronectin can be duplicated by small synthetic fragments of the molecule. Nature **309:** 30-33.

Ramon y Cajal, S. (1890) Sur l'origine et les ramifications des fibres nerveuses de la moelle embryonaire. Anat. Anz. **5:** 609-613; 631-639.

Rickmann, M., J.W. Fawcett, and R.J. Keynes (1985) The migration of neural crest cells and the growth of motor axons through the rostral half of the chick somite. J. Embryol. exp. Morph. **90:** 409-414.

Rogers, S.L., P.C. Letourneau, S.L. Palm, J. McCarthy, and L.T. Furcht (1983) Neurite extension by peripheral and central nervous system neurons in response to substratum-bound fibronectin and laminin. Dev. Biol. **98:** 212-220.

Rogers, S.L., J.B. McCarthay, S.L. Palm, L.T. Furcht, and P.C. Letourneau (1985) Neuron-specific interactions with two neurite-promoting fragments of fibronectin. J. Neurosci. **5:** 369-378.

Rogers, S.L., K.J. Edson, P.C. Letourneau, and S.C. McLoon (1986) Distribution of laminin in the developing nervous system of the chick. Dev. Biol. **113:** 429-435.

Schachner, M., G. Schoonmaker, and R.O. Hynes (1978) Cellular and subcellular localization of LETS protein in the nervous system. Brain Res. **158:** 149-158.

Timpl, R., J. Engel, and G.R. Martin (1983) Laminin - A multifunctional protein of basement membranes. Trends Biochem. Sci. **8:** 207-209.

Tosney, K.W., and L.T. Landmesser (1984) Pattern and specificity of axonal outgrowth following varying degrees of chick limb bud ablation. J. Neurosci. **4:** 2518-2527.

Tosney, K.W., and L.T. Landmesser (1985) Development of major pathways for neurite outgrowth in the chick hindlimb. Dev. Biol. **109:** 193-214.

Trinkaus, J.P. (1984) *Cells into Organs*. Prentice-Hall, Inc., Englewood Cliffs, N.J.

Wartiovaara, J., I. Lievo, and A. Vaheri (1980) Matrix glycoproteins in early mouse development and in differentiation of teratocarcinoma cells. In *The Cell Surface: Mediator of Developmental Processes*, S. Subtelney and N.K. Wessells, eds. pp. 305-324. Academic Press, New York.

THE ROLE OF BASAL LAMINA IN AXON REGENERATION

M. Berry[1], S.M. Hall[1], E.L. Rees[1], P. Yiu[1], and J. Sievers[2]

[1] **Anatomy Department**
Guy's Hospital Medical School
London, Great Britain

[2] **Anatomy Department**
University of Kiel
Kiel, Federal Republic of Germany

Abstract. Normal and acellular peripheral nerves of adult rats have been anastomosed to the optic nerves, disc and retina. At 30 days post lesion, axons regenerated into normal PN grafts but none grew into acellular PN grafts. Similarly, central axons did not regenerate into muscle or optic nerve grafts. Treatment of a PN graft with a mitotic arrest agent (mitomycin C) markedly reduced the number of fibres regenerating therein. No regenerating optic fibres were found in muscle grafts or in acellular optic nerve grafts. The results show that central axons grow only into Schwann cell populated grafts and not into either acellular PN, muscle or optic nerves even in the presence of a laminin rich basal lamina. These findings do not support the contention that laminin per se is neurotrophic in vivo but, instead, suggest that central axons regenerate into PN because Schwann cells elaborate a chemotactic factor which attracts central axons into the basal lamina tubes of the PN. It is not clear if the substrate for growth of axons within the basal lamina tubes is the laminin rich lamina lucida or the plasmalemma of the Schwann cell.

Introduction

The faculty of substrate adhesion is an important determinant of neurite elongation (Weiss, 1955). Adhesion enhances neurite survival (Letourneau, 1979) and this may be related to the binding of trophic molecules to the substrate (Collins, 1978; Berg, 1984). The glycoprotein laminin appears both to promote neurite outgrowth and survival (Baron Van Evercooren et al., 1982; Manthorpe et al., 1983; Rogers et al., 1983; Edgar et al., 1984; Liesi et al., 1984; Smalheiser et al., 1984; Hopkins et al., 1985) and provide an adhesive substrate for growth cones (Timpl et al., 1983; Davis et al., 1985; Hammarback et al., 1985; Lander et al., 1985). Laminin is a constituent of basal lamina throughout the body (Timpl et al., 1979; Yamada, 1983), including that of skeletal muscle (Sanes, 1983) and peripheral nerve (PN) (Bunge and Bunge, 1983), and is localised (Tohyama and Ide,

NATO ASI Series, Vol. H5
Mesenchymal-Epithelial Interactions in Neural Development
Edited by J. R. Wolff et al.

1984; Carbonetto, 1984; Tohyama, 1985), although not necessarily accessible (Yourchenco et al., 1986), within the lamina lucida. Except around blood vessels, over the glia limitans externa and within the olfactory nerve (Liesi, 1985), laminin is absent from the mammalian and avian central nervous system (CNS). However, in the piscine and amphibian brain laminin is continuously expressed by astrocytes (Liesi, 1985; Hopkins et al., 1985 and, in the injured mammalian brain laminin is transiently produced by reactive astrocytes (Bernstein et al., 1985; Liesi et al., 1984) which also secrete the glycoprotein into the newly refashioned glia limitans externa (Berry et al., 1983; Bernstein et al., 1985; Krüger et al., 1986).

The differential distribution of laminin across genera, and within species, is correlated with the regeneration of axons after injury, which readily takes place in the mammalian olfactory nerve and peripheral nervous system (PNS), and in amphibian and piscine CNS, but is classically described as abortive in the mammalian and avian CNS (Cajal, 1928; Björklund and Stenevi, 1979; Kiernan, 1979; Berry, 1979, 1983, 1985). It has therefore been suggested that an absence of laminin in the mammalian and avian CNS may account for the regenerative failure of most CNS axons in these groups (Hopkins et al., 1985; Liesi, 1985). Indeed, such a thesis gains support from PN transplantation studies which show that rat CNS axons will grow in a Schwann cell environment where basal lamina is abundant, (Benfey and Aguayo, 1982; Aguayo et al., 1984; Richardson et al., 1980, 1984; David and Aguayo, 1985; Kromer and Cornbrooks, 1985). The route taken by regrowing axons within a PN graft is invariably on the inner aspect of the basal lamina tubes of Schwann cells (Ide et al., 1976), over the surface of the laminin rich lamina lucida (Carbonetto, 1985; Yourchenco et al., 1986). Within denervated muscle, regenerating axons also follow the lamina lucida of the basal lamina of the sarcolemma (Sanes et al., 1978; Nitkin et al., 1983; Keynes et al., 1984; Glasby et al., 1986). We have tested the hypothesis that laminin is a growth promoting molecule in vivo by suturing PN and optic nerve grafts to the cut ends of both CNS and PNS axons having pretreated one group of grafts with a freezing/thawing procedure which kills all cellular elements but leaves the basal lamina intact (Berry et al., 1986a, b, c). According to the laminin hypothesis (Liesi, 1985), regenerating CNS axons should penetrate both cellular and acellular PN grafts since basal lamina is retained in both.

Experimental Procedures

(a) Transplants

WAG albino rats (250 g) of either sex were used throughout, PN transplants were either segments of intact normal, acellular, predegenerate or mitomycin C treated sciatic or obturator nerves. In the acellular preparations all Schwann cells were killed in the sciatic nerve graft, prior to transplantation, by repeatedly freezing (x 5 to -25°C) and thawing (Ide et al., 1976). Predegenerate nerves were first isolated in the animal by severing the nerve and suturing the cut ends. After 5 days, the isolated nerves were removed from the donor animal and transplanted. Mitomycin C treated nerves were injected with a solution (400 μg/ml in physiological saline) using the technique of Hall and Gregson (1975). Injections were sited at one end of the nerves and this area was then transected ready for transplantation.

Optic nerve transplants consisted of normal and acellular nerves. The latter were prepared in the same way as the PN. Each experimental group of animals contained 6 rats.

(b) Transplantation of PN to optic nerve (Fig. 1)

Complete sectioning of the optic nerve was achieved by removing a 3 mm length, 2-3 mm from the globe. A 15-20 mm segment of normal or acellular sciatic nerve was then anastomosed to the distal optic nerve stump, as previously described by us (Berry et al., 1986a, b). At 30 days post-lesion (dpl), 5 μl of a 4% solution of HRP-WGA (Sigma L 2384) was slowly infused into a crushed region of the sciatic nerve grafts, 10 mm from the site of anastomosis, and the retinae examined 36 hrs later for the presence of HRP-WGA filled ganglion cells. Animals were perfused with 4% paraformaldehyde and the anastomotic site of these animals was also embedded in polyester wax and alternate, 5 μm, sections were used for immunocytological demonstration of either laminin using rabbit antisera to mouse laminin (Bethesda Research Laboratories - BRL) or axons, using a mouse monoclonal anti-neurofilament antibody (dil 1/3, Clone No. 97 producing antibody against 200,000 dalton sub-unit of rat brain neurofilament - Dr. B. Anderton, St. George's Hosp. London). The immunofluorescence demonstrated in the rat with laminin antisera, used at 1/100 dilution, was completely suppressed by preincubation with purified laminin (BRL) from EHS mouse sarcoma.

The orthograde tracer rhodamine B isothiocyanate (Thanos and Bonhoeffer, 1983) (RITC-B, Sigma; R-1755) was injected into the vitreous of the operated eye, at 5, 10, 15 and 30 dpl, to study regeneration across approximated stumps of the optic nerve and also into the different PN grafts.

(c) Transplantation of PN to the optic disc (Fig. 1)

Normal, acellular and predegenerate sciatic nerve grafts were sutured to the sclera at the point where the optic nerve joins the eye ball. The optic nerve was removed by transection at the lamina cribrosa. The cut end of each PN graft was placed against the lamina and the epineurium sutured to the sclera. The application of HRP-WGA and the anti-laminin and anti-neurofilament immuno-fluorescence were carried out as described above.

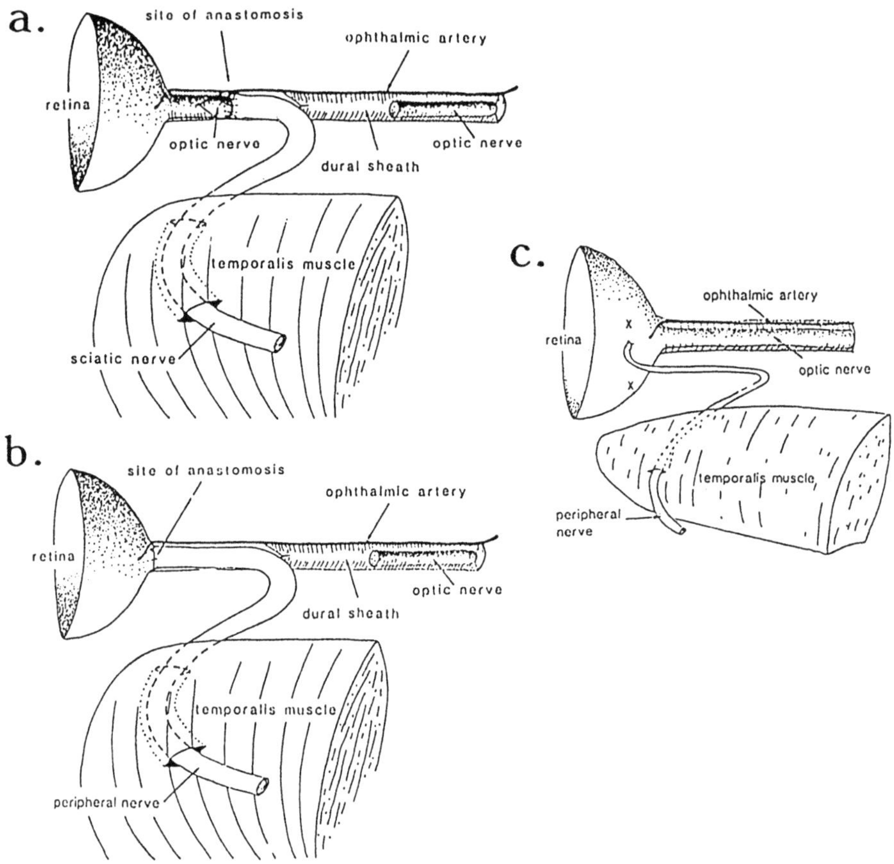

Figure 1: Diagrammatic illustration of the operative technique employed for anastomosis of (a) peripheral nerve (PN) to optic nerve (b) PN to optic disc (c) PN to retina - the crosses mark the positions of independent retinal lesions juxtaposed to the site of implantation of the PN. In each case HRP-WGA was subsequently injected into the crush site of the PN over the temporalis muscle outside the orbit.

(d) Transplantation of PN into the retina (Fig. 1)

Normal, acellular, predegenerate and mitomycin C treated obturator nerves were implanted into the retina near the disc by making a 'T' shaped incision through the sclera, (each arm of the 'T' being 0,5 mm long). One mm of the PN, stripped of epineurium, was inserted into the retina and sutured to the sclera with 10/0 nylon. The retina was lesioned at varying distances from the point of insertion of the PN in the retina. HRP-WGA was subsequently applied to the PN as described above.

(e) Transplantation of optic nerve to retina and to PN implants

Normal and acellular optic nerve segments were implanted into the retina using the technique described above. Optic nerve segments were also anastomosed, end to end, to normal PN grafts implanted into the retina 30 days previously. The PN grafts were trimmed, to within 5 mm of the sclera, before suturing the acellular optic nerves in place with 10/0 nylon.

(f) Control animals

Complete optic nerve section both at the disc and 2 mm from the disc followed by re-suturing of the cut ends, together with simple needle (25 g) lesions of the retina provided the surgical controls for the experiments. Controls for diffusion of HRP-WGA were provided by grafting normal sciatic nerve segments to the optic nerve and injecting a crush site, 10 mm from the site of anastomosis, with HRP-WGA, at the time of operation and killing 36 hrs later.

(g) Ganglion cell counts

The total retinal ganglion cell (RGC) population and their size distribution were estimated in the normal retina by immediately infusing HRP-WGA into normal optic nerve at the point of crush 2-3 mm from the globe. Animals were also perfused with 4% paraformaldehyde and their retinae whole mounted onto gelatin subbed slides and stained with cresyl fast violet for RGC counting (Berry et al., 1986a, b). The numbers of HRP-WGA filled RGCs in each size group, arbitrarily set at 5 μm^2 intervals, were subtracted from the total Nissl stained retinal count to obtain the number of displaced amacrine cells (Perry, 1981) at each size interval. This mean size distribution of amacrine cells was subtracted from the mean total Nissl stained retinal count to obtain the mean total of RGCs/glial cells and their size profile in each experimental group. The contralateral eye in each HRP injected rat acted as a control for systemic diffusion.

(h) Electron microscopy

Animals were perfused with 2.5% gluteraldehyde in 0.1 M phosphate buffer at 5, 10, 15 and 30 dpl. The site of anastomosis was post-fixed in 1% osmium tetroxide, divided into 2 mm segments in a proximo-distal sequence and sections examined using a Hitachi HU 12 A electron microscope.

Results

(a) Transplantation of PN to optic nerve

RGC axons regenerated into normal PN transplants anastomosed to the optic nerve. The first signs of regeneration were seen at 10 dpl and, by 30 dpl, large numbers of axons crossed the anastomosis (Fig. 2). The density of axons decreased with distance along the PN graft, indicating that regenerating RGC axons were leaving the transplant along its

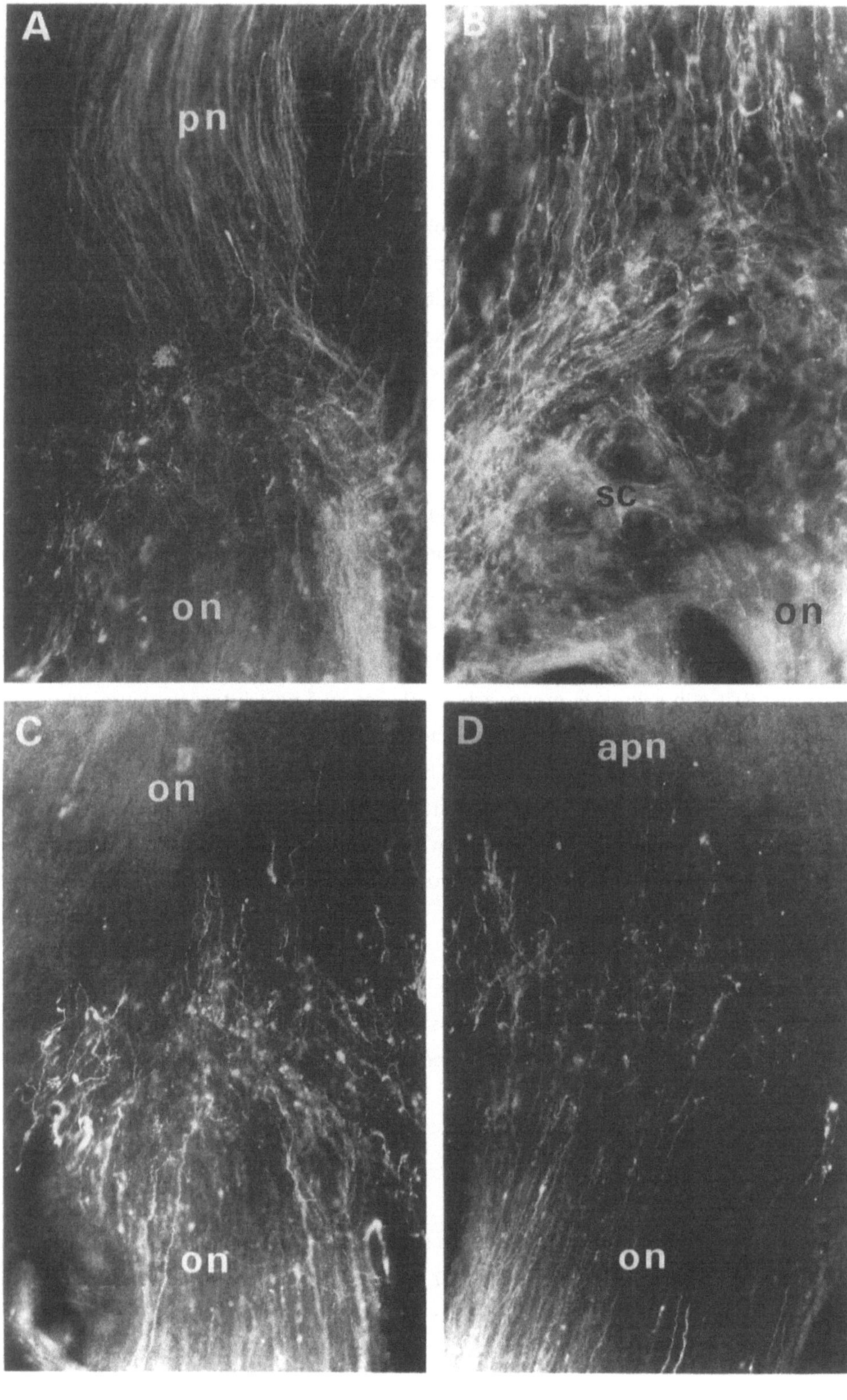
A
pn
on
B
sc
on
C
on
on
D
apn
on

Figure 2 (facing page): RITC-B orthograde tracing of RGC axons at 30 days post lesion (dpl) in the optic nerve (on) into a normal peripheral nerve (pn) graft (A) and into the scar about the proximal segment of the transected optic nerve (C), and at the site of anastomosis of a length of acellular peripheral nerve (apn). Note that, in C and D, no RGC axons reinvade neural tissue. Detail of the anastomosis between optic nerve and normal PN graft is shown in B, illustrating how the orientation of regenerating RGC axons is perturbed as they move through the scar (sc) fromt he optic nerve (on) before becoming aligned by Schwann cell basal lamina tubes within the peripheral nerve graft. A and C x220; B and D x500.

length. No axons crossed the anastomosis at any time after operation in the optic nerve section experiments, although many ramified within the scar (Fig. 2). In the acellular PN grafts, a small number penetrated the implants at 10 and 15 dpl but, by 30 dpl, only a few beaded axons remained at the margins of the grafts (Fig. 2).

Anti-laminin immunocytology detected laminin in both the normal and the acellular PN transplants (Fig. 3) within the basal lamina tubes of Schwann cells. Electron microscopical examination of both types of PN graft confirmed the presence of basal lamina tubes. In normal PN grafts, these tubes contained small diameter axons and Schwann cell processes, first detected at 8 dpl at the anastomotic site. No Schwann cells invaded the optic nerve from the PN graft. A few axons were present in the collapsed basal lamina tubes in the acellular grafts at 15 dpl at the site of the anastomosis (Fig. 4), but these did not advance into the graft by 30 dpl. In both PN grafts, myoblasts invaded from the surrounding orbital muscles.

Estimates of the numbers of regenerating RGC axons into the PN implants were achieved by injecting HRP-WGA into the grafts 10 mm from the site of anastomosis. HRP-WAG was found in RGCs only after injection of normal PN grafts (Fig. 5). The mean total number of HRP-WGA filled RGC in the normal retina was 111,206 ± 3,290 (S.E.) whilst the number of RGCs filling after normal PN grafting was 1,529 ± 125 (Table 1). Thus, 1,4% of the normal ganglion cell population regenerate their axons 10 mm into the normal PN grafts but none regenerate into the acellular graft.

RGC survival was estimated by first calculating the size of the displaced amacrine cell population in the normal retina. This was achieved by subtraction of the total RGC count (number of HRP-WGA filled cells) from the total Nissl stained cell count. The mean total number of amacrine cells in the normal retina was 108,045 ± 1.764 (Table 1). RGC survival was computed by subtracting the normal amacrine count from the total number of Nissl stained cells in the retina in each group, assuming that the normal amacrine population is not affected by optic nerve transection.

The number of RGCs surviving after optic nerve section, grafting an acellular PN and transplanting a normal PN is 10,443 ± 4,058, 27,851 ± 3,794, 37,658 ± 6,329 respectively. Representing regeneration in terms of surviving RGCs, 4.0% of surviving RGCs

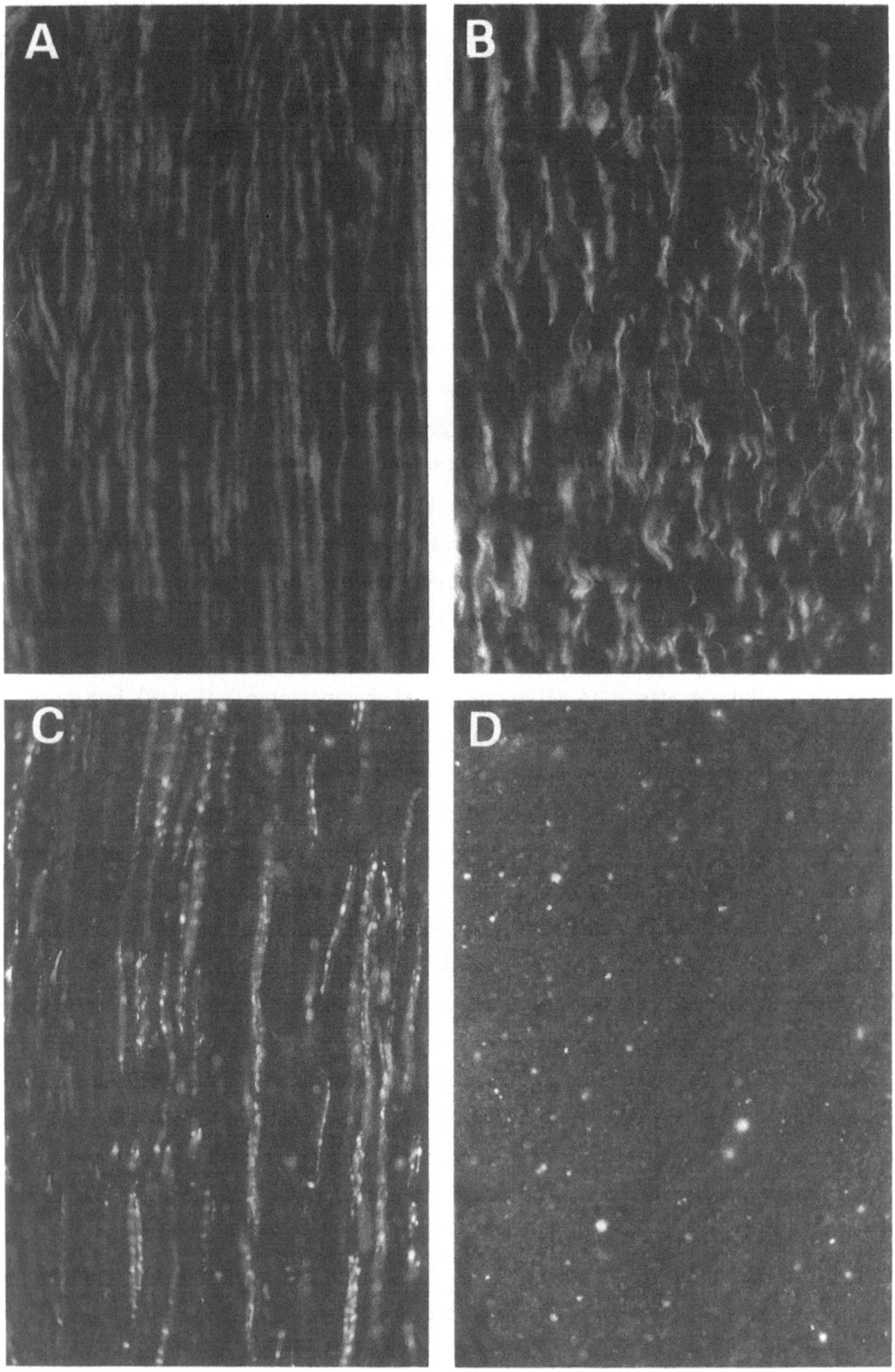

Figure 3

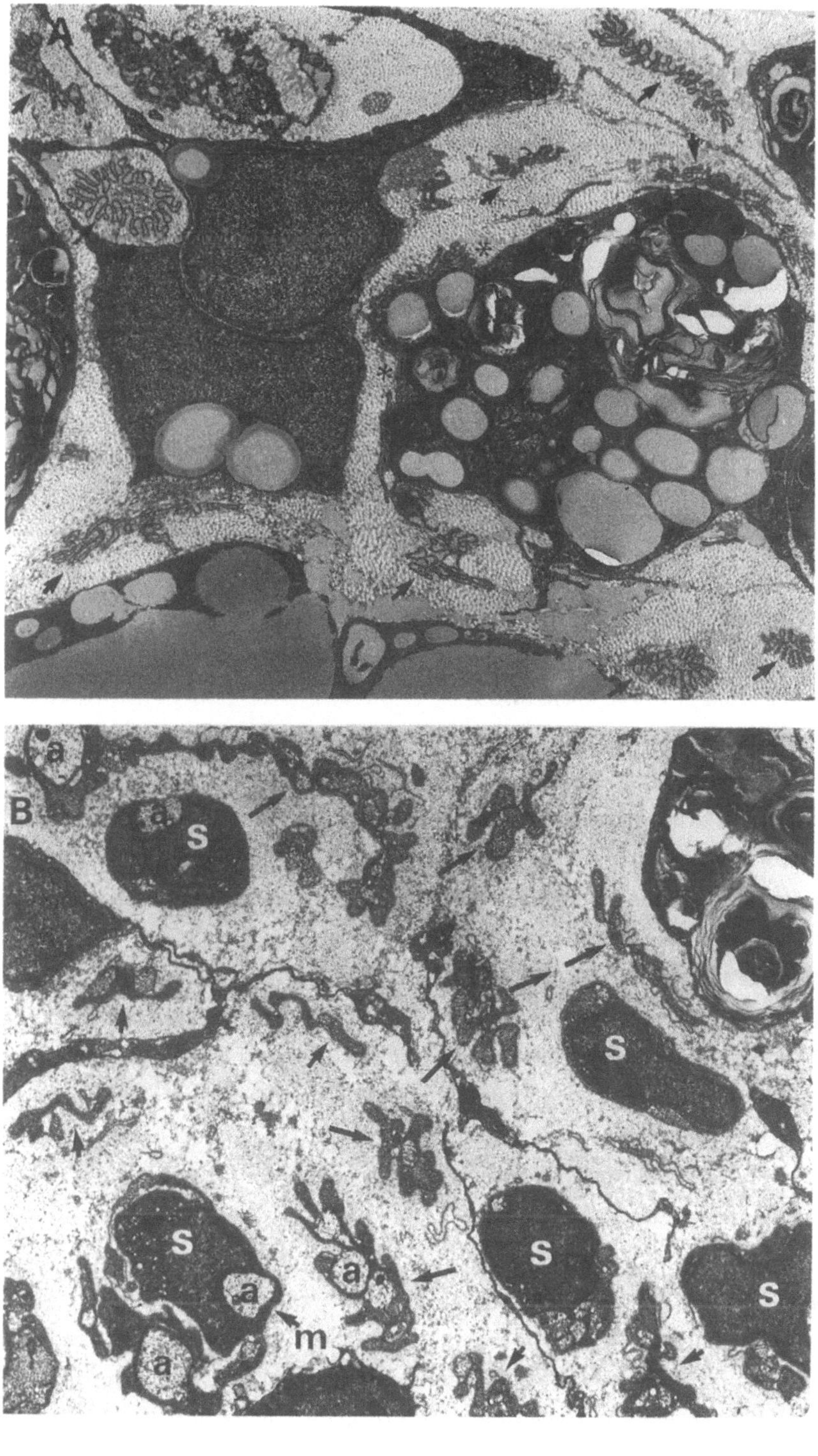

Figure 4

Figure 3 (previous page): Normal (A) and acellular (B) PN grafts at 30 days post lesion stained with anti-laminin antibody showing the presence of laminin positive Schwann cell basement membrane tubes in both specimens. Tubes in normal PN graft are normally aligned and some contain regenerating axons (C) stained with anti-neurofilament antibody, but those in the acellular graft are collapsed, disorganised and do not contain axons (D). The punctuate anti-neurofilament staining in D represents persistant axonal debris.
Figure 4 (previous page): A: Transverse section through an acellular PN graft, 0.5 cm distal to the point of anastomosis with the optic nerve, at 15 days post lesion (dpl), showing endoneurium containing a number of crenulated, collapsed and empty Schwann cell basal lamina tubes (arrows). Debris is present in macrophages, one of which is associated with a breach of the tubal wall (asterisk). Negatively stained collagen is abundant between these structures. x 8500.
B: Transverse section through a normal PN graft, 0.5 cm distal to the point of anastomosis with the optic nerve, at 15 dpl, showing endoneurium containing numerous basal lamina tubes in which Schwann cells (S) or Schwann cell processes (arrowed) and large (a) and small RGC axon sprouts are present. One neurite is undergoing remyelination (m). x 6300.

regenerated their axons 10 mm along the normal PN graft (Table 1).

(b) Transplantation of PN to optic disc

The results of experiments in which PNs were grafted to the optic disc replicated those of PN/optic nerve anatomosis (Fig. 5, Table 1). 2,332 .. 835 axons regenerated into the normal intact PN graft from the retina, but none were found in an acellular graft. RGC axonal regrowth was enhanced into predegenerate PN grafts where a mean of 4,777 ± 524 axons regenerated (Table 1). Effects on RGC survival were also similar since grafts of acellular, normal and predegenerate PN all saved respectively 26,062 ± 1,369, 39,976 ± 4,216 and 31,195 ± 3,442 RGCs (Table 1), and the percentage RGCs regenerating their axons along the graft was, respectively, 0%, 6% and 15%. Antilaminin immunofluorescence showed the presence of laminin in all grafts at 5, 10, 15 and 30 dpl, including the acellular PN grafts, whilst electron microscopical studies confirmed the presence of empty basal lamina tubes in the latter at these times. In normal and predegenerate grafts, basal lamina tubes were packed with Schwann cell processes and axons, some of which were already myelinated at 30 dpl (Fig. 6).

(c) Transplantation of PN to retina

RGC axons respond to injury in the fibre layer of the retina by growing randomly about the wound (Fig. 7). Growth continues for at least as long as 250 dpl (the time when our experiments were terminated) when the entire retina is packed with a tangle of randomly growing axons. Most of the regenerated axons are in either the fibre layer or the innermost strata of the inner plexiform layer.

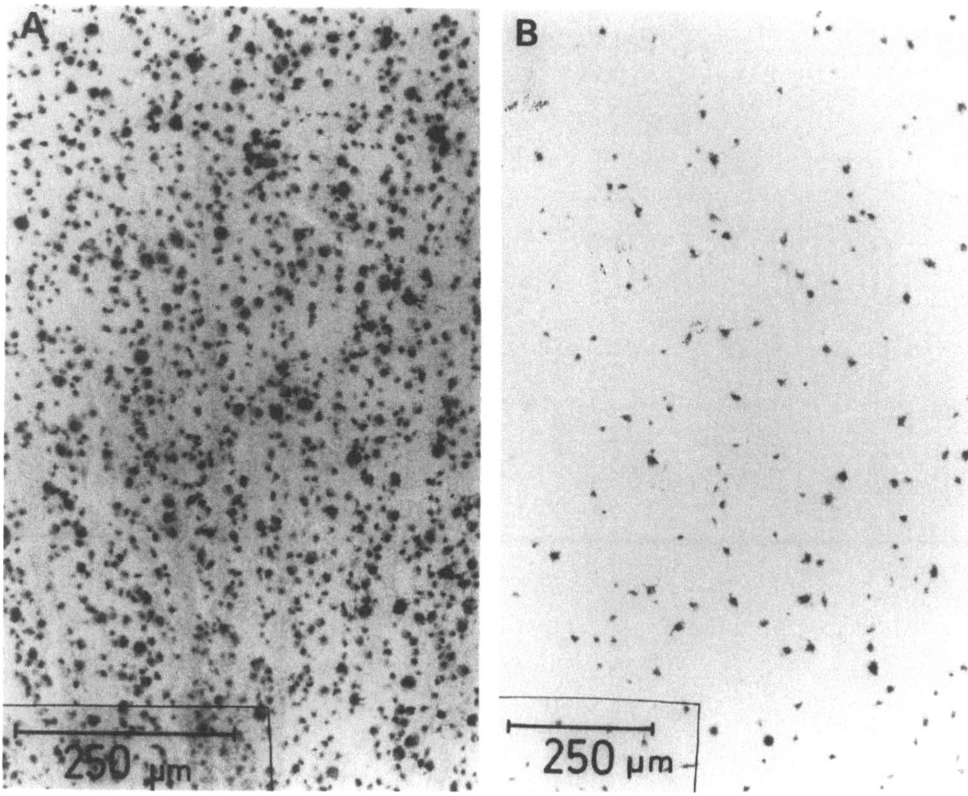

Figure 5: A: HRP-WGA filled RGCs in normal retina 36 hrs after injection of the crush site in the optic nerve 2 mm from the disc. B: HRP-WGA filled RGCs in the retina of a rat 36 hrs after injection of the crush site in a PN, 10 mm from the site of anastomosis to the optic disc. Anastomosis was performed 30 days previously.

When a normal PN is sutured into a retinal lesion, severed RGC axons regenerate into the grafts directly, without deviation (Fig. 8) HRP-WGA studies show RGC filling in the expected sectors of the retina (Fig. 9). By measuring the density of filled RGCs in the sector and correcting for the total area of the retina, the local effects of PN implants are scaled-up and expressed in terms of RGC axon regeneration in the retina as a whole, thus offering a means of comparison with the results of disc and optic nerve grafting. Normal PN implants into the retina allow the equivalent of 2,743 ± 305 axons to regenerate over the entire retina and predegenerate PN grafts permit 5,359 ± 1,162 (Table 1). No RGC axons enter the acellular PN grafts, by 30 dpl (Fig. 8), in which lami-

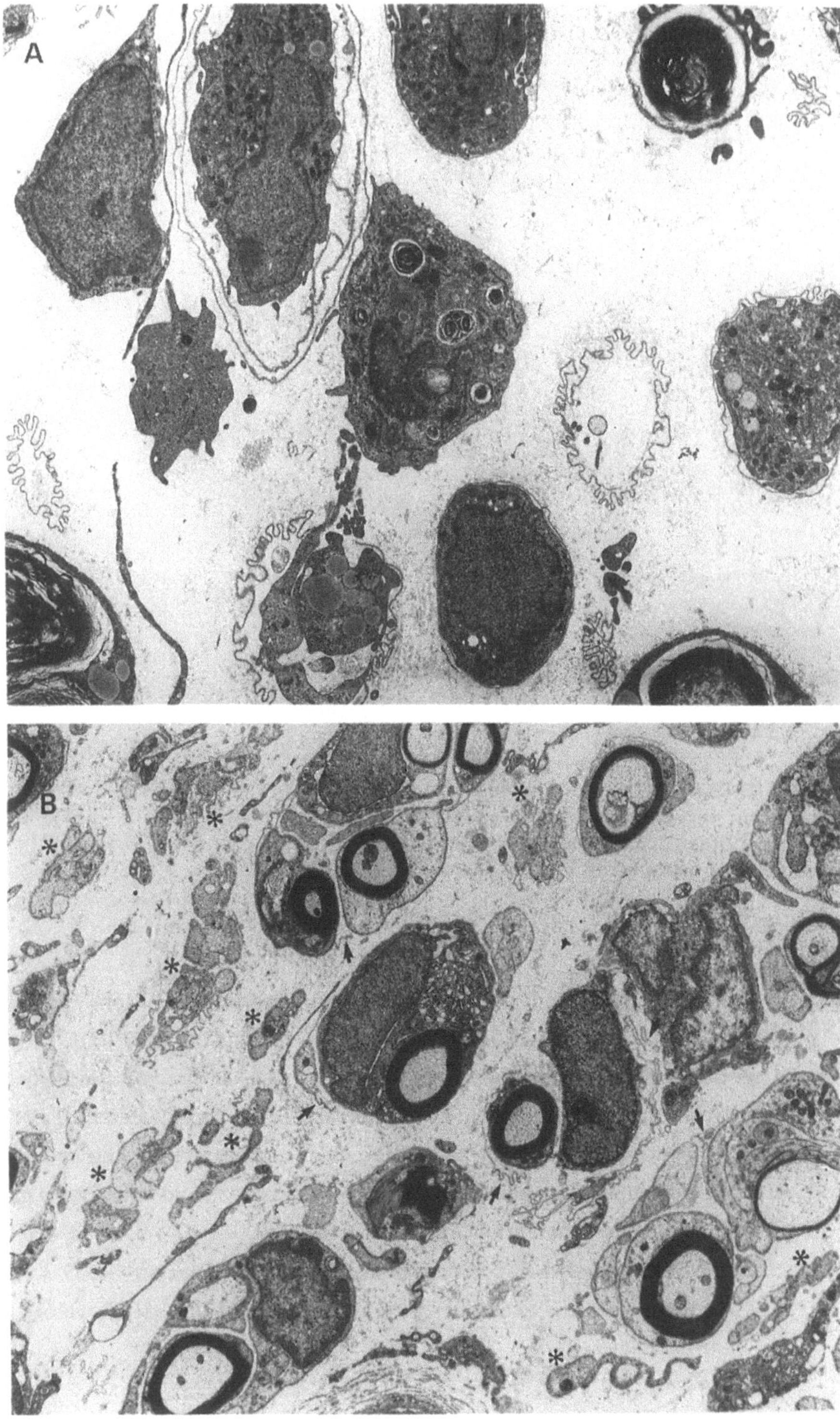

Figure 6

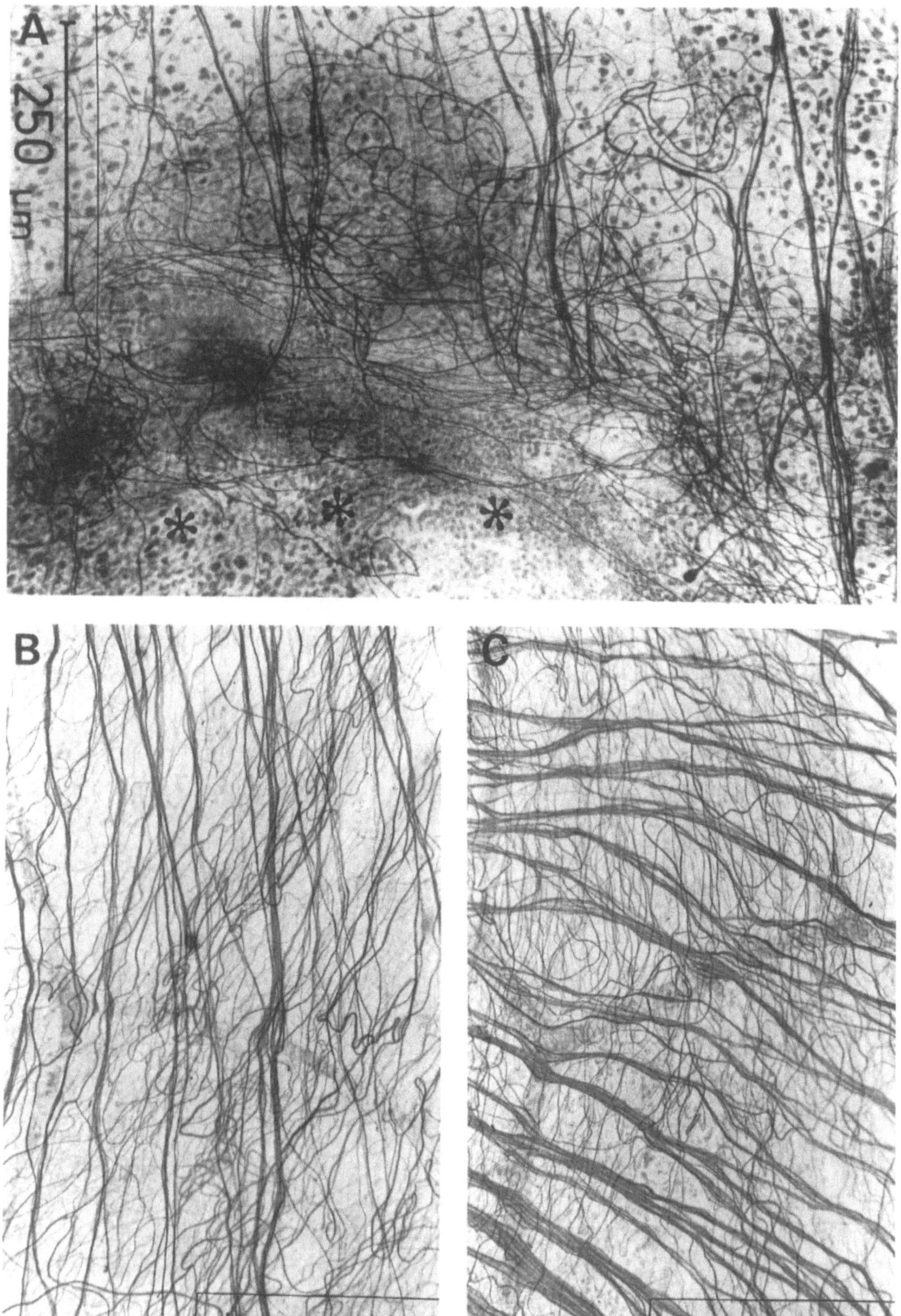

Figure 7

Figure 6 (previous page): A: Transverse section through an acellular PN graft, 0.5 mm distal to the point of anastomosis with the optic disc, at 30 days post lesion (dpl), showing endoneurium containing collapsed, crenulated Schwann cell basal lamina tubes which are either empty or containing macrophages. x 9000.
B: Transverse section through a normal PN graft, 0.5 mm distal to the point of anastomosis with the optic disc, at 30 dpl, showing endoneurium containing Schwann cells and myelinated axons herniating free from the old basal lamina tubes (arrowed). These Schwann cells now have their own basal lamina. Old basal lamina tubes persist packed with Schwann cell processes with very fine axons (asterisks). x 6300.
Figure 7 (previous page): A: Random growth of RGC axons about the site of a retinal lesion (asterisks) 30 dpl. These fibres are growing in the fibre and inner plexiform layers - optic disc beyond the lower edge of the picture.
B and C: Randomly orientated regeneration of RGC axons in the inner plexiform and fibre layers of an area of the retina far removed from the original site of injury 250 dpl. Marker = 250 µl, silver stained retinal whole mounts, optic disc beyond the lower edge in B and left hand edge in C, normal radial fascicules of RGC axons coursing in the fibre layer are clearly seen in A, B and C.

nin positive basal lamina tubes persist. Instead, retinal axons grow randomly about the stump of the acellular implant within the fibre layer and internal plexiform layer of the retina. No estimates of the number of surviving RGCs within affected retinal sectors were made. The tips of some PN grafts were treated with mitomycin C before implantation into the retina. In these animals, the equivalent of 161 ± 49 HRP-WGA labelled RGCs were found in the retina (Table 1), at 30 dpl, after injecting the PN graft 5 mm from the site of implantation. Some axons were seen to grow randomly about these amitotic implants.

Another group of normal PN grafts were implanted into the retina and lesions were placed 1, 2, 3 and 4 mm respectively from the implants. RGC axons regenerated from the lesion site into the implant when separated by distances up to 3 mm but not beyond (Fig. 9). In only two animals, in which lesions were 1 or 2 mm apart, were axons seen to traverse the distance through the retina. In all other cases, axons passed in the scar, left by the original needle tract in the vitreous, from lesion to PN implant, over the posterior surface of the lens capsule.

In another set of experiments, an acellular PN was anastomosed to the cut end of a 5 mm length of normal PN, implanted into the retina 30 days previously. After a further 30 days, HRP-WGA was applied to the acellular segment 5 mm from the anastomotic site and the equivalent of 210 ± 106 (n = 6) RGC were filled in the retina (Table 1).

(d) Transplantation of muscle to optic nerve, disc and retina

Transplantation of pieces of gluteus maximus to the optic nerve, disc or retina did not result in any regeneration of RGC axons into the grafts, despite the persistence of sarco-

Table 1:

	Number of RGCs*	Number of RGCs regenerating axons	% normal regererating	% surviving RGCs regenerating
Normal	111,206 (3,290)	-		
ON section	10,743 (4,058)	-		
Normal PN to ON	37,658 (6,329)	1,529 (125)	1.4	4.0
Acellular PN to ON	27,851 (3,794)	-		
Disc section	14,212 (3,945)	-		
Normal PN to disc	39,976 (4,216)	2,332 (835)	2.1	6.0
Acellular PN to disc	26,062 (1,369)	-		
Predegenerate PN to disc	31,195 (3,442)	4,777 (524)	4.3	1.5
Normal PN to retina		2,743 (305)		
Acellular PN to retina		-		
Predegenerate PN to retina		5,359 (713)	5.0	
ON to retina		-		
Acellular ON to retina		-		
Acellular ON to PN to retina		-		
Acellular PN to PN to retina		210 (106)	0.2	
PN M/C to retina		161 (49)		
Muscle to ON		-		
Muscle to disc		-		
Muscle to retina		-		

* Total cells in retina less glia and endothelial cells 219,251 (5,938)
Total HRP filled RGCs 111,206 (3,290) Amacrine population 108,000

Numbers of RGCs surviving and RGC axons regenerating in the grafting experiments. The number of RGCs in the retinae of grafted rats was calculated by subtracting an estimate of the number of amacrine cells from the Nissl stained count (in the latter, glia and endothelial cells were excluded). The normal amacrine population was estimated by subtracting the number of HRP labelled RGCs from the total Nissl count in normal retinae. The numbers in brackets are the standard error of the mean. M/C = mitomycin C.

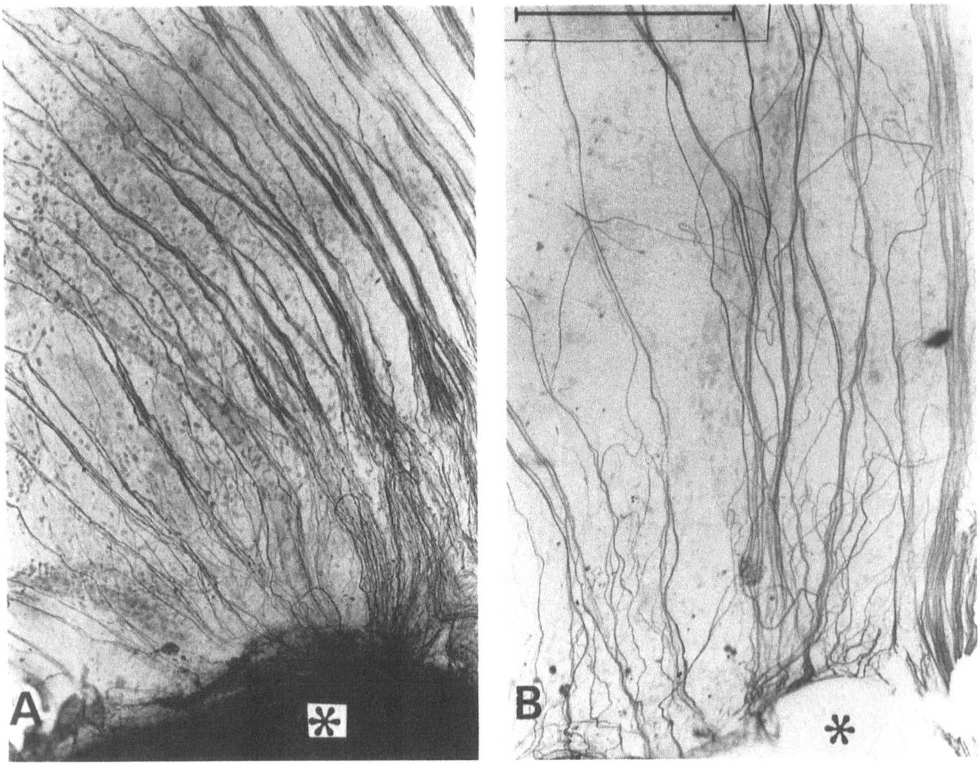

Figure 8: A: RGC axons regenerating directly into a PN implanted into the retina (asterisk) 30 days previously.
B: RGC axons regenerating randomly in the inner plexiform and fibre layers proximal to the site of implantation of an acellular PN (site of implant - asterisk) 30 days previously. Silver stained retinal whole mount; marker = 250 μm, in both figures the optic disc is beyond the lower edge.

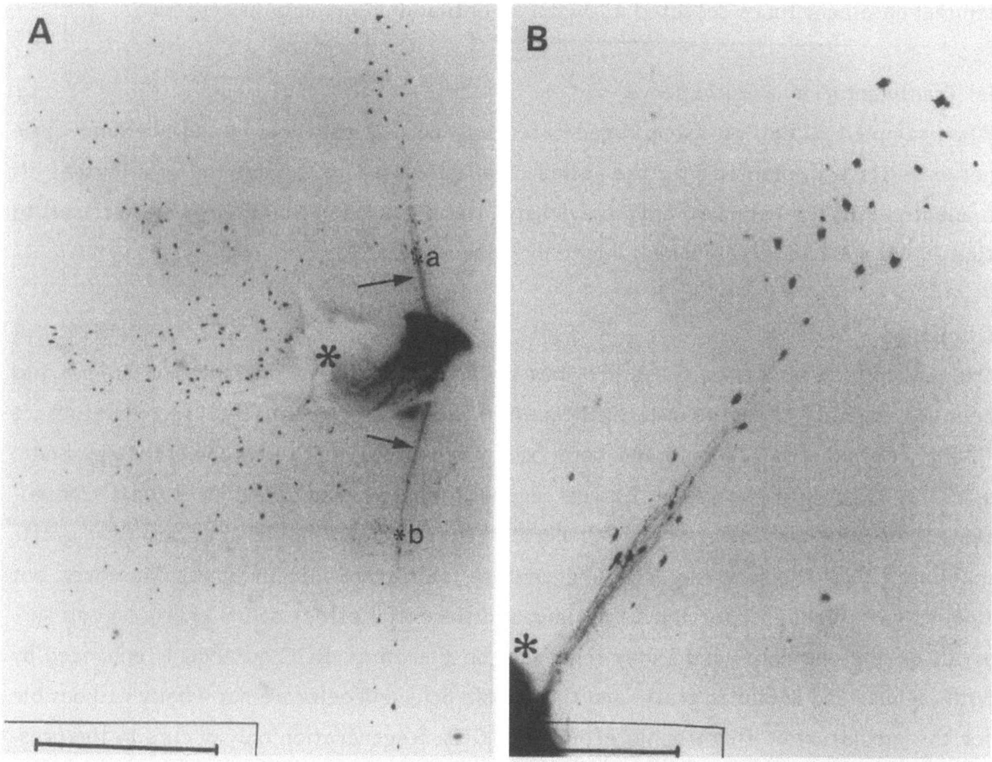

Figure 9: A: Site of implantation of a PN to the retina (large asterisk) 30 dpl - optic disc is off the right hand side of the picture opposite the PN implant. HRP-WGA has been injected into the PN 10 mm from the implantation site. Filled RGCs are seen in a large sector of the retina corresponding to the position of those RGCs whose axons were injured during implantation and which have subsequently regenerated into the PN. At the time of implantation of the PN, two lesions (small asterisks) were made 1 mm (a) and 2 mm (b) from the PN nerve stump. These lesioned RGC axons have also regenerated into the PN. Their cell bodies are filled with HRP-WGA within the appropriate retinal sectors and their axons (arrows) still contain the tracer, marking the course of regeneration into the PN which protrudes into the vitreous and is also heavily labelled with HRP-WGA (marker = 2 mm).
B: Detail of RGC axons and cell bodies associated with lesion (a) (Fig. 9A) made 1 mm from the PN implant (asterisk). Note that the axons from both lesions (a) and (b) in Fig. 9A are orthogonal to the intact RGC fascicles passing towards the optic disc. Retinal whole mount preparation, marker = 1 mm.

lemmal basement tubes detected at 30 dpl by antilaminin immunofluorescence.

(e) Transplantation of optic nerve

When sampled 30 dpl, no axons regenerated into normal optic nerve and acellular optic nerve grafts transplanted into the retina or anastomosed to the cut end of a normal PN implanted into the retina 30 days previously. The acellular optic nerves were difficult to find 30 dpl and had largely been cleared by macrophages.

Discussion

One interpretative difficulty of previous results of PNS/CNS implantation studies has been the inability to distinguish regenerating CNS axons from collateral sprouts of intact fibres. The experiments reported here have been designed to overcome that problem, since regrowth into the grafts is true regeneration following complete axonal transection. The results also quantify the regenerative response and show that at best (disc grafting) only 15% of the surviving RGCs regenerate their axons into the graft. Moreover, our findings also highlight, for the first time, a differential effect of PN grafts on cell survival, on the one hand, and axons regrowth, on the other. RGC survival is enhanced by both cellular and acellular grafts and thus viable Schwann cells are not wholly responsible for the mediation of this trophic effect on RGCs. Regeneration only occurs in the presence of live Schwann cells; basal lamina per se does not promote regeneration of axons but the lamina lucida does act as a preferential substrate for axon regrowth.

RGC axons consistently fail to regenerate into the laminin-rich basal lamina tubes which persist in all acellular PN grafts implanted into the optic nerve, optic disc and retina. Basal lamina may be essential for the pioneering outgrowth of axons in order to establish early connections in the developing optic pathway of the fish (Easter et al., 1981, 1984) and chick (Cohen, this volume), but appears irrelevant to the regrowth of injured axons in mature fish (Easter, this volume) and birds (Cohen, this volume). In mammals, axon regeneration fails despite the creation of a laminin rich environment by reactive astrocytes within the wound (Bernstein et al., 1985; Liesi et al., 1984).

During regrowth within either acellular grafts or regeneration chambers (Nageotte, 1917; Pollard and McLeod, 1981; Lundborg et al., 1981; Anderson et al., 1983; Williams et al., 1984; Hall, 1986a, b) Schwann cells either co-migrate with or precede the growing neurites; isolated axons are never found within basal lamina tubes of acellular PN grafts. It is therefore reasonable to postulate that the migration of Schwann cells from the intact stump into the graft is a primary event in the regenerative response, Schwann cell movement occurring preferentially over the inner aspect of the basal lamina. Axon invasion is secondary, possibly organised by the secretions of the pioneering population of Schwann cells that fill the tubes. Treatment of the proximal stump of a PN with mitomycin C (Hall and Gregson, 1975) significantly delays regeneration into the distal stump (Hall, 1986b) presumably because the production of postmitotic migratory Schwann cells is retarded. Such experiments are replicated by the CNS acellular graft paradigm where no Schwann cells are available from the CNS to repopulate the acellular PN graft and thus regeneration does not occur, even though a laminin adhesive substrate is present. Results of experiments have been reported in which a few axons enter pre-existing basal lamina scaffolds in acellular muscle (Keynes et al., 1984; Glasby et al., 1986) and nerve (Haftek and Thomas, 1968; Ide et al., 1976; Hall, 1986a, b; Bunge, 1982; Anderson et al., 1983). They do so exclusively along the inner aspect of the basal lamina in association with Schwann cells which presumably migrate into the graft from the proximal PN stump. Our negative findings after muscle and optic nerve grafting into the optic system, and with mitomycin C treatment of the PN grafts show that an association with Schwann cells is the significant factor for successful axonal regrowth rather than any possible neurotrophic influence of the basal lamina per se (Liesi, 1985; Hopkins et al., 1985).

Our findings also suggest that Schwann cells actively secrete a chemotactic factor which attracts CNS axons over distances of up to 3 mm. Thus, lesioned retinal fibres, which normally regenerate, but lack direction (McConnell and Berry, 1982), preferentially grow into a PN graft when the retinal lesion and implant are separated by less than 3 mm. Similar conclusions have been reported by Politis et al. (1982) and Anderson and Turmaine (1986) in the PNS using 'Y' shaped tubes.

One reason for grafting PN segments to different sites in the optic pathway was to test the axon growth inhibitory hypothesis (Berry, 1983, 1985) which assumes that oligodendrocytes and the products of degenerating CNS myelin permanently inhibit CNS regeneration in the immediate post-injury period. Our findings do not support this thesis since neither the number of axons regenerating into the normal PN grafts from the optic nerve, nor the numbers of RCCs surviving are significantly different at the different sites. Nonetheless, regeneration is significantly enhanced when transplantation of the graft is delayed for 5 days following isolation, when mitosis in the isolated PN is maximal (Sunderland, 1978). Twice the number of axons normally regenerating into a PN graft are found in the predegeneratic nerve. This effect is achieved presumably because the greater number of Schwann cells present in predegenerate PNs secrete higher titres of neurotrophic factor at an optimal time in the immediate post-injury period. It should be noted that, although the predegenerate PN grafts enhance regeneration, they do not increase RGC survival significantly above the number saved by a normal PN implant.

Anderson and Turmaine (1986) have shown that PN axons will penetrate an acellular optic nerve graft but will not invade a living optic nerve, an observation also reported by others (Aguayo et al., 1978, 1982; Weinberg and Spencer, 1979). Our experiments show that CNS axons will not invade acellular optic nerve grafts nor do CNS axons already growing within a PN enter such grafts. We assume that Schwann cells migrating ahead of regenerating PN axons penetrate the acellular optic nerve graft in the experiments of Anderson and Turmaine (1986) and that this is crucial to regenerative success into the graft. Without this bridge, CNS axons, in our experiments, fail to grow directly from the CNS into acellular optic nerve grafts. Perhaps the small number of optic fibres in our retinal PN implants makes it improbable that any will enter the acellular optic nerve graft since regenerative paths, defined by migrating Schwann cells moving over the lamina lucida, are limited to the basal lamina of both the glia limitans externa and effete blood vessels. The fact that the substrate for Schwann cell migration into the CNS is normally limited to the lamina lucida may account for the poor penetration of regenerating axons from PNS into CNS both at the root/cord interface (Lugaro, 1906; Tower, 1931; Ten Cate, 1932; Paskind, 1936, Westbrook and Tower, 1940; Kimmel and Moyer, 1947; Moyer et al., 1953; McCough, 1955; Gamble, 1976; Stensaas et al., 1979) and after implantation of both ends of a PN graft into the CNS (David and Aguayo, 1981; Aguayo et al., 1984).

References

Aguayo, A.J., R. Dickson, J. Trecarten, M. Attiwell, G.M. Bray, and P. Richardson (1978) Ensheathment and myelination of regenerating PNS fibres by transplanted optic nerve glia. Neurosci. Lett. **9**: 97-104

Aguayo, A.J., S. David, P. Richardson, and G.M. Bray (1982) Axonal elongation in peripheral and central nervous system transplants. In *Advances in Cellular Neurobiology*, S. Fedoroff and L. Hertz, eds, vol. 3, pp. 215-234. Academic Press, New York.

Aguayo, A.J., U. Björklund, U. Stenevi, and T. Carlsted (1984) Fetal mesencephalic neurons survive and extend long axons across peripheral nervous system grafts inserted into the adult rat striatum. Neurosci. Lett. **45**: 53-58.

Anderson, P.N., J. Mitchell, D. Mayor, and V.V. Stauber (1983) An ultrastructural study of the early stages of axonal regeneration through rat nerve grafts. Neuropath. Appl. Neurobiol. **9**: 455-466.

Anderson, P.N., and M. Turmaine (1986) Peripheral nerve regeneration through grafts of living and freeze-dried CNS tissue. J. Anat., in press.

Baron-Van Evercooren, A., H. Kleinman, S. Ohno, P. Marangos, J. Schwartz, and M. Dubois-Dalcq (1982) Nerve growth factor, laminin and fibronectin promote neurite growth in human fetal sensory ganglia cultures. J. Neurosci. Res. **8**: 179-193.

Benfey, M., and A.J. Aguayo (1982) Extensive elongation of axons from rat brain into peripheral nerve grafts. Nature **296**: 150-152.

Berg, D.K. (1984) New neuronal growth factors. Ann. Rev. Neurosci. **7**: 149-170.

Bernstein, J.J., R. Getz, M. Jefferson, and M. Kelemen (1985) Astrocytes secrete basal lamina after hemisection of rat spinal cord. Brain Res. **327**: 135-141.

Berry, M. (1979) Regeneration in the central nervous system. Rec. Adv. Neuropath. **1**: 67-111.

Berry, M. (1983) Regeneration of axons in the central nervous system. Progr. Anat. **3**: 213-233.

Berry, M. (1985) Regeneration and plasticity in the CNS. In *Scientific Basis of Clinical Neurology*, M. Swash and C. Kennard, eds, pp. 658-679. Churchill Livingstone, Edinburgh.

Berry, M., W.L. Maxwell, A. Logan, A. Mathewson, P. McConnell, D.E. Ashhurst, and G.H. Thomas (1983) Deposition of scar tissue in the central nervous system. Acta

Neurochirurgica Suppl. **32:** 31-53.

Berry, M., L. Rees, and J. Sievers (1986a) Regeneration of axons in the mammalian visual system. In *Processes of Recovery from Neural Trauma*, G.M. Gilad, A. Gorio, and G.W. Kreutzberg, eds., Exp. Brain Res. Suppl. **13**, pp. 18-33. Springer-Verlag, Berlin.

Berry, M., L. Rees, and J. Sievers (1986b) Unequivocal regeneration of rat optic nerve axons within sciatic nerve isografts. In *Neural Transplantation and Regeneration*, G.D. Das and R.B. Wallace, eds., pp. 63-79. Springer-Verlag, New York.

Berry, M., L. Rees, S. Hall, P. Yui, and J. Sievers (1986c) Optic axons regenerate into sciatic nerve isografts only in the presence of Schwann cells. Nature, submitted.

Björklund, A., and U. Stenevi (1979) Regeneration of monoaminergic and cholinergic neurons in the mammalian central nervous system. Physiol. Rev. **59:** 62-100.

Bunge, R.P. (1982) Contributions of tissue culture studies to our understanding of basic processes of peripheral nerve regeneration. In *Posttraumatic Peripheral Nerve Regeneration. Experimental Basis and Clinical Implications*, A. Gorio, H. Millesi, and S. Migrino, eds., pp. 105-113. Raven Press, New York.

Bunge, R.P., and M.B. Bunge (1983) Interrelationship between Schwann cell function and extracellular matrix production. Trends Neurosci. 6: 499-505.

Cajal, Ramon Y.S. (1928) *Degeneration and Regeneration in the Nervous System*. Oxford University Press, London.

Carbonetto, S. (1984) The extracellular matrix of the nervous system. Trends Neurosci. **7**: 382-387.

Collins, F. (1978) Induction of neurite outgrowth by a conditioned-medium factor bound to the culture substratum. Proc. Natl. Acad. Sci. (USA) **75:** 5210-5213.

David, S., and A.J. Aguayo (1981) Axonal elongation into peripheral nervous system bridges after central nervous system injury in adult rats. Science **214:** 931-933.

David, S., and A.J. Aguayo (1985) Axonal regeneration after crush injury of rat central nervous system fibres innervating nerve grafts. J. Neurocytol. 14: 1-12.

Davis, G.E., M. Manthorpe, E. Engvall, and S.J. Varon (1985) Isolation and characterization of rat Schwannoma neurite-promoting factor: Evidence that the factor contains laminin. J. Neurosci. **5**: 2662-2671.

Easter, S.S. Jr., A.C. Rusoff, and P.E. Kish (1981) The growth and organisation of the optic nerve and tract in juvenile and adult goldfish. J. Neurosci. **1**: 793-811.

Easter, S.S., Jr., B. Bratoon, and S.S. Scherer (1984) Growth-related order of the retinal fiber layer in goldfish. J. Neurosci. **4**: 2173-2190.

Edgar, D., R. Timpl, and H. Thoenen (1984) The heparin-binding domain of laminin is responsible for its effects on neurite outgrowth and neuronal survival. EMBO J. **3**: 1463-1468.

Gamble, H.J. (1976) Spinal and cranial nerve roots. In *The Peripheral Nerve*, J. Landon, ed., pp. 330-354. Chapman and Hall, London.

Glasby, M.A., J.I. Hitchcock, and L.-H. Huang (1986) The dependence of nerve regeneration through muscle grafts on the availability and orientation of basement membrane in the rat. Muscle and Nerve, in press.

Haftek, J., and P.K. Thomas (1968) Electron-microscope observations on the effects of localized crush injuries on the connective tissues of peripheral nerve. J. Anat. **103**: 233-243.

Hall, S.M. (1986a) Regeneration in cellular and acellular autografts in the peripheral nervous system. Neuropath. Appl. Neurobiol. **12**: 27-46.

Hall, S.M. (1986b) The effect of inhibiting Schwann cell mitosis on the re-innervation of acellular autografts in the peripheral nervous system. Neuropath. Appl. Neurobiol., in press.

Hall, S.M., and N.A. Gregson (1977) The effects of mitomycin C on the process of remyelination in the mammalian peripheral nervous system. Neuropath. Appl. Neurobiol. **1**: 65-78.

Hammarback, J.A., S.L. Palm, L.T. Furcht, and P.C. Letourneau (1985) Guidance of neurite outgrowth by pathways of substratum-absorbed laminin. J. Neurosci. Res. **13**:

213-220.
Hopkins, J.M., T.S. Ford-Holsvinski, J.P. McCoy, and B.W. Agranoff (1985) Laminin and optic nerve regeneration in the goldfish. J. Neurosci. **5:** 3030-3038.
Ide, C., K. Tohyama, R. Yokota, T. Nitatori, and S. Onodera (1976) Schwann cell basal lamina and nerve regeneration. Brain Res. **288:** 61-75.
Keynes, R.J., W.G. Hopkins, and C.L.-H. Huang (1984) Regeneration of mouse peripheral nerves in degenerating skeletal muscle: Guidance by residual muscle fibre basement membrane. Brain Res. **295:** 275-281.
Kiernan, J.A. (1979) Hypotheses concerned with axonal regeneration in the mammalian nervous system. Biol. Rev. **54:** 155-197.
Kimmel, D.L., and E.K. Moyer (1947) Dorsal roots following anastomosis of the central stumps. J. Comp. Neurol. **87:** 289-319.
Kromer, L.F., and C.J. Cornbrooks (1985) Transplants of Schwann cell cultures promote axonal regeneration in the adult mammalian brain. Proc. Natl. Acad. Sci. (USA) **82:** 6330-6334.
Krüger, S., J. Sievers, C. Hansen, M. Sadler, and M. Berry (1986) Three morphologically distinct types of interface develop between adult host and fetal brain transplants: Implications for scar formation in the adult central nervous system. J. Comp. Neurol., in press.
Lander, A.D., D.K. Fujii, and L.F. Reichardt (1985) Laminin is associated with the neurite outgrowth-promoting factors found in conditioned media. Proc. Natl. Acad. Sci. (USA) **82:** 2183-2187.
Liesi, P., D. Dahl, and A. Vaheri (1984) Neurons cultured from developing rat brain attach and spread preferentially to laminin. J. Neurosci. Res. **11:** 241-251.
Liesi, P. (1985) Laminin-immunoreactive glia distinguish regenerative adult CNS systems from non-regenerative ones. EMBO J. **4:** 2505-2511.
Liesi, P., S. Kakkola, D. Dahl, and A. Vaheri (1984) Laminin is induced in astrocytes of adult brain by injury. EMBO J. **3:** 683-686.
Letourneau, P.C. (1979) Cell-substratum adhesion of neurite growth cones, and its role in neurite elongation. Exp. Cell Res. **124:** 127-138.
Lugaro, E. (1906) Sulla presenta rigenerazione autogena delle radici posteriori. Rev. pat. nerv. ment. **11:** 337-340.
Lundborg, G., L.B. Dahlin, N.P. Danielsen, H.A. Hansson, and K. Larsson (1981) Reorganisation and orientation of regenerating nerve fibres, perineurium, and epineurium in preformed mesothelial tubes - An experimental study on sciatic nerve grafts. J. Neurosci. Res. **4:** 265-281.
Manthorpe, M., E. Engvall, E. Ruoslahti, F. Longo, G. Davis, and S. Varon (1983) Laminin promotes neurite regeneration from cultured peripheral and central neurons. J. Cell Biol. **97:** 1882-1890.
McConnell P., and M. Berry (1982) Regeneration of ganglion cell axons in the adult mouse retina. Brain Res. **241:** 362-365.
McCouch, G.P. (1955) Comments on regeneration of functional connections. In *Regeneration in the Central Nervous System*, W.F. Windle, ed., pp. 171-175. Charles C. Thomas, Springfield, Illinois.
Moyer, E.K., D.C. Kimmel, and L.W. Winborne (1953) Regeneration of sensory nerve roots in young and in senile rats. J. Comp. Neurol. **98:** 283-308.
Nageotte, J. (1917) Sur la greffe des tissues mants et au particulies sur la reperation pertes de substances des nerfs a l'aide de greffons nerveux conserves dans l'alcohol. Comp. ren. des sean. soc. de biol. **80:** 925-933.
Nitkin, R.M., B.G. Wallace, M.E. Spira, E.W. Godfrey, and U.J. McMahan (1983) Molecular components of the synaptic basal lamina that direct differentiation of regenerating neuromuscular junctions. Cold Spring Harbor Symp. **48:** 653-665.
Paskind, H.A. (1936) Regeneration of posterior root fibres in cat. Arch. Neurol. Psychiat. **36:** 1077-1084.
Perry, V.H. (1981) Evidence for an amacrine cell system in the ganglion cell layer of the rat retina. Neuroscience **6:** 931-944.

Pollard, J.D., and J.G. McLeod (1981) Fresh and predegenerate nerve allografts and isografts in Trembler mice. Muscle and nerve **4**: 274-281.

Politis, M., K. Ederle, and P.S. Spencer (1982) Tropism and nerve regeneration in vivo. Attraction of regenerating axons by diffusible factors derived from cells in distal stumps of transected peripheral nerves. Brain. Res. **278**: 229-231.

Richardson, R.M., V.M. McGuiness, and A.J. Aguayo (1980) Axons from CNS neurones regenerate into PNS grafts. Nature **284**: 264-265.

Richardson, P.M., V.M.K. Issa, and A.J. Aguayo (1984) Regeneration of long axons in the rat. J. Neurocytol. **13**: 165-182.

Rogers, S.L., P.C. Letourneau, S.L. Palin, J. McCarthy, and L.T. Furcht (1983) Neurite extension by peripheral and central nervous system neurons in response to substratum-bound fibronectin and laminin. Dev. Biol. **98**: 212-220.

Sanes, J.R. (1983) roles of extracellular matrix in neural development. Ann. Rev. Physiol. **45**: 581-600.

Sanes, R.J., L.M. Marshall, and U.J. McMahan 81978) Reinnervation of muscle fiber basal lamina after removal of myofibers. Differentiation of regenerating axons at original synaptic sites. J. Cell. Biol. **78**: 176-198.

Scherer, S.S., and S.S. Easter (1984) Degenerative and regenerative changes in the trochlear nerve of goldfish. J. Neurocytol. **13**: 519-565.

Smalheiser, N.R., S.M. Crain, and L.M. Reid (1984) Laminin as a substrate for retinal axons in vitro. Dev. Brain Res. **12**: 136-140.

Stensaas, L.J., P.K. Burgess, and K.W. Horch (1979) Regenerating dorsal root axons blocked by spinal cord astrocytes. Soc. Neurosci. Abstr. **5**: 684.

Sunderland, S. (1978) Degeneration of the axon and associated changes. In *Nerve and Nerve Injuries*, 2nd edn., pp. 90-91. Churchill-Livingstone, Edinburgh.

Ten Cate, J. (1932) Befunde nach der experimentellen Isolierung eines Rückenmarksabschnittes. Arch. neerl. phys. l'homme annim. **17**: 149-238.

Thanos, S., and F. Bonhoeffer (1983) Investigations on the development and topographic order of retinotectal axons: Anterograde and retrograde staining of axons and perikarya with rhodamine in vivo. J. Comp. Neurol. **219**: 420-430.

Timpl, R., H. Rohde, P.G. Robey, S.I. Rennard, J.-M. Foidart, and G.R. Martin (1979) Laminin - A glycoprotein from basement membranes. J. Biol. Chem. **254**: 9933-9937.

Timpl, R., J. Engel, and G.R. Martin (1983) Laminin - A multifunctional protein of basement membranes. Trends Biochem. Sci. **6**: 207-209.

Tohyama, K., and C. Ide (1984) The localisation of laminin and fibronectin on Schwann cell basal lamina. Arch. Histo. Jap. **47**: 519-532.

Tohyama, K. (1985) The localisation of lectin-binding sites on Schwann cell basal lamina. J. Neurocytol. **14**: 49-61.

Tower, S.S. (1931) A search for trophic influences of posterior spinal roots on skeletal muscle, with a note on the nerve fibres found in the proximal stumps of the roots after excision of the root ganglion. Brain **54**: 99-110.

Weinberg, E.L., and P.S. Spencer (1979) Studies on the control of myelinogenesis. 3. Signalling of oligodendrocyte myelination by regenerating peripheral axons. Brain Res. **162**: 273-279.

Weiss, P. (1955) Special vertebrate organogenesis. Nervous system. In *Analysis of Development*, B.H. Willier, P.A. Weiss, and V. Hamburger, eds. Saunders, Philadelphia.

Westbrook, W.H.L., and S.S. Tower (1940) An analysis of the problem of migrant fibres in the posterior spinal root dealing with the rate of growth of extraneous fibres into the roots after ganglionectomy. J. Comp. Neurol. **72**: 383-397.

Williams, L.R., H.C. Powell, G. Lundorg, and S. Varon (1984). Competence of nerve tissue as digital insert promoting nerve regeneration in a silicone chamber. Brain Res. **293**: 201-211.

Yamada, K.M. (1983) Cell surface interactions with extracellular materials. Ann. Rev. Biochem. **52**: 761-799.

Yourchenco, P.D., E.C. Tsilibary, A.S. Charonis, and H. Furthmayr (1986) Models for the self-assembly of basement membrane. J. Histochem. **34**: 93-102.

RETINAL AXONS AND THE BASAL LAMINA

Stephen S. Easter, Jr.
Natural Science Building
Department of Biology
University of Michigan
Ann Arbor, MI
U.S.A.

Fish and amphibians continue to add new neurons for years after they hatch and begin to live independently. This post-hatching neurogenesis has been studied mostly in the visual system, particularly in the retina. The continuous addition of new neurons to a retina that was already functioning adequately presents a number of interesting developmental questions, among which is how the axons of the retinal ganglion cells find their way to the brain. In the process of asking that question, in fish, my collaborators and I discovered that the new axons all grow along the basal lamina that surrounds the central nervous system. This physical association between a nerve cell and a part of the extracellular matrix is a very appropriate one to be discussed in this particular workshop.

The paper is organized as follows. First, post-hatching neurogenesis will be discussed, and related to the outgrowth of optic axons. Secondly, the structures of two nerves (cranial nerves II and IV, the optic and trochlear) will be compared, with particular attention to the basal lamina and its relation to axons. Thirdly, the regeneration of optic and trochlear axons will be compared, with special emphasis on the relationship of the axons with the basal lamina.

Post-hatching Neurogenesis

The retina develops according to a similar plan in all vertebrates. The presumptive diencephalon bulges out to form the bilateral optic vesicles which then invaginate to form two-layered, cup-shaped presumptive eyes. The surface facing laterally becomes the neural retina, and the adjacent medial surface, the pigmented epithelium. Our attention will be restricted to the neural retina. At early stages of development, it is histologically similar to the neural tube from which it arose; that is, it is a pseudostratified columnar epithelium populated with mitotically active cells that will become neurons and glia. Very early, within a few days after fertilization in both amphibians and fish, the first

NATO ASI Series, Vol. H5
Mesenchymal-Epithelial Interactions in Neural Development
Edited by J.R. Wolff et al.

cells drop out of the mitotic cycle and begin to differentiate (Hollyfield, 1968, 1972; Straznicky and Gaze, 1971; Sharma and Ungar, 1980). These are all in central retina, the part near the optic stalk, which will provide the avenue along which the axons of the ganglion cells will pass toward the brain. Before long, nearly all of the cells in the central retina are post mitotic, and the remaining proliferative zone is ring-shaped, out in the peripheral retina near the boundary with the pigmented epithelium.

A centro-peripheral developmental gradient exists in all vertebrates, and the division into postmitotic center and proliferative periphery is most striking in fish and amphibians. But the most striking difference between anamniotes (fish and amphibia), on the one hand, and amniotes (or at least birds and mammals, since no one has investigated reptiles), on the other, is in the duration of the proliferative phase. In fish and amphibians, it continues for years (Johns and Easter, 1977; Johns, 1977; Meyer, 1978; Reh and Constantine-Paton, 1983). The relic population of proliferating cells on the edge of the retina continues to produce neurons, glia, and more proliferative cells. The post mitotic cells stay where they are, and differentiate, and the proliferative cells remain on the edge. Thus, the proliferative zone continually recedes, and the retina grows. New rings of retina are added, and the distance of a cell from the center of the retina correlates with its age; the more peripheral the cell, the younger it is. Hundreds of new retinal neurons are added daily (Easter et al., 1981).

For our purposes, the ganglion cell is of particular interest, because its axon travels through the optic nerve to the brain. In large animals, this distance is of the order of a centimeter or more; how do the axons make it?

We asked the question with anatomical methods. In goldfish, as in many other animals, the optic axons run across the innermost retinal layer (that is, the layer closest to the vitreous) in a direct line toward the optic disk, the retinal exit. When the retina is prepared for electron microscopy and sectioned perpendicular to these spoke-like nerve bundles, the result is shown in Fig. 1. The axons are clustered together, separated by glial cells from the ganglion cell layer below, from the inner limiting membrane above, and from other bundles of axons on either side. There is a gradient of axonal diameters - larger ones deep, smaller ones superficially. Moreover, most of the large ones are wrapped in a loose "myelin." These observations suggest that the most superficial axons are the youngest. This was supported by the observations that near the periphery, <u>only</u> non-myelinated axons were seen, and that when axons were severed very peripherally in the retina, degenerating axons in central retina were superficial. The most compelling evidence came from images such as the one in Fig. 2. The flattened profile at the top (gc) is part of a growth cone. Note that it has pushed the glial-end feet aside to contact the basal lamina directly. This apposition was quite extensive, as the reconstruction in Fig. 3 illustrates (Easter et al., 1984).

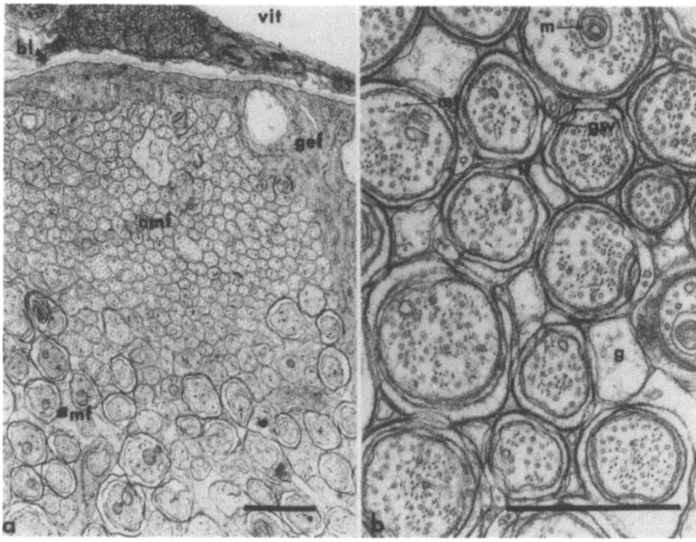

Figure 1: Transverse sections through the retinal fiber layer midway between the retinal margin and the optic nervehead. a: Most of one fascicle is shown. It contains non-myelinated fibers (nmf) superfically and myelinated fibers (mf) deeply. It is separated from the vitreous (vit) by a cell associated with a blood vessel, the basal lamina (bl), and glial end-feet (gef). b: A more highly magnified view of the myelinated fibers, with intermediate filaments (if), microtubules (mt), and mitochondria (m) inside, and glial wrapping (gw) and processes (g) outside. Calibrations: 1 μm. (From Easter et al., 1984).

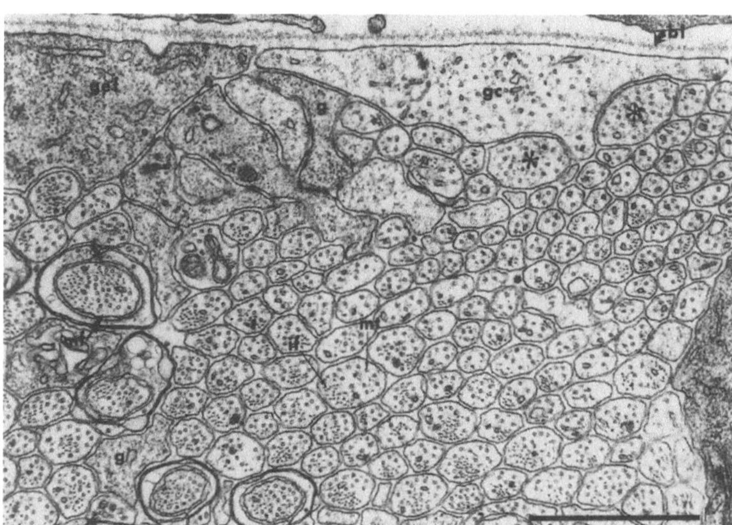

Figure 2: Transverse section through the retinal fiber layer midway between the retinal margin and the optic nerve head. The growth cone (gc) directly contacts the basal lamina (bl). Asterisks mark presumed young axons. Other abbreviations as in Fig. 1. Calibration: 1 μm. (From Easter et al., 1984).

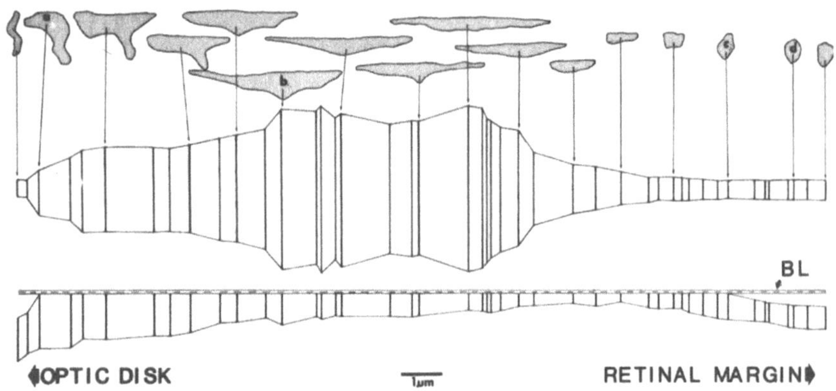

Figure 3: A reconstruction, from 45 quasi-serial sections, of a retinal growth cone. The hatched contours above show the shapes of certain processes in cross-section. The middle sketch shows the process as viewed from the vitreous, with each vertical line scaled to show that maximum width of the process. The lower figure shows this same process, viewed from the side in relation to the basal lamina (bl). (From Easter et al., 1984).

The data just described lead to the conclusion that individual axons emerge from the newly differentiated cell bodies at the retinal edge, and squeeze in between their most recent predecessors below, the basal lamina above, and the glial end-feet on either side. Thus, each axon, in one of the fascicles of Fig. 1, once lay on the most superficial lamina, but was gradually separated from that level by the intervention of newer axons.

The direct apposition to the basal lamina has been seen only in teleost fish. In other animals (chick: Krayanek and Goldberg, 1981; mouse: Silver and Robb, 1979), the new axons occupy the innermost layer, but the space remains enveloped by glial end-feet. What feature of this environment is so suitable for axonal elongation that the axons in all of these species grow there? Three structures bound the space; which is the more influential?

The basal lamina has been shown to be conducive to axonal growth in the peripheral nervous system (see below), and compounds extracted from it, particularly laminin, promote neuritic outgrowth in vitro. More recently, my colleagues at the University of Michigan (Hopkins et al., 1985) have shown that laminin-like immunoreactivity is increased in the regenerating optic nerve of goldfish, but not in the non-regenerating optic nerve of a rat. One is tempted to conclude that laminin plays an essential role in regeneration, and, by extrapolation, in de novo development, too. We resist the temptation to speculate that the basal lamina and laminin are the important elements, for two reasons. The first is that avian and mammalian optic fibers grow without contacting the basal lamina; indeed Williams and Rakić (1985), prompted by out work, specifically searched for such contacts in the developing optic nerve of the monkey, and found none.

The second is given two paragraphs below.

The glial end-feet have been favored as guidance structures for growing axons by Silver and Rutishauser (1984), who showed immunoreactivity to neural cell adhesion molecule (N-CAM) in this region, and suggested that the molecule lay in the glial membranes. Without questioning their observations, we doubt that the glial end-feet play a very significant role as a substrate for axonal outgrowth in fish, since the axons push them aside, and contact them minimally. It is generally accepted that axons spread out on the substrate to which they adhere (Letourneau, 1982), and the growth cone in Fig. 2 certainly does not spread on the glial end-feet.

The other fibers are probably the most important structures because the new growth cones were always on a pre-existing bundle of axons. The bundles are generally spaced many bundle widths apart, so if the growth cones were to seek out either basal lamina or glial end-feet, they have plenty of both that are not associated with a bundle of axons. Moreover, regenerating growth cones also favor other recently produced axons in the optic nerve (see below). Thus, we are led to the tentative conclusion that new axons prefer to grow with other axons, and the most superficial layer is probably selected for reasons other than the proximity of the basal lamina.

We continued our electron microscopic investigation from the retina into the nerve, and concluded that the growth cones occupied a similar location, immediately below the pial basal lamina. In this case, the astrocytes had been pushed aside. More recently, using light microscopy, Bernhardt and I (1986) carried the analysis through the tract and into the tectum, with a similar result. In summary, the vast majority of optic axons in goldfish appear to have followed this same route.

Optic and Trochlear Nerves

These two nerves, the former a part of the central nervous system, and the latter, peripheral, have quite different structures.

Close to the retina, the optic nerve is a cylinder, surrounded by the basal lamina that envelopes the astroglial end-feet. This glia limitans, continuous with the inner limiting membrane of the retina, provides the boundary between CNS and somatic space. The boundary is more complex at more proximal locations in the nerve, where it is partitioned into a number of individual bundles of axons, one of which is seen in the electron micrograph of Fig. 4. These bundles appear isolated from one another in a cross section such as this one, but in three dimensions, they are linked. The compartmentalization has been created by local indentations and invaginations of the cylindrical pia, with the result that each of the compartments is surrounded by a glia limitans, and outside that, the basal lamina. The bundle in Fig. 4 contains a few hundred myelinated axons, no unmyelinated axons, and glia. The oligodendrocytes provide the myelin, and the astrocytes provide the glia limitans around the outside, and fill the space between the axons inside, Fig. 5 is

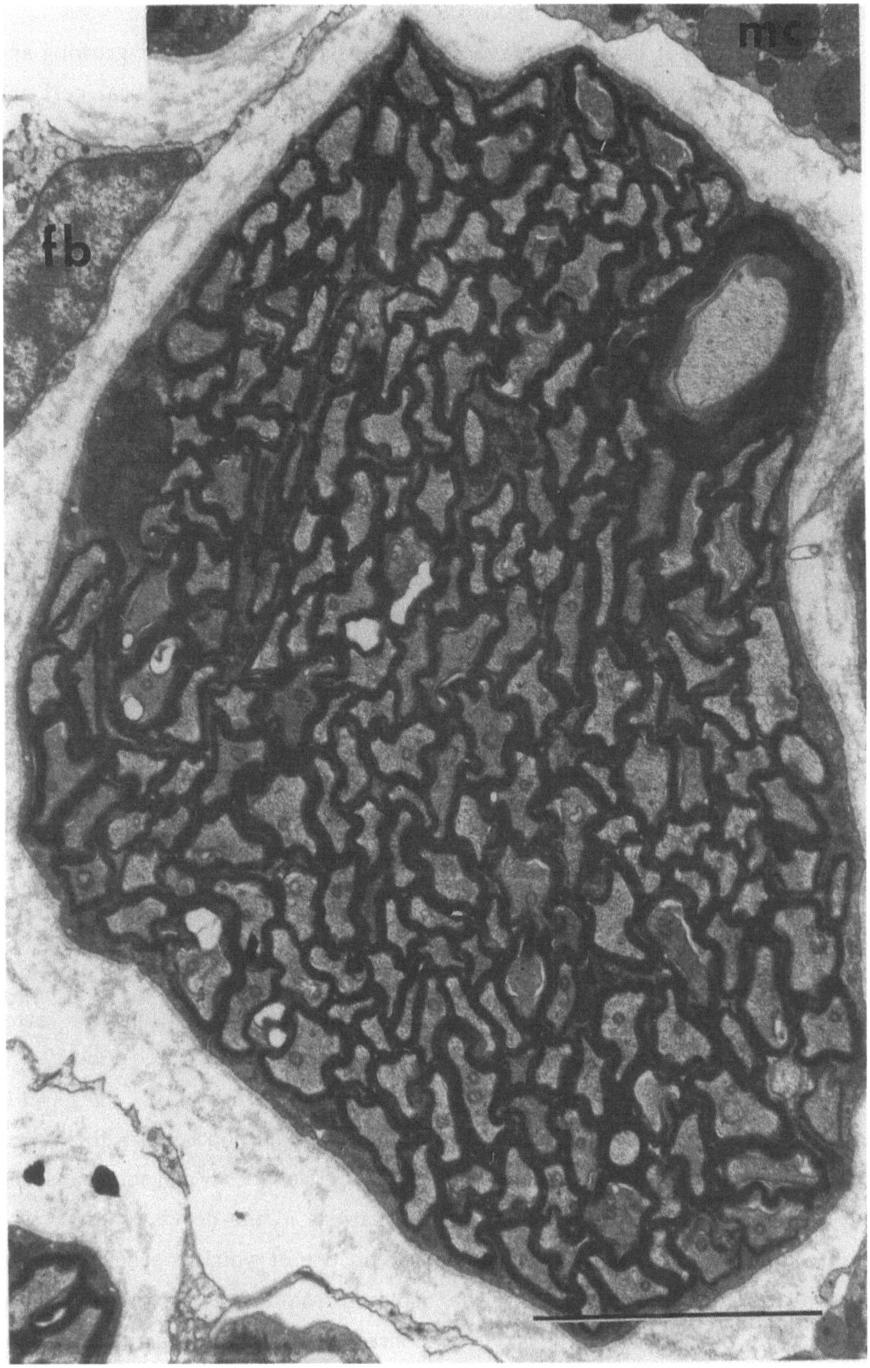

Figure 4: A transverse section through one bundle of axons in the optic nerve. A granular mast-like cell (mc) is seen in the top right, and a fibroblast (fb) top left. Note that most axons are far from the outer boundary of the fascicle. Calibration: 5 μm.

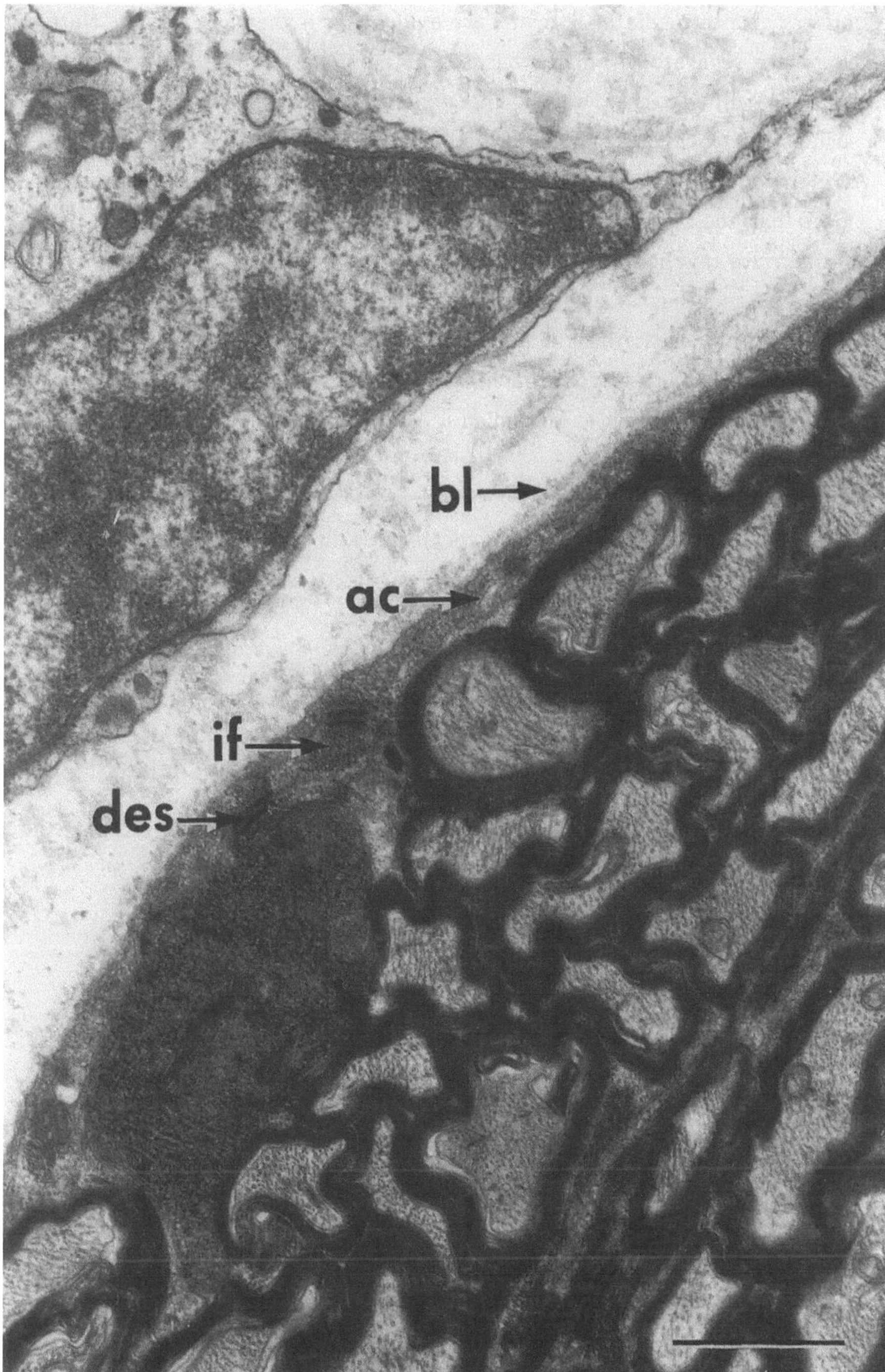

Figure 5: A detail of the top left of Fig. 4. The bundle of axons is wrapped with a basal lamina (bl) over a ring of astrocytic cytoplasm (ac) recognizable by its characteristic dense bundles of intermediate filaments (if) and demosomes (des). Calibration: 1 μm.

a more highly magnified view of the outside of the fascicle. The desmosomes and the dense clusters of intermediate filaments are evident, as is the basal lamina. Note that all of the axons are separated from the basal lamina by the astrocytic envelope, and most are separated by other axons, as well. This contrasts with the arrangement in peripheral nerves such as the trochlear, to be described next.

An electron micrograph of the trochlear nerve appears in Fig. 6. It contains many fewer fibers than the optic (typically 80 vs. more than 100,000), but like the optic, most or all of them are myelinated. In this case, the Schwann cell provides the myelin, and each myelinated axon has its own individual basal lamina. For our purposes, this feature is most important; whereas the individual trochlear nerve fibers are separated from the basal lamina by a thin layer of Schwann cell cytoplasm, the optic fibers are in general much farther from the nearest basal lamina.

Axonal Regeneration

Peripheral axons regenerate well in all vertebrates, including mammals, and work by many people, especially McMahan and his colleagues, has shown that the basal lamina

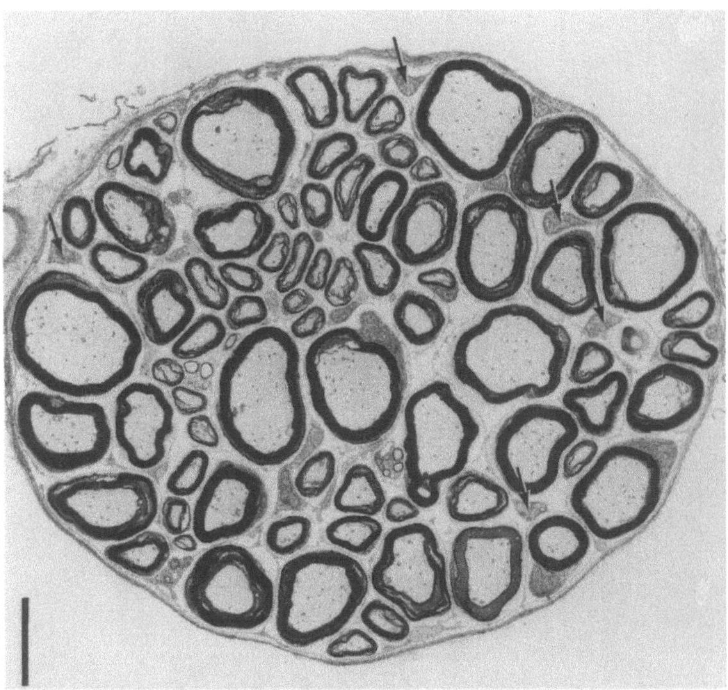

Figure 6: Transverse section of a trochlear nerve inside the orbit. The perineurium (arrowheads) surrounds the nerve, which contains 95 axons, (83 myelinated, 12 unmyelinated) and abundant endoneurial fibroblasts (arrows). Calibration: 10 μm. (From Scherer and Easter, 1984).

surrounding the "Schwann tube" is a very effective substrate for growth. The same is true of the trochlear nerve of goldfish. Fig. 7 shows a growth cone regenerating along the basal lamina (outside) having displaced the Schwann cell cytoplasm internally toward the degenerating axon. In the study from which these illustrations are taken (Scherer and Easter, 1984) all growth cones that were seen were in immediate contact with the basal lamina.

Given the proximity of the new optic axons to the pial basal lamina and the demonstrated affinity that regenerating peripheral axons have for the basal lamina, it was tempting to suppose that regenerating optic axons might grow in along the basal lamina. We pursued this idea electron microscopically, and have concluded that it is not true.

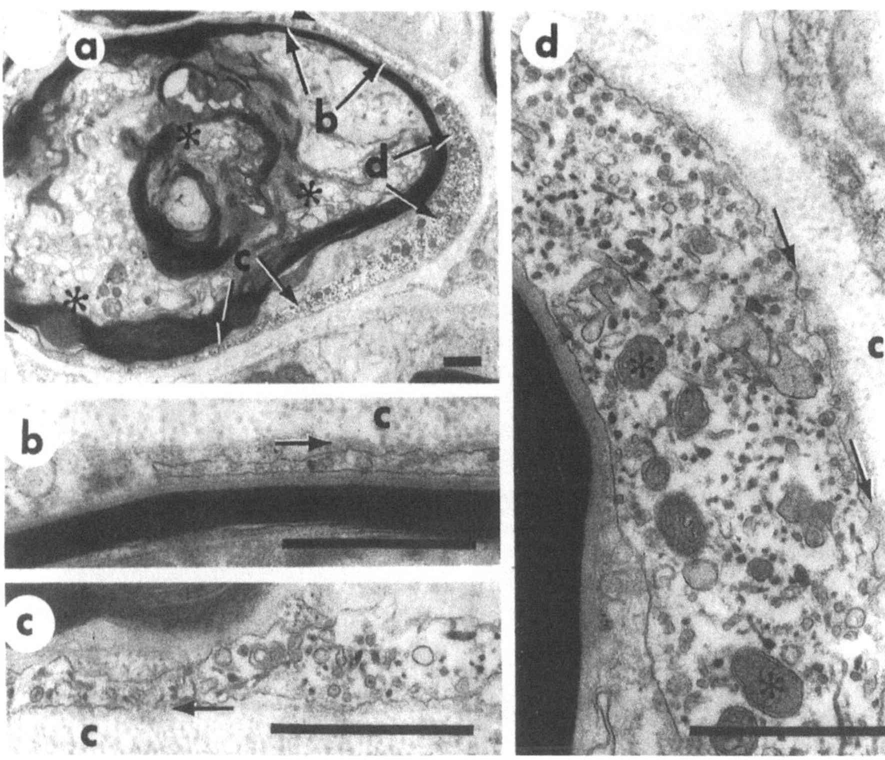

Figure 7: A growth cone inside a Schwann tube of a regenerating trochlear nerve. a: The thin growth cone extends around roughly two-thirds of the inner surface of the basal lamina (between the arrowheads). The inside contains myelin debris (asterisks). b,c,d: Details of the regions indicated by the arrows in a. The basal lamina is indicated by arrows. Asterisks: mitochondria. C: collagen fibers. Calibrations: 1 μm. (From Scherer and Easter, 1984).

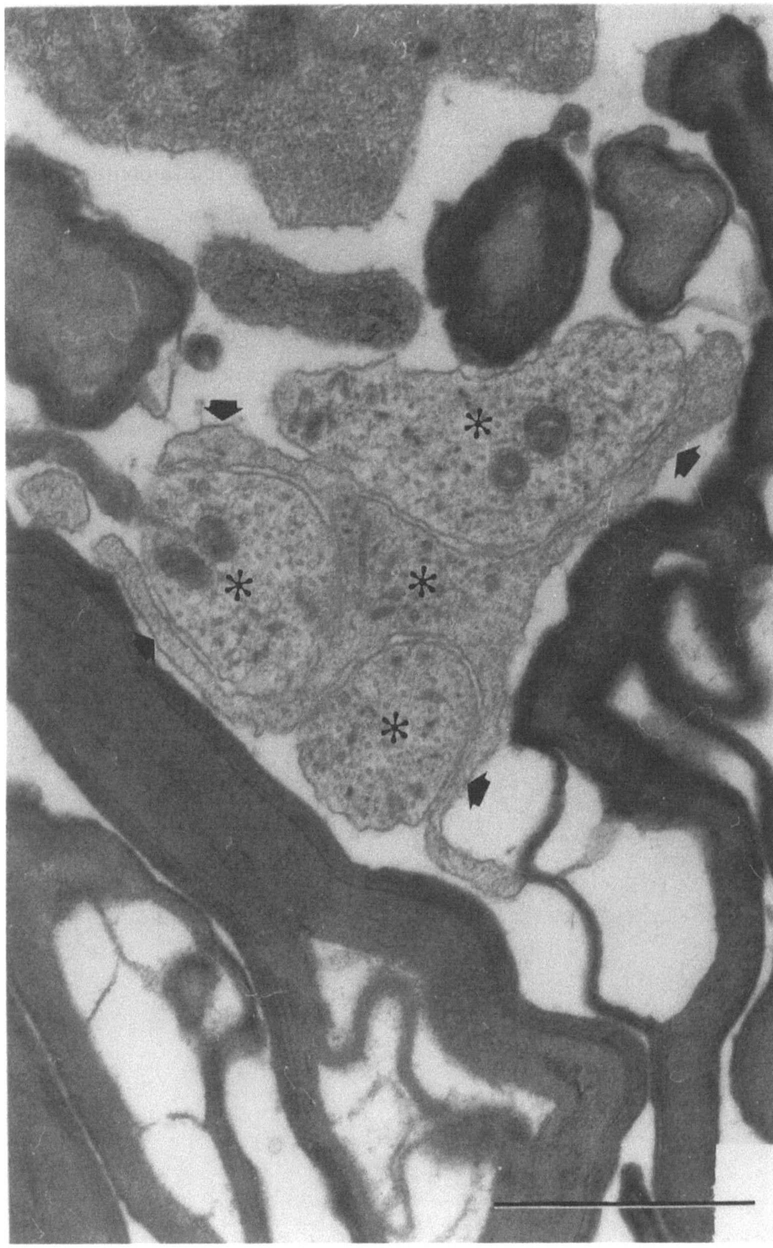

Figure 8: A bundle of four regenerating axons (asterisks), in a transverse section of the optic nerve ten days after crush, about 1 mm on the brain side of the lesion site. The central axon extends four lamellipodia (arrows) which contact extensively the other three axons, and make scarcely any contact with any other structures. This is number 299 in a series of 420 serial sections. The lamellipodia-bearing axon had lamellipodia from section 246 to the end of the series, and presumably beyond, closer to the brain. Calibration: 1 µm.

We made retinal lesions by piercing the retina through the sclera with a microknife. Four days later, the retina was fixed for electron microscopy, the lesion site was located, and sections perpendicular to the fibers were cut downstream from the lesions. In all cases, the new fibers, and what putative growth cones we could identify, were all deep in the fiber layer; none directly contacted the basal lamina.

We crushed the optic nerve in other fish and then examined cross sections of the nerve downstream from the lesion (between the lesion and the brain). The regenerating axons were unmyelinated, and therefore easily distinguished from the degenerates. With rare exceptions, the regenerates were inside the compartments bound by astrocytes, but no growth cones or regenerating axons were seen between the glia limitans and the basal lamina. It is possible that all the bundles of regenerates were pioneered by axons that initially grew in along the underside of the basal lamina, and then dropped off to occupy the center of the fascicle, but this seems unlikely. A more likely explanation is that they grew in the center from the beginning.

As I write this, I have not seen any pioneering growth cones. But in those section containing small bundles of regenerates (2-50 axons), growth cones have been noted. They were always in immediate contact with the axons, and sometimes extensively so, as Fig. 8 shows. I beleive that it is significant that these follower growth cones did not associate with glia or degenerating axons nearly as much as with other axons, slightly more advanced. This supports the inference, made earlier, that axons provide the preferred substrate for other axons.

These electron microscopic descriptions should be reconciled with the immunocytochemical data of Hopkins et al. (1985), who showed that laminin-like immunoreactivity increased in the disconnected stump of the goldfish optic nerve. Normally, the only binding of anti-laminin is by the basal laminae surrounding the fascicles, but within days after nerve crush or eye removal, binding extended across the width of the fascicles. The fact that it occurred even after eye removal indicates that it is not contingent on the presence of regenerating axons. It seems likely that the laminin was incorporated into the membranes of either glia or degenerating debris or both, and the pioneer growth cones adhered to, and grew along, this substrate. The electron microscopic investigation continues in an effort to learn which of the available substrates is favored by the pioneering growth cones.

References

Bernhardt, R., and S.S. Easter (1986) The map of retinal position onto the cross section of the optic pathway of goldfish. Submitted.

Easter, S.S., B. Bratton, and S.S. Scherer (1984) Growth-related order of the retinal fiber layer in goldfish. J. Neurosci. 4: 2173-2190.

Easter, S.S., A.C. Rusoff, and P.E. Kish (1981) The growth and organization of the optic nerve and tract in juvenile and adult goldfish. J. Neurosci. 1: 793-811.

Hollyfield, J.G. (1968) Differential addition of cells to the retina in Rana pipiens tadpole. Dev. Biol. **18:** 163-179.
Hollyfield, J.G. (1972) Histogenesis of the retina in the killifish, Fundulus heteroclitus. J. Comp. Neurol. **144:** 373-380.
Hopkins, J.M., T.S. Ford-Holevinski, J.P. McCoy, and B.W. Agranoff (1985) Laminin and optic nerve regeneration in the goldfish. J. Neurosci. **5:** 3030-3038.
Johns, P.R. (1977) Growth of the adult goldfish eye. III. Source of the new retinal cells. J. Comp. Neurol. **176:** 343-357.
Johns, P.R., and S.S. Easter (1977) Growth of the adult goldfish eye. II. Increase in retinal cell number. J. Comp. Neurol. **176:** 331-342.
Krayanek, S., and S. Goldberg (1981) Oriented extracellular channels and axonal guidance in the embryonic chick retina. Dev. Biol. **84:** 41-50.
Letourneau, P.C. (1982) Nerve fiber growth and its regulation by extrinsic factors. In *Neuronal Development*, N.C. Spitzer, ed., pp. 213-254. Plenum, New York.
Meyer, R.L. (1978) Evidence from thymidine labeling for continuing growth of retina and tectum in juvenile goldfish. Exp. Neurol. **59:** 99-111.
Reh, T.A., and M. Constantine-Paton (1983) Qualitative and quantitative measures of plasticity during the normal development of the Rana pipiens of retinotectal projection. Dev. Brain Res. **10:** 187-200.
Scherer, S.S., and S.S. Easter (1984) Degenerative and regenerative changes in the trochlear nerve of goldfish. J. Neurocytol. **13:** 519-565.
Sharma, S.C., and F. Ungar (1980) Histogenesis of the goldfish retina. J. Comp. Neurol. **191:** 378-382.
Silver, J., and R.M. Robb (1979) Studies on the development of the eye cup and optic nerve in normal mice and in mutants with congenital optic nerve aplasia. Dev. Biol. **68:** 175-190.
Silver, J. and U. Rutishauser (1984) Guidance of optic axons in vivo by a preformed adhesive pathway on neuroepithelial endfeet. Dev. Biol. **106:** 485-499.
Straznicky, K., and R.M. Gaze (1971) The growth of the retina in Xenopus laevis: An autoradiographic study. J. Embryl. Exp. Morphol. **26:** 67-79.
Williams, R.W., and P. Rakić (1985) Dispersion of growing axons within the optic nerve of the embryonic monkey. Proc. Natl. Acad. Sci. (USA) **82:** 3906-3910.

Section 6

INTRACEREBRAL VASCULARIZATION

THE DEVELOPMENT OF INTRACEREBRAL BLOOD VESSELS INTERACTS WITH ASTROCYTE DEVELOPMENT AND NEURON POSITIONING IN THE RAT NEOCORTEX

J.R. Wolff
Department of Anatomy
University of Göttingen
and
Institute for Advanced Study Berlin
Federal Republic of Germany

The nervous system of vertebrates and invertebrates contains neuroglial cells which completely separate neurons from the meninges. This "membrana limitans gliae" (Held, 1909) expands considerably where blood vessels enter the nervous tissue (Table 1). It is, therefore, not surprising that, in vertebrates, the number of non-radial or astrocyte-like glial cells increases with the expansion of intracerebral vascularization, while the number of radial or ependymal glial cells decreases (Sarnat et al., 1975). The correlation, however, is not absolute. Some non-radial glial cells occur in the absence of intracerebral vascularization (e.g., in the avascular spinal cord of Amphioxus; Bone 1960) and radial or ependymal glial cells ("Tanycytes"; Horstmann, 1954) can also form perivascular sheaths (e.g., Müller cells in vascularized retina). The data surveyed here may explain this lack of absolute correlation.

This review will focus on three topics: 1) adult structure of glial-vascular interface and glial cells in the rat cerebral cortex; 2) relations between intracortical vascularization, gliogenesis and neuron positioning; 3) pathophysiological relations between astrocytes and intracortical vessels. Because of limited space, our own publications, which contain references to the work of other authors, will be emphasized.

Available data suggest that on astrocytes and ependymal glial cells two types of glial-vascular interactions have to be distinguished, which are mediated by glial end-feet and en passant contacts, respectively.

Structural Characteristics of the Glial-Vascular Interface in Adult Neocortex

Intracortical blood vessels show three characteristic features: 1) perivascular spaces are narrow or absent, 2) polarized glial cells (mainly astrocytes) form perivascular sheaths,

NATO ASI Series, Vol. H5
Mesenchymal-Epithelial Interactions in Neural Development
Edited by J. R. Wolff et al.

Table 1: Relation between the meningeal and vascular fractions of the mesenchymal-glial interface in cerebral cortex*.

species	men.S[cm²]	V[cm³]	$L_V[\frac{mm}{cm^3}]$	$\bar{d}$[µm]	vasc.S[cm²]	men.S / tot.S
man	1,600	480	1.35	5.8	110,000	1:70
rat	4	0.5	1.1	4.2	72	1:20

men.S : meningeal surface area of cerebral cortex
V : total cortex volume
L_V : capillary length per unit of cortex volume
d : estimated average diameter of all intra cortical vessels
vasc.S : surface area of intracortical blood vessels
tot.S : total glial-mesenchymal (meningeal + vascular) interface

* All data were taken from Blinkov and Glezer (1968) and Eulner (1980).

and 3) the structure of glial-vascular interfaces varies in different parts of the neocortical angioarchitecture.

Capillaries (diameter 8 µm) lack an open perivascular space. They are surrounded by an unusually thick basal lamina (Wolff, 1963, 1977). Serial section reconstructions (unpublished), show that the pericapillary glial sheath consists almost exclusively of en passant contacts with astrocyte processes (Fig. 1). These processes carry lamellar excrescences, which are attracted by neuronal components and synapses (Wolff, 1965, 1970), but also cover 60% of the capillary walls (Wolff and Bär, 1976).

At distal portions of stem vessels and at their preterminal branches, the basal lamina is often split by reticulin fibrils intervening between the glial and vascular part (Wolff, 1963). The attachment of astrocytic lamellae is less prominent than on capillaries, and some astrocytic end-feet can be seen.

The proximal parts of stem vessels are usually surrounded by an open perivascular space. This so-called Virchow-Robin space contains an amorphous extracellular matrix, collagen fibrils and some fibroblast-like cells. Like meningeal cells, these cells are not covered by basal lamina, they are capable of phagocytosis (see below), and they accumulate neurotransmitters, such as monoamines (see Sievers, this volume) and amino acids (e.g., GABA, unpublished observations). The glial sheath is composed of numerous astrocytic end-feet and some en passant contacts with astrocytic processes (Fig. 1).

Thus, at least in the cerebral cortex, the structure of glial-vascular interfaces varies at different sites of the vascular system. In other parts of the CNS, capillaries

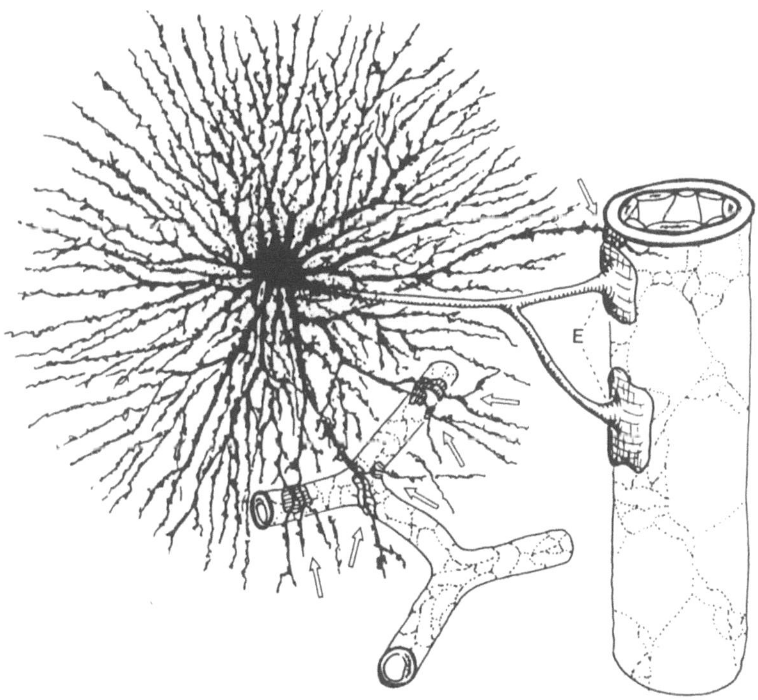

Figure 1: Isolated astrocyte with two types of processes: one leads to end-feet (E) on a stem vessel. The other ones penetrate the surrounding tissue and make en passant contacts (open arrows) mainly on capillaries, but also the large vessels. Dotted lines delineate the borders on vessel walls between astrocytic end-feet and en passant contacts of astrocytic processes.

with open perivascular spaces have been found (e.g., medulla; Wolff and Nemecek, 1968). Since in periventricular organs the existence of open pericapillary spaces is not closely correlated with the absence of blood-brain-barrier (see Leonhardt, 1980), other functional and/or morphogenetic conditions may be responsible for the differentiation of glial-vascular interfaces.

Vascularization of the Cerebral Cortex

In the rat neocortex, as in other parts of the vertebrate CNS, vascularization can be divided into at least 3 phases corresponding to the formation of characteristic structures: 1) meningeal plexus, 2) intracerebral stem vessels and their preterminal branches, and 3) capillaries.

The meningeal plexus corresponds to the epiparenchymal vascular plexus of other epithelial organs (Wolff et al., 1975). It initially consists of sinusoidal vessels with irregularly shaped endothelial cells, which are covered by filopodia and incomplete basal lamina (Bär and Wolff, 1972a). Meningeal vessels are always separated from the complete basal lamina of the adjacent neural tissue. The intervening space contains connective tissue, including fibroblast-like or meningeal cells (for cerebellar meninges, see J. Sievers, this volume). The vascular walls show high mitotic activity which leads to sprouting (forming new meshes of the plexus and intracerebral stem vessels), to dilation or elongation of preexisting vessels or to formation of pericytes or smooth muscle cells. Postnatally, while the intracortical angioarchitecture is maturing, parts of the meningeal plexus degenerate, and persisting vessels transform to pial arteries and veins (Bär et al., 1986). This remodelling establishes a spatial correlation between venous drainage fields and functional areas of the cortex (Ambach et al., 1986).

Intracortical stem vessels develop from sprouts of the meningeal plexus. They penetrate the cortical anlage and form a subventricular plexus (Strong, 1964). Sprouts are not primarily covered by a basal lamina or connective tissue. Only after embryonic day (E) 15, patches of amorphous, weakly osmiophilic and fuzzy material appears, where endothelial cells or pericytes become closely apposed to perivascular cells (Bär and Wolff, 1972a). Such perivascular cells have been identified as non-radial glial cells by their characteristic polarity and, by E18, by the expression Cl- and GFAP-antigens (Rickmann and Wolff, 1985). Postnatally, a continuous lamina densa appears when the perivascular space diminishes and glial sheaths become complete (Bär and Wolff, 1972a). It is unclear when Virchow-Robin spaces appear and whether extracellular matrix components and fibroblast-like cells are locally formed or shift from the meninges into the cortex.

Until postnatal day (P)8, three further sets of radial stem vessels are added in a characteristic spatio-temporal pattern (Fig. 2A). Younger vessels terminate progressively in more superficial layers of the cortex (Wolff, 1976; Wolff et al., 1978; Eulner, 1980; Bär, 1980). Thus, the basic plan of cortical angioarchitecture emerges during the pre- and perinatal period. Local proliferation of vascular wall cells proceeds for about 2 weeks after birth (Mareš and Brückner, 1978), permitting elongation and dilation of stem vessels. Glial cells and processes are concomitantly added to expand the perivascular sheath, coinciding with the main peak of glial proliferation (Fig. 2C).

The formation of pre- and postcapillary vessels begins with the branching of stem vessels. Stem vessels become interconnected and establish the subventricular plexus and at least three further plexi within the cortex. Intracortical sprouts are primarily devoid of basal lamina and glial sheaths (Bär and Wolff, 1972a) and spread from vessel stems toward those parts which have most recently been formed. Finally, arterioles and venules dilate during the beginning of capillarization and reach their adult surface area at about P10 (Fig. 2B) when the proliferation of astrocytes terminates (Fig. 2C).

During capillary formation (Fig. 2B; Bär and Wolff, 1973; Bär, 1980), sprouts con-

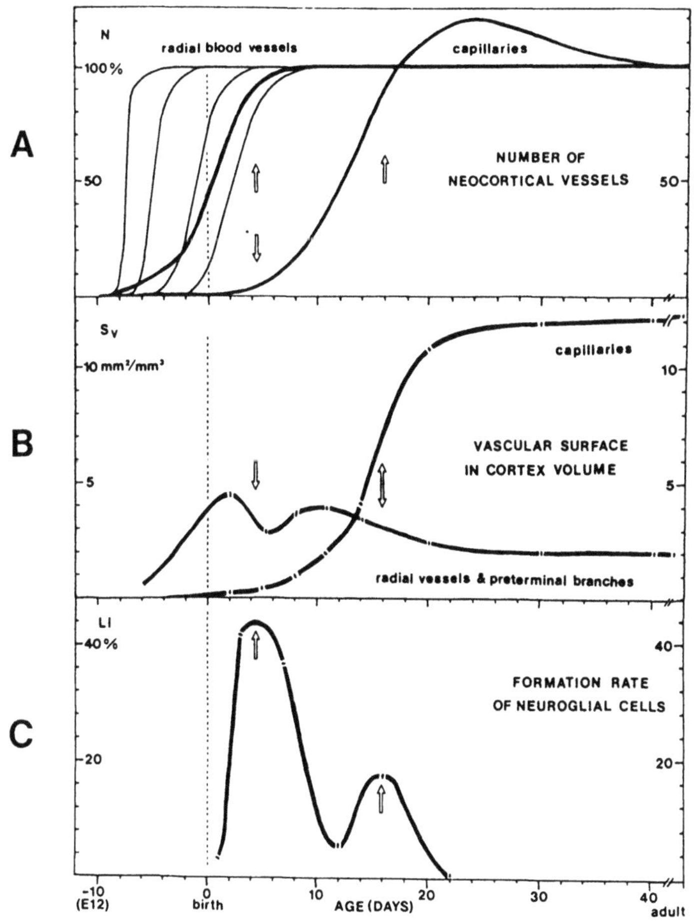

Figure 2: Comparison between the time courses of vascularization and gliogenesis. A: The number of radial vessels per whole cortex (N as % of adult values) is plotted against age for four sets of stem vessels (thin lines) and all stem vessels (thick lines; data from Eulner, 1980). The capillary network is represented by the number of branching points in the whole cortex (calculated from data of Bär and Wolff, 1973; Bär, 1980; and Eulner, 1980). B: The vascular surface area in the whole cortex volume (S_V) was calculated from the data of Bär and Wolff (1973c); capillary length), Bär and Wolff (1972b; vascular diameter), Bär (1980; thickness of cortex) and Eulner (1980; development of neocortical surface area in the rat). The separation of capillaries from radial vessels and preterminal branches was based on fractions of vessels with diameters smaller or larger than 8 μm, respectively. C: the formation rate of "ectodermal glial cells" is based on the labelling index (LI) of nuclei after ^{3}H-thymidin-pulse labelling (data of Mareš and Brückner, 1978). Open arrows indicate the position of gliogenetic peaks. Broken lines label the day of birth.

sist of a single endothelial cell, rather than of several cells, as do those formed at earlier stages. Consequently, the lumen does not form by opening intercellular clefts, but by penetrating the cell leading to "seamless" endothelia (Wolff and Bär, 1972). Apart from these markers of capillaries, growth of the capillary network can be described by the density of branchings. This density increases rapidly between P5 and P25, and then decreases toward adulthood, indicating a loss of about 20% of the branching points, i.e., 10% of the capillaries, within the meshwork (Fig. 2A). Thereafter, capillary length increases by a telescope-like mechanism. This causes a delayed increase in the surface area of capillaries (Fig. 2B; Bär and Wolff, 1972b, 1973, 1976; Bär, 1980).

The glial sheath covers about 65% of the capillary surface at P2, 80% at P6, 90% at P12 and becomes complete between P20 and P30. The final 40% of the glial sheath consisting of astrocytic lamellae are formed between P8 and P20, i.e., during the rapid phase of capillarization (Wolff and Bär, 1976).

These observations indicate that there is a characteristic difference in the formation of glial sheaths between stem vessels and capillaries which lead to a differential distribution of end-feet and en passant contacts of astroglial processes. Correspondingly, basal lamina formation is discontinuous on stem vessels at the beginning, while it is more or less continuous on capillaries and spares only the tips of sprouts.

Conditions for the Formation of Basal Lamina

In the meninges and Virchow-Robin-spaces around proximal stem vessels, basal laminae forms separately along plasma membranes of glial cells and vascular wall cells. This is in the presence of connective tissue containing extracellular matrix and fibroblast-like cells.

On intracortical stem vessels, basal lamina formation is generally delayed until perivascular spaces diminish by apposition of what appears as glial precursor cells, while neighbouring neurons do not seem to provide proper conditions. Between glial cells or their processes and vascular wall cells, fuzzy and weakly osmiophilic material appears at first, which gradually condenses and forms two thin layers of lamina-dense-like structures attached to the adjacent plasma membranes. Two sublayers can often be discerned even in thick and continuous basal laminae of the adult cortex whether or not collagen fibrils intervene between them. This indicates that for establishing lamina densa, though not necessarily for its maintenance, the material needs anchoring points on the adjacent plasma membranes (Bär and Wolff, 1972a).

The characteristic thickening of pericapillary basal lamina occurs relatively late during development (Stage 4, 3rd to 4th postnatal week). The thickening occurs synchronously with the extension of astrocytic lamellae within the glial sheath. And it is greater on the arterial than on the venous side of the microvasculature. Thickening might depend on specific interactions between glial elements and endothelial cells, because it is often absent between pericytes, and endothelial cells or astrocytes (Wolff, 1963).

These observations suggest that basal lamina can be formed on vascular wall cells in the absence of glial cells, if connective tissue is present. In the absence of connective tissue, however, the intracerebral vessels do not form basal lamina before glial processes approach them.

The Fate of Basal Lamina after Vascular Degeneration

Several parts of the primary cortical angioarchitecture undergo degeneration, but in each case the fate of basal laminae is different, in the meningeal network many anastomoses degenerate postnatally (Bär et al., 1986). This includes the complete degradation of basal laminae.

According to Strong (1964), many primary stem vessels degenerate by E18. In later stages no remnants of basal lamina can be seen in their respective positions. Although Strong's observations still need confirmation, this may suggest that either the basal lamina was degraded or the degenerating vessels had not yet been covered by basal lamina material.

Large parts of the subventricular plexus probably degenerate, since in later developmental stages there are only localized vascular arrays directly below the ependyma. In this position, basal lamina labyrinths (for references, see Leonhardt, 1980) have been found which represent duplications of basal lamina bordered by processes of ependymal cells and astrocytes. These labyrinths seem to develop relatively late (about P30). They are said to grow as expansions of persisting vessel walls with which they remain connected. It is not clear whether they can also develop from degenerating vessels of the subventricular plexus.

Finally, when capillaries degenerate, their basal laminae are apparently preserved and form "intercapillary bridges" (Guseo and Gallyas, 1974). Electron microscopy reveals them to be collapsed tubes of basal lamina covered by glial processes. Similar intercapillary bridges accumulate in glial scars after neuronal degeneration (Wolff and Bär, 1976). Another condition has been found in which basal lamina covered with astrocytic processes can be preserved after degeneration of vascular wall cells. This is the case with vessel fragments in cortical explant cultures (Wolff et al., 1974).

These observations indicate that basal lamina in contact with glial processes can be maintained (but probably not formed) in the absence of mesodermal derivatives such as fibroblasts, endothelia and pericytes. At present, it cannot be ruled out that essential factors diffuse from persisting vessel walls into the "empty" basal lamina tubes.

Gliogensis in Neocortex

The development of the ependymo-astrocytic cell line proceeds essentially in two phases. Prenatally the position of primary glioblasts is determined, while the main number of astrocytes is formed postnatally.

Gliogenesis begins when and where radial glial cells appear. These cells have to antedate the positioning of neurons and glioblasts in the cortical anlage, because they represent migration guides (Rakic et al., 1974). Radial glial cells are polarized between two processes. The apical process makes contact with the ventricular surface, while the basal process ends with end-feet either at the meningeal surface or on intracortical blood vessels (corresponding to "long" and "short" radial glial cells; Rickmann and Wolff, 1985, Fig. 3).

Non-radial glioblasts lose the ventricular contact, while they preserve or regain contact with basal lamina. These contacts have the form of end-feet which are located either along the meningeal surface (precursors of marginal astrocytes in lamina I) or contact intracortical stem vessels (Fig. 3). Recently, it has been demonstrated that both types of non-radial glial cells appear very early in the neocortex anlage (see Rickmann, this volume).

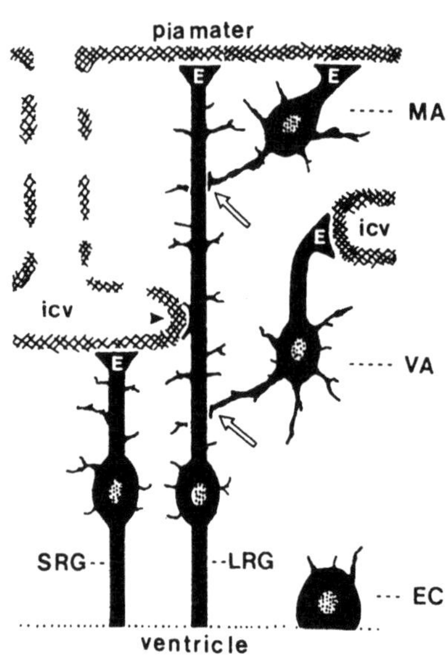

Figure 3: Schematic representation of various forms of immature ependymo-glial cells with different contact relationships to the ventricle, to other glial cells (open arrows), and to basal lamina. The latter are medited by end-feet (E) or en passage contact (arrow head). EC ependymal cell, icv intracerebral blood vessel, LRG long radial glial cell, MA marginal astrocyte, SRG short radial glial cell, VA astrocyte with vascular contact.

During the early postnatal period, cortical gliogenesis shows two peaks and then terminates almost completely three weeks after birth (Fig. 2C; Mareš and Brückner, 1978). The first peak coincides with the end of stem vessel formation. The second (smaller) peak occurs when capillarization begins and preterminal vessels establish their final size (compare Fig. 2A to C). The characteristic pattern of astrocyte processes develops after their last mitosis. This coincides with the sprouting of capillaries.

These data provide circumstantial evidence for the following developmental sequence of the glial-vascular interface: radial glia and marginal astrocytes are added and form end-feet during the extension of the meningeal surface of the cortex. This is a pre- and postnatal process. The predominant fraction of astrocytes also establishes basal lamina contacts via the perikaryon or end-feet which are the primary processes. This, however, is correlated with the dilation of stem vessels and preterminal branches. Secondary processes of astrocytes do not show any orientation toward blood vessels, but form en passant contacts with capillary sprouts. Finally, lamellar excrescences appear on astrocytic processes when capillaries elongate. Basal lamina becomes thicker when a major part of the capillary surface is covered by lamellae.

Prenatal Gliogenesis Determines Neuron Positioning

In the rat neocortex, the currently accepted mode of neuron positioning is the inside-outside layering (Berry and Rogers, 1965). This pattern of positioning evolves when post-mitotic neurons migrate from the ventricular proliferation zone along radial glial cells up to the border of lamina I and cortical plate (Rakic et al., 1974). It is not clear why the neurons stop migrating at this border instead of at the marginal surface. Since lamina I neurons do pass this border, there might be some kind of stop signal which can be recognized at least by pyramidal cells (see Rickmann, this volume).

Recently, non-radial glial cells have been found which are formed prenatally and form end-feet contacting stem vessels. These cells have a second process which makes a specialized contact with radial glial processes. Such contacts can be formed at various levels of the cortical anlage including lamina I and subplate. Since neurons are not committed to migration along radial glial cells (e.g., cerebellar granule cells on Bergmann glial cells, see J. Sievers, this volume), it is conceivable that while passing such contacts, neurons might choose non-radial glial cells for further migration and stop near stem vessels. This possibility may explain why there is a second mode of neuron positioning in the neocortex. This so-called diffuse positioning mode (Wolff et al., 1978, 1983) continuously provides all cortical layers including lamina I and the subcortical white matter with non-pyramidal neurons throughout the period of neuron production and migration.

Thus, neurons use the radial as well as non-radial glial cells as migration guides. In

Figure 4 (facing page): Removal of degeneration products from the rat neocortex at 7 days (A), 10 days (B) and 20 days following callosotomy (C, D). Astrocytes accumulate degeneration products of degenerating axons and synapses first in their processes (arrow heads in A and C, "a" in D). Then, residual bodies appear in perivascular end-feet (arrow-head in B) and outline the wall of stem vessels (white arrows in C). Finally, degeneration products accumulate in flattened cells within the glial vascular interface (between broken lines in D). The arrow in D points to the nucleus surrounded by stained granules. Note unlabeled endothelial cells ("e") of the stem vessel and unlabelled capillaries ("c" in D). Bär represents the length of 50 µm (A, c) and 25 µm (B, D). A and B dark field. Silver staining of Gallyas et al. (1980).

combination, these provide the neocortex with two characteristic populations of neurons: pyramidal cells and non-pyramidal neurons.

Intracellular Transport in Ependymo-astrocytic Cells

The central nervous tissue can cope with degeneration products in two ways. First, macrophages and microglial cells can incorporate such material, digest it partly in phagolysosomes and transport the residual bodies out of the tissue by migration. This reaction of migratory cells is predominant when there are large defects including lesions in the vascular system.

A second type of reaction is observed when anterograde axonal degeneration is induced by axotomy. Degenerating axon terminals and synapses are engulfed by swelling lamellae and incorporated in the processes of astrocytes. The next step observed is an accumulation of residual bodies in the perikaryon. Later, residual bodies accumulate in end-feet and then in what appears as the fibroblast-like cells along stem vessels and some preterminal branches (Fig. 4A-D; Wolff et al., 1981, Holzgraefe et al., 1981). Similar sequences of phagocytosis and intracellular transfer of lysosomes in astrocytes can be observed after local application of horseradish peroxidase to brain tissue in situ and in explant cultures of brain tissue. In these cultures, many neurons degenerate after some time in vitro, while astrocytes survive and keep or re-build perivascular endfeet-like structures. These astrocytes incorporate degeneration products, transport them intracellularly, and deposit debris at the vessel walls, although there is no blood flow (Wolff et al., 1974).

These observations suggest that astrocytes have an intracellular transport system that can be directed from sites of phagocytosis in processes through the pericaryon to perivascular end-feet but not to en passage contacts with vessels. This transport is not dependent on blood flow. It is not yet clear whether this intracellular transport of astrocytes is only induced under pathological conditions or is also a physiological phenomenon (e.g., in tanycytes of the median eminence, see Leonhardt, 1980).

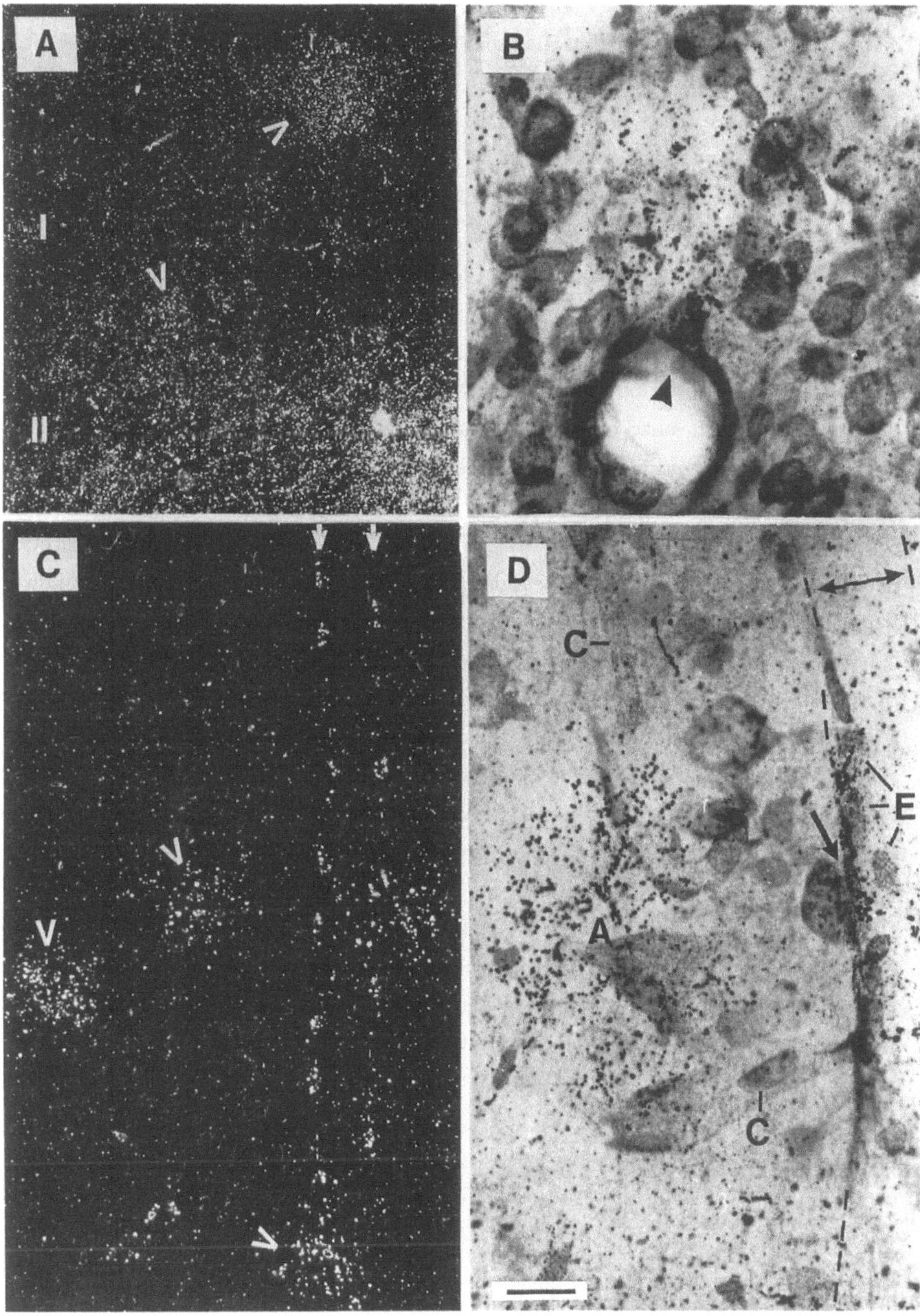
A
B
C
D
I
II
C
A
C
E

Conclusions

1. Astrocytes are involved in the glial sheath towards both meningeal tissue and intracerebral vessel walls. However, the vast majority of astrocytes contact vascular surfaces. This may explain why astrocytes and intracerebral vessels co-vary roughly, although there is no strict correlation.
2. Intracortical vascularization and development of astrocytes are interdependent processes. Formation of perivascular basal lamina requires the vicinity of glial cells, while glioblasts seem to depend on basal lamina contact for proliferation.
3. Stem vessels and preterminal branches develop in conjunction with the genesis of glioblasts and astrocytes and their primary processes establish end-feet on vascular walls. Capillary formation coincides with the differentiation of secondary processes of astrocytes which make en passant contacts with vessel walls. Correspondingly, the distribution of astrocytic end-feet and the structure of glial vascular interfaces differ between stem vessels and capillaries.
4. Astrocytes show an intracellular transport of lysosomes, which is selectively directed towards their end-feet at stem vessels but excludes en passant contacts (capillaries) from accumulation of residual bodies. This discrimination shows a characteristic polarization of astrocytes which is based between different types of basal lamina contact.
5. Pyramidal neurons, following radial glial cells as migration guides, are positioned within the neocortex according to the "inside-outside layering." In contrast, non-pyramidal neurons are continuously added to all layers ("diffuse positioning mode"). This difference may be due to specific contacts of radial glial cells with non-radial glial cells which are already anchored on intracortical vessels during the prenatal period.
6. Thus, intracortical vascularization is not only induced by neural tissue (see Risau, this volume), but it also regulates gliogenesis and determines the polarization of astrocytes, and may in turn influence the distribution of non-pyramidal interneurons, including those serving GABAergic inhibition. The role of extracellular matrix components in glial-vascular interactions is yet unknown.

Acknowledgements. Our studies were supported by grants from the Deutsche Forschungsgemeinschaft (SFB33: E3, E5, Wo 279/6-1). The author is grateful to Drs. J. Adler and G. Stent for critical comments and improvement of the English text.

References

Ambach, G., J. Toldi, O. Fehér, F. Joó, and J.R. Wolff (1986) Spatial correlation between sensory regions and the drainage fields of pial veins in rat cerebral cortex. Exp. Brain Res. **61**: 540-548.

Bär, Th. (1980) *The Vascular System of the Cerebral Cortex*. Adv. Anat. Embryol. Cell Biol. **59**: 1-89. Spinger, Berlin, Heidelberg, New York.

Bär, Th., A. Miodoński, and A.W. Budi-Santoso (1986) Postnatal development of the vascular pattern in the rat telencephalic pia-arachnoid. SEM study. Anat. Embryol. **174**, in press.
Bär, Th., and J. Wolff (1972a) The formation of capillary basement membranes during internal vascularization of the rat's cerebral cortex. Z. Zellforsch. **133**: 231-248.
Bär, Th., and J.R. Wolff (1972b) On the vascularization of the rat's cerebral cortex. Bibl. Anat. **11**: 515-519.
Bär, Th., and J.R. Wolff (1973) Quantitative Beziehungen zwischen der Verzweigungsdichte und Länge von Capillaren im Neocortex der Ratte während der postnatalen Entwicklung. Z. Anat. Entw.gesch. **141**: 207-221.
Bär, Th., and J.R. Wolff (1976) Development and adult variations of the wall of brain capillaries in the neocortex of rat and cat. In *The Cerebral Vessel Wall*, Cervós-Navarro, J. et al., eds., pp. 1-6. Raven Press, New York.
Berry, M., and A.W. Rogers (1965) The migration of neuroblasts in the developing cerebral cortex. J. Anat. (Lond.) **99**: 691-709.
Blinkov, S.M., and IL. Glezer (1968) *Das Zentralnervensystem in Zahlen und Tabellen*. 458p. VEB G. Fischer, Jena.
Bone,Q. (1960) The central nervous system of Amphioxus. J. Comp. Neurol. **115**: 27-51
Eulner, S. (1980) Beziehungen zwischen Wachstum und Vaskularisation der Großhirnrinde. Eine morphometrische Studie an der Albinoratte. Thesis, University of Göttingen, pp. 1-88.
Gallyas, F., J.R. Wolff, H. Böttcher, and L. Zaborszky (1980) A reliable and sensitive method to localize terminal degeneration and lysosomes in the central nervous system. Stain Technol. **55**: 299-306.
Guseo, A., and F. Gallyas (1974) Intercapillary bridges and the development of brain capillaries. In *Pathology of Cerebral Microcirculation*. Cervós-Navarro et al., eds., pp. 448. De Gruyter, Berlin.
Held, H. (1909) Über die Neuroglia marginalis der menschlichen Großhirnrinde. Mschr. Psychol. Neurol. **26 (Erg.Heft)**: 360-416.
Horstmann, E. (1954) Die Faserglia des Selachiergehirns. Z. Zellforsch. **39**: 588-617.
Holzgraefe, M., G. Teuchert, and J.R. Wolff (1981) Chronic isolation of visual cortex induces reorganization of cortico-cortical connections. In *Lesion-induced Neuronal Plasticity in Sensorimotor Systems*, H. Flohr and W. Precht, eds., pp. 351-359. Springer, Berlin, Heidelberg, New York.
Leonhardt, H. (1980) Ependym und circumventrikuläre Organe. In *Handbuch der Mikroskopischen Anatomie des Menschen, vol. IV/10: Neuroglia I*, A. Oksche, ed., pp. 177-665. Springer Verlag, Berlin, Heidelberg, New York.
Mareš, V., and G. Brückner (1978) Postnatal formation of non-neuronal cells in the rat occipital cerebrum: An autoradiographic study of the time and space pattern of cell division. J. Comp. Neurol. **177**: 519-528.
Rakic, P., L.J. Stensaas, E.P. Sayre, and R.C. Sidman (1974) Computer-aided three-dimensional reconstruction and quantitative analysis of cells from serial electron microscopic montages of foetal monkey brain. Nature **250**: 31-34.
Rickmann, M., and J.R. Wolff (1985) Prenatal gliogenesis in the neopallium of rat. Adv. Anat. Embryol. Cell Biol. **93**: 1-104. Springer, Berlin, Heidelberg, New York.
Sarnat, H.B., J.F. Campa, and J.M. Lloyd (1975) Inverse prominence of ependyma and capillaries in the spinal cord of vertebrates: A comparative histochemical study. Amer. J. Anat. **143**: 439-450.
Strong, L.H. (1964) The early embryonic pattern of internal vascularization of the mammalian cerebral cortex. J. Comp. Neurol. **123**: 121-138.
Wolff, J. (1963) Beiträge zur Ultrastruktur der Kapillaren in der normalen Großhirnrinde. Z. Zellforsch. **60**: 409-431.
Wolff, J. (1965) Elektronenmikroskopische Untersuchungen über die Struktur und Gestalt von Astrocytenfortsätzen. Z. Zellforsch. **66**: 811-828.
Wolff, J. (1970) Quantitative aspects of astroglia. Proc. 6th Intern. Congr. Neuropath., Mason, Paris, pp. 327-352.

Wolff, J.R. (1976) An ontogenetically defined angioarchitecture of the neocortex. Drug Res. **26/6:** 12.

Wolff, J.R. (1977) Ultrastructure of the terminal vascular bed as related to function. In: *Microcirculation*, G. Kaley and B.M. Altura, eds., vol. I, pp. 95-130. Univ. Park Press, Baltimore.

Wolff, J.R., and Th. Bär (1972) Seamless endothelia in brain capillaries during development of the rat's cerebral cortex. Brain Res. **41:** 17-24.

Wolff, J.R., and Th. Bär (1976) Development and adult variations on the pericapillary glial sheath in the cortex of rat. In *The Cerebral Vascular Wall*, J. Cervós-Navarro et al., eds., pp. 7-13. Raven Press, New York.

Wolff, J.R., B.M. Chronwall, and M. Rickmann (1978) Morphogenetic relations between cell migration and synaptogenesis in the neocortex of rat. In *Proc. Europ. Soc.*, Neurochem. V. Neuhoff, ed., pp. 158-173. Verlag Chemie, Weinheim, New York.

Wolff, J.R., B.M. Chronwall, and M. Rickmann (1983) "Diffuse deposition mode" provides visual cortex with non-pyramidal and GABA-ergic neurons. Intern. J. Devel. Neurosci. **1:** 246.

Wolff, J.R., S. Eins, M. Holzgraefe, and L. Zaborszky (1981) The temporo-spatial course of degeneration after cutting cortico-cortical connections in adult rats. Cell Tissue Res. **214:** 303-321.

Wolff, J.R., Ch. Goerz, Th. Bär, and F.-H. Güldner (1975) Common morphogenetic aspects of various organotypic microvascular patterns. Microvasc. Res. **10:** 373-395.

Wolff, J., and S.T. Nemeček (1968) Über kollagenhaltige perivasculäre Räume an "Kapillaren" in der Medulla des Rhesusaffen. Experientia **24:** 930.

Wolff, J.R., K.T. Rajan, and W. Noack (1974) The fate and fine structure of fragments of blood vessels in CNS tissue cultures. Cell Tiss. Res. **156:** 89-102.

POSTNATAL DEVELOPMENT OF γ-GT ACTIVITY IN RAT BRAIN MICROVESSELS. A SEMIQUANTITATIVE HISTOCHEMICAL APPROACH.

Th. Bär and A.W. Budi Santoso
Max-Planck-Institut für Systemphysiologie, Dortmund, and
Zentrum Anatomie, Universität Göttingen, Göttingen
Federal Republic of Germany

The membrane bound glycoprotein enzyme γ-glutamyl-transpeptidase (γ-GT) is highly active in mature endothelial cells (EC) of cerebral capillaries of different mammals as demonstrated by histochemical as well as biochemical techniques. γ-GT takes part in the transfer of large neutral amino acids and appears to be related to the blood brain barrier, a specific property of differentiated capillary endothelium in the brain tissue. The activity of γ-GT varies within the pre- and postnatal developmental stages and in different organs. At the dedifferentiation stage, γ-GT activity is at the level usually found in early embryonic development of the particular organ. In brain tissue γ-GT activity usually disappears when the EC loses its contact to the adjacent astroglial cells, i.e., during processes leading to migration and proliferation of EC in vitro and in vivo. The correlation between differentiation of EC and the presence of γ-GT activity prompted us to examine developmental changes of γ-GT in capillaries isolated from rat cerebral cortex at different ontogenetic stages (i.e., on days 2, 7, 11, 14, 21 and 60 after birth). The cerebral cortex of the rat represents a suitable model for studies of vascular growth and differentiation because major changes occur in the capillary bed after birth. For instance, detailed morphometric studies document a phase of vigorous proliferation in capillary length followed by the functional maturation of the capillary bed (Bär, 1980).

γ-GT activity is determined by measuring the enzyme kinetics of the histochemical reaction on isolated capillaries using a flying spot microscope densimeter (for details of the methods see Budi Santoso and Bär, 1986). Enzyme activity is expressed as an increase in relative optical density (at 500 um) in arbitrary units/min/μm^2 during the two minutes immediately after starting the histochemical reaction. This method is suitable to detect low levels of γ-GT activity and offers further advantages (when compared with common biochemical methods), because the capillary fragments, which are expected to exhibit γ-GT activity, can be selectively measured for their enzyme kinetic properties.

NATO ASI Series, Vol. H5
Mesenchymal-Epithelial Interactions in Neural Development
Edited by J. R. Wolff et al.

The activity of γ-GT detected in isolated cortical microvessels is restricted to the capillary EC. The reaction product is mostly localized at the luminal side. The brain capillary fragments from 2 day old animals have large round endothelial cell nuclei lying close together. After incubation for γ-GT, the cytoplasm of the EC is stained intensively. The nuclei are not stained. However, a decrease in γ-GT activity is observed in capillaries from 7 day old animals. Thereafter, γ-GT activity increases and reaches adult values on day 60. Thus, a biphasic change of γ-GT activity takes place. The postnatal period, between the 2nd and 12th day, is characterized by a significant decrease in γ-GT activity, coinciding with the onset of the rapid mitotic proliferation of EC in the cerebral cortex. Between the 12th and the 21st day, a fast increase of γ-GT activity is measured. It is suggested that γ-GT activity remains low as long as the majority of the plasma membranes of the newly formed endothelial tubes is in a state of low differentiation during the rapid postnatal growth. The activity of some membrane bound enzymes may be positively correlated to the incorporation of poly-unsaturated fatty acids in the plasma membrane. Indeed, γ-GT activity in the plasma membrane of lymphocytes is changed together with the content of linoleic and arachidonic acid, after stimulation with concanavalin A (Szamel and Resch, 1981).

For a continuous induction of γ-GT an interaction between astroglial cells and EC is required (De Bault and Cancilla, 1980). Co-culturing mouse EC with C6 rat glioma cells can induce the re-expression of γ-GT activity in EC. In the cerebral cortex, the fraction of the capillary surface adjacent to astroglial endfeet and lamellae increases from 66% to 84% during the first postnatal week. At the same time, the proliferation of the astroglial cells reaches maximum values in the cerebral cortex. Thus, the mitotic proliferation of EC and of astroglial cells may be correlated with the observed decrease of γ-GT activity in brain capillaries during the early postnatal period.

Our study supports evidence that the presence of astroglial cells is important for the expression of γ-GT activity, and for the structural and functional maturation of the blood brain barrier.

References

Bär, Th. (1980) The vascular system of the cerebral cortex. In Adv. Anat. Embryol. Cell Biol. **59**, A. Brodal, W. Hild, J. van Limborgh, R. Ortmann, T.H. Schiebler, G. Töndury, E. Wolff, eds., pp. 1-62, Springer-Verlag, Berlin-Heidelberg-New York.

Budi Santoso, A.W., and Th. Bär (1986) Postnatal development of γ-GT activity in rat brain microvessels corresponds to capillary growth and differentiation. Int. J. Dev. Neurosci. In press.

De Bault, L.E. and P.A. Cancilla (1980) γ-glutamyl-transpeptidase in isolated brain endothelial cells: Induction by glial cells in vitro. Science **27**: 653-655.

Szamel, M. and K. Resch (1981) Modulation of enzyme activities in isolated lymphocyte plasma membranes by enzymatic modification of phospholipid fatty acids. J. Biol. Chem. **256**: 11618-11623.

DEVELOPMENT AND DIFFERENTIATION OF THE BLOOD-BRAIN BARRIER

Werner Risau
Max-Planck-Institut für Entwicklungsbiologie
Tübingen
Federal Republic of Germany

Specialized properties of capillaries in different organs are crucial for special organ functions (e.g., ultrafiltration in kidney glomeruli; absorption and transport in the liver). The morphological appearance of capillary endothelial cells in different organs is often indicative of their function. Fenestrated endothelium, for instance, occurs in the kidney glomerulus, sinusoidal endothelium in liver, spleen and bone marrow, continuous endothelium in muscle and brain (for review see Majno, 1965). Brain endothelium is unique because it forms the blood-brain barrier (BBB).

The existence of a BBB has remained controversial since its discovery by Ehrlich (1885), but has been confirmed by many experiments during the last 30 years (for review see Bradbury, 1979, 1984). One of the most important experiments for our studies on the development and differentiation of the BBB was performed by Reese and Karnovsky (1967). They injected peroxidase intravenously into mice and found that the tight junctions of brain endothelial cells prevented the passage of peroxidase from the blood to the brain. The problem of BBB formation is therefore a problem of endothelial cell differentiation in the brain.

Development of the Brain Vascular System

The vascular system originates early in embryogenesis from so-called blood islands (hemangioblasts). They differentiate from the splanchopleuric mesoderm, and contain the precursor cells of blood cells as well as endothelial cells. Proliferation and fusion of blood islands give rise to capillary plexus (i.e., those of the yolk sac), which later differentiate through a complex series of regressions and enlargegments to form an arteriovenous vascular network (Evans, 1909; Romanoff, 1960).

The neural tube is invested by a perineural vascular plexus, which in a 2-day chick embryo covers the entire surface of the neural tube, and later differentiates into the leptomeningeal arteries and veins (Evans, 1909; Bär, 1980). From the perineural plexus vascular sprouts penetrate the proliferating neuroectoderm after 4 days of embryonic

NATO ASI Series, Vol. H5
Mesenchymal-Epithelial Interactions in Neural Development
Edited by J. R. Wolff et al.

development in the chick. The radially invading vessels are called stem vessels which do not penetrate the ependyma, but rather give off manifold branches in the deeper layers of the developing brain (Bär, 1980; Roncali et al., 1985; see Fig. 1). Several waves of penetration of stem vessels have been observed during development. The maturation of the blood vessels later in development involves the disappearance of pseudopodia-like processes, the formation of a basement membrane, and the differentiation of the vascular wall (pericytes, smooth muscle cells, and glial end-feet) (Wolff, 1963; Bär and Wolff, 1972; Roncali et al., 1985).

Angiogenesis of the Brain

There are in principle two explanations of how the differentiated state of a capillary endothelial cell could be determined in a specific organ. Either endothelial cells differentiate in situ, from precursor cells of the organ itself or they invade the organ and differentiate into a specific endothelium under the influence of the tissue environment.

Very early in development, most blood vessels arise in situ from blood islands. Later in development, capillary sprouts have been observed invading developing organs and tissues. This angiogenic event occurs in the brain (Bär, 1980; Stewart and Wiley, 1981), kidney (Ekblom et al., 1982), and limb bud (Joutereau and LeDouarin, 1978; Wilson, 1983). The external origin of vascular endothelium in these tissues has been clearly established using the chick-quail nuclear marking technique of LeDouarin (1973).

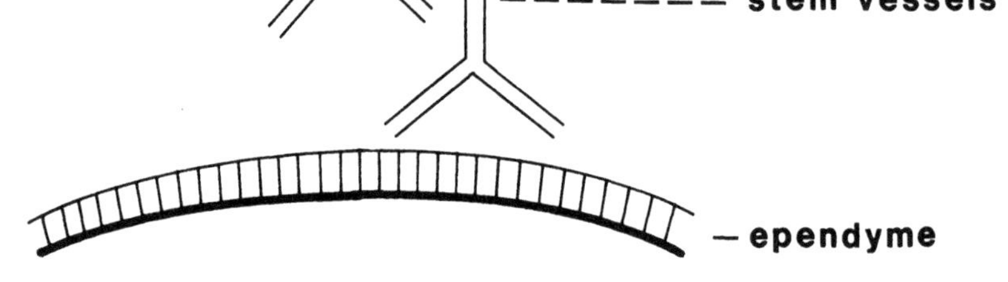

Figure 1: Scheme of the development of brain blood vessels (after Bär, 1980).

These results prompted us to investigate whether embryonic organs produce factors which are able to stimulate the ingrowth of blood vessels. One way to approach this question is to analyze these tissues for the presence of endothelial cell growth factors. Embryonic chick brain indeed produces a growth factor which is present early in development at a high specific activity. Total activity increased 700-fold during embryonic development. The growth factor has a molecular weight of 17 kDa, binds to heparin sepharose and is angiogenic in vivo in the rabbit cornea and chick chorioallantoic membrane (CAM) (Risau, 1986). The biochemical characteristics of this factor are very similar to fibroblast growth factor (FGF) - like angiogenic factors. Since the amino acid sequences of both anionic and cationic FGF have been published (Gimenez-Gallego et al., 1985; Esch et al., 1985) we will hopefully know soon what the relationship is between FGF and the embryonic brain derived angiogenesis factor.

Embryonic brain development is different in different species. The chick is a precocious animal, because its brain is much more mature at the end of embryogenesis than the rat brain (Horstmann, 1959). Robertson et al. (1985) have shown that in the rat, maximal capillary proliferation occurs between 5 and 9 postnatal days.

Adult bovine brain has been the major source for endothelial growth factors. The role of these growth factors in adult brain is not known, there is essentially no turnover of capillaries in adult tissues. Since both embryonic and adult endothelial growth factors are equally active in vitro and perhaps also in vivo, the regulation of angiogenesis during development and in the adult is one of the major problems of endothelial cell biology. It is also important with respect to pathological conditions, many of which are characterized by abnormal vessel growth (i.e., tumor growth, inflammation, wound healing).

The pattern of blood vessels in the brain is very characteristic and reproducible (Bär, 1980; see Fig. 1). How this pattern is established is an open question. One possibility is that a locally produced chemotactic factor for endothelial cells might guide the capillaries to their point of branching in the brain.

Development of the BBB

Capillaries are present very early in the embryonic brain, in the chick at day 4 (Feeney and Watterson, 1946), and in the rat at day 11 (Phelps, 1972). The BBB as defined by impermeability to peroxidase, however, develops considerably later, in the chick around embryonic day 13 (Wakai and Hirokawa, 1978), in the mouse around day 16 (Risau et al., 1986). In addition, the barrier does not develop abruptly but rather forms successively at different times in different regions of the brain. A gradient of impermeability from the periphery (pial surface) to the ependyma has been observed (Wakai and Hirokawa, 1978).

According to our hypothesis that the tissue environment would somehow induce the differentiation of invading endothelium, we would suppose that the BBB inducing principle in the brain is active at the time of BBB formation and not at the time of invasion of capillaries.

Differentiation of BBB Endothelium

Apart from the barrier characteristics of brain endothelium, a number of morphological as well as physiological and biochemical differences have been observed in brain endothelium as compared to other endothelial cells in the body. Table 1 summarizes many unique characteristics of brain endothelial cells which could possibly be used as markers of BBB endothelium.

Using the biochemical markers, we recently correlated the expression of proteins in brain endothelium with the development of the BBB in chick, quail, and mouse embryos (Risau et al., 1986).

These proteins which are possibly involved in carrier or metabolic activities of the BBB might therefore be good markers for BBB endothelium.

We have recently characterized a monoclonal antibody, called HT7, which specifically reacts with chick brain endothelial cells but with no other endothelial cells. Ex-

Table 1: Unique properties of BBB endothelium.

Morphology	Complex tight junctions	Coomber and Stewart (1985) Saunders and Møllgard (1984)
	Low number of pinocytotic vesicles	Coomber and Stewart (1985)
	Investment by glial end-feet	Wolff (1963)
	Large number of seamless endothelial cells	Bär et al. (1984)
Physiology	High electrical resistance	Crone and Olesen (1981)
	Specific carrier mechanisms	Pardridge (1984) Fenstermacher (1985)
Biochemistry	Gamma-glutamyl transpeptidase	Albert et al. (1966)
	Alkaline phosphatase	Landow et al. (1942)
	Cholinesterase	Brightman and Albers (1959)
	Transferrin receptor	Jefferies et al. (1984)

pression of the HT7 antigen on brain endothelium begins on embryonic day 10. It is therefore induced early on brain endothelium. Since it is not present in choroid plexus endothelium which represents permeable fenestrated endothelium, it may also be a marker for the BBB.

Induction of the BBB

Stewart and Wiley (1981) found that capillaries which had invaded chick brain tissue, transplanted into the quail embryo, not only were derived from the quail but also had acquired BBB characteristics. These studies and our data on the correlation of marker expression on brain endothelium and BBB development strongly suggest that mechanisms exist in the brain that control the expression of BBB-specific molecules. Furthermore, using species-specific HT7 antibodies we found that embryonic mouse brain transplanted onto the chick CAM induced the expression of the BBB-specific molecule on chick capillaries that had invaded the transplanted mouse brain.

DeBault and Cancilla (1980) provided evidence that the enzyme gammaglutamyltranspeptidase can be reinduced in brain endothelial cells in vitro by glioma cells. This has been the first approach to study the expression of brain-specific endothelilal cell markers in vitro. More studies of this kind, however, are needed to confirm the inductive capacity of glioma cells, because glioma blood vessels in vivo do not always possess a BBB (Groothuis and Vick, 1982; Coomber and Stewart, 1985).

Glial cells are nevertheless good candidates for BBB-inducing cells in the brain, because the presence of glial end-feet on capillaries is a characteristic feature of brain blood vessels. Indeed, in the chick (Wechsler, 1965) and rat (Phelps, 1972), the development of glial end-feet coincides with BBB formation.

Raff and co-workers (Williams et al., 1985) have shown that two types of astrocytes exist in the brain, which are morphologically and biochemically distinct, and develop on different time schedules. Type 1 astrocytes differentiate (with respect to defined cell surface markers) in the embryo at a time when the BBB develops, whereas type 2 astrocytes develop much later. Both types are, however, apparently able to form vascular glial end-feet (Miller and Raff, 1984). The availability of purified populations of astrocytes may provide a means by which the induction of BBB characteristics in endothelial cells could be studied.

Summary

The formation of the BBB is a differentiated function of brain endothelial cells in higher vertebrates. Brain endothelium invades brain tissue early in embryonic development, probably stimulated by a soluble brain-derived angiogenesis factor. Considerably later the BBB, as defined by impermeability to peroxidase, develops. The formation of the BBB can be correlated with the expression of specific proteins in brain endothelium. Transplanted brain tissue can induce the expression of BBB-specific endothelial cell proteins in invading endothelial cells derived from non-barrier endothelium. Brain cells (possibly the end-feet of astrocytes) may play a role in the induction of BBB characteristics in endothelial cells. A long-range (diffusible angiogenesis factor) as well as a local cell-cell interaction (glial end-feet) between invading mesenchymal cells (endothelium) and neuroectoderm-derived cells therefore seems to be crucial for BBB development and differentiation.

References

Albert, Z., M. Orlowski, Z. Rzucidlo, and J. Orlowska (1966) Studies on γ-glutamyl transpeptidase activity and its histochemical localization in the central nervous system of man and different animal species. Acta Histochem. **25:** 312-320.

Bär, Th. (1980) The vascular system of the cerebral cortex. Adv. Anat. Embryol. Cell Biol. **59:** 1-62.

Bär, Th., and J.R. Wolff (1972) The formation of capillary basement membranes during internal vascularization of the rat's cerebral cortex. Z. Zellforsch. **133:** 231-248.

Bär, Th., F.-H. Güldner, and J.R. Wolff (1984) "Seamless" endothelial cells of blood capillaries. Cell Tiss. Res. **235:** 99-106.

Bradbury, M.W.B. (1979) *The Concept of a Blood-Brain Barrier.* Wiley, Chichester, U.K.

Bradbury, M.W.B. (1984) The structure and function of the blood-brain barrier. Federation Proc. **43:** 186-189.

Brightman, M.W., and R.W. Albers (1959) Species differences in the distribution of extraneuronal cholinesterase within the vertebrate central nervous system. J. Neurochem. **4:** 244-250.

Coomber, B.L., and P.A. Stewart (1985) Morphometric analysis of CNS microvascular endothelium. Microvasc. Res. **30:** 99-115.

Crone, C., and S.P. Oleson (1981) The electrical resistance of brain capillary endothelium. J. Physiol. **316:** 53P-54P.

DeBault, L.E., and P.A. Cancilla (1980) γ-Glutamyl transpeptidase in isolated brain endothelial cells: Induction by glial cells in vitro. Science **207:** 653-655.

Ekblom, P., H. Sariola, M. Karkinen, and L. Saxén (1982) The origin of the glomerular endothelium. Cell Diff. **11:** 35-39.

Esch, F., A. Baird, N. Ling, N. Ueno, F. Hill, L. Denoroy, R. Klepper, D. Gospodarowicz, P. Böhlen, and R. Guillemin (1985) Primary structure of bovine pituitary basic fibroblast growth factor (FGF) and comparison with the amino-terminal sequence of bovine brain acidic FGF. Proc. Natl. Acad. Sci. (USA) **82:** 6507-6511.

Evans, H.M. (1909) On the development of the aortae, cardinal and umbilical veins, and the other blood vessels of vertebrate embryos from capillaries. Anat. Rec. **3:** 498-519.

Feeney, J.F. Jr., and R.L. Watterson (1946) The development of the vascular pattern within the walls of the central nervous system of the chick embryo. J. Morph. **78:** 231-303.

Fenstermacher, J.D. (1985) Current models of blood-brain transfer. TINS **8:** 449-453.

Gimenez-Gallego, G., J. Rodkey, C. Bennett, M. Rios-Candelore, J. DiSalvo, and K. Thomas (1985) Brain-derived acidic fibroblast growth factor: Complete amino acid sequence and homologies. Science **230:** 1385-1388.

Groothuis, D.R., and N.A. Vick (1982) Brain tumors and the blood-brain barrier. TINS **5:** 232-235.

Horstmann, E. (1959) Die postnatale Entwicklung der Kapillarisierung im Gehirn eines Nesthockers (Ratte) und eines Nestflüchters (Meerschweinchen). Anat. Anz. **106/107:** 405-410.

Jefferies, W.A., M.R. Brandon, S.V. Hunt, A.F. Williams, K.C. Gatter, and D.Y. Mason (1984) Transferrin receptor on endothelium of brain capillaries. Nature **312:** 162-163.

Jotereau, F.V., and N.M. LeDouarin (1978) The developmental relationship between osteocytes and osteoclasts: A study using the quail-chick nuclear marker in endochondral ossification. Dev. Biol. **63:** 253-265.

Landow, H., E.A. Kabat, and W. Newman (1942) Distribution of alkaline phosphatase in normal and in neoplastic tissues of nervous system: Histochemical study. Arch. Neurol. Psychol. **48:** 518-530.

LeDouarin, N. (1973) A biological cell labeling technique and its use in experimental embryology. Dev. Biol. **30:** 217-222.

Majno, G. (1965) Ultrastructure of the vascular membrane. In *Handbook of Physiology, Circulation*, W.F. Hamilton, ed. Washington, D.C.: Am. Physiol. Soc., sect. 2, vol. III, chapt. 64, pp. 2293-2375.

Miller, R.H., and M.C. Raff (1984) Fibrous and protoplasmic astrocytes are biochemically and developmentally distinct. J. Neurosci. **4:** 585-592.

Pardridge, W.M. (1984) Transport of nutrients and hormones through the blood brain barrier. Federation Proc. **43:** 201-204.

Phelps, C.H. (1972) The development of glio-vascular relationships in the rat spinal cord. An electron microscopic study. Z. Zellforsch. **128:** 555-563.

Reese, T.S., and M.J. Karnovsky, (1967) Fine structural localization of a blood-brain barrier to exogenous peroxidase. J. Cell Biol. **34:** 207-217.

Risau. W. (1986) Developing brain produces an angiogenesis factor. Proc. Natl. Acad. Sci. (USA) **83,** in press.

Risau, W., R. Hallmann, and U. Albrecht (1986) Differentiation-dependent expression of proteins in brain endothelium during development of the blood-brain barrier. Dev. Biol., in press.

Robertson, P.L., M. du Bois, P.D. Bowman, and G.W. Goldstein (1985) Angiogenesis in developing rat brain: An in vivo and in vitro study. Dev. Brain Res. **23:** 219-223.

Romanoff, A.L. (1960) *The Avian Embryo*. Macmillan, New York.

Roncali, L., D. Ribatti, and G. Ambrosi (1985) Ultrastructural basis of the vessel wall differentiation in the chick embryo optic tectum. J. Submicrosc. Cytol. **17:** 83-88.

Saunders, N.R., and K. Møllgard (1984) Development of the blood-brain barrier. J. Dev. Physiol. **6:** 45-57.

Stewart, P.A., and M.J. Wiley (1981) Developing nervous tissue induces formation of blood-brain barrier characteristics in invading endothelial cells: A study using quail-chick transplantation chimeras. Dev. Biol. **84:** 183-192.

Wakai, S., and N. Hirokawa (1978) Development of the blood-brain barrier to horseradish peroxidase in the chick embryo. Cell Tiss. Res. **195:** 195-203.

Wechsler, W. (1965) Die Entwicklung der Gefäße und perivasculären Gewebsräume im Zentralnervensystem von Hühnern (Elektronenmikroskopischer Beitrag zur Kenntnis der morphologischen Grundlagen der Bluthirnschranke während der Ontogenese). Z. Anat. Entwickl. Gesch. **124:** 367-395.

Williams, B.P., E.R. Abney, and M.C. Raff (1985) Macroglial cell development in embryonic rat brain: Studies using monoclonal antibodies, fluorescence-activated cell sorting, and cell culture. Dev. Biol. **112:** 126-134.

Wilson, D. (1983) The origin of the endothelium in the developing marginal vein of the chick wing bud. Cell Diff. **13:** 63-67.

Wolff, J. (1963) Beiträge zur Ultrastruktur der Kapillaren der normalen Großhirnrinde. Z. Zellforsch. **60:** 409-431.

SUBJECT INDEX

NATO ASI Series H

Vol. 1: **Biology and Molecular Biology of Plant-Pathogen Interactions.**
Edited by J.A. Bailey. 415 pages. 1986.

Vol. 2: **Glial-Neuronal Communication in Development and Regeneration.**
Edited by H.H. Althaus and W. Seifert. 865 pages. 1987.

Vol. 3: **Nicotinic Acetylcholine Receptor: Structure and Function.**
Edited by A. Maelicke. 489 pages. 1986.

Vol. 4: **Recognition in Microbe-Plant Symbiotic and Pathogenic Interactions.**
Edited by B. Lugtenberg. 449 pages. 1986.

Vol. 5: **Mesenchymal-Epithelial Interactions in Neural Development.**
Edited by J.R. Wolff, J. Sievers, and M. Berry. 428 pages. 1987.

MIX
Papier aus verantwortungsvollen Quellen
Paper from responsible sources
FSC® C105338

If you have any concerns about our products,
you can contact us on
ProductSafety@springernature.com

In case Publisher is established outside the EU,
the EU authorized representative is:
Springer Nature Customer Service Center GmbH
Europaplatz 3, 69115 Heidelberg, Germany

Printed by Libri Plureos GmbH
in Hamburg, Germany